Geothermal Energy

Geothermal Energy: Engineering Fundamentals, Principles, and Applications is a much-needed book that offers a timely, practical, and comprehensive treatment of geothermal energy from the perspective of engineers and applied scientists. Bridging the gap between academic theory and real-world application, the book serves as a technology-transfer tool, providing essential technical information, design methodologies, and illustrative examples for students and professionals alike. Covering topics from basic thermodynamic principles to advanced geothermal applications and environmental considerations, the book is structured to support both learning and on-the-job training. Current and relevant, with an emphasis on pragmatic, calculation-based content in a clear, accessible style, this resource is an indispensable guide for anyone involved in the evolving field of geothermal energy or other sustainable energy systems.

Coverage includes:

- a comprehensive coverage of geothermal energy fundamentals and principles;
- a review of environmental/economic considerations associated with the various geothermal energy technologies;
- a practical, engineer-focused approach with real-world applications, design methods, and calculation tools;
- a review of legal and regulatory concerns;
- an examination of the potential role of artificial intelligence in the future of this technology;
- a review of potential investment opportunities in the geothermal field; and
- illustrative examples and predictive algorithms to support system design and analysis.

Geothermal Energy

Engineering Fundamentals, Principles, and Applications

Marybeth Reynolds, R. Ryan Dupont,
Walter Matystik, and Louis Theodore

CRC Press
Taylor & Francis Group
Boca Raton London New York

CRC Press is an imprint of the
Taylor & Francis Group, an informa business

Designed cover image: Shutterstock

First edition published 2027
by CRC Press
2385 NW Executive Center Drive, Suite 320, Boca Raton FL 33431

and by CRC Press
4 Park Square, Milton Park, Abingdon, Oxon, OX14 4RN

CRC Press is an imprint of Taylor & Francis Group, LLC

ISBN: 978-1-041-13035-2 (hbk)
ISBN: 978-1-041-13034-5 (pbk)
ISBN: 978-1-003-66775-9 (ebk)

DOI: 10.1201/9781003667759

Typeset in Times
by Newgen Publishing UK

Contents

PART 3 Geothermal Energy Applications

PART 4 Epilogue

Preface

As its title *Geothermal Energy: Engineering Fundamentals, Principles, and Applications* implies, this book – more than anything else – provides information on geothermal energy from an engineers' and applied scientists' perspective. The need and requirement to evaluate and design geothermal-based energy systems have led to the writing of this book. Concise and easy to use, it brings together a wealth of general information on geothermal energy systems and applications in one compact book. It offers technical information, calculation details that include design and predictive algorithms, and illustrative examples.

The book has been written not only for students but also for technical individuals, particularly, engineers and applied scientists who work in energy fields. It is a tool that may be used whenever and wherever information about geothermal energy is sought. Thus, it is a must for those working in the energy arena and those involved peripherally with this field. The authors feel that no one source covers all of the information on these topics in the manner presented in this book. It is hoped that this book will serve to fill the growing information required – both academically and professionally – in this field.

Here are the facts concerning geothermal energy – the subject title of this book:

1. Planet Earth contains an inexhaustible quantity of energy that is referred to as geothermal energy. It exists in both low-temperature and high-temperature states. The former state has application in space heating and cooling, while the later state is primarily applicable for electricity generation.
2. Ground source heat pumps have been employed for heating and cooling purposes. So-called advanced/enhanced geothermal technology is employed for electricity generation.
3. Hydrothermal geothermal energy employs heated, underground water as the source of energy to generate electricity while an external heat transfer fluid can extract high-temperature geothermal energy at depths significantly displaced from the Earth's outer surface.
4. Capital costs associated with geothermal power production continue to decrease at near exponential rates, leading to a conclusion that geothermal energy will almost certainly become the primary energy option of the future, displacing fossil fuels, hydropower, and nuclear energy as the primary base-load energy source.

A brief description of the five parts of the book's content is as follows:

Part 1 introduces the reader to a host of energy issues. Part 2 is concerned with the engineering fundamentals and principles associated with energy and energy-related topics. Part 3 topics focus on geothermal applications that include the role of artificial intelligence, investment opportunities, and illustrative examples. Peripheral engineering energy topics are featured in Part 4. The book concludes with Part 5, the Index.

Reasonable care has been taken to assure the accuracy of the information contained in this book. However, the authors and the publisher cannot be responsible for erroneous omissions in the information presented or for any consequences arising from the use of the information published in the book. Accordingly, reference to original sources is encouraged. Reporting of any errors or omissions is solicited in order to assure that appropriate changes may be made in future editions.

Last but not least, the authors wish to thank the following individuals for their contributions to this work: Emma Parente, Mary K. Theodore, and Ronnie Zaglin.

About the Authors

Marybeth Reynolds is a researcher and author specializing in clean energy and climate solutions. With more than 20 years of experience integrating engineering, public policy, and technical communication, she focuses on environmental impacts, regulatory frameworks, and sustainable pathways for decarbonization.

Ms. Reynolds holds a Bachelor's of Science in Chemical Engineering from Manhattan College, a Master's of Public Policy from Georgetown University, and is currently engaged in graduate studies in climate, energy, and environmental sustainability at Johns Hopkins University. Born and raised in the Bronx, she resides in the New York metro area with her husband and two children.

Most recently, Ms. Reynolds worked on geothermal energy research and technology-transfer initiatives supporting the deployment of geothermal sources for electricity generation. Previously, at the Natural Resources Defense Council (NRDC), she managed digital production and developed policy and advocacy resources for a full portfolio of environmental priorities, including renewable energy, environmental health, biodiversity protection, and equitable climate policy. She is a contributing author to publications on environmental considerations and barriers to deployment for hydrogen energy, and served as technical editor for research and market studies focused on energy efficiency standards and policies.

Ms. Reynolds began her career at BOC Edwards, managing knowledge-sharing initiatives to leverage internal engineering competencies and environmental, health, and safety resources to serve the semiconductor industry. She has spearheaded research and communications projects across a number of nonprofit and philanthropic organizations, including the Verizon Foundation.

R. Ryan Dupont has more than 40 years of experience teaching and conducting applied and basic research in environmental engineering at the Utah Water Research Laboratory at Utah State University. His main research areas have addressed soil and groundwater bioremediation, stormwater management via green stormwater infrastructure, field remediation technology demonstration and treatment system performance verification, and wastewater and biosolids reuse affected by contaminants of emerging concern (pharmaceuticals, personal care products, and polyfluorinated alkyl substances (PFAS) compounds).

He received a B.S. degree in Civil Engineering, and M.S. and Ph.D. degrees in Environmental Health Engineering from the University of Kansas, Lawrence. Dr. Dupont has been a Full Professor of Civil and Environmental since 1995. He served as the Head of the Environmental Engineering Division for 10 years, was instrumental in establishing an Undergraduate Degree in Environmental Engineering at USU, and has been responsible for attracting more than $\$5.8 \times 10^6$ in extramural funding through the Water Research Lab since joining the faculty in 1982.

Dr. Dupont is a member of Sigma Xi, Tau Beta Pi, Chi Epsilon, the American Society of Civil Engineers, the Water Environment Federation, and the AWMA. Dr. Dupont was recognized as an Outstanding Young Engineering Educator by the

ASEE, and was a 2015 recipient of the Richard I. Stessel Waste Management Award, for "distinguished achievement as an educator in the field of waste management" from the AWMA.

Dr. Dupont was also recognized as an endowed Cazier Professor of Civil and Environmental Engineering in 2021 for lifetime academic achievement and public service by Utah State University. One of the university's most prestigious honors, the award recognizes senior faculty members who have demonstrated more than 20 years of consistent excellence in their academic careers.

Walter Matystik, M.E., J.D., started his 45-year academic career in 1972 upon earning a Bachelor of Engineering degree in Chemical Engineering from the acclaimed industry-sponsored program at Manhattan College. He was an undergraduate research assistant in the Department of Energy-sponsored Particulate Solids Research Institute at the College's Paulian Labs under the direction of Dr. Frederick Zenz, innovator of fluidization technology for the petrochemical industry. He earned an EPA Training Fellowship and entered the prestigious Environmental Engineering Program at Manhattan College. He completed his Master of Engineering in Environmental Engineering in 1974, and stayed at the College to work with Dr. Thomann as a research engineer under an EPA grant to address eutrophication in the Great Lakes. Dr. Matystik decided to pursue a part-time law degree, earning a Juris Doctor from New York Law School in 1981 and admitted to the NY Bar in 1982 and the U.S. District Court in 1984.

Manhattan College appointed Dr. Matystik as Assistant Provost in 1992 and later as Associate Provost in 2010. Wanting to share his engineering and legal expertise with undergraduate and graduate students, Walter became a member of the College's adjunct faculty rising quickly to full professor and developing and teaching courses such as Legal Aspects of Engineering; Environmental Law; Hazardous Waste Management; Environmental Impacts of Construction Projects; and Environmental Impact Analysis. He is the recipient of EPRI's Distinguished Service Award.

Retiring from the College (now Manhattan University) in 2018, Dr. Matystik continues the pro-bono environmental law work he started while at the College assisting organizations such as the Bronx Council for Environmental Quality, the Bronx River Alliance, and the Environmental Law and Justice Project. He still serves as Secretary and a board member of the Academic Federal Credit Union since 2008.

Dr. Matystik's interest in geothermal energy combines his chemical and environmental engineering background along with his concern for the environmental impacts of technology plus his expertise with regulatory compliance. His current interest in how artificial intelligence can inform/impact future exploration and exploitation led him to this collaboration.

Louis Theodore, born and raised in Hell's Kitchen by poor immigrant Greek parents, received the degrees of M.Ch.E and Eng.Sc.D from New York University and B.Ch.E. from The Cooper Union. Dr. Theodore was a successful educator for over 50 years at Manhattan College (holding the rank of Full Professor of Chemical Engineering), Graduate Program Director, researcher, professional innovator, and technical communicator. During this period, he was primarily responsible for his program achieving a No. 2 ranking by the *U.S. News & World Report* and particularly successful in placing students in internships, jobs, and graduate schools.

Dr. Theodore is an internationally recognized lecturer who has provided more than 200 seminars/courses to industry, government, and technical associations. He also developed and served as the Principal Moderator/Lecturer for U.S. Environmental Protection Agency (U.S. EPA) courses on air pollution control equipment, consulted for several industrial companies in the field of air pollution control equipment, pollution prevention and environmental management, and served as a consultant/expert witness for the U.S. EPA and U.S. Department of Justice. He currently serves as a part-time consultant to Theodore Tutorials and is actively involved in air pollution control equipment, geothermal energy, and potable water/desalination activities.

Dr. Theodore is the author of more than 130 text/reference books ranging from air pollution control equipment to pollution prevention to hazardous waste incineration to engineering and environmental ethics. He has also served as section editor of *Environmental Management* for the last four editions of McGraw-Hill's *Perry's Chemical Engineers' Handbook*.

Dr. Theodore is the recipient of the Air & Waste Management Association's (AWMA) prestigious Ripperton Award that is "presented to an outstanding educator who through example, dedication and innovation has so inspired students to achieve excellence in their professional endeavors." He was also the recipient of the American Society of Engineering Education's (ASEE) AT&T Foundation Award for "excellence in the instruction of engineering students."

In addition to providing invited testimony to a Presidential (Ford) Crime Commission Hearing, Dr. Theodore was honored at Madison Square Garden in 2008 for his contributions to basketball and youth of America. He is also the author of *Basketball Coaching 101* (2015) and *Winning at Casinos: The Definitive Guide* (2024).

Dr. Theodore is a member of Phi Lambda Upsilon, Sigma Xi, Tau Beta Pi, American Chemical Society, ASEE, AWMA, and The Royal Hellenic Society.

Part 1

Prologue

The objective of the first part of this book is to introduce the reader to several key topics in the geothermal energy field. Chapter numbers and their accompanying titles are listed below:

Chapter 1. Energy History
Chapter 2. U.S. Department of Energy
Chapter 3. U.S. Energy Regulations
Chapter 4. Energy Resources: Fossil Fuels
Chapter 5. Energy Resources: Renewables

DOI: 10.1201/9781003667759-1

1 Energy History

1.1 INTRODUCTION

The concept of *energy* developed slowly over a period of several centuries and culminated in the establishment of the general principle of conservation of energy and its application. The availability of energy, which for many years was just a thermodynamic term to engineers and scientists, has now become a familiar matter to many others in the United States and elsewhere. Because of demand, energy availability has become a matter of concern to every household in the United States. Allocations of fuel oil for heating, even the remote possibility of institution of regulated distribution of fuels, reflect the current state of demand. Many factors are involved, some of which are: (1) energy requirements brought about by technology advance, (2) domestic local political atmosphere relative to the development of energy resources on an orderly basis and even on an emergency basis, (3) worldwide energy reserves and their distribution, (4) political and economic developments within countries that have significant energy reserves, and (5) the international political-economic situation related to energy resources (Elrod et al. 1973).

It must be recognized that with regard to fossil fuels – that is, coal, oil, and natural gas – the quantity available is not unlimited in the United States or, in fact, in the world. However, tremendous quantities of energy are consumed due to demand, and there are many possibilities for limiting or regulating its use. The effects of unwise use of energy and irreplaceable natural resources, and inadequate consideration of future needs and development of natural resources could have very drastic effects on future generations.

The operation of today's technological society depends upon both the production and use of large amounts of energy. Many of the world's present problems are closely related to energy distribution, dwindling fossil fuel supplies, and environmental effects of various methods of energy extraction, production, and utilization. The demand for energy worldwide has been increasing at over 5% per year and predictions are for this trend to continue. With approximately 6% of the world's population, the United States consumes nearly 30% of the world's total energy demand. This growth in the demand for energy is primarily a function of two factors:

DOI: 10.1201/9781003667759-2

1. Increased demand per person
2. Steady increases in the population

Industry, another major user of energy, is much more complex and is intimately related to the economy of the nation and its population. In the best-case scenario, reductions of energy demand could ideally come through improvements in operations and processes without a significant effect on economic growth. However, this is a real challenge and is intertwined with the competition in a free enterprise system. It should be remembered that a large part of nonautomobile transportation provides support to industry and has developed as a result of a large variety of social and economic pressures.

Better utilization of "waste" energy suggests recovery of energy that, up to now, has often been released to the local environment. In this and raw materials reclamation, a cooperative arrangement among industrial concerns may prove feasible in some instances. Utilizing some of the waste heat from the electric utilities for commercial and industrial heating is possible and is already occurring in limited instances across the globe.

Improvements in technology with emphasis on reduced energy requirements, reduced pollution, and conserving raw materials are needed in all use categories, particularly in industry, electric utilities, and transportation. More conscientious application of existing technology in lighting, heating, and cooling systems in building design to optimize their application can also produce substantial reductions in energy demands. Advanced technologies are generally associated with improved energy efficiency; however, the high energy demands of new artificial intelligence (AI) technologies are testing this notion, and the spread of AI technologies in virtually all economic sectors will put increasing pressure on future energy production demands.

In recent years, there have been dramatic changes in the supply and demand patterns of the basic "end use fuels" of oil, natural gas, propane, coal, etc. Since pricing is directly associated with the supply and demand for these products, it is certain that future energy demand trends will change.

Section numbers and their accompanying titles for this introductory chapter are listed:

1.2 Early History
1.3 The First Humans
1.4 The Industrial Revolution
1.5 Recent History
1.6 Conservation Practices
1.7 Concluding Remarks
References

1.2 EARLY HISTORY

BANG! The Big Bang. In 1948 physicist George Gamow proposed the Big Bang theory of the origin of the universe. He believed that the universe was created in a gigantic explosion as all mass and energy were created in an instant of time. Current estimates of the age of the universe are approximately 14 billion years, with 3–5 billion years often mentioned as the age of planet Earth (Theodore and Theodore 1995).

The Big Bang occurred in a split second and within a minute the universe was approximately a trillion miles wide and expanding at an unbelievable rate. Several minutes later all the matter known to humanity had been produced. This universe as it is known today was in place (Theodore and Theodore 1995).

Gamow further believed that the various elements observed today were produced within the first few minutes after the Big Bang, when near infinitely high temperatures fused subatomic particles into the chemical elements that now comprise the universe. More recent studies suggest that hydrogen and helium were the primary products of the Big Bang, with heavier elements being produced later within stars. The extremely high density within the primeval atom caused the universe to expand rapidly. As it expanded, the hydrogen and helium cooled and condensed into stars and galaxies. This perhaps explains the expansion of the universe and the physical basis for planet Earth.

Man has been benefitting from the energy of the sun since the birth of dawn. Some of it is stored in organic materials on the surface of the Earth. It was absorbed over its outer surface areas, and after billions of years, this solar energy was transformed by converting carbon into fossil fuels. Solar energy continues to be transformed into the kinetic energy of air (windmills) and water (waterfalls).

Fast forward to the present. More than any other time in history, the 21st century will likely be a turning point for human civilization. Human beings are facing pending ecological disasters that could affect one's ability to survive. These crises have forced many countries to reexamine the "modern" industrial energy value systems that have governed their lives for the past centuries.

1.3 THE FIRST HUMANS

Environmental problems have bedeviled humanity since the first person discovered fire. The earliest humans appear to have inhabited a variety of locales within a tropical and semitropical belt stretching from Ethiopia to southern Africa about 1.9 million years ago. These first humans provided for themselves by a combination of gathering food and hunting animals. Humans, for the majority of their two million years' existence, lived in this manner. It is believed that the steady development and dispersion of these early humans were largely due to an increase in their brain size. This led to the ability to think abstractly, which was vital in the development of technology and the ability to speak. This in turn led to cooperation and more elaborate social organization (Gordon and Suzuki 1991). These capabilities allowed early humans to develop techniques to overcome their hostile environment and ultimately led to the expansion of these first human settlements.

With the use of primitive tools and skins of animals for clothing, the first humans moved outside Africa about one and a half million years ago. This migration led them into the frost-free zones of the Middle East, India, southern China, and parts of Indonesia. The humans at this time could only adapt to those ecosystems found in the semitropical areas that contained a wide variety of vegetation and small, easily hunted animals to supplement their diet. Despite relatively easy access, Europe was not settled for a long period of time due to the deficient ecosystem, which was later overcome by technology. The first evidence of human settlement in Europe is dated to about 730,000 years ago with the settlement of towns that served as marketplaces. That meant clearing more land, cutting more timber, planting more crops, and raising more cattle, all at a rate that could be sustained only at a cost of significant damage to the land (Ponting 1991).

In the United States, the deforestation of New England and the disappearance of beaver in the East are but two dramatic examples of how the demands of the energy market could deplete abundant resources in short order. By the time of the American Revolution, the wilderness along the eastern seaboard had been tamed. While some pockets of forest remained, the 13 colonies were largely covered with farms, dotted with villages, and punctuated by a few cities, notably Boston, New York, Philadelphia, and Charleston. All of this occurred even though the population of the United States at the time of the American Revolution was only a small fraction of what it is today (Ponting 1991).

The first energy shortage may have occurred when man's hunting techniques were so well developed that large animals eventually disappeared. Because of this shortage, man probably developed methods of agriculture that led to the discovery of fire. Wood was the first and, for most of human history, the major source of energy. It was readily available because extensive forests grew in many parts of the world, and the amount of wood needed for heating and cooking was relatively modest. Certain other energy sources, found only in localized areas, were also used in ancient times including asphalt, coal, and peat from surface deposits and oil from seepages of underground deposits.

This situation changed when wood began to be used during the Middle Ages to make charcoal, which in turn was used to reduce ores to metals. As forests were cut and wood supplies dwindled at the onset of the Industrial Revolution, charcoal was replaced by coke from coal in the reduction of ores. Coal, which also began to be used to drive steam engines, became the dominant energy source soon after the arrival of the Industrial Revolution.

1.4 THE INDUSTRIAL REVOLUTION

Early in the 19th century, an awesome new force was gathering strength in Europe. The term *Industrial Revolution* was coined by the French as a metaphor of the affinity between technology and the great political revolution of modern times. Soon exported to the United States, the Industrial Revolution swept away any visions of America being an agrarian society. The steam engine, the railroad, the mechanical thresher, and hundreds of other ingenious artifacts that increased man's ability to transform the natural world and put it to use would soon be puffing and clattering and roaring in all corners of the nation. The new machines not only swiftly accelerated the consumption of raw materials from farms, forests, and mines but also increased energy demand.

Lumbering became the most important industry in the United States in the late 18th century. Wood was the most widely used raw material for fuel, houses, barns, and shops; the same can be said for ships, furniture, railroad ties, factories, and paper-making. The supply seemed inexhaustible since the forest still darkened huge parts of the country. The forest melted away before the axes of the advancing Americans. The settlers never thought of their axes work as deforestation, but as the progress of civilization.

Meanwhile, the big cities and growing wealth of the East were creating a more rapidly expanding market for wheat, corn, beef, and other cash crops. New roads and canals, the steamboat, and the locomotive made domestic and foreign markets increasingly accessible to farms in the center of the continent. Eli Whitney's cotton gin, Cyrus McCormick's reaper, Benjamin Holt's combine, and other ingenious inventions encouraged the development of a highly productive and efficient agriculture that sharply reduced the biological diversity of the land. Mining both preceded and quickly followed the settlement of the interior. Gold in California, lead in Illinois, coal and oil in Pennsylvania, iron ore in Minnesota, and copper in Montana attracted fortune hunters and job seekers from all over the world (Ponting 1991).

Although petroleum had been used in small quantities for centuries for purposes as diverse as medicine and building, the modern petroleum era began when a commercial well was brought into production in Pennsylvania in 1859. The American oil industry expanded rapidly as refineries sprang up to make oil products from crude oil. The oil companies soon began exporting their principal products.

1.5 RECENT HISTORY

In recent years, the United States oil companies have been enormously successful in expanding in the rest of the world. By 1955, the five major U.S. oil companies produced two-thirds of the oil for the world oil market (not including North America and the Soviet bloc). The world price at that time was about (get this!) a dollar a barrel, and during the mid-1950s, the United States was largely self-sufficient, with its imports limited by a quota.

Two series of events coincided to change this secure supply of cheap oil into an insecure supply of expensive oil. In 1960, enraged by cuts in oil prices by the seven major oil companies, the governments of the major oil-exporting countries – Venezuela and four countries around the Persian Gulf – formed the Organization of Petroleum Exporting Countries or OPEC to try to prevent further cuts in the price they received for their oil. As a result of the 1973 Arab–Israeli war, the Arab oil-producing countries cut back oil production and embargoed oil shipments to the United States and other countries. Arab OPEC nations proclaimed an oil embargo after a decision by the United States to resupply the Israeli military. The Arab OPEC group in effect was able to raise the price of oil from approximately $3 per barrel to roughly $13 per barrel by the end of the embargo.

In 1979 and 1980, a second oil crisis occurred as a result of the revolution that drove the Shah of Iran from his throne. Iranian oil production and exports dropped to negligible levels at that time. Because Iran had been a major exporter, consumers again panicked. The very high oil prices that resulted caused a worldwide recession and gave energy conservation a big push; as oil demand slackened, the world oil market slumped temporarily. From 1979 to 1982, oil prices rose from $15 per barrel to almost $40 per barrel. Today (at the time of the preparation of the chapter), oil is still relatively expensive, ranging roughly from $60 to $80 per barrel for West Texas Intermediate crude, and supplies are not guaranteed.

An expanded presentation of energy resources is provided later in this part. However, no discussion of energy issues would be complete without presenting a list of the present major sources of energy available in the United States. The ten key resources are:

1. Coal
2. Oil
3. Natural gas
4. Oil shale
5. Tar sands
6. Nuclear
7. Solar
8. Hydroelectric
9. Wind
10. Geothermal

Some may also consider hydrogen (an energy carrier) as an additional energy resource (DelGatto et al. 2025).

1.6 CONSERVATION PRACTICES

Three types of possible energy conservation practices may be described. The first type is curtailment, that is, doing without – for example, closing factories or staying home instead of taking trips. The second type is overhaul, that is, changing drastically the way people live – for example, outlawing further suburbanization. The third type involves the more efficient use of energy, that is, adjusting to higher energy costs – for example, making cars that go farther per unit of fuel, capturing waste heat in factories, and insulating houses. This last alternative is most readily acceptable to governments and to society in general.

By 1980 many people had come to recognize that increased energy efficiency could help the world energy balance in the short and middle term, and that productive conservation should be considered as no less an energy alternative than the energy sources that have been described previously. Substantial energy savings began to occur in the 1970s with the start of the modern environmental movement, and further savings of 25% to 40% appear possible without dramatically affecting the way people live.

A number of obstacles stand in the way, however. One major roadblock to productive conservation is its highly fragmented and unglamorous character; it requires hundreds of millions of people to do mundane things such as turn off lights and keep tires properly inflated. Another barrier has been the lack of emphasis given to conservation in national energy and energy-related debates. A third obstacle is the large subsidies given in some countries – particularly the United States and Canada – that encourage the overconsumption of energy. Most subsidies have been in the form of price controls that hold the price of conventional energy sources far below their true economic value, for example, artificial cost controls on natural gas in the United States. A fourth obstruction is the lack of information and adequate money for energy consumers to make energy conservation investments. A final impediment is the fundamental misconception about the relationship of energy use to economic growth.

In the face of increasing evidence to the contrary, many people continue to believe that increases in energy use are necessary for economic growth, thereby weakening the commitment of governments to pursue strategies to improve energy efficiency. Major areas for significant energy conservation improvements include transportation, industries, and buildings.

1.7 CONCLUDING REMARKS

The world's energy supply represents the combined resources by which the nations of the world attempt to meet their energy needs. Energy is the basis of industrial civilization; without it, modern life would cease to exist. During the 1970s the world began a painful adjustment to new realities of the conflict between energy demand and energy supply and availability. In the long run, conservation practices may provide the time to sort out the range of available or potential supply choices and technological options. While that process occurs, however, the world will likely continue to be vulnerable to disruptions in the supply of oil, which, after World War II, became, and remains, the most favored energy source for its growing economies.

The United States is at a turning point in the use of its energy resources. The cost of energy will no longer enjoy its coveted position of the past; it will cost more, much more, and be harder to obtain. Fossil fuels (the most abundant energy resource in the United States) and their utilization will almost certainly have to fill both the demand and the price gap in the near term if the United States is to continue as a growing industrialized nation.

In the final analysis, one can either accept or reject the grim projections for the future obtained by extending the consumption patterns and trends of the past, which establish the basis for defining "energy demand." Once it has been determined that the demand exists, the choice among the various means of energy conversion systems, either available at present or in some stage of development, can be made. This requires an evaluation, for each means of power generation, of the available fuel resources, the environmental implications of their use, the advances in waste energy recovery, technology improvements, and the relation to relevant economic and social issues. However, all of these considerations are themselves influenced by assumptions regarding future demands for power, and these, too, must be reexamined. For example, by analyzing the various components that presently constitute energy demand, various alternatives may be devised to maximize the long-term social return per unit of energy consumed. In turn, such alternatives may have important implications for the economic system, for social processes, and for lifestyles. These considerations define the energy issues and concerns of the future.

Enter geothermal energy … and the writing of this book.

REFERENCES

DelGatto, V.J., Theodore, L., Dupont, R.R., and Ogwa, M.C. 2025. *Hydrogen Energy: Principles and Applications.* Hoboken, NJ: Wiley.

Elrod, W., Wreglit, H., and Zabaneycz, S. 1973. *Unlimited Energy, Unlimited Control, Ultimate Disaster, First National Conference on Energy and the Environment.* Dayton, OH: Dayton AIChE Section.

Gordon, A., and Suzuki, D. 1991. *It's a Matter of Survival.* Cambridge, MA: Harvard University Press.

Ponting, C. 1991. *A Green History of the World.* New York, NY: St. Martin's Press.

Theodore, M.K., and Theodore, L. 1995. *Major Environmental Issues Facing the 21st Century.* contributing author (Chapter 1): A. Meier. East Williston, NY: Theodore Tutorials (originally published by Simon & Schuster).

2 U.S. Department of Energy

2.1 INTRODUCTION

The Department of Energy (DOE) is a cabinet-level branch of the U.S. government created by an act of Congress nearly 50 years ago in 1977. The DOE is headed by the Secretary of Energy, who is appointed by the President with the advice and consent of the Senate.

The DOE was established by Congress to deal with urgent national problems of dwindling supplies of coal, oil, and natural gas, and the increasing dependence of the U.S. economy on foreign sources of fossil fuels, particularly oil. It is responsible for overseeing the management of the fuel and energy resources of the United States and for fostering more efficient technologies for the utilization of those resources. The DOE is also involved with the marketing of electric power, regulation of energy production and use, pricing and allocation, energy conservation, and the nuclear weapons program. Of major concern are consumer interests, the protection of the environment, expanding the conversation on energy issues, and upholding open government principles of transparency, participation, and collaboration. One of the key ways DOE seeks to accomplish this is through the use of "social media," that is, social networks, blogs, wikis, podcasts, online videos, etc.

Section numbers and their accompanying titles for this chapter covering the DOE are listed:

2.2 DOE OVERVIEW

DOE has its headquarters in Washington, D.C., and several operations offices throughout the United States. It has assumed control of the facilities and personnel of five regional administrations for the transmission and marketing of electric power and

DOI: 10.1201/9781003667759-3

of strategic reserves of oil, shale, and uranium. Many research laboratories, including, for example, Brookhaven National Laboratory in Upton, NY, and the nuclear research facilities at Oak Ridge, TN, and Idaho Falls, ID, are also part of the DOE.

DOE was created by the Department of Energy Organization Act. It stipulated that it work closely with other departments of the government to reduce energy waste in federal programs. The DOE is also required to work with other federal agencies such as the Environmental Protection Agency (EPA), the Bureau of Mines, and the Nuclear Regulatory Commission, all of which have direct authority over the establishment of air and water pollution standards and standards for the safe design of nuclear power plants, respectively.

The DOE is *supposedly* required to prepare and submit to Congress biennially a National Energy Policy Plan, outlining objectives for energy use in light of other aims, such as full employment, price stability, national security, economic growth, environmental quality and protection, control of dangerous nuclear materials, and efficient utilization of publicly and privately owned energy resources.

The Federal Energy Regulatory Commission (FERC) was established within the DOE to assume many of the responsibilities of the former Federal Power Commission, including the right to establish and monitor rates charged for electricity and for the transportation of oil and gas by pipeline. However, FERC is an independent regulatory agency; its head is responsible only to the U.S. President.

The Economic Regulatory Administration (ERA), a department within the DOE, is responsible for administering many of the directives proposed by FERC. One of its first and earliest activities was to hold hearings in 1978 to set rates for oil transported through the Trans-Alaska Pipeline. The ERA also historically had the power to order electric utility companies to convert their operations to coal instead of the once scarcer oil and natural gas. The United States has since moved in the opposite direction, with most of the coal-fired power plants closed or closing for air and environmental quality concerns.

Another department, the Energy Information Administration (EIA), collects, analyzes, and publishes data on energy reserves, production, consumption, and other energy-related matters. Extensive fieldwork is carried out by EIA to ensure the accuracy of these data. The DOE took over all the programs of the Energy Research and Development Administration, which no longer exists. These programs, now distributed among a number of departments within the DOE, are concerned mainly with scientific and engineering research. They aim to develop better methods for extracting and combusting traditional fossil fuels, such as coal, oil, and natural gas, and also to develop new sources of energy, such as solar power, geothermal energy, and nuclear fusion.

The Advanced Research Projects Agency – Energy is a wing of the DOE that funds and directs the discovery of outlier energy technologies that are strategic to U.S. energy security. DOE has provided support for technologies such as coal liquefaction and coal gasification, fluidized bed combustion of coal, and improvement of methods for extracting oil from shale. Other concepts being explored in recent years are the conversion of animal wastes into methane gas, the burning of solid wastes as fuel, wind energy, and ocean thermal energy. Most of the solar power research is being pursued in connection with efforts to design residential commercial buildings that will use energy more efficiently. The most promising devices to date are solar collectors for water heating, solar mirrors, and advanced photovoltaics.

Finally, the DOE is responsible for research on laser-fusion and isotope-separation technology. The department also promotes peacetime uses of nuclear energy and is attempting to solve the difficult (if not impossible) problems involved in disposing of nuclear waste materials. Projects to develop other new types of reactors are being investigated, so "if not impossible today," they may, in fact, be possible in the future.

2.3 THE FEDERAL ENERGY REGULATORY COMMISSION

The FERC plays an important role within the DOE. The following provides some details on the scope of its activities. FERC is an independent agency that regulates the interstate transmission of electricity, natural gas, and oil. FERC also reviews proposals to build liquefied natural gas (LNG) terminals and interstate natural gas pipelines as well as licensing hydropower projects. The Energy Policy Act of 2005 gave FERC additional responsibilities as outlined in FERC's Top Initiatives and updated Strategic Plan. As part of that responsibility, FERC:

1. regulates the transmission and wholesale sales of electricity in interstate commerce;
2. reviews certain mergers and acquisitions and corporate transactions by electricity companies;
3. regulates the transmission and sale of natural gas for resale in interstate commerce;
4. regulates the transportation of oil by pipeline in interstate commerce;
5. approves the siting and abandonment of interstate natural gas pipelines and storage facilities;
6. reviews the siting application for electric transmission projects under limited circumstances;
7. ensures the safe operation and reliability of proposed and operating LNG terminals;
8. licenses and inspects private, municipal, and state hydroelectric projects;
9. protects the reliability of the high-voltage interstate transmission system through enforcement of mandatory reliability standards;
10. monitors and investigates energy markets;
11. enforces FERC regulatory requirements through imposition of civil penalties and other means;
12. oversees environmental matters related to natural gas and hydroelectricity projects and other matters; and
13. administers accounting financial reporting regulations and conduct of regulated companies.

Many areas outside FERC's jurisdictional responsibility are dealt with by state public utility commissions. Areas considered outside FERC's responsibility include:

1. Regulation of retail electricity and natural gas sales to consumers.
2. Approval for the physical construction of electric generation facilities.
3. Regulation of activities of municipal power systems, federal power marketing agencies like the Tennessee Valley Authority, and most rural electric cooperatives.
4. Regulation of nuclear power plants by the Nuclear Regulatory Commission.
5. Issuance of state water quality certificates.
6. Oversight for the construction of oil pipelines.
7. Abandonment of service related to oil facilities.
8. Mergers and acquisitions related to natural gas and oil companies.
9. Responsibility for pipeline safety or for pipeline transportation on or across the outer continental shelf.
10. Regulation of local distribution pipelines of natural gas.
11. Development and operation of natural gas vehicles.
12. Reliability problems related to failures of local distribution facilities.
13. Tree trimmings near local distribution power lines in residential neighborhoods.

2.4 THE ENERGY INFORMATION ADMINISTRATION

The U.S. EIA is the statistical and analytical agency within DOE. Some discussion of the EIA is warranted since it maintains the primary public repository of data for the United States and the world on energy resources. The Department of Energy Organization Act of 1977 established the EIA as the primary federal government authority on energy statistics and analysis, building upon systems and organizations first established in 1974 following the oil market disruption of 1973. Located in Washington, D.C., the EIA is an organization of approximately 250 federal employees, with an annual budget in Fiscal Year 2024 of $135 million.

The EIA collects, analyzes, and disseminates independent and impartial energy information to formulate sound energy policy and programs, efficient markets, and public understanding of energy and its interaction with the economy and the environment. The EIA is the nation's premier source of energy information and, by law, its data, analyses, and forecasts are independent of approval by any other officers or employees of the U.S. government. The EIA conducts a comprehensive data collection program that covers the full spectrum of energy sources, end users, and energy flows. It also prepares informative energy analyses, monthly short-term forecasts of energy market trends, and long-term U.S. and international energy outlooks. The EIA disseminates its data, analyses, and other products primarily through its website and a customer contact center.

The EIA provides the technical community with energy information. The information is disseminated in different ways, including reports, web products, press releases, databases, and maps. The information is issued weekly, monthly, annually, and periodically as needed or requested. Some of the products deal with specific energy industries or fuels. Some products contain mostly data; some also have analysis and forecasts. Other products present an integrated view over different fuels or energy uses. The EIA's Information Center also responds to customer inquiries by e-mail at InfoCtr@eia.gov, with an average response time of three business days.

The EIA's data and analyses are widely used by federal and state agencies, industries, media, researchers, consumers, and educators.

Industry looks to the EIA for official estimates on energy demand, supply, prices, markets, and financial indicators. Media and the general public rely on the EIA for the most comprehensive source of current and historical data and information on all aspects of U.S. energy. The international community relies on the EIA's products for timely information on the world energy supply and demand, and primary and secondary school educators and students turn to the EIA Kid's Page (www.eia.gov/kids/) for a wealth of energy-related information and activities.

All of EIA's products can be accessed through its website, www.eia.gov, which averages approximately two million visits per month. The EIA also distributes information and data on specific topics through 39 e-mail subscription lists and 7 RSS (Really Simple Syndication) feeds.

2.5 THE DOE GEOTHERMAL TECHNOLOGY OFFICE

The DOE Geothermal Technology Office (GTO) focuses on realizing the potential to generate electricity and produce heating and cooling for U.S. homes from domestic geothermal resources. To do so, GTO works in partnership with industries, academia, DOE's national laboratories, and others on research, development, and demonstration activities focused on these areas:

1. *Enhanced Geothermal Systems (EGS)* – Advancing the commercial viability of EGS (human-made geothermal energy) with a focus on reservoir characterization, enhancement, and sustainability. Initiatives include *Frontier Observatory for Research in Geothermal Energy* (www.energy.gov/eere/geothermal/forge) in Milford Utah; *Wells of Opportunity* (www.energy.gov/eere/geothermal/wells-opportunity), exploring both well retrofitting and co-production to help establish the commercial viability of geothermal energy production from existing hydrocarbon fields; and *EGS Pilot Demonstrations*, an initiative to demonstrate EGS in a variety of geographic locations, geologic formations, and subsurface conditions at field scale.
2. *Hydrothermal Resources* – Advancing technologies to expand electricity generation using naturally occurring geothermal resources and value-added opportunities like lithium extraction. Initiatives include *Hidden Systems* (www.energy.gov/eere/geothermal/hidden-systems), which addresses the importance of discovering hidden geothermal systems, or geothermal sources without obvious markers – like hot springs – on Earth's surface. This initiative sponsors integrated research, development, and deployment focused on moving hidden resources through the phases of project development where the risk is highest. These phases strongly influence a developer's accuracy in well targeting and the probability of success during more advanced drilling phases; *GeoFlight* (www.energy.gov/eere/geothermal/geoflight) seeks to capture surface and near-surface geologic data in the Salton Sea region to identify new geothermal energy prospects and areas of highest potential for mineral recovery, especially lithium. It is a joint effort between the GTO and

the U.S. Geological Survey (USGS); and the *Geothermal Drilling Technology Demonstration Campaign* (*Drilling RD&D*, www.energy.gov/hgeo/geothermal/drilling-demonstrations-initiative) to advance geothermal drilling technology and methods, helping to transfer research from laboratory settings into the field and, ultimately, the marketplace.

3. *Low-Temperature and Coproduced Resources* – Improving the efficiency and expanding the utility of low-temperature (<300°F) geothermal systems such as geothermal heat pumps and district heating and cooling systems. Programs include *Community Geothermal Heating and Cooling Design and Deployment* (https://publications.mygeoenergynow.org/grc/1035012.pdf), which is currently supporting five communities to install geothermal heating and cooling systems where coalitions selected project sites, assessed the geothermal resource and permitting needs, conducted feasibility analysis and local engagement, and identified workforce and training needs. The five selected communities include Ann Arbor, MI; Chicago, IL; Framingham, MA; Hinesburg, VT; and Shawnee, OK; and *Federal Geothermal Partnerships* (www.energy.gov/eere/ geothermal/federal-geothermal-partnerships), which is focused on the federal government's energy consumption (nearly 1% of all end-use energy in the United States). Approximately 450 federal sites comprise more than 75% of the federal government's energy use. A team of national laboratories, universities, a state agency, and industry partners were selected to provide technical assistance and help expand deployment of geothermal heating and cooling technology at federal sites. Two project sites have been selected for this initiative, which include the U.S. Army Detroit Arsenal and the West Point Military Academy.
4. *Data, Modeling, and Analysis* – Addressing nontechnical barriers to geothermal deployment through environmental and resource assessments, data stewardship, and analytical tools that advance geothermal exploration and development. Initiatives include the *GTO Multi-Year Program Plan* (www.energy.gov/eere/geothermal/articles/geothermal-technologies-office-multi-year-program-plan-fy-2022–2026), which provides a high-level technology plan that focuses on key areas of GTO research to support the growth and long-term contribution of geothermal energy to the U.S. electricity grid and American homes and buildings; the *U.S. Geothermal Power Production and District Heating Market Report* published in 2021 by the National Renewable Energy Lab in Golden, CO (https://docs.nrel.gov/docs/fy21osti/78291.pdf) captures domestic capacity and usage for geothermal power production and district heating and cooling, the impacts of state and federal policies on geothermal energy development, current research on geothermal development, and future opportunities for the domestic geothermal market and industry; *Permitting for Geothermal Power Development Projects* (www.energy.gov/eere/geothermal/permitting-geothermal-power-development-projects); and the *Geothermal Data Repository* (https://gdr.openei.org), which provides open access to geothermal data and information.

2.6 THE EPA'S ENERGY-RELATED REGULATORY RESPONSIBILITY

No discussion of the DOE's role in geothermal energy development would be complete without the mention of EPA's energy-related regulatory responsibilities. EPA was established by Congress in 1970. It strives to formulate and implement actions that lead to a balance between human activities and ability of natural systems to support and nurture life. The EPA also regulates energy use to protect human health and the environment. Some key energy-related regulations are the following:

1. *Transportation* – The EPA regulates air pollution from motor vehicles, engines, and the fuels used to operate them.
2. *Stationary sources* – The EPA regulates nonmoving sources – that is, fixed-site producers of pollution such as power plants, chemical plants, oil refineries, manufacturing facilities, and other industrial facilities. The key energy-related regulations that apply to stationary sources include (Stander and Theodore 2012):
   - *New Source Performance Standards*
   - *New Source Review*
   - *Prevention of Significant Deterioration*
   - *Clean Air Interstate Rule*
   - *Clean Air Mercury Rule*
3. *Clean air markets* – The EPA offers various market-based regulatory programs designed to improve air quality by reducing outdoor concentrations of fine particles, sulfur dioxide, nitrogen oxides, and mercury.
4. *Nuclear power radiation protection standards* – The EPA develops radiation protection standards to safeguard human health and the environment.
5. *Underground injection control* – The EPA regulates the construction, operation, permission, and closure of injection wells that place fluids underground for storage or disposal.

The EPA works with many partners to promote the responsible use of energy resources and enhances the nation's energy security. Some of the more important energy programs, agencies, groups, etc., are listed:

1. Stewardship, partnership, and voluntary programs – numerous EPA programs including ENERGY STAR, the Green Power Partnership, and the SmartWay Transport Partnership address energy consumption and conservation.
2. Office of Atmospheric Programs – these programs focus on climate change issues, protecting the ozone layer, and improving regional air quality by addressing problems such as acid rain.
3. Office of Transportation and Air Quality – EPA actively promotes travel options that reduce emissions and regulate air pollution from motor vehicles, engines, and fuel use.
4. Regional programs – EPA's ten regions provide energy programs for one's local area.

5. Energy Strategies Group – this program develops regulations and evaluates energy strategies to secure the nation's energy future.
6. Oil Program – EPA works to prevent, prepare for, and respond to oil spills that occur in and around inland waters of the United States.
7. Oil and Gas Sector Program – EPA's Sector Strategies Program for the oil and gas sector achieves performance improvement and burden reduction in a collaborative setting.
8. Resource Conservation Challenge – EPA's Office of Solid Waste promotes waste reduction and energy conservation.
9. Smart Growth Program – EPA's outreach programs help communities grow in ways that conserve energy.
10. Sustainable Infrastructure for Water and Wastewater – EPA identifies approaches to integrate energy-efficient practices for the water sector.
11. Underground Storage Tank Program – EPA oversees a regulatory program for underground systems that store petroleum products.
12. Underground Injection Control Program – this program oversees the underground injection of wastes, including wastes from oil and gas operations.

The EPA also offers the following energy-efficiency and energy-related programs:

1. *Combined Heat and Power Partnership (CHP)* – This program promotes the use of CHP to reduce the environmental impact of power generation.
2. *ENERGY STAR* – This joint program with the U.S. DOE helps save money and protect the environment through energy-efficient products and practices.
3. *Federal Electronics Challenge* – The EPA works with federal facilities to purchase greener electronic products, reduce impacts of these products during use, and manage obsolete electronics in an environmentally safe way.
4. *Sustainable Infrastructure for Water and Wastewater* – The EPA helps water utilities improve their energy efficiency.
5. *Design for the Environment* and *Safer Choices* – These programs work with industries that combine the potential for chemical risk reduction with improvements in energy efficiency.
6. *Responsible Appliance Disposal Program* – Partners recover ozone-depleting chemicals from old refrigerators, freezers, air conditioners, and humidifiers.
7. *Green Power Partnership* – This program encourages organizations such as Fortune 500 companies; small- and medium-sized businesses; local, state, and federal governments; and colleges and universities to buy green power.
8. *Clean Energy Programs* – The EPA's Clean Energy Programs are working with state and local policy makers, electric and gas utilities, energy customers, and other key stakeholders to develop clean energy policy and technology solutions.
9. *Best Workplaces for Commuters* – This national program recognizes employers who provide commuter benefits such as telework programs, carpooling matching, and van pool subsidies.

10. *Clean School Bus USA* – The EPA works to reduce children's exposure to diesel exhausts and fosters the deployment of innovative technologies.
11. *SmartWay Transport Partnership* – This innovative collaboration between the EPA and the freight industry increases energy efficiency while significantly reducing greenhouse gases and air pollution.

Theodore and Theodore (2010), Stander and Theodore (2012), and Skipka and Theodore (2014) provide extensive details on the EPA.

REFERENCES

Skipka, K., and Theodore, L. 2014. *Energy Resources Availability, Management and Environmental Impacts.* Boca Raton, FL: CRC Press/Taylor & Francis Group.

Stander, L., and Theodore, L. 2012. *Environmental Regulatory Calculations Handbook.* Boca Raton, FL: CRC Press/Taylor & Francis Group.

Theodore, M.K., and Theodore, L. 2010. *Introduction to Environmental Management.* Boca Raton, FL: CRC Press/Taylor & Francis Group.

3 U.S. Energy Regulations

3.1 INTRODUCTION

It should now be obvious that one of the objectives of this book is to lay the foundation of a national energy policy – or is it a national energy plan? Webster defines the two words in this manner (Merriam-Webster 2025a,b):

Policy: a high-level overall plan embracing the general goals and acceptable procedures, especially of a governmental body.

Plan: a method for achieving an end …an often customary method of doing something… a detailed formulation of a program of action.

In addition, Matystik (2011) has defined policy as:

> a general framework or plan for a course of action embodying desired principles. Policy can form the basis for law (e.g., a nondiscrimination policy can form the basis for civil rights legislation). Laws can directly flow from policy or be the rationale for legislation. Additionally, in some cases the law is the policy (e.g., U.S. Energy Policy Act). Policy can also stand-alone without a supporting law, e.g., foreign policy while setting a general direction for diplomacy.

Confusing? Perhaps. Obviously, there is some overlap in Webster's definitions, but in the energy field, the two words are essentially synonymous. For this text, it appears that "plan" might be the more appropriate word since "policy" would more appropriately describe an agency's interpretation of the law. The following describes the approaches that have been used in the United States. Similar approaches have been used in developed countries.

As noted previously, the "policy" or "plan" must first be developed and sent to Congress. Congress then passes a law that (hopefully) reflects the objectives of the policy. In the process, Congress must identify an agency or regulatory body to oversee and implement the law. The agency or regulatory body then may pass a regulation (or regulations) to ensure that the intent of the law is achieved. However – and here is where it gets tricky – the regulators may set a policy without going through the normal process associated with passing a regulation.

DOI: 10.1201/9781003667759-4

Matystik (2011) has further summarized the difference between laws, regulations, and policy as they apply to the environment in the following manner:

1. *Laws* provide vision, scope, and authority. They are usually enacted by a legislative body and serve as a basis for administrative implementation.
2. *Regulations* establish general requirements that must be met by a regulated community. These requirements generally apply at a national, state, or local level and are usually adopted by an administrative entity that has received an authorization from a law. These regulations generally interpret the enabling legislation and explain how the law is to be implemented.
3. *Policy* is developed by an administrative entity to explain and further interpret how laws and regulations are to be implemented and to resolve issues and conflicts of interpretation.

The material that follows addresses a host of regulatory-related topics, including:

3.2 The Regulatory System
3.3 Laws and Regulations: The Differences
3.4 The Role of the States
3.5 Environmental Regulations
3.6 The 2015 New York State Energy Plan
3.7 Future Outlook: U.S. Energy Policy and Energy Independence
References

3.2 THE REGULATORY SYSTEM

Over the past five decades, environmental regulations have become a system in which laws, regulations, and guidelines have become interrelated. For example, the history and development of the environmental regulatory system has led to laws that focus principally on only one environmental medium (i.e., air, water, or land). Some environmental managers feel that more needs to be done to manage all of the media simultaneously in an integrated fashion. Hopefully, the environmental regulatory system will eventually evolve into a truly multimedia management framework in the future.

Federal laws are the product of Congress. Regulations written to implement the law are promulgated by the executive branch of government, but until judicial decisions are made regarding the interpretations of the regulations there may be uncertainty about what regulations mean in real situations.

Until recently, environmental protection groups were most frequently the plaintiffs in cases brought to court seeking interpretation of the law. Today, industry is becoming more active in this role. Enforcement approaches for environmental regulations are environmentally management-oriented in that they seek to remedy environmental harm, rather than simply a specific infraction of a given regulation. All laws in a legal system may be used in enforcement to prevent damage or threats of damage to the environment or human health and safety. Tax laws (e.g., tax incentives) and business regulatory laws (e.g., product claims, liability disclosure) are examples of laws not directly focused on environmental protection but that may also be used to encourage compliance and discourage noncompliance with environmental regulations.

Common law also plays an important role in environmental management. Common law is the set of rules and principles relating to the government and security of persons and property. Common law authority is derived from the usages and customs that are recognized and enforced by the courts. In general, no infraction of the law is necessary when establishing a common law court action. A common law "civil wrong" (e.g., environmental pollution) that is brought to court is called a tort. Environmental torts may arise because of nuisance, trespass, or negligence.

Laws tend to be general and contain uncertainties relative to the implementation of principles and concepts they contain. Regulations derived from laws may be more specific, but are also frequently, too broad to allow clear translation into technology practice. Permits may be used to bridge this gap and prescribe specific technical requirements concerning the discharge of pollutants or other activities carried out by a facility that may impact the environment.

Most major federal environmental laws provide for citizen lawsuits. This empowers individuals to seek compliance or monetary penalties when these laws are violated and regulatory agencies do not take enforcement action against the violator.

3.3 LAWS AND REGULATIONS: THE DIFFERENCES

The following is a listing of some of the major differences between a federal law and a federal regulation, as briefly described in the previous section:

1. A law (or act) is passed by both houses of Congress and signed by the president. A regulation is issued by a government agency such as the DOE, U.S. EPA, or the Occupational Safety and Health Administration (OSHA).
2. Congress can pass a law on any subject it chooses. It is only limited by the restrictions in the Constitution. A law can be challenged in court only if it violates the Constitution. It may not be challenged if it is merely unwise, unreasonable, or even silly. If, for example, a law were passed that placed a tax on sneezing, it could not be challenged in court just because it was unenforceable. A regulation can be issued by an agency only if the agency is authorized to do so by a law passed by Congress. When Congress passes a law, it usually assigns an administrative agency to implement that law. A law regarding radio stations, for example, may be assigned to the Federal Communications Commission. Sometimes a new agency is established to implement a law. This was the case with the Consumer Product Safety Commission. OSHA is authorized by the Occupational Safety and Health Act to issue regulations that protect workers from exposure to the hazardous chemicals they use in industry and manufacturing processes.
3. Laws can include, for example, a Congressional mandate directing the EPA to develop a comprehensive set of regulations. Regulations, or rule-makings, are issued by an agency, such as the EPA, and translate the general mandate of a statute into a set of requirements for the agency and the regulated community to follow to achieve the intent of Congress.

4. Regulations are developed by agencies in an open and public manner according to an established process. When a regulation is formally proposed, it is published in an official government document called the *Federal Register* to notify the public of the agency's intent to create new regulations or modify existing ones. The agency then provides the public, which includes the potentially regulated community, with an opportunity to submit comments. Following an established comment period, EPA may revise the proposed rule based on both an internal review process and public comments (Stander and Theodore 2012).
5. The final regulation is published, or promulgated, in the Federal Register. Included with the regulation is a discussion of the agency's rationale for the regulatory approach, known as preamble language. Final regulations are compiled annually and incorporated in the *Code of Federal Regulations* (CFR) according to a highly structured format based on the topic of the regulation. This latter process is called codification, and each CFR title corresponds to a different regulatory authority. For example, DOE's regulations are primarily in Title 10 of the CFR. The codified DOE Worker Safety and Health Program regulations can be found in Title 10 of the CFR, Part 851, while Power and Transmission Rates regulations are codified in Title 10 of the CFR, Part 903. These regulations are cited as 10 CFR with the part listed afterward, that is, 10 CFR Part 903, or with the part and section referenced as 10 CFR §903.1.
6. A regulation may be challenged in court on the basis that the issuing agency exceeded the mandate given to it by Congress. If the law requires the agency to consider costs versus benefits of the regulation, the regulation could be challenged in court on the basis that the cost/benefit analysis was not correctly or adequately performed. If OSHA issues a regulation limiting a worker's exposure to a hazardous chemical to 1 part per million, OSHA could be called upon to prove in court that such a low limit was needed to prevent a worker from being harmed. Failure to prove this would mean that OSHA exceeded its mandate under the law, as OSHA is charged to develop standards only as stringent as those required to protect worker health and provide worker safety.
7. Laws are usually brief and general. Regulations are usually lengthy and detailed. The Hazardous Materials Transportation Act, for example, is approximately 20 pages long. It speaks in general terms about the need to protect the public from the dangers associated with transporting hazardous chemicals and identifies the Department of Transportation (DOT) as the agency responsible for issuing regulations implementing the law. The regulations issued by the DOT are several thousand pages long and are very detailed, down to the exact size, shape, design, and color of the warning placards that must be used on trucks carrying any of the thousands of chemicals regulated by the act.
8. Generally, laws are passed infrequently. Often years pass between amendments to an existing law. A completely new law on a given subject already addressed by an existing law is unusual. Laws are published as a "Public Law #__-_" and are eventually codified into the United States Code.

9. Regulations are issued and amended frequently. Proposed and final new regulations and amendments to existing regulations are published daily in the *Federal Register*. Final regulations have the force of law when published.

3.4 THE ROLE OF THE STATES

States have the ability to develop their own sets of laws and regulations similar to those of the federal government. Resource Conservation and Recovery Act (RCRA), for example, like most federal environmental legislation, encourages states to develop and run their own waste management programs as an alternative to EPA management. Thus, in a given state, the waste regulatory program may be run by the EPA or by a designated state agency. For a state to have jurisdiction over its waste management programs, they must receive approval from the EPA by showing that their programs are at least as stringent as those of the federal EPA.

States that are authorized to operate RCRA hazardous waste management or other programs oversee the hazardous waste tracking system in their state, operate the permitting system for hazardous waste facilities, and act as the enforcement arm in cases where an individual or a company practices illegal hazardous waste management. If needed, the EPA steps in to assist the states in enforcing the law. The EPA can also act directly to enforce RCRA or other laws in states that do not yet have or desire to have authorized programs. The EPA and the states currently act jointly to implement and enforce the regulations (U.S. EPA 1997).

State agencies do not operate in a vacuum. Indeed, most trace their very existence to the passage of major federal laws. While federal environmental legislation, such as the Clean Air Act (CAA), for example, established national ambient air quality standards (NAAQS), it went on to direct that these standards be met by state implementation plans (SIPs). The National Pollutant Discharge Elimination System, set up by the Clean Water Act, even anticipated a gradual handing off of control to states to administer their own permitting systems once they evidenced adequate authority. Thus, a mix of federal and state statutory laws evolved along with the appropriate bureaucracies to implement them – for example, DOE (discussed in the previous chapter) and other agencies at the federal level along with separate, parallel state regulatory agencies.

One can blame this patchwork quilt of laws, regulations, and standards on the notion of federalism on which the U.S. system of governance was built. Federalism is basically a sharing of power between the states and national (federal) government. In 1787, the new federal government only exercised limited or enumerated powers granted to it by the Constitution, such as making treaties and printing money. The Tenth Amendment to the Constitution clarified that all other powers belonged to the states –"The powers not delegated to the United States by the Constitution, nor prohibited by it to the states, are reserved to the states respectively, or to the people."

3.5 ENVIRONMENTAL REGULATIONS

Environmental regulations in a general sense are laws that attempt to protect the environment, not only within the borders of the United States but also throughout the entire planet. These regulations are aimed at protecting the air, land, and water that are subjected to society's private and commercial interactions.

In the United States, environmental regulations can be dated back to 1898 with the River and Harbor's Appropriations Act. This law was directed at the removal of logs and other debris from the nation's navigable waterways in order to maintain clear passage. The amount of environmental regulation has not only increased since then but also accelerated and broadened in scope. Environmental regulations within the United States have expanded the Federal Water Pollution Control Act of 1948 which was simply aimed at maintaining water quality, to the CAA Amendments of 1990 that mandates and enforces air pollution reduction, issues plans to meet air quality standards, and specifies compliance dates.

Historically, California was the first state to take steps against air pollution as a result of deteriorating air quality, especially in the Los Angeles area. As it became apparent that other urban areas were experiencing a decline in air quality, it was realized that federal intervention would become necessary. Although air pollution legislation was promulgated between 1955 and 1970, it was not until the CAA of 1970, combined with the creation of the EPA, that the framework for the current regulations was developed

The legislative approval process for environmental and energy law is the same for any law. As noted previously, the legislation must be approved by the U.S. Congress and sent to the president for approval and signature. If approved by the president, the legislation is signed into law. EPA is the branch of the federal government that is empowered to set and enforce environmental quality standards. The EPA is required by law to set national air and water quality and solid waste management standards and ensure that the standards are met. Stander and Theodore (2012) provide additional details on the EPA.

The environmental regulation most pertinent to energy production and consumption in the United States is the 1970 CAA. Although air pollution has been around for a long time, it has only been during the past 55 years that the governments of this country and many others have taken active roles in controlling it. In the United States, the first efforts at regulating air quality were principally at the individual community level in cities such as Cincinnati, Ohio, Pittsburgh, Pennsylvania, and Los Angeles. State requirements for controlling air pollution on a statewide basis came later.

At the federal level, the CAA defines the national policy for air pollution abatement and control in the United States. It establishes goals of protecting health and natural resources and delineates what is expected of federal, state, and local governments to achieve those goals. From its modest beginnings as an effort to determine the causes of illnesses and deaths related to air pollution that occurred in such places as Donora, PA, or in New York City, to the comprehensive document that exists today, the evolution of the CAA is an example of the country's efforts to protect people and the environment from adverse effects from air pollution, much of which comes from energy production and the consumption of various fuels.

The CAA of 1970 was enacted because of the slow process in developing ambient standards and the continuing deterioration of air quality across the country. These amendments required the newly formed EPA to adopt uniform NAAQS to protect health and welfare and required that these standards be attained and maintained across the country. Other aspects of the 1970 amendments included a significant federal enforcement authority; the development of SIPs to achieve the ambient air quality standards; more stringent motor vehicle emission standards; requirements for EPA to establish standards of performance for new and modified sources of air pollution; requirements for establishing national emission standards for hazardous air pollutants; and provisions for citizen suits.

In 1977, the CAA was again amended to enable EPA to make some corrections in the country's efforts to achieve and maintain ambient standards. Additional time was allowed for achieving the NAAQS in the most polluted areas, and the concept of Prevention of Significant Deterioration was made part of the law. Additionally, EPA was given the flexibility to allow for some growth in "dirty" areas.

The 1990 Amendments to the CAA are the most recent changes in effect. Since 1977, the country had been unsuccessful in its efforts to achieve the ambient standards for many pollutants and EPA had been slow in identifying and requiring controls of hazardous air pollutants. The 1990 Amendments to the CAA provided a listing of hazardous air pollutants (Title III) and required EPA to establish emission standards for certain industrial and commercial sources of these pollutants.

The CAA (as amended in 1990) is composed of six subchapters, each of which establishes requirements for a portion of the nation's air pollution control effort. Subchapter I requires controls for stationary sources (including sources of toxic air pollutants as modified by Title III of the 1990 amendments to the CAA) to establish more stringent control requirements in those areas that continue to have polluted air. Subchapter II establishes requirements for mobile sources to reduce emissions from motor vehicles. Subchapter III includes definitions and administrative provisions for the development of regulations and for public involvement. The acid rain provisions are prescribed in Subchapter IV reducing emissions of sulfur dioxide from power plants. Subchapter V (referred to as Title V) covers the state's operating permits program, and Subchapter VI provides requirements for protection of stratospheric ozone and implements the Montreal Protocol, an international agreement to reduce emissions and ultimately eliminate the use of chlorofluorocarbons and other ozone-depleting chemicals.

The following five major rules were promulgated to achieve significant improvement in air quality, health, and quality of life.

1. *Clean Air Interstate Rule* (70 FR 25161, May 12, 2005). The Clean Air Interstate Rule (CAIR) provided states with a solution to the problem of power plant pollution that drifts from one state to another. The rule uses a cap and trade system to reduce the target pollutants by 70%.
2. *Mercury Rule* (70 FR 28605, May 18, 2005). EPA issued the Clean Air Mercury Rule on March 15, 2005. This rule builds on the CAIR to reduce mercury emissions from coal-fired power plants, the largest remaining domestic source of human-caused mercury emissions. Issuance of the Clean Air Mercury Rule marked the first time EPA regulated mercury emissions from utilities, and made the United States the first nation in the world to control emissions from this major source of mercury pollution.

3. *Nonroad Diesel Rule* (69 FR 38957, May 11, 2004). The Clean Air Nonroad Diesel Rule changes the way diesel engines function to remove emissions and the way diesel fuel is refined to remove sulfur. The Rule is one of EPA's Clean Diesel Programs, which were promulgated to produce significant improvements in air quality from the transportation sector.
4. *Ozone Rules* (www.epa.gov/ground-level-ozone-pollution/ozone-national-ambient-air-quality-standards-naaqs). The Clean Air Ozone Rules (dealing with 8-hour ground-level ozone designation and implementation) designated those areas whose air did not meet the health-based standards for ground-level ozone. The ozone rules classified the seriousness of the problem and required states to submit plans for reducing the levels of ozone in areas where the ozone standards were not being met.
5. *Fine Particle Rules* (www.epa.gov/pmdesignations/). The Clean Air Fine Particles Rules designated those areas whose air does not meet the health-based standards for fine-particulate pollution. This rule required states to submit plans for reducing the levels of particulate pollution in areas where the fine-particle standards are not met.

3.6 THE 2015 NEW YORK STATE ENERGY PLAN

Energy planning does not stop with the federal government in the United States. Several states have proposed or set forth formal energy plans. An abbreviated outline of the 2015 energy plan for New York State is provided here. The 2015 New York State Energy Plan ("Plan") (https://energyplan.ny.gov), updated in 2020, when it was amended to align with the objectives of the 2019 Climate Leadership and Community Protection Act (Climate Act), is a comprehensive roadmap to build a clean, resilient, and affordable energy system for all New Yorkers. It provides broad policy and program direction to guide energy-related decision-making in the public and private sectors within New York State. The Plan (New York State 2025b) addresses meeting future energy needs, while balancing reliability, cost, environmental and public health impacts, and economic growth. The topic areas developed in the Plan are guided by statutory requirements (Article 6 of the Energy Law). The statute requires that the Plan seeks to:

1. Insulate consumers from volatility in market prices, emphasizing energy affordability;
2. Animate clean energy markets to transition to a clean energy economy;
3. Improve the reliability and resiliency of the state's energy systems;
4. Reform the energy regulatory burden in the state;
5. Minimize public health and environmental impacts, particularly those related to climate change, and particularly those on minority and low-income communities;
6. Guide the development and implementation of clean, reliable transportation systems in the state.

The primary plan objective is initially to provide an integrated overview of the state's energy systems including historic, current, and forecasted demand and prices, disaggregated by fuel and customer types and followed by the evaluation of future energy and infrastructure requirements and costs, supply options, and system reliability needs. Then, the Plan is to assess system changes, technology development, economic growth, and environmental impacts, with focus on the potential contribution of energy efficiency, renewable energy, and distributed sources. And, finally, the Plan examines current energy prices for New York State customers, particularly for low-income consumers.

The Plan is further divided into seven interrelated categories of energy initiatives (New York that include:

1. Renewable Energy
2. Buildings and Energy Efficiency
3. Clean Energy Financing
4. Sustainable and Resilient Communities
5. Energy Infrastructure Modernization
6. Innovation and R&D
7. Transportation

The New York Energy Plan 2030 targets include a 40% reduction in greenhouse gas emissions from 1990 levels; 50% electricity generation from renewable energy sources; and a 600 Trillion Btu increase in statewide energy efficiency. From an evaluation perspective, a comprehensive plan integrating all of these analyses and activities into a concise energy plan will be a major challenge.

3.7 FUTURE OUTLOOK: U.S. ENERGY POLICY AND ENERGY INDEPENDENCE

Energy is central to all current and future human activities, and, as such, this nation is approaching a crisis stage. Historically, the primary impetus for the emergence of the United States as a global leader has been the ability to satisfy energy needs independently. Without a strategy for determining the most beneficial path to achieving future independence, the future is resigned to be at the mercy of those that will control world energy supplies.

The current energy policy being pursued by the United States is disjointed, random, and fraught with vested interests. Achieving energy independence by pursuing current approaches will be costly, inefficient, and disruptive, especially as resources diminish. In addition, environmental impacts will become major concerns, and the consequences of continuing on the current course may be surprisingly difficult to correct.

The facts on present-day energy consumption are universally accepted. Even the projections for raw material reserves of oil, coal, gas, and uranium cause little argument. But, consensus on all other aspects of energy policy is nonexistent. In the broadest sense, many cannot agree whether there is presently a crisis or a problem. In any event, a number of measures must be taken to assure that where energy problems exist, they will not worsen. To better appraise the magnitude of these measures, one must set short-term and long-term goals, both of which are briefly discussed next (Theodore and Theodore 2010; Skipka and Theodore 2014).

3.7.1 Short Term

Because of the long lead times required to substantially improve a nation's energy position, the only answer to present energy shortages is to allow the continued importation and production of crude oil, distillate, and residuals. Because of the serious strain that such imports can place on the balance of payments, it is critical that conservation steps be initiated.

3.7.2 Long Term

In the United States, a comprehensive, coordinated national energy policy must be formulated. This policy should be subject to continuing review and adjustment to address such needs as depleting resources, new technological breakthroughs, conservation measures, etc. In short, the policy should address the following (baker's dozen) categories:

1. Establish a comprehensive cabinet-level agency dealing with all forms of energy. The DOE could assume this role.
2. Systematize and refine air and water pollution guidelines and establish land utilization policies to ease the siting of refineries, power plants, and petrochemical operations.
3. Stimulate increases in domestic oil and gas production by deregulating gas prices or regulating them with realistic price guidelines.
4. Stimulate increases in domestic oil and gas production by increasing the size of offshore leases as well as the frequency of lease sales.
5. Change Internal Revenue guidelines to favor exploration in the United States rather than favoring international exploration, and production abroad by allowing royalties paid to foreign governments to be written off against federal tax owed the U.S. government.
6. Foster research on improved coal mining and new reclamation techniques.
7. Sponsor, fund, and encourage coal gasification projects to a degree that would lead to possible commercialization of the process within this and the next decade.
8. Foster additional research on alternative energy sources (solar, nuclear, tidal, wind, biofuel, and geothermal energy).
9. Provide regulations and/or incentives to allow the use of coal as a petroleum substitute in certain industrial operations.
10. Accelerate and systematize the environmental review process of proposed energy projects.
11. Coordinate all energy research to take advantage of breakthrough technologies with the end goal being energy independence.

12. Foster additional research on the environmental problems involved in the production of both geothermal and nuclear energy.
13. Analyze and implement energy conservation measures whenever such a need arises.

Similar options can be considered in other developed or developing nations. In a very real sense, one of the objectives of the above list is to develop and propose an energy policy to lead the country into a future of energy independence and clean and inexpensive energy availability for all citizens.

REFERENCES

Matystik, W. 2011. *Personal Communication*. East Williston, NY.

Merriam-Webster. 2025a. Plan. In *Merriam-Webster.com dictionary*. Retrieved June 2, 2025, from www.merriam-webster.com/dictionary/plan.

Merriam-Webster. 2025b. Policy. In *Merriam-Webster.com dictionary*. Retrieved June 2, 2025, from www.merriam-webster.com/dictionary/policy.

New York State. 2025a. *New York Energy Plan*. https://energyplan.ny.gov/Plans/2015-Energy-Plan. (accessed June 2, 2025).

New York State. 2025b. *State Energy Plan. Volume 1. Energy to Lead*. https://energyplan.ny.gov/-/media/Project/EnergyPlan/files/2015-state-energy-plan.pdf. (accessed June 2, 2025).

Skipka, K., and Theodore, L. 2014. *Energy Resources: Availability, Management and Environmental Impacts*. Boca Raton, FL: CRC Press/Taylor & Francis Group.

Stander, L., and Theodore, L. 2012. *Environmental Regulatory Calculations Handbook*. Boca Raton, FL: CRC Press/Taylor & Francis Group.

Theodore, M.K., and Theodore, L. 2010. *Introduction to Environmental Management*. Boca Raton, FL: CRC Press/Taylor & Francis Group.

U.S. Environmental Protection Agency. 1997. *RCRA: Reducing Risk from Waste*. EPA 530-K-97-004, Solid Waste and Emergency Response. Washington, D.C.: U.S. EPA. https://archive.epa.gov/epawaste/inforesources/web/html/risk.html.

4 Energy Resources
Fossil Fuels

4.1 INTRODUCTION

This chapter contains information on fossil fuels; it provides an expanded treatment of material introduced earlier in Chapter 1. The subject matter varies from coal to tar sands.

Section numbers and their accompanying titles are listed as follows:

The reader should note that an expanded treatment of environmental issues as they relate to the fossil fuels is addressed later in this text.

4.2 COAL

Coal is a combustible black or brownish-black sedimentary rock normally occurring in layers or veins referred to as coal beds or coal seams. It is composed primarily of elemental carbon along with variable quantities of other elements – chiefly hydrogen, with smaller quantities of sulfur, and even smaller quantities of oxygen and nitrogen.

Coal was formed from the energy that plants absorbed from the sun millions of years ago. These living plants converted solar energy to living plant materials through photosynthesis. This stored energy is usually released when the plants decay. Under conditions favorable to coal formation, however, this decay process was interrupted, preventing the release of the plant's stored solar energy; in effect, the energy became locked into the coal.

Over time, the dead plant matter was deposited into swampy water and a thick layer of dead plants decayed at the bottom of the swamps. As the surface and climate

DOI: 10.1201/9781003667759-5

of the Earth changed, more swamp seeped above the plants, halting the decay process. The mass of the top layer of swamp water packed down the lower layers of dead plants. Under both heat and the increased pressure, the plant matter underwent chemical reactions and physical changes. This resulted in leaving rich hydrocarbon deposits; in effect, the plants gradually turned into coal. As the preceding physical and chemical processes continued, the progressive changes were from wood and vegetation to peat, which in turn was changed by heat and pressure to brown coals and lignite, sub-bituminous coals, and finally to anthracite and graphite.

The chemical and physical process that transformed plant debris to coal is referred to as coalification. The first product of this process was the aforementioned peat. However, peat was progressively transformed to lignite, which eventually became anthracite, etc., depending on the progression of geological changes. These chemical changes affected an increase in carbon content plus loss of water, oxygen, and hydrogen. The degree of coalification is determined by Skipka and Theodore (2014):

1. Varying types of vegetation from which the coal originated
2. Depths of burial
3. Temperatures and pressures at those depths
4. Length of time the coal has been forming in the deposit

The history of coal varies with the literature chosen for review. One of the earliest references to the use of coal as a fuel is from the geological treatise *On Stones* by the Greek scientist Theophrastus (Mattusch 2008):

> Among the materials that are dug because they are useful, those known as coals are made of earth, and, once set on fire, they burn like charcoal. They are found in Liguria...and in Elis as one approaches Olympia by the mountain road; and they are used by those who work in metals.

Others claim the earliest reference was by the Chinese during the time of Marco Polo.

Coal reserves in the Unites States have been estimated from drillings, outcroppings, gorges cut by rivers, and explorations for minerals and oil. The two great coal-producing regions in the Unites States are the Appalachians including the states of Pennsylvania, West Virginia, Ohio, western Maryland, eastern Kentucky, Virginia, Tennessee, and Alabama; and the Central States region including Illinois, Indiana, western Kentucky, Iowa, Missouri, western Kansas, Oklahoma, and Arkansas. The Unites States is the largest user of coal in the world. However, Russia, Great Britain, Poland, and West Germany also use considerable amounts of coal resources. The reserves of coal available are variously estimated to be from 8 to 16 trillion tons. Coal generated 23% of the electricity used in the United States in 2023 (EIA 2025a). Other energy sources, especially renewables and nuclear energy, will increase their rate of generation of electricity but it is expected that coal will continue to play an important role in the U.S. energy sector during the transition to more sustainable energy sources, especially as a result of the 2024 U.S. election.

4.3 OIL

For over a century the terms oil and petroleum have been used interchangeably by industry. However, there are some who refer to petroleum as the crude oil prior to any treatment/refining, with oil reserved as the definition of "any of the various kinds of greasy, combustible substances obtained from animal, vegetable, and mineral sources, and liquid at ordinary temperature." Both terms are employed interchangeably in this chapter as a matter of convenience. The typical petroleum reservoirs are mostly sandstone or limestone formations containing oil. The viscosity of the oil in these formations may be as thin as gasoline or as thick as tar. It may be almost clear or black.

Unlike coal, in which the stored energy actually takes the form of rock, oil is an entirely separate substance from the rock in which it was formed. Oil, in its liquid form, is very mobile and tends to move upward as the heat in the Earth's crust causes it to expand. Rising oil frequently penetrates surrounding rock formations and continues its upward journey until it meets a layer of rock that is so tightly compressed that passage is not possible. Such a barrier might be produced by movements in the Earth's crust that have caused hard, impenetrable rock to lie against the porous layers through which oil is moving. When this happens, the oil becomes trapped in a formation that is defined today as an *oilfield* (Skipka and Theodore 2014).

Although oil deposits have been formed by geological processes similar to those that created coal seams, the origin of oil itself is more difficult to discover. It probably derives from stores of carbon and hydrogen that built up in the plants and tiny animals that lived in the shallow seas covering many parts of the Earth 400–500 million years ago. The remains of these creatures eventually sank and mixed with mud on the sea bottom. As new layers of silt were washed into the seas by rivers, the mixture was sealed off from the oxygen necessary for aerobic decomposition to take place and release the stored energy back into the atmosphere in the form of heat. Instead, the pressure and heat caused by the buildup of new upper layers of material gradually turned the mud into a firm rock and, over millions of years, forced out the carbon and hydrogen. These components were then transformed by heat and pressure to become the fossil fuel referred to as oil.

Oil, like coal, is a fossil hydrocarbon fuel. The particular combination of hydrogen and carbon determines whether the fuel takes the solid form of coal or the liquid form of oil or the gaseous form of natural gas. Oil has a higher proportion of hydrogen than coal and this makes it lighter in weight. Due to its chemical composition, 1 ton of oil has about the same energy content as 1.5 tons of coal.

It is important to note that fossil fuels are *finite* in supply. The increased use of petroleum around the world has caused supply and demand problems that have resulted in price increases everywhere. The Unites States continues to be reliant on oil as a primary fuel for transportation, as well as a supply for the industrial production of chemicals and other materials.

Oil is discovered by small and large companies by means of geological exploration and the drilling of exploratory wells. A total of 918,068 oil-producing wells were identified in 2023 (EIA 2025b). Drilling technology has steadily improved in part due to interest in geothermal energy. Although many exotic drilling techniques have been proposed and are being investigated, the rotary drilling technique still remains the most efficient and economic method for oil-well drilling in the Earth's crust. One recent innovation reported by EIA (2025b) is the expansion of horizontal well drilling technology that has increased from 8% to 21% of the active well inventory between 2013 and 2023.

From a processing perspective, an oil refinery turns crude oil into gasoline, diesel oil, lubricating oil, kerosene, and jet fuel. Gasoline is the chief product of a refinery (approximately 45%) and fuel oil of various grades ranks second in importance. Fuel oil is used to heat homes and buildings and to supply power for factories and railroads. Jet fuel contains a mixture of gasoline, kerosene, and oils with low freezing points. Asphalt for roads is another product of crude oil. Petrochemicals are chemicals produced from petroleum; examples are plastics. nylon, fertilizers, and medical drugs.

4.4 NATURAL GAS

Natural gas consists of a mixture of different gases. The main component is methane (CH_4), which is odorless, colorless, and tasteless. Companies add a chemical odorant called mercaptan (with an odor like rotten eggs) so that any escaping gas can be detected. This process is employed as a safety measure (Theodore and Dupont 2012).

Natural gas is a nonrenewable fossil fuel, because most scientists and engineers believe that natural gas (as with oil) was formed from the remains of sea animals and plants that died 300–400 million years ago. As noted earlier, when these sea animals and plants died, they sank to the bottom of the oceans where they were buried by layers of sediment that turned into rock. Over the millennia, these layers increased in height to several thousands of feet, subjecting the remains to high pressure, which, when combined with the high temperature of the Earth, converted this matter into not only oil but also natural gas. Over time, high concentrations of natural gas became trapped in the layers of rock.

As with other fossil fuels, natural gas was formed in the Earth's crust, possessing the ability of rising toward the Earth's surface. It built up in pockets when its progress was blocked by an upper layer of solid impervious rock. This process has been described as being analogous to trapping water with a sponge. The gas accumulation can be reached by drilling wells in a manner similar to how oil is recovered from an oil field. It should be mentioned that natural gas is produced on all continents except Antarctica.

Natural gas is often found with oil. The oil fields of the Middle East once contained about one-fifth of the world's natural gas reserves. In the past, large quantities of natural gas recovered along with oil were burned because there was no use for this energy source locally. Today, this gas is recovered and liquefied for transport to users. The leading producers of natural gas are Russia, the United States, Canada, and Western Europe. Prior to the commercialization of natural gas, the gas was manufactured from coal and oil. However, this manufactured gas had only about half the heating value of natural gas.

Natural gas ranks second, after oil, in energy consumption in the United States and approximately 36% of the energy consumed is supplied by this fossil fuel (EIA

2025a). Best estimates of natural gas consumption by sector in 2021 are the following (EIA 2025a):

Electricity generation: 40%
Industry: 32%
Residential: 14%
Commercial use: 10%
Transportation: 4%

Natural gas is a relatively inexpensive fuel and has become an important part of the U.S. energy portfolio because it is relatively nonpolluting and is widely available throughout the country. Natural gas is the cleanest and most flexible fossil fuel, but it is more difficult to store and transport than liquid or solid fuels, especially when the distribution system must extend from one continent to another. The annual consumption of natural gas in the Unites States continues to increase.

4.5 SHALE OIL

Oil deposits are not always buried deep down in the Earth's crust. In some places oil has been formed under just a thin layer of soil and rock. This layer has been too light to compress the oil-bearing sands into the firm rock typical of deep underground oil reservoirs. Even so, there has been sufficient heat at these shallow depths in the Earth's crust for hydrogen and carbon from decomposing plant and marine life to combine over millions of years into crude oil similar to that found at much greater depths (Skipka and Theodore 2014).

Shales are geologic formations composed of thinly layered sedimentary rocks, and these are widely dispersed on the Earth. In certain shale deposits, organic material has been laid down and intermixed with the inorganic material. It has partially decomposed and metamorphosed over the ages until it now exists as a mixture of organic substances locked within the shale. The organic material contained in shale is called *kerogen* (Skipka and Theodore 2014).

The shale itself contains no oil as it is known today. However, if the shale is pretreated and fractionated at elevated temperatures, the kerogen decomposes and part of it is discharged as an oily vapor, which can be condensed into a viscous liquid referred to as shale oil. Unlike typical crude oil, this shale oil contains large amounts of compounds of nitrogen and large amounts of sulfur. These must be removed before a product similar to crude oil results.

Vast new energy resources are being discovered every year across North America. One major source of these "vast new resources" comes from the development of deep shale natural gas formations. Deep shale natural gas is typically found thousands of feet below the Earth's surface in tight, low-permeability shale formations. Experts have known for years that natural gas deposits existed in deep shale formations, but until recently, the vast quantities of natural gas in these formations were thought to be unrecoverable.

Today, through the use of a technique called *hydraulic fracking*, combined with sophisticated horizontal drilling techniques, extraordinary amounts of natural gas from deep shale formations across the United States are being produced (Skipka and Theodore 2014).

The oil shale industry is an industry focused on mining and processing of oil shale from which liquid hydrocarbons can be obtained. The industry was first developed in Estonia and has expanded to the United States, Germany, Israel, Russia, China, Brazil, and to a growing extent in Canada. Several other countries are currently investigating the potential of their oil shale reserves.

Following the 1973 oil crisis, the oil shale industry was resurrected in several countries, including the United States. However, in the 1980s, when oil prices fell, the industry again declined. The industry grew in full force at the start of the 21st century. The oil shale development program was initiated in the United States and later led to the commercial leasing program for oil shale. Interestingly, Estonia is actively involved in the exploitation of oil shale as oil shale and oil shale gas accounted for 54% of its total domestic energy generation in 2020. Due to the negative environmental impacts of oil shale mining, processing, and use, however, Estonia has set a target of stopping oil shale power generation by 2035 and all uses of oil shale in its energy sector by 2040 at the latest (Wikipedia 2025a).

4.6 TAR SANDS

Tar sands is a unique term. The word tar has been defined as oily, dark-colored bituminous products obtained by the distillation of various petroleum feedstocks, including crude oil. Tars are typically rich in organic compounds that are related to benzene. High-molecular tars (or pitches) are used in the manufacture of roofing papers and for making coal briquettes. Alternately, the term sand has been defined as a loose, incoherent mass of mineral materials in a finely granular condition, usually consisting of quartz (silica), with small quantities of mica, feldspar, magnetite, and other resistant minerals possibly present. It is the product of the chemical and mechanical disintegration of rocks under the condition of weathering and abrasion. When formed, the particles are usually angular and sharply pointed, becoming smaller and more rounded by attrition. It is also a constituent of most soils and is extremely abundant as a surface deposit along the courses of rivers, on the shores of lakes and oceans, and in arid regions.

Tar sands (also alternately referred to as bituminous or oil sands) are located in certain beds. They contain a somewhat dense, viscous form of petroleum, with the assigned technical term of bitumen. The term tar sands was first employed around the turn of the 20th century. The interstices between the grains of sand are largely filled with a thick tarry hydrocarbon mixture that can be extracted. This extract can be processed to yield 65%–75% synthetic crude oil. The sand was initially mined and transported to a separation plant. Only open-pit mining was practical, and since most of the deposits were buried too deeply for this activity, there were also relatively few places where commercial, large-scale operations were possible at that time. One of many of these deposits is the Athabasca Tar Sands, located in northern Alberta, Canada, along both sides of the Athabasca River.

The use of tar sands dates back to Paleolithic times. The first known use of bitumen was by Neanderthals, approximately 40,000 years ago. Humans later used bitumen for construction of buildings and waterproofing. Bitumen was primarily a Mesopotamian raw material used by the Babylonians and also those residing in Persia (Iran). Early Canadian settlers used bitumen from the Athabasca deposits to waterproof canoes.

The presence of the Athabasca tar sands in Canada has been known for many years. These tar sand deposits cover an area of approximately 9,000 square miles in the province of Alberta. The tar sands are beds, or layers, of a mixture of sand, water, and bitumen, with the water and bitumen forming a film around each grain of sand. When the sand is compressed, it excretes an oil residue. Significant bitumen deposits have been reported in Canada, Kazakhstan, Russia, and Venezuela. The estimated worldwide deposits of oil are more than 2 trillion barrels with approximately 71% residing in Alberta. Canada (Wikipedia 2025b).

It is estimated that 90% of U.S. tar sand deposits are located in eastern Utah. This amount is dwarfed, however, by the massive Athabasca tar sand deposits in Canada, where commercial activity is beyond the developmental stage. Both surface mining and in situ recovery are now being used to obtain oil from tar sands. Thus, considerable efforts are being made today to further develop processes to exploit these considerable reserves. However, it is difficult to estimate the percentage of these reserves that are economically recoverable while complying with environmental regulations. In addition, the characteristics of the Utah deposits are significantly different from the Athabascan deposits as these U.S. tar sands are "hydrocarbon wet," while the Canadian tar sands are "water wet."

4.7 ENVIRONMENTAL ISSUES

This section briefly examines environmental concerns associated with the five fossil fuel resources reviewed in the previous sections of this chapter.

4.7.1 Coal

The four source categories (as they relate to coal) for which performance standards have been established are listed as follows:

1. Fossil-fuel-fired steam generators
2. Electric utility steam-generating units
3. Industrial/commercial/institutional steam-generating units
4. Small industrial/commercial steam-generating units

The coal industry's largest specific environmental challenge today is removing organic sulfur. When coal is combusted, the sulfur is released and combines with oxygen to form sulfur dioxide (and to a lesser extent, sulfur trioxide). Sulfur dioxide has been shown to have adverse effects on humans and the ecosystem. Scrubbers are employed to remove sulfur-bearing gases; as a water and limestone mixture reacts with the sulfur dioxide to form a high-sulfur-content sludge. These scrubbers can operate at efficiencies of 98%. The coal industry has made significant improvements in reducing sulfur emissions through the application of this control technology. Since 1989, coal-fired plants in the Unites States have lowered sulfur dioxide emissions per ton by a fourth and have increased efficiency significantly. Mercury and other heavy metals in coal flue gas are also air contaminants of human health and environmental concern. Coal plants recycle millions of tons of fly ash (viewed as a coal by-product) into useful products, such as road-building materials, cement additives, and, in some cases, pellets to be used in rebuilding oyster beds.

Another potential air emission problem is carbon dioxide. It is released when coal is combusted, just as it is released from the human body during respiration. Carbon dioxide combines with other gases, such as those emitted from automobiles, to form a shield that allows the sun's light through the atmosphere but does not let the heat that is produced out of the atmosphere.

This phenomenon is called the *greenhouse* effect. Without this greenhouse effect, the Earth would be too cold to support life. There is concern that the combustion of coal is contributing to major changes in greenhouse gas levels in the Earth's atmosphere that are responsible for recent changes in the Earth's climate. Interestingly, Theodore has shown that CO_2 emissions per unit of energy generated for coal are 50% greater than those for oil and nearly double those for natural gas.

The capture of both SO_2 and the (fly)ash from coal has also created another problem. Current SO_2 regulations have encouraged the use of low-sulfur coals. However, most electrostatic precipitators (ESPs) normally used for particulate capture are *not* capable of achieving sufficiently high collection efficiencies (Theodore 2008; Richardson and Theodore 2025). This has limited the use of coals located east of the Mississippi River (in addition to their lower heating value). In effect, it is a double-edged sword: low-sulfur content and heating values in the West versus higher-sulfur content and heating values of the East. The problem has existed for years and has yet to be resolved, and no solution is expected in the immediate future.

Suffice it to say that the potential impacts on water are significant with the extraction, processing, and transportation of coal from the mine to its end use point. These can be equally, if not more, significant than the impacts on air resources; however, the impacts will depend on specific activities or accidents that would cause environmental harm. The same is true for solid wastes associated with the use of coal as an energy resource in the form of coal ash.

4.7.2 Oil

As with coal, environmental concerns will continue to exist for the petroleum industry. Drilling for and transporting oil can endanger wildlife and the environment if it spills into rivers or oceans. Leaking underground storage tanks (as at the home of one of the authors) can pollute groundwater and create noxious fumes. Processing oil at the refinery can contribute to air and water pollution. Burning gasoline to fuel cars contributes to air pollution. Even the careless disposal of waste oil drained from a car can pollute rivers, lakes, and estuaries. Power plants burning all the various grades of oil produce SO_2 (with trace quantities of SO_3). The combustion of any carbon-bearing compound also produces CO_2, the primary greenhouse gas. The oxides of nitrogen, lead, any unburned hydrocarbon, and other criteria and noncriteria pollutants are also of concern.

Many advances have been made in protecting the environment since the passage of the Clean Air Act in 1970 and its 1990 amendments. Oil companies have redesigned their refineries and motor vehicle manufacturers have redesigned engines to reduce emissions into the air and water. Gasolines have also been reformulated to burn cleaner, dramatically reducing the levels of lead, nitrogen oxide, carbon monoxide, and hydrocarbons released into the air.

The oil industry, including the production, transportation, distribution, and consumption of petroleum products, is regulated to minimize the negative effects on the environment. Yet, environmentalists continue to object to virtually all of the activities of the oil industry. No discussion on the environment can avoid mention of both the *Exxon Valdez* oil spill of 1987 and the 2010 BP *Deepwater Horizon* explosion in the Gulf of Mexico. While some spills may have involved more oil, no oil spill received as much attention as the BP fiasco and had as much impact on the local ecosystem. Fortunately, the combination of the cleanup efforts and nature appears to have prevented some of the worst predictions of unrecoverable damage. Theodore and Dupont (2012) have provided information and illustrative examples on the BP fiasco. Although it is too early to draw any conclusions regarding the environmental implications associated with the failed oil rig, many have described the incident as the worst environmental disaster in U.S. history.

4.7.3 Natural Gas

There are three primary fossil fuels – coal, petroleum, and natural gas – that release pollutants into the atmosphere when burned. However, of these primary fuels, natural gas is the most environmentally friendly fossil fuel. Combusting natural gas produces less oxides of sulfur, carbon, and nitrogen than other fossil fuels. Natural gas also emits little to no ash (particulate) into the environment.

The main sources of environmental concern related to natural gas may be placed in these four categories:

1. Pipeline leakage (release of CH_4 as a more potent greenhouse gas than CO_2)
2. Liquefied natural gas (LNG) transport risks
3. Combustion: CO_2 emissions, SOx, NOx
4. Health and safety concerns

Items 1 and 2 are generally not considered major problems. The aforementioned combustion issues (Reynolds, Jeris and Theodore 2004; Theodore and Ricci 2010; Theodore 2011) raised early in this section have yet to be resolved to the satisfaction of environmentalists. Health and safety remain a concern for any industrial process.

To summarize, the five major environmental pollutants of concern in using primary fossil fuels are the following:

1. Carbon dioxide, CO_2
2. Particulates, PM
3. Oxides of nitrogen, NOx

4. Oxides of sulfur, SOx
5. Unburned hydrocarbons, CH_4 (greenhouse gas), other nonmethane HCs for ozone production

4.7.4 Shale Oil

Environmental concerns with oil shale are broad and encompass not only the industrial complex but also the effect on the surrounding socioeconomic area and region. There is a considerable amount of air pollution, including the release of particulate, SO, NOx, CO_2, and hydrocarbons from any of the processes discussed in the previous section (Theodore and Theodore 2021). Blasting, mining, crushing, and transporting the oil shale all contribute to air quality particulate levels, and the transportation and storage of spent shale contribute even more. The *in situ* process minimizes these types of emissions, but both processes emit substantial amounts of pollutants from the retort operations themselves. Control techniques, such as wetting agents and covered conveyors, are commonly used to minimize fugitive emissions, but air and water pollution will still be a major obstacle to oil shale development.

It is estimated that slightly more than three barrels of water is required for every barrel of oil produced from shale oil. Oil shale processing involves a number of environmental impacts; they are more pronounced in surface mining than in underground mining. They include acid drainage induced by the sudden rapid exposure and subsequent oxidation of formerly buried materials; the introduction of both organics and metals into surface water and groundwater; and increased erosion, sulfur emissions, particulate matter, and other forms of air pollution. The processes can also damage the biological and recreational value of land and the ecosystem in the area. The extensive use of water in arid regions is an extremely sensitive environmental issue as well.

4.7.5 Tar Sands

The magnitude of the environmental problems associated with the extraction, processing, and producing oil from tar sands remains a major concern. Environmental problems include the siltation of streams and lakes from the destruction of natural drainage patterns; destruction of salmon spawning areas near tar sands resources; the contamination of groundwater; and air pollution. Any sulfur and (to a lesser extent) nitrogen in the tar sand materials contribute further to air pollution. The sulfur is distributed in organic and inorganic forms in the tar sands. Heavy metals, including lead, mercury, chromium, cadmium, arsenic, etc., can also be present in tar sands. The disposal of spent tar sands after they have been processed through the bitumen extraction process is another major problem.

The *in situ* process discussed earlier offers advantages in that the disposal of spent sands is essentially eliminated. However, both surface mining and *in situ* processes can lead to groundwater contamination and loss of land for recreation or other uses. It is fair to say that the environmental issues discussed for shale oil also apply to tar sands. Many of the problems associated with coal and to a lesser extent oil also apply.

4.8 FUTURE PROSPECTS AND CONCERNS

4.8.1 Coal

Future prospects for the role of coal in energy production are both positive and uncertain. On the positive side are relatively new processes involving coal:

1. *Coking of coal* generates a solid carbonaceous residue from the burning of bituminous coal. It is used as a reducing agent in iron ore smelting operations and as a new fuel.
2. *Gasification of coal* can be used to produce syngas, a valuable mixture of both carbon monoxide (CO) and hydrogen (H_2) gas.

$$(\text{Coal}) + O_2 + H_2O \rightarrow H_2 + 3CO \tag{4.1}$$

   This syngas can then be converted into transportation fuels through the Fischer-Tropsch process (Theodore, Ricci, and VanVliet 2009). In addition, the hydrogen obtained from gasification can be used for various purposes, such as powering a hydrogen economy, making ammonia, or upgrading fossil fuels. If more hydrogen is desired, the CO can be converted via the water gas shift reaction (Theodore, Ricci, and VanVliet 2009) into more hydrogen.

$$CO + H_2O \rightarrow CO_2 + H_2 \tag{4.2}$$

3. *Liquefaction* occurs when coal is converted into liquid fuels, such as gasoline or diesel by several different processes. In the direct liquefaction processes, the coal is either hydrogenated or carbonized. Hydrogenation processes include several solvent refined coal processes and the NUS Corporation hydrogenation process. In the process of low-temperature carbonization, coal is coked at temperatures between 360°C (680°F) and 750°C (1380°F). These temperatures optimize the production of coal tars richer in lighter hydrocarbons than normal coal tar. The coal tar is then further processed into fuels. Alternatively, coal can be converted into a gas first, and then into a liquid by using the Fischer-Tropsch process. An overview of coal liquefaction and its future potential is available in U.S. EPA literature (U.S. EPA 2025). Coal liquefaction methods involve CO_2 emissions in the conversion process. If coal liquefaction is accomplished without employing either carbon capture and storage technologies or biomass blending, the result is life cycle greenhouse gas footprints that are generally greater than those released in the extraction and refinement of liquid fuel production from crude oil. For most future synthetic fuel projects, CO_2 sequestration has been proposed to avoid releasing it into the atmosphere. Sequestration will, however, add significantly to the cost of production.
4. *Refined coal* is the product of a coal-upgrading technology that removes moisture and certain pollutants from lower rank coals such as sub-bituminous and lignite coals. It is one form of several precombustion treatment processes that alter the characteristics of coal.

Five key areas of uncertainty that affect the coal industry today, from an air pollution perspective, are the following:

1. CO_2 emissions
2. SO_2 emissions
3. NOx emissions
4. Mercury
5. Other air toxics

Despite coal once generating approximately half of the electricity used in the Unites States, the cost of complying with air and other regulations will test the ability of the coal industry to survive in the competitive market with oil, natural gas, and renewables under mounting regulatory and societal pressures (Stander and Theodore 2008).

4.8.2 Oil

The increasing U.S. dependence on petroleum presents a continuing challenge for the future. This dependence must be balanced by the growing demand for petroleum products, increased costs, the depletion of this irreplaceable natural resource, and the protection of the environment. As most are aware, while the Unites States is producing a record amount of crude oil (~13.4 million barrels per day, the highest rate in the world), not all of this crude oil can be efficiently utilized in U.S. refineries (AFPM 2025). In addition, U.S. refineries need about 16.5 million barrels per day to maintain current production levels at roughly 90% utilization (AFPM 2025). About 60% of the crude oil that runs through U.S. refineries is U.S. sourced. Many refineries need heavier crude oil to maximize flexibility of gasoline, diesel, and jet fuel production. Most crude oil produced in the United States is light and is not a good replacement for the heavy crude oil sourced from Canada and Mexico. Nearly 70% of U.S. crude oil imports come from Canada (60%) and Mexico (7%), keeping refineries efficient, keeping costs down, and maintaining energy security.

Re-tooling refineries to process solely U.S. crude oil (light crude) would cost billions – a risky investment that would take decades to permit, construct, and eventually pay off. In addition, the infrastructure (like pipelines) needed to cost effectively supply U.S crude oil and refined products to every region is lacking. Even if the economics of re-tooling facilities worked, it can take close to a decade to permit and build pipelines in the United States. Based on statements of the leading trade association for the sector, the American Fuel and Petroleum Manufacturers Association (AFPM 2025), the ability to import and export crude oil and refined products is good for consumers and safeguards U.S. energy security.

There are steps that need to be taken in the future to help ensure oil energy resources and reduce high oil prices. Some believe the most important step is to decrease demand for oil through increased conservation (e.g., reducing the oil used by increasing the efficiency of vehicles and transportation). Others believe the Unites States should increase homeland oil production, particularly in the Arctic National

Wildlife Refuge in northern Alaska and in offshore areas. Others argue for an increase in the use of other transportation fuels. Most agree that in the future the Unites States must increase efficiency and continue development of nonpetroleum transportation fuels.

4.8.3 Natural Gas

Imports and exports are an important part of the natural gas market in the United States. They provide additional sources of supply for U.S. consumers and additional customers for U.S. producers. Sustained growth in domestic natural gas production in the United States during the past two decades has reduced demand for natural gas imports and created opportunities to export natural gas. Today, the United States is a net exporter of natural gas and one of the top exporters of LNG in the world.

Domestic dry natural gas production in 2023 totaled 37.8 trillion cubic feet (Tcf), or 103.6 billion cubic feet per day (Bcf/d), which is 4.3% higher than the 2022 level of 99.3 Bcf/d. For the seventh consecutive year, the Unites States was a natural gas net exporter (exports minus imports) in 2023, exporting 4,681 Bcf (12.8 Bcf/d) more natural gas than it imported for the year (EIA 2025c).

There are significant benefits to natural gas development. When combusted it is cleaner than coal or oil and releases fewer emissions. Advancements in drilling and extraction techniques such as employing fracturing fluids have reduced the problem of meeting increasing demand for natural gas. Since 1985, natural gas prices have been set by the marketplace. The federal government sets the price of transportation for gas that crosses state lines. State public utility commissions will continue to regulate natural gas utility companies, just as they regulate electric utilities. These commissions regulate how much utilities may charge and monitor the utilities' policies. The policy for the future is uncertain.

Changes have occurred in the U.S. natural gas industry over the past decade as a result of the deregulation of many aspects of the industry, and significant future changes are anticipated. Currently, interstate pipeline companies are reassessing their markets and positioning themselves for the future. The strategic position that a pipeline company could take will depend on the market position of its parent company and any physical relationships to suppliers and consumers in existing and potential markets. Other challenges in the future include the outcome of electric utility deregulation and the market for electric vehicles. Also, many major cities are converting buses, garbage trucks, mail trucks, and other vehicles to natural gas. Electric utilities are installing natural gas-fueled power plants to comply with air regulations. Thus, natural gas usage is likely to increase.

4.8.4 Oil Shale

In the future, it may be possible to produce oil from shale without costly excavation and processing. Some pyrolysis processes have been recommended and are currently under study. One of the processes involves combining a fluidized bed and a circulated fluidized bed for the purpose of burning the byproducts of pyrolysis. Perhaps the most promising process is a pressurized fluid bed combustion process; however, these new processes are still in the early stages of development.

4.8.5 Tar Sands

The deposits of tar sands are no longer restricted to only a few regions. A systematic, worldwide exploration for these deposits has taken place since the start of this century. Even the reserves of some known deposits have been expanded.

Earlier, when the overburden was too deep, economic considerations discouraged surface mining. Research continues to expand the feasibility of both surface and *in situ* production methods. It is anticipated that gases containing CO_2, CO, and CH_4 will continue to be produced in the *in situ* process. Although the *in situ* process offers advantages in that the disposal of spent sands is a minor problem for that option, both surface mining and *in situ* processes have led to the formation of "boom towns" (e.g., Calgary).

The economics of oil extraction from tar sands have become more attractive in the past decades. Support for this once unfavorable alternate energy source continues to increase. Further support is expected in both the near and distant future. It is now evident that considerable efforts will be made in the future to systematically explore the considerable potential of tar sands and to develop technologies for society's utilization. It is also now evident that considerable efforts should and must be made in the future to continue to explore the potential of oil shale, shale gas, and tar sands and to develop technologies for their utilization by mankind. However, environmental activists and indigenous peoples have consistently protested against both the shale oil and gas and tar sands industries due to their large environmental footprint, right behind mountain top removal mining, and impacts to indigenous lands. Mitigation methods for their environmental impacts should also be a significant focus of future resource development.

REFERENCES

American Fuel & Petroleum Manufacturers (AFPM). 2025. How Much Oil Does the United States Import (and Why)?www.afpm.org/newsroom/blog/how-much-oil-does-united-states-import-and-why. (accessed June 6, 2025).

Energy Information Agency. 2025a. U.S. Energy Facts Explained. U.S. Energy Consumption by Source and Sector, 2023. www.eia.gov/energyexplained/us-energy-facts/. (accessed June 6, 2025).

Energy Information Agency. 2025b. Petroleum and Other Liquids, U.S. Oil and Natural Gas Wells by Production Rate, Figure 1. Total U.S. Wells by Production Rate Brackets. www.eia.gov/petroleum/wells/. (accessed June 6, 2025).

Energy Information Agency. 2025c. Natural Gas, Natural Gas Annual. www.eia.gov/naturalgas/annual/. (accessed June 6, 2025).

Mattusch, C. 2008. Networking and Tools. In *The Oxford Handbook of Engineering and Technology in the Classical World*, ed. J. Oleson. Oxford: Oxford University Press.

Reynolds, J., Jeris, J., and Theodore, L. 2004. *Chemical and Environmental Engineering Calculations Handbook*. Hoboken, NJ: Wiley.

Richardson, J., and Theodore, L. 2025. *Optimizing Air Pollution Control Equipment Performance*. Hoboken, NJ: Wiley.

Skipka, K., and Theodore, L. 2014. *Energy Resources: Availability, Management and Environmental Impacts.* Boca Raton, FL: CRC Press/Taylor & Francis Group.

Stander, L., and Theodore, L. 2008. *Environmental Regulatory Calculations Handbook.* Hoboken, NJ: Wiley.

Theodore, L. 2008. *Air Pollution Control Equipment Calculations*. Hoboken, NJ: Wiley.

Theodore, L. 2011. *Heat Transfer for the Practicing Engineer*. Hoboken, NJ: Wiley.

Theodore, L., and Dupont, R.R. 2012. *Environmental Health Risk and Hazard Risk Assessment: Principles and Calculations*. Boca Raton, FL: CRC Press/Taylor & Francis Group.

Theodore, L., and Ricci, F. 2010. *Mass Transfer Operations for the Practicing Engineer*. John Hoboken, NJ: Wiley.

Theodore, L., Ricci, F., and VanVliet, T. 2009. *Thermodynamics for the Practicing Engineer*. Hoboken, NJ: Wiley.

Theodore, M.K., and Theodore, L. 2021. *Introduction to Environmental Management, 2nd edition*. Boca Raton, FL: CRC Press/Taylor & Francis Group.

U.S. Environmental Protection Agency. 2025. Science Inventory. Environmental Assessment of Coal Liquefaction. https://cfpub.epa.gov/si/si_public_record_report.cfm?Lab=ORD&dirEntryId=46636. (accessed June 6, 2025).

Wikipedia Contributors. 2025a. Oil Shale in Estonia. *Wikipedia, The Free Encyclopedia*. https://en.wikipedia.org/w/index.php?title=Oil_shale_in_Estonia&oldid=1287501389. (accessed June 6, 2025).

Wikipedia Contributors. 2025b. Oil Sands. *Wikipedia, The Free Encyclopedia*. https://en.wikipedia.org/w/index.php?title=Oil_sands&oldid=1293940235. (accessed June 6, 2025).

5 Energy Resources
Renewables

5.1 INTRODUCTION

Interest in all forms of energy resources is growing. Municipal, commercial, and industrial energy users are all expanding their activities in the wind, solar, nuclear, and other energy resources. The previous chapter provided information on fossil fuel resources including coal, oil, natural gas, shale, and tar sands. This chapter reviews a host of other energy sources in the renewable energy category. Resources reviewed with their accompanying section numbers include:

5.2 Solar Energy
5.3 Nuclear Energy
5.4 Hydroelectric Energy
5.5 Wind Energy
5.6 Hydrogen Energy
5.7 Biomass-Derived Energy
5.8 Hydrokinetic Energy
5.9 Environmental Issues
5.10 Future Prospects and Concerns
References

Obviously this list does not include geothermal energy, the subject title of this book. This material on geothermal energy receives detailed treatment in Part III of this book.

5.2 SOLAR ENERGY

Solar energy is the first of the renewable resources to be reviewed. It cannot be depleted (at least for millions of years). Solar energy is radiant energy that is produced by the Sun. Every day the Sun radiates an enormous amount of energy that maintains the planet Earth at an acceptable temperature to sustain life as is known today.

Only a small portion of the energy radiated by the Sun into space strikes the Earth – approximately one part in two billion. Yet, this amount of energy is enormous. Enough solar energy strikes the United States every day to supply the nation's

DOI: 10.1201/9781003667759-6

energy needs for one-and-a-half years. Approximately 15% of the energy that strikes the Earth is reflected back into space. Another 30% evaporates water, and some is absorbed by plants, the land, and the oceans. The remainder is available to supply society's energy needs (Skipka and Theodore 2014).

In a very real sense, almost all the radiant energy impacting the surface of the Earth can be attributed to solar energy. This chapter covers what has been traditionally defined as solar energy. Solar energy is not a single energy technology, but rather a term that in reality covers a diverse set of renewable energy technologies. Solar energy applications can be divided into two groups – heating and cooling, and electricity generation.

In terms of solar energy, it should be noted that, for most locations on the Earth, the Sun does not shine 24 hours a day. This period of darkness is unavailable when compared to a conventional power plant that is not subjected to such constraints. The angle of incidence of radiation also depends on the motion of the Sun and upon the geographic position of the Earth. The angle of incident radiation becomes more pronounced the farther away a receptor site is located from the position where the Sun is directly overhead.

In terms of energy strength, the Sun constantly delivers 1.36 kW (1,360 J/s) of power per square meter (430 Btu/h-ft^2 or 123 W/ft^2) to the Earth. However, some of this power is absorbed by the atmosphere so that, even under ideal conditions, a receptor on the Earth may receive less than 93 W/ft^2 (1,000 W/m^2 or 0.1 W/cm^2). From an overall global perspective, of the total annual radiation reaching the Earth's atmosphere (1.5×10^{18} kWh) (Skipka and Theodore 2014):

- 50% is reflected back into space by the clouds
- 27.8% is absorbed in evaporating water
- 15% is reflected back into space by the Earth's surface
- 5.3% is absorbed by bare soil
- 1.7% is absorbed by marine vegetation
- 0.2% is absorbed by land/vegetation

5.3 NUCLEAR ENERGY

Energy that is obtained from the nucleus of an atom is defined as nuclear energy. The energy is released from an atom through one of two processes: (1) nuclear *fusion* and (2) nuclear *fission*.

Energy is released in nuclear fusion when atoms are combined or fused together (fusion). This is how energy is generated in the Sun. In nuclear fission, energy is released when atoms are split apart. Nuclear fission is the only method currently used by nuclear plants to generate electricity. However, the terms nuclear fusion and nuclear fission are referred to synonymously as nuclear energy. Both involve changes in (nuclear) mass.

Regarding fission, uranium is the heaviest of the 92 naturally occurring elements. Since it is also one of the few elements that is easily fissioned, it is the fuel of choice used by today's nuclear power plants. This element was formed when the Earth was created and is usually found in rocks. Rocks that contain significant quantities of uranium are referred to as uranium ore, or pitchblende. Two forms (isotopes) of uranium are found in these rocks: uranium-235 and uranium-238. These three-digit numbers refer to the number of neutrons and protons in each atom. Uranium-235 is the form normally used for energy production because, unlike uranium-238, its nucleus splits more easily when bombarded by a neutron. During fission, the uranium-235 atom absorbs a bombarding neutron, causing its nucleus to split apart into two atoms of lighter weight (See Figure 5.1). The fission reaction simultaneously releases energy as both heat and radiation. It also releases more neutrons. These released neutrons go on to bombard other uranium atoms, and the process repeats itself. This repetitive process is referred to as a *chain reaction*.

When a uranium atom is split apart (fissioned), the mass of the *fragments* is less than the mass of the original atom. The energy corresponding to this loss of mass is defined as fission energy. It is represented in equation form by Einstein's equation:

$$E = mc^2 \tag{5.1}$$

where E = energy, J; m = mass, kg; and c = speed of light, 3 x 10^8 m/s.

One can show that each kilogram of mass converted liberates approximately 10^{17} J. This thermal energy is enough to run a large turbine generator (a 1-GW unit) for a year at 35% thermal efficiency and is equivalent to the energy possessed by 2.5 million tons of coal (Skipka and Theodore 2014).

Similarly, if two or more light elements (e.g., hydrogen) are combined (*fused*) to form a heavier one, the resulting mass of the atom is less than the sum of the two original atoms. The difference (or loss) in mass is converted to energy, as with the fission process. This energy is defined as *fusion* energy.

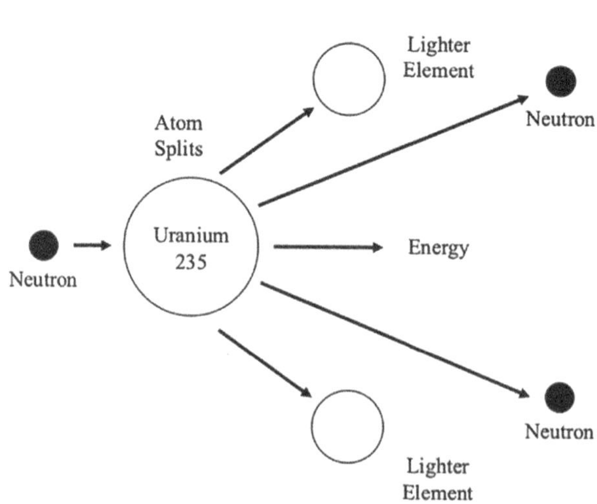

FIGURE 5.1 The fission process.

The operation of nuclear reactors depends upon various types of interaction of neutrons with atomic nuclei. An atom consists of a positively charged nucleus surrounded by a number of negatively charged electrons, so that the atom as a whole is electrically neutral. Chemical energy, such as that obtained from the combustion of coal or oil, results from a rearrangement of the atoms caused by a redistribution of the electrons. In the processes taking place in a nuclear reactor, leading to the release of atomic energy, only the atomic nuclei are involved; the electrons may be neglected.

The building or planning of nuclear reactors is presently directed toward one of the following objectives: (1) the production of heat and power, (2) the production of plutonium, uranium-233, or radioisotopes, (3) the production of neutrons for research, or the experimental development of reactors. Nuclear reactors can also be classified according to type of fuel used or to the energy of the neutrons at the time they are captured by the fuel to induce fission (fast, thermal, or intermediate).

Electricity generation from commercial nuclear power plants in the United States began in 1958. As of August 1, 2023, the United States had 93 operating commercial nuclear reactors at 54 nuclear power plants in 28 states. The average age of these nuclear reactors is about 42 years. The oldest operating reactor, Nine Mile Point Unit 1 in New York, began commercial operation in December 1969. The newest reactor to enter commercial service is Unit 3 at the Alvin W. Vogtle Electric Generating Plant in Georgia, which began commercial operation on July 31, 2023, and is the first reactor to come online since Watts Bar 2 was commissioned in 2016 (EIA 2025a).

The number of operating U.S. nuclear reactors peaked at 112, and their combined net summer electricity generation capacity was 99,624 MW. The number of operating reactors declined to 104 in 1998 and remained there through 2013. The number declined to 92 operating reactors in 2022. Total U.S. nuclear net summer electricity generation capacity peaked in 2012 at about 102,000 MW and declined to 94,765 MW in 2022. Although the number of reactors has declined since 2012, power plant upgrades – modifications to increase capacity – at individual nuclear power plants have made it possible for the entire operating nuclear reactor fleet to maintain high capacity-utilization rates. These relatively high capacity rates helped nuclear power to provide 19%–20% of total annual U.S. electricity generation from 1990 through 2021. Some reactors also increased annual electricity generation by shortening the length of time reactors are offline for refueling (EIA 2025a).

A major innovation in nuclear power is the development of small modular reactors (SMRs). SMRs are nuclear reactors that have a power capacity of up to 300 MW(e) per unit, which is about one-third of the generating capacity of traditional nuclear power reactors. Since they are designed to be modular units, it is possible for systems and components to be factory-assembled and transported as a unit to a location for installation, standardizing their design and reducing their overall cost and start-up time (IAEA 2025).

NuScale Power, LLC (www.nuscalepower.com) is the only U.S. company to have received design approval from the U.S. Nuclear Regulatory Commission for their pressurized water-cooled reactor technology. Their 77 MW module is designed to provide power for electrical generation, district heating, desalination, commercial-scale hydrogen production, and other process heat applications, and is being considered by a number of utilities for generation of carbon-free power to meet future energy needs.

5.4 HYDROELECTRIC ENERGY

The fall and movement of water is part of a continuous natural cycle called the *water cycle*. The cycle consists of four basic steps (Theodore and Theodore 2021):

1. Energy from the Sun evaporates water in the Earth's oceans and rivers and draws it upward as water vapor.
2. When the water vapor reaches cooler air in the atmosphere, it condenses and forms clouds.
3. The moisture eventually falls to the Earth as rain or snow, replenishing the water in the oceans and rivers.
4. Gravity then drives the water, moving it from high elevation to lower elevation.

Hydroelectricity is the term referring to electricity generated by this moving water. The production of electrical power "arises" through the use of the gravitational force of falling or flowing water. It is the most widely used form of renewable energy across the globe. Once a hydroelectric complex is constructed, the project produces no direct waste and has a significantly lower output level of greenhouse gases, that is, CO_2, than any of the fossil-fuel-powered energy plants. There was an installed capacity of 4,236 GW hydroelectricity supplied worldwide in 2023. This represented approximately 14.5% of the world's electricity and accounted for about 47% of electricity from renewable sources (EIA 2025b).

The major advantage of hydroelectricity is the elimination of the cost of fuel. The cost of operating a hydroelectric plant is not affected by increases in the cost of fossil fuels such as oil, natural gas, or coal, and no imports (to the United States) are required. These plants also have long lives, with some plants still in service after 50+ years. Operating labor cost is also usually low since many of the plants are automated and have few personnel onsite during normal operation.

The equation describing electric power production at a hydroelectric plant is (Abulencia and Theodore 2009):

$$P = (\rho q g \Delta h)E \tag{5.2}$$

where P = power, W; ρ = density of water, 1,000 kg/m^3; q = flow rate, m^3/s; g = gravitational acceleration, 9.8 m/s^2; Δh = height of water drop across the hydroelectric power plant, m; and E = fractional energy conversion efficiency, unitless, 0–1, normally assumed 0.3.

Note that the annual electric energy production depends on the available water supply. The water flow rate, q, in some installations can vary by a factor of 10:1 during the course of a year.

Hydropower is the most economical way to generate electricity today. No other energy source, renewable or nonrenewable, is comparable to it. It costs approximately 1.1 cent per kilowatt-hour (kWh) to produce electricity at a typical hydro plant. In comparison, it costs coal plants about 3.8 cents per kilowatt-hour and nuclear plants about 2.7 cents per kilowatt-hour to generate electricity (EIA 2025c). Producing electricity from hydropower is inexpensive because, once a dam or the equivalent has been built and the equipment installed, the energy source – flowing water – is free. Hydroelectric plants also produce power at a minimal cost due to their sturdy structures and simple equipment design.

5.5 WIND ENERGY

Wind can be defined as air in motion. It is produced by the uneven heating of the Earth's surface by energy from the Sun. Since the Earth's surface is made of very different types of land and water, it absorbs the Sun's radiant energy at different rates. Much of this energy is converted into heat as it is absorbed by land areas, bodies of water, and the air over these formations.

The Earth is not uniformly heated by the sun, as the poles receive less energy from the sun than the Equator. Along with this, dry land heats up (and cools down) more quickly than do the bodies of water. Solar insolation causes temperature difference driving forces between the equator and the poles. These create a global atmospheric convection system reaching from the Earth's surface to the stratosphere, which acts as a virtual ceiling. On a local scale, the warm air over the land expands and rises, and the heavier, cooler air over the water rushes in to take its place, creating localized winds.

Wind power is the conversion of this air movement into a useful form of energy. This is accomplished by using wind turbines to make electricity, windmills to produce mechanical power, wind pumps for water pumping or drainage, or sails to propel ships. The total amount of usable power available from the wind is estimated to be considerably more than present human power use from all other sources. Continental U.S. wind potential of 43,000 TWh/yr exceeds 2022 U.S. electricity use of 4,000 TWh by more than an order of magnitude (Center for Sustainable Systems 2024).

Wind energy is perhaps the most prominent and fastest growing source of renewable energy. And while there are some limitations and constraints, wind power is expected to continue to grow worldwide. Annual global onshore wind installations surpassed 100 GW for the first time in 2023, while the United States experienced a slowdown. A total of 10.8 GW of offshore wind capacity was added worldwide in 2023, a 24% increase from 2022, bringing global offshore wind capacity to 75.2 GW (Center for Sustainable Systems 2024).

U.S. wind capacity grew from 45 GW in 2010 to 147 GW in 2023, a 10% average increase/year. Texas (40,151 MW), Iowa (12,783 MW), and Oklahoma (12,222 MW) are the leading states in installed wind capacity. Texas generated the most wind electricity of any U.S. state, while Iowa generated 62.4% of its electricity from wind, the largest share of any state in 2022 (Center for Sustainable Systems 2024).

5.6 HYDROGEN ENERGY

Hydrogen is the most abundant gas in the universe and the source of all the energy received from the Sun. The Sun is basically a giant ball of hydrogen and helium gases. In a process called fusion (see Section 5.3), four hydrogen atoms combine to form one helium atom, releasing energy as radiation. This radiant energy is Earth's most abundant energy source. It provides light and heat and makes plants grow. It also causes the wind to blow and the rain to fall. And, it is stored as chemical energy in fossil fuels. Most of the present sources of energy originally came from the Sun.

Since hydrogen is not a basic energy resource (except in the Sun), it must be supplied by using some other basic energy resource to generate hydrogen from water or other hydrogen-containing chemical compounds. Unlike carbon-based fuels, hydrogen used directly as a fuel produces only water as an end product at low combustion temperatures. At higher combustion temperatures, as in all combustion processes requiring air, nitrogen oxides form and must be vented (Skipka and Theodore 2014).

Hydrogen is the lightest in weight of all gases and has been used for the inflation of balloons and dirigibles. It ignites very easily – a small spark causing it to burn. Several dirigibles, including the infamous Hindenburg, have been destroyed by hydrogen fires. Today, helium, which has 92% of the lifting power of hydrogen and is not inflammable, is used in place of hydrogen whenever possible for these activities.

Hydrogen fuel is an ecologically friendly fuel that uses electrochemical cells or combusts in internal engines to power vehicles and electric devices. It is also used in the propulsion of spacecraft and can potentially be mass produced and commercialized for passenger vehicles and aircraft.

In a flame of pure hydrogen gas (combusting in air), the hydrogen (H) reacts with oxygen (O_2) to form water (H_2O) and liberates heat. It does not produce other chemical by-products, except for a small amount of nitrogen oxides at elevated temperatures. Hence, a key feature of hydrogen as a fuel is that it is relatively nonpolluting.

Pure hydrogen does not occur naturally; it takes energy to manufacture it. Once manufactured, it is an energy carrier (i.e., a store for energy first generated by other means). The energy is eventually delivered as heat, a radiant emission, when the hydrogen is burned. DelGatto et al. (2025) repeatedly expressed concern with the future of hydrogen energy.

Dorf (1978) listed the following characteristics of a desirable energy carrier:

1. High efficiency of conversion of original energy source to the carrier fuel.
2. Availability of a method of transporting and distributing the carrier fuel with low attendant energy losses and at low cost per unit distance.
3. Availability of a method of storing the carrier fuel for relatively long periods of time and at low cost.
4. Availability of several methods of utilizing the carrier fuel for transportation and industry.
5. Safety in handling and storage.

In recent years, hydrogen has been touted as the answer to many of society's energy problems. It is the perfect clean fuel. It has multiple applications through direct combustion and conversion to electricity in fuel cells. Fuel cell technology in vehicles has also been identified as an opportunity to address and solve petroleum and urban air pollution from transportation in the future.

5.7 BIOMASS-DERIVED ENERGY

Biomass is any organic matter – wood, crops, seaweed, animal waste, etc. – that can be used as an energy source. Biomass is probably the oldest source of energy after the Sun. People have burned wood for thousands of years to heat homes and cook food.

When humans changed from nomadic life and began living together in organized settlements, they began to organize available resources. Primitive humans discovered that certain wastes that were being disposed of on land could be burned to provide heat. These two methods, land disposal and incineration, remain the most popular forms of waste disposal today.

Biomass-derived fuels include several different forms, including wood and alcohol fuels. Wood is still a major fuel in some developing countries, and high oil prices have caused a resurgence of interest in wood in some industrialized countries. Approximately one-half of the households in Vermont, for example, are estimated to heat with wood (Skipka and Theodore 2014).

Biomass gets its energy from the Sun; in fact, all organic matter contains stored energy from the Sun. During a process called photosynthesis, sunlight gives plants the energy they need to convert water and carbon dioxide into oxygen and sugars. These sugars, called carbohydrates, supply plants and the animals that eat plants with energy. Foods rich in carbohydrates are also a good source of energy for humans.

Biomass-derived fuels classified by the Energy Information Agency (EIA 2025d) include biofuels, wood, and waste materials. *Biofuels* are classified as liquid fuels that blend components produced from biomass feedstocks, such as corn or soy. Biofuels include fuel ethanol, biodiesel, renewable diesel fuel, and other biofuels consumption as well as losses and coproducts from fuel ethanol and biodiesel production. *Wood and wood products* used as fuels include round wood (cord wood), limb wood, wood chips, bark, saw dust, forest residues, charcoal, pulp waste, spent pulping liquor, densified biomass (which includes wood pellets), and other wood-based solids and liquids. *Waste* biomass-derived fuels include organic nonfossil material of biological origin that is a byproduct or a discarded product. Waste includes municipal solid waste from biogenic sources; landfill gas; sludge; agricultural crop byproducts; straw; and other biomass solids, liquids, and gases. Waste does not include wood and wood-derived fuels (including black liquor), biofuels feedstock, biodiesel, and fuel ethanol. In addition, EIA biomass waste data include energy crops grown specifically for energy production, which would not normally be considered waste (EIA 2025d).

As of 2024, the EIA (2025d) indicated that biofuels, and wood and waste represented 33% and 26% of renewable energy consumption in the United States, respectively. A total of 98% of the industrial use of renewable energy was as biomass-derived fuels, 100% of renewable energy in the transportation sector was in the form of biomass-derived fuels, 58% of residential and commercial renewable energy consumption was in the form of biofuels, wood and waste combustion, and 9% of electrical power generation from renewables came from wood and waste combustion. Biomass-derived fuel consumption is an important energy source in the United States that has grown by more than fivefold over the last two decades (EIA 2025e).

5.8 HYDROKINETIC ENERGY

5.8.1 Tidal Energy

The gravitational pull of the Moon and the Sun, along with the rotation of the Earth creates tides in the oceans. In some places, tides cause water levels near the shore to rise and fall up to 40 feet. People in Europe harnessed this movement of water to operate grain mills more than a 1,000 years ago. Today, tidal energy systems generate electricity. Producing tidal energy economically requires a tidal range of at least 10 feet (EIA 2025f).

The United States does not have any commercially operating tidal energy power plants, although several demonstration projects are in various stages of development. Two places in the United States with potential for tidal power are the Cook Inlet of Alaska, which has the second-highest tidal range in North America, and several places in Maine. There are three types of tidal energy systems being considered that include *tidal barrages*, *tidal turbines*, and *tidal fences*.

Tidal barrages use a structure similar to a dam called a barrage. The barrage is installed across an inlet of an ocean bay or lagoon that forms a tidal basin. Sluice gates on the barrage control water levels and flow rates to allow the tidal basin to fill on the incoming high tides and to empty through an electricity turbine system on the outgoing ebb tide. A two-way tidal power system generates electricity from both the incoming and outgoing tides. Several tidal power plants operate around the world. The largest is the Sihwa Lake Tidal Power Station in South Korea, at 254 MW of electricity-generation capacity. The oldest and second-largest operating tidal power plant is in La Rance, France, with 240 MW of electricity-generation capacity. Smaller tidal power plants are in Canada, China, Russia, and South Korea (EIA 2025f).

Tidal turbines are similar to wind turbines because they both have blades that turn a rotor to power a generator. They can be placed on the sea floor where the tidal flow is strong. Because water is about 800 times denser than air, tidal turbines have to be much sturdier and heavier than wind turbines. Tidal turbines are more expensive to build than wind turbines but can capture more energy with the same size blades. The United States does not have any commercially operating tidal power plants, but several demonstration projects have been implemented or are in various stages of development including the Roosevelt Island Tidal Energy Project Pilot in the East River of New York, and the Western Passage Tidal Energy Project and the Cobscook Bay Tidal Energy Project, both in Maine (EIA 2025f).

A tidal fence has vertical-axis turbines mounted on the sea bed in a fence or row, similar to tidal turbines. Water passing through the turbines generates electricity. In 2023, no tidal fence projects were operating in the United States (EIA 2025f).

5.8.2 Wave Energy

Waves form as wind blows over the surface of open water in oceans and lakes. Ocean waves contain tremendous energy. The theoretical annual energy potential of waves off the coasts of the United States is estimated to be as much as 2.64 trillion kWh, which is equal to about 63% of total U.S. utility-scale electricity generation in 2023 (EIA 2025g). The west coasts of the United States and Europe and the coasts of Japan and New Zealand have potential sites for harnessing wave energy. Many different methods and technologies for capturing and converting wave energy to electricity are under development. These methods include placing devices on or just below the surface of the water and anchoring devices to the ocean floor.

The United States has no commercially operating wave energy projects, but several research projects are underway or planned. The U.S. Department of Energy's (DoE) Marine and Hydrokinetic Technology Database (https://openei.org/wiki/Marine_and_Hydrokinetic_Technology_Database) provides information on marine and hydrokinetic renewable energy in the United States and around the world.

5.9 ENVIRONMENTAL ISSUES

This section briefly examines environmental concerns associated with a number of the renewable energy sources reviewed in the previous sections of this chapter.

5.9.1 Solar Energy

Solar collectors covering a large area are necessary for solar energy systems, and therefore large quantities of material are required for the construction of these devices. As with all energy systems utilizing equipment for power production, it is the extraction, production, processing, and transportation of the materials used in equipment manufacturing that will continue to have the greatest environmental impact. Environmental effects will differ somewhat for three basic solar energy technologies: solar photovoltaic energy, solar thermal energy conversion, and solar heating and cooling.

One of the most obvious environmental effects of a photovoltaic energy system is the large land requirement and the resulting visual (aesthetic) impact. Some farmland or grazing land may be taken out of use or production by centralized photovoltaic solar plants or thermal conversion plants, but these plants are usually located in the desert and semidesert of the southwestern United States. There is some air pollution as a result of the energy used in the production of solar equipment and related materials. Quantities of particulates SOx, and NOx, released to the atmosphere result from the production of the power required and the production of materials (such as carbon steel), glass glazing, and aluminizing back panels. Primary aluminum production results in emissions of particulates and gaseous fluorides. Air pollutants produced in photovoltaic solar cells are cadmium sulfide, silicon, and gallium arsenide. There are also three potential areas of significant water pollution caused by the production of solar photovoltaic materials (Skipka and Theodore 2014):

1. Surface water degradation and groundwater disturbance due to coal, zinc, copper, and bauxite mining.
2. Water pollution related to ore roasting, smelting, and refining.
3. Wastewater discharges from CdS, Si, and GaAs production and device fabrication.

5.9.2 Nuclear Energy

Questions about environmental issues and safety in the nuclear energy industry have created numerous battles. The advocates, comprising primarily industry and business leaders plus some in the scientific/engineering communities, believe that the most realistic alternative to fossil fuels is nuclear power. Although some problems remain, they maintain that solutions will be found in the near future. Nuclear opponents, on the other hand, emphasize a number of unanswered questions about nuclear power and the environment, including:

1. Waste management and disposal
2. Plant accidents/safety
3. Radiation effects

Each of these is discussed in the following sections.

5.9.2.1 Waste Management and Disposal

Disposal of nuclear waste is often said to be the main environmental concern of both society and industry. Presently, waste is usually stored at individual reactor sites where radioactive material continues to accumulate. Experts agree that centralized underground repositories that are well managed, guarded, and monitored are presently the most cost-effective approach to this problem.

Dorf (1978) has outlined a typical waste management program. In a typical reprocessing plant, the spent fuel elements are received and placed in a storage pool. Later, they are chopped and dissolved in nitric acid. The tubes that contain the fuel pellets are not dissolved by the acid but are buried. An organic solvent dissolves the uranium and plutonium, which are later separated, purified, and further converted into plutonium oxide and uranium hexafluoride. The nitric acid containing the fission-product solution is sent to an evaporator where the acid, along with some fission products, is boiled off. The remaining high-level waste is condensed and placed in storage tanks where it is cooled.

The waste sites are divided into three categories:

1. Those referred to as low level in terms of their radioactivity are piped directly into surface ponds on the site.
2. Intermediate-level wastes are treated more cautiously and are emptied into concrete-covered trenches known as cribs; the cribs are open to the soil at the bottom, and the water in the wastes gradually seeps downward, taking the radioactive isotopes with it.
3. The most radioactive wastes, known as high level, are buried in steel-lined concrete tanks in the ground; the storage of these wastes is currently the major challenge to the nuclear industry.

Most scientists believe the safest way to store nuclear waste is in rock formations deep underground – called *geological repositories*. In 1982, Congress passed the Nuclear Waste Policy Act. This law directed the DoE to site, design, construct, and operate America's first repository by 1998. Unfortunately, this did not happen. The same law also established the Nuclear Waste Fund to pay for a permanent repository. People who use electricity from nuclear power plants pay 1/10 of a cent for each kilowatt-hour of electricity they use. An average household using about 7,500 kilowatt-hours a year contributes $750 a year to the fund. In 2002, the U.S. Congress voted to designate Yucca Mountain in Nevada as the proposed national repository for used fuel. Congress ended federal funding for the site in 2011 during the Obama administration.cite_note-3 The DoE stopped collecting annual NWF fees in 2014 following an order by the U.S. Court of Appeals for the District of Columbia Circuit, which found that the DoE failed to justify the continued imposition of the fee following the suspension of the Yucca Mountain repository project.

The project has encountered many difficulties and was highly contested by the public, the Western Shoshone peoples, and many politicians. The project also faces strong state and regional opposition. The Government Accountability Office stated that the closure was for political, not technical or safety reasons (Wikipedia 2025a). This leaves the U.S. government (which disposes of its transuranic waste from nuclear weapons production at the Waste Isolation Pilot Plant (WIPP) in New Mexico) and American nuclear power plants without any designated long-term storage for their high-level radioactive waste (spent fuel) stored onsite in steel and concrete casks at 76 reactor sites in 34 states.

To make waste storage matters worse, an accident occurred at WIPP in February 2014, resulting from a truck fire deep in the mine spreading soot and disabling the air monitoring system. A subsequent chemical reaction of sWheat Scoop cat litter used to absorb liquids in a waste drum resulted in drum failure and release of stored radioactive waste, which lead to a three-year shutdown of the facility and an estimated $2 billion in clean up, fine payment, and waste management costs (Bulletin of Atomic Scientists 2019). This incident has put into question long-term nuclear waste management practices and has resulted in an impasse in finding a permanent solution for commercial spent nuclear fuel disposal that is yet to be resolved.

5.9.2.2 Plant Accidents/Safety

It should again be noted that nuclear power plants are designed and constructed to a very high standard of safety. To meet such standards and assure a high-quality product, strict quality assurance in design and quality control in the manufacture of components and in the plant construction are required. Inherent in the design are such features as redundancy of components and systems, diversity and backup of plant shutdown systems, multiple fission product release barriers, and the application of the fail-safe principle. Additionally, safeguard systems are installed to mitigate the consequences of severe incidents not expected to occur but which contribute to the design bases of the plant.

Nuclear reactors are designed so that, in the case of component failure, the reactor can still be safely shut down. However, when several components fail, as well as the safety system, an accident can occur. The probability of such accidents occurring can be calculated through the use of event-tree and fault-tree methods as well as application of the Weibull probability distribution; Theodore and Dupont (2012) provide numerous illustrative examples, plus two case studies in this area.

Despite these safeguards being put into place, there have been more than 100 nuclear plant accidents and incidents throughout the world between 1952 and 2019 (Wikipedia 2025b), perhaps the most well-known being at Three Mile Island in 1979, Chernobyl in 1986, and Fukushima in 2011.

5.9.2.3 Radiation Effects

Perhaps the greatest potential risk from nuclear power plants is the release of high-level radiation and radioactive material either from waste management incidents or from plant releases. Plants in the United States are specifically designed to contain radiation and radioactive material in the unlikely case of an accident. Emergency plans are in place to alert and advise nearby residents if there is a release of radiation into the local environment.

Natural sources of radiation include cosmic rays from outer space, minerals in the ground, and radon in the air. Manmade sources of radiation include the x-ray equipment used by doctors, smoke detectors, color television sets, and luminous dial clocks. Nuclear waste is another kind of manmade radiation that usually contains higher than natural concentrations of radioactive atoms.

5.9.3 Hydroelectric Energy

Hydropower can potentially cause environmental problems. Ten of these problems are listed here:

1. Damming rivers may permanently alter river systems and wildlife habitats.
2. Fish may no longer be able to swim upstream.
3. It may affect water quality by churning up dissolved metals that may have been deposited by discharges from industrial operations.
4. It may increase silting, change water temperatures, and lower the levels of dissolved oxygen in the downstream water.
5. It may destroy biologically rich and productive lowland and river valley forests, marshland, and grasslands.
6. It can be disruptive to surrounding aquatic ecosystems – both upstream and downstream of the plant site.
7. Water exiting a turbine usually contains very little suspended sediment which can lead to scouring of river beds and riverbanks downstream of the dam.
8. Flowing water has the ability to transport particles more dense than itself downstream.
9. Hydroelectric dams may result in the relocation of the people living where the reservoirs are planned or located.
10. Dam failures due to poor construction, terrorism, or other causes (hurricane Katrina is but one example) can be catastrophic to downriver settlements and infrastructure. Dams are an obvious target for wartime attack, sabotage, and terrorism.

It should be noted that most of these problems can be and have been successfully managed by the technical community.

5.9.4 Wind Energy

Compared to the environmental impact of traditional energy sources, the environmental impact of wind power is relatively minor. Wind power consumes no fuel and emits no air pollution, unlike fossil fuel power sources. The energy consumed to manufacture and transport the materials used to build a wind power plant is equal to the new energy produced by the plant within a few months. While a wind farm may cover a large area of land, many land uses, such as agriculture, are compatible with only small areas of turbine foundations and infrastructure made unavailable for use.

Most wind power technologies, as presently envisioned, would have a benign effect on the environment. The specific effects of wind power systems will depend upon the method of wind energy extraction, the peripheral equipment used for energy conversion, energy storage, and transmission and the ultimate method of their use. Some people would regard visual (aesthetic) pollution caused by wind machines and transmission lines to feed a power network to be the severest wind power system impact. To some, windmills on the horizon are an eyesore, while to others they are a beautiful example of technology at work.

Although noise is an area of concern, newer turbines avoid the sound used when downwind blades flex as they pass behind their towers. In addition, newer blades have improved aerodynamic characteristics that make them much quieter. There has also been some token concern associated with the impact of wind turbines on aviation crashes/collisions and the effect of tall buildings in and around wind turbines. Overall, environmental issues are relatively minor concerns related to this renewable energy technology.

5.9.5 Hydrogen Energy

Hydrogen is an ecologically friendly fuel. It does not produce harmful chemical by-products, except for a small amount of nitrogen oxides (at elevated temperatures). Hence, a key feature of hydrogen as a fuel is that it is relatively nonpolluting. With regard to safety from unwanted explosions, hydrogen fuel in automotive vehicles is probably as safe as gasoline. The technical community now realizes that the safe handling of hydrogen presents them with a challenge to work out the methods necessary to utilize hydrogen for commercial, residential, and transportation uses. However, the risks of ignition or spreading flames are actually less than those for gasoline since in open air and well-ventilated places, leaks or spills diffuse so rapidly (because hydrogen is the lightest of all elements). The basic risk management requirements are appropriate ventilation, leak prevention, and elimination of ignition sources. The Hindenburg dirigible accident is a constant reminder of the need for safety precautions when using hydrogen.

5.9.6 Biomass-Derived Energy

The pollutant species of concern during the combustion of biomass, like and carbon-based fuel, can include (Theodore 2008):

1. Nitrogen oxides (NOx)
2. Carbon monoxide (CO)
3. Acid gases (SO_2, HCl, HF, etc.)
4. Particulate matter

5. Metals (As, Be, Cd, Pb, Hg, Ni, etc.)
6. Toxic organics

5.9.7 Hydrokinetic Energy

A potential disadvantage of tidal power is the effect a tidal station can have on plants and animals in estuaries of the tidal basin. Tidal barrages can change the tidal level in the basin and increase turbidity. They can also affect navigation and recreation.

Wave energy is generally considered a clean and sustainable energy source, but it does have potential environmental impacts that include disruptions to marine habitats, potential impacts on marine animal migration and behavior, and changes to wave dynamics and coastal erosion.

5.10 FUTURE PROSPECTS AND CONCERNS

5.10.1 Solar Energy

Improved solar systems are being developed. For example, concentrated photovoltaic (CPV) systems employ sunlight concentrated onto photovoltaic surfaces for the purpose of electrical power production. Solar concentrators of all varieties may be used, and these are often mounted on a solar tracker in order to keep the focal point upon the cell as the Sun moves across the sky. Luminescent solar concentrators (when combined with a PV-solar cell) can also be regarded as a CPV system. Luminescent solar concentrators are useful as they can drastically improve performance of PV-solar panels. In addition, thermoelectric, or thermovoltaic devices convert a temperature difference between dissimilar materials into an electric current. Finally, space-based solar power (SBSP) is a theoretical design at this time for the collection of solar power in space for use on the Earth. SBSP differs from the usual method of solar power collection in that the solar panels used to collect the energy would reside on a satellite in orbit, often referred to as a solar power satellite (SPS), rather than on the Earth surface improving solar incident light potential and overall energy production capacity.

5.10.2 Nuclear Energy

Today, the nuclear industry in most countries is bogged down in political, environmental, and economic controversy. This growth in opposition has now much to do with public alarm over the persistent and complex nature of radioactive emissions, wastes, and waste disposal resulting from fuel reprocessing. Few orders for new plants are being placed and although many plants currently under construction are being completed, delays abound and many orders for new plants have been cancelled. As might be expected, estimates of future energy that will be obtained from nuclear power vary widely. Some now worry that in the 21st century, it will not be the panacea its advocates once foresaw. In effect, nuclear power could increase or decrease in the years ahead.

The recent innovation in nuclear power of modular SMRs that can be factory-assembled and transported as a unit to a location for installation, standardizing their design and reducing their overall cost and start-up time (IAEA 2025), could be a potential boon to the nuclear power industry as utilities look for carbon-free baseload power options, the choices of which are extremely limited.

5.10.3 Hydroelectric Energy

It is clear how hydroenergy can be exploited. The technical community needs to increase the conversion of hydroenergy into electrical energy in the future. Hydropower plans should attempt to:

1. Utilize the total flow available
2. Increase the total height of the waterfall
3. Reduce any heat losses

In effect, existing plants need to be modernized with turbine and generator upgrades, operational improvements, and an increase in generating capacity. In addition, many flood-control dams not equipped for electricity production could be retrofitted with generating equipment. However, it should be noted that the most economical sites for hydropower dams have already been developed; the development of big hydro plants beyond those planned in the United States and elsewhere is unlikely.

Finally, the procedure for licensing and relicensing dams is presently a lengthy and expensive process. Many environmental impact studies must be undertaken and it takes anywhere from 3 to 7 years to obtain a license to build a hydroelectric dam or a relicense to continue operations. Hopefully, this time period will be reduced in the future.

5.10.4 Wind Energy

Wind power will not solve all of the energy problems of the United States. However, it might provide a significant fraction of the energy needs in the future at costs that do not appear to be excessive. And, as noted previously, wind power produces no greenhouse gas emissions during operation and uses relatively little land. The overall cost per unit of energy produced has been comparable to the cost for new fossil fuel installations. In addition, wind energy received government or other support prior to Donald Trump's second administration to encourage its development. Wind power in the United States received a tax credit prior to 2025 for each kilowatt hour produced; the credit had a yearly inflationary adjustment. Another tax benefit is accelerated depreciation. Many states still provide incentives, such as exemption from property tax, mandated purchases, and additional markets for "green credits." Countries such as Canada and Germany also provide some credits. It is assumed that some of these incentives will continue or increase in the near future.

5.10.5 Hydrogen Energy

The goal of the U.S. DoE's Hydrogen Program is for hydrogen to provide 10% of the total energy demand by the Year 2030. Hydrogen has the potential to provide clean, renewable energy for the future. However, before one can judge whether hydrogen is the right choice as a primary energy system, several issues need to be analyzed. Hydrogen could make a significant contribution to the U.S. energy picture, but many new systems would need to be designed and built. There must be large production and storage facilities plus a distribution system. And, consumers must have access to the technology and a desire to use it.

And then there is Murphy's law. A hydrogen economy is not without its pitfalls. The key question: Where is all this hydrogen going to come from? It is true that hydrogen is the most abundant element on this planet; however, it is tied up in a very stable molecule called water and requires a large amount of energy to separate hydrogen from oxygen. The old "catch-22" is back.

5.10.6 Biomass-Derived Energy

It appears that the future of wood burning will remain relatively unchanged. The future of ethanol and biodiesel is positive, however. The federal government mandated in 2012 that 12 billion gallons of renewable fuels be produced per year in the United States. Most of this fuel was to be ethanol. For comparative purposes, the United States consumed over 137 billion gallons of gasoline in 2023. While the total volume of biomass-based biofuels is relatively small, capacity to produce biofuels increased 7% in the United States during 2023, reaching 24 billion gallons per year at the start of 2024, led by a 44% increase in the category *renewable diesel and other biofuels* (EIA 2025h). *Other biofuels* include renewable heating oil, renewable jet fuel (also known as sustainable aviation fuel), and renewable naphtha and gasoline. Given continued state and federal tax incentives, regulatory policies, plant expansions, and projected new plant construction, the EIA expects U.S. biofuels production capacity to continue increasing in the future.

5.10.7 Hydrokinetic Energy

Marine energy resources, such as wave, tidal, and ocean currents, are abundant, geographically diverse, energy dense, predictable, and complementary to other renewable energy sources. Marine energy therefore has the potential to contribute to a carbon pollution-free power sector in an environmentally just and sustainable way. Utility-scale marine energy technologies are at an early stage of development compared to other renewable energy technologies due to the fundamental challenges of generating power from dynamic, low-velocity, and high-density waves and currents, while surviving in corrosive marine environments. These challenges are intensified by high costs and lengthy permitting processes associated with in-water testing (DoE 2022).

Even though deploying marine energy to power the electric utility grid remains challenging, there are near-term opportunities for smaller-scale marine energy systems to enhance the resilience of coastal communities, including through offsetting a significant amount of fossil fuels, and to power systems in the ocean where other energy sources are challenging to access. DoE's Water Power Technology Office is supporting the Powering the Blue Economy initiative to address the energy needs of the rapidly growing "blue economy" and ocean-based activities with marine renewable energy to enable sustainable aquaculture, to power observations that expand the understanding of the ocean and track the changing climate, to power desalination systems, and more (DoE 2022).

REFERENCES

Abulencia, P., and Theodore, L. 2009. *Fluid Flow for the Practicing Engineer*. Hoboken, NJ: Wiley.

Bulletin of Atomic Scientists. 2019. What Really Went Wrong at WIPP: An Insider's View of Two Accidents at the only US Underground Nuclear Waste Repository. https://thebulletin.org/2019/06/what-really-went-wrong-at-wipp-an-insiders-view-of-two-accidents-at-the-only-us-underground-nuclear-waste-repository/. (accessed June 10, 2025).

Center for Sustainable Systems. 2024. *Wind Energy Factsheet*. Pub. No. CSS07-09. Ann Arbor, MI: University of Michigan. https://css.umich.edu/sites/default/files/2025-02/CSS07-09.pdf.

DelGatto, V., Theodore, L., Dupont, R.R., and Ogwa, M. 2025. *Hydrogen Energy: Principles and Applications*. Hoboken, NJ: Wiley.

Department of Energy (DoE). 2022. *Multi-Year Program Plan*. Water Power Technologies Office. Oak Ridge, TN: U.S. DoE Office of Scientific and Technical Information. www.energy.gov/sites/default/files/2022-03/wpto-mypp-2022-1.pdf. (accessed June 10, 2025).

Dorf, R. 1978. *Energy Resources and Policy*. Reading, MA: Addison-Wesley.

Energy Information Agency. 2025a. Nuclear Explained. U.S. Nuclear Industry. www.eia.gov/energyexplained/nuclear/us-nuclear-industry.php. (accessed June 9, 2025).

Energy Information Agency. 2025b. International. Electricity. Electricity Generation. www.eia.gov/international/data/world. (accessed June 9, 2025).

Energy Information Agency. 2025c. Table 8.4. Average Power Plant Operating Expenses for Major U.S. Investor-Owned Electric Utilities, 2013 through 2023 (Mills per Kilowatthour). www.eia.gov/electricity/annual/html/epa_08_04.html. (accessed June 9, 2025).

Energy Information Agency. 2025d. U.S. Renewable Energy Consumption by Source and Sector, 2024. www.eia.gov/totalenergy/data/monthly/pdf/flow/renewable_spaghettichart_2024.pdf. (accessed June 10, 2025).

Energy Information Agency. 2025e. Total Energy. Table 10.1. Renewable Energy Production and Consumption by Source. www.eia.gov/totalenergy/data/browser/ index.php?tbl=T10.01#/?f=A&start=200001. (accessed June 10, 2025).

Energy Information Agency. 2025f. Hydropower Explained. Tidal Power. www.eia.gov/energyexplained/hydropower/tidal-power.php. (accessed June 10, 2025).

Energy Information Agency. 2025g. Hydropower Explained. Wave Power. www.eia.gov/energyexplained/hydropower/wave-power.php. (accessed June 10, 2025).

Energy Information Agency. 2025h. Today in Energy. U.S. Capacity to Produce Biofuels Increased 7% in 2023. www.eia.gov/todayinenergy/detail.php?id=63465. (accessed June 10, 2025).

International Atomic Energy Agency. 2025. What Are Small Modular Reactors (SMRs)? www.iaea.org/newscenter/news/what-are-small-modular-reactors-smrs. (accessed June 9, 2025).

Skipka, K., and Theodore, L. 2014. *Energy Resources: Availability, Management and Environmental Impacts*. Boca Raton, FL: CRC Press/Taylor & Francis Group.

Theodore, L. 2008. *Air Pollution Control Equipment Calculations*. Hoboken, NJ: Wiley.

Theodore, L., and Dupont, R.R. 2012. *Environmental Health and Hazard Risk Assessment: Principles and Calculations*. Boca Raton, FL: CRC Press/Taylor & Francis Group.

Theodore, M.K., and Theodore, L. 2021. *Introduction to Environmental Management, 2nd edition*. Boca Raton, FL: CRC Press/Taylor & Francis Group.

Wikipedia Contributors. 2025a. Yucca Mountain Nuclear Waste Repository. *Wikipedia, The Free Encyclopedia.* https://en.wikipedia.org/w/index.php?title=Yucca_Mountain_nuclear_waste_ repository&oldid=1287542576. (accessed June 10, 2025).

Wikipedia Contributors. 2025b. Nuclear and Radiation Accidents and Incidents. *Wikipedia, The Free Encyclopedia.* https://en.wikipedia.org/w/index.php?title=Nuclear_and_radiation_accidents_and_incidents&oldid=1294421920. (accessed June 10, 2025).

Part 2

Energy Fundamentals and Principles

The objective of the second part of this book is concerned with energy engineering and related subject matter that arises in geothermal energy technology and applications. Chapter numbers and their accompanying titles are listed as follows:

Chapter 6. Introduction to Engineering Fundamentals and Principles
Chapter 7. Conservation Laws
Chapter 8. Thermodynamics
Chapter 9. Heat Exchangers
Chapter 10. Prime Movers
Chapter 11. Steam: The Quintessential Medium in Energy Applications
Chapter 12. Air Conditioning, Refrigeration, and Cryogenics
Chapter 13. Power Plant Thermodynamic Cycle Analysis

DOI: 10.1201/9781003667759-7

6 Introduction to Engineering Fundamentals and Principles

6.1 INTRODUCTION

Basic concepts of units and dimensions, some physical and chemical properties of substances, the Ideal Gas Law (IGL), the Reynolds number, and phase equilibria all play a role in energy-related calculations. Because many of these topics are unrelated to each other, this chapter admittedly lacks the cohesiveness that chapters covering a single topic might have. This is usually the case when basic material from such widely differing areas of knowledge as physics, chemistry, and engineering is surveyed. Though these topics are widely divergent and covered with varying degrees of thoroughness, all of them will find later use in this text. If additional information on these review topics is needed, the reader is directed to the literature in the reference section of this chapter.

Topics reviewed with their accompanying section numbers for this introductory engineering chapter include:

6.2 FUNDAMENTALS OF MEASUREMENT

6.2.1 Units and Dimensions

The units used in this text are consistent with those adopted by the engineering profession in the United States. For engineering work, SI (*Systeme International*) and *English* units are most often employed. In the United States, the *English engineering* units are generally used, although efforts are still underway to obtain universal adoption of SI units for all engineering and science applications. SI units have the advantage of being based on the decimal system, which allows for more convenient

DOI: 10.1201/9781003667759-8

conversion of units within the system. There are other systems of units; some of the more common of these are shown in Table 6.1. Although English engineering units are primarily used, Tables 6.2 and 6.3 present units for both the English and SI systems, respectively. Some of the more common prefixes for SI units are given in Table 6.4.

6.2.2 Conversion of Units

Converting a measurement from one unit to another can conveniently be accomplished by using *unit conversion factors*; these factors are obtained from the simple equation that relates the two units numerically. For example, from

$$1 \text{ foot (ft)} = 12 \text{ inches (in)} \tag{6.1}$$

the following conversion factor can be obtained:

$$12 \text{ in}/1 \text{ ft} = 1 \tag{6.2}$$

Since this factor is equal to unity, multiplying some quantity (e.g., 18 ft) by the factor cannot alter its value. Hence

$$18 \text{ ft } (12 \text{ in}/1 \text{ ft}) = 216 \text{ in} \tag{6.3}$$

Note that in Equation 6.3 the old units of *feet* on the left-hand side cancel out leaving only the desired units of *inches*.

Physical equations must be dimensionally consistent. For the equality to hold, each term in the equation must have the same dimensions. This condition can be and should be checked when solving engineering problems. Throughout the text great

TABLE 6.1
Common Systems of Units

System	Length	Time	Mass	Force	Energy	Temperature
SI	meter	second	kilogram	Newton	Joule	Kelvin, degree Celsius
Egs	centimeter	second	gram	dyne	erg, Joule, or calorie	Kelvin, degree Celsius
Fps	foot	second	pound	poundal	foot poundal	degree Rankine, degree Fahrenheit
American Engineering	foot	second	pound	pound (force)	British thermal unit, horsepower-hour	degree Rankine, degree Fahrenheit
British Engineering	foot	second	slug	pound (force)	British thermal unit, foot pound (force)	degree Rankine, degree Fahrenheit

TABLE 6.2
English Engineering Units

Physical quantity	Name of unit	Symbol for unit
Length	foot	ft
Time	second, minute, hour	s, min, hr
Mass	pound (mass)	lb
Temperature	degree Rankine	°R
Temperature (alternative)	degree Fahrenheit	°F
Moles	pound mole	lbmol
Energy	British thermal unit	Btu
Energy (alternative)	horsepower-hour	hp-hr
Force	pound (force)	lb_f
Acceleration	foot per second squared	ft/s^2
Velocity	foot per second	ft/s
Volume	cubic foot	ft^3
Area	square foot	ft^2
Frequency	cycles per second, Hertz	cycles/s, Hz
Power	Horsepower, Btu per second	hp, Btu/s
Heat capacity	British thermal unit per (pound mass-degree Rankine)	Btu/lb-°R
Density	pound (mass) per cubic foot	lb/ft^3
Pressure	pound (force) per square inch	psi
	pound (force) per square foot	psf
	atmospheres	atm
	bar	bar

care is exercised in maintaining the dimensional formulas of all terms and the dimensional homogeneity of each equation. Equations will generally be developed in terms of specific units rather than general dimensions (e.g., ft, rather than length). This approach should help the reader to more easily attach physical significance to the equations presented in these chapters.

6.2.3 Significant Figures and Scientific Notation

Significant figures provide an indication of the precision with which a quantity is measured or known. The last digit represents, in a qualitative sense, some degree of doubt. For example, a measurement of 8.32 inches implies that the actual quantity is somewhere between 8.315 and 8.325 inches. This applies to calculated and measured quantities; quantities that are known exactly (e.g., pure integers) have an infinite number of significant figures.

The significant digits of a number are the digits from the first nonzero digit on the left to either (a) the last digit (whether it is nonzero or zero) on the right if there is a decimal point, or (b) the last nonzero digit of the number if there is no decimal point. For example:

TABLE 6.3
SI Units

Physical Unit	Name of Unit	Symbol for Unit
Length	meter	m
Mass	kilogram, gram	kg, g
Time	second	s
Temperature	Kelvin	K
Temperature (alternative)	degree Celsius	$^{\circ}C$
Moles	gram mole	gmol
Energy	Joule	J, kg-m^2/s^2
Force	Newton	N, kg-m^2/s^2, J/m
Acceleration	meters per second squared	m/s^2
Pressure	Pascal, Newton per square meter	Pa, N/m^2
Pressure (alternative)	bar	bar
Velocity	meters per second	m/s
Volume	cubic meters, liters	m^3, L
Area	square meters	m^2
Frequency	Hertz	Hz, cycles/s
Power	Watt	W, kg-m^2-s^3, J/s
Heat capacity	Joule per kilogram-Kelvin	J/kg-K
Density	kilogram per cubic meter	kg/m^3
Angular velocity	radians per second	rad/s

TABLE 6.4
Prefixes for SI Units

Multiples and Submultiples		Prefixes	Symbols
1 000 000 000 000	10^{12}	tera	T
1 000 000 000	10^{9}	giga	G
1 000 000	10^{6}	mega	M
1000	10^{3}	kilo	k
100	10^{2}	hecto	h
10	10^{1}	deka	da
Base unit 1	10^{0}		
0.1	10^{-1}	deci	d
0.01	10^{-2}	centi	c
0.001	10^{-3}	milli	m
0.000 001	10^{-6}	micro	µ
0.000 000 001	10^{-9}	nano	n
0.000 000 000 001	10^{-12}	pico	P
0.000 000 000 000 001	10^{-15}	femto	f
0.000 000 000 000 000 001	10^{-18}	atto	a

370	has two significant figures
370.	has three significant figures
370.0	has four significant figures
28,070	has four significant figures
0.037	has two significant figures
0.0370	has three significant figures
0.02807	has four significant figures

Whenever quantities are combined by multiplication and/or division, the number of significant figures in the result should equal the lowest number of significant figures of any of the quantities. In long calculations, the final result should be rounded off to the correct number of significant figures. When quantities are combined by addition and/or subtraction, the final result cannot be more precise than any of the quantities added or subtracted. Therefore, the position (relative to the decimal point) of the last significant digit in the number that has the lowest degree of precision is the position of the last permissible significant digit in the result. For example, the sum of 3702., 370, 0.037, 4, and 37. should be reported as 4110 (without a decimal). The least precise of the five numbers is 370, which has its last significant digit in the tens position. The answer should also have its last significant digit in the tens position.

Unfortunately, engineers and scientists rarely concern themselves with significant figures in their calculations. However, it is recommended that the readers attempt to follow the calculation procedure set forth in this section.

In the process of performing engineering calculations, very large and very small numbers are often encountered. A convenient way to represent these numbers is to use *scientific notation*. Generally, a number represented in scientific notation is the product of a number ($<$ 10 but $>$ or = 1) and 10 raised to an integer power. For example,

$$28{,}070{,}000{,}000 = 2.807 \times 10^{10}$$
$$0.000\ 002\ 807 = 2.807 \times 10^{-6}$$

A positive feature of using scientific notation is that only the significant figures need appear in the number.

Zeros are significant when they occur *between* significant digits. In the following examples, all zeros are significant: 10001, 402, 1.1001, and 500.09 (five, three, five, and five significant figures, respectively). *Zeroes are not significant when they are used as place holders*. When used as a holder, a zero simply identifies where a decimal is located. For example, each of the following numbers has only one significant digit: 1000, 500, 60, 0.09, and 0.0002. In the numbers 1200, 540, and 0.0032, there are two significant digits, and the zeroes are not significant. Once again, when *zeros follow a decimal and are preceded by a significant digit, the zeros are significant*. In the following examples, all zeroes are significant: 1.00, 15.0, 4.100, 1.90, 10.002, and 10.0400. For 10.002, the zeroes are significant because they fall between two significant digits. For 10.0400, the first two zeroes are significant because they fall between two significant digits; the last two zeroes are also significant because they follow a decimal and are preceded by a significant digit.

When truncating (removing final, unwanted digits), rounding is normally applied to the last digit to be kept. Thus, *if the value of the first digit to be discarded is less than 5, one should retain the last kept digit with no change. If the first digit to be discarded is 5 or greater, increase the last kept digit's value by one.* Assume, for example, only the first two decimal places are to be kept for 25.0847 (the 4 and 7 are to be dropped). The number is then 25.08. Since the first digit to be discarded (4) is less than 5, the 8 is not rounded up. If only the first two decimal places are to be kept for 25.0867 (the 6 and the 7 are to be dropped), it should be rounded to 25.09. Since the first digit to be discarded (6) is 5 or more, the 8 is rounded up to 9.

6.3 CHEMICAL AND PHYSICAL PROPERTIES

6.3.1 Temperature

Whether in the gaseous, liquid, or solid state, all molecules possess some degree of kinetic energy, that is, they are in constant motion – vibrating, rotating, or translating. The kinetic energies of individual molecules cannot be measured, but the combined effect of these energies in a very large number of molecules can. This measurable quantity is known as *temperature*; it is a macroscopic concept only and as such does not exist on the molecular level.

Temperature can be measured in many ways; the most common method makes use of the expansion of a liquid (usually encased inside a glass capillary tube) with increasing temperature, or through the employment of thermocouples or thermistors. The two most commonly used temperature scales are the Celsius (or Centigrade) and Fahrenheit scales. The Celsius scale is based on the boiling and freezing points of water at 1 atm pressure; to the former, a value of 100°C is assigned, and to the latter, a value of 0°C is assigned. On the older Fahrenheit scale, these temperatures correspond to 212°F and 32°F, respectively. Equations 6.4 and 6.5 show the conversion from one scale to the other.

$$°F = 1.8\,(°C) + 32 \tag{6.4}$$

$$°C = (°F - 32)/1.8 \tag{6.5}$$

where °F = temperature on the Fahrenheit scale, and °C = temperature on the Celsius scale.

Experiments with gases at low-to-moderate pressures (up to a few atmospheres) have shown that, if the pressure is kept constant, the volume of a gas and its temperature are linearly related (Charles' Law) and that a decrease of 0.3663% or (1/273) of the initial volume is experienced for every temperature drop of 1°C. These experiments were not extended to very low temperatures, but if the linear relationship were extrapolated, the volume of the gas would *theoretically* be zero at a temperature of approximately – 273°C or – 460°F. This temperature has become known as *absolute zero* and is the basis for the definition of two absolute temperature scales.

An absolute scale is one which does not allow negative quantities. These absolute temperature scales are the kelvin (K) and Rankine (°R) scales; the former is defined by shifting the Celsius scale by 273°C so that 0 K is equal to –273°C; Equation 6.6 shows this relationship:

$$K = °C + 273 \tag{6.6}$$

The Rankine scale is defined by shifting the Fahrenheit scale 460°, so that

$$°R = °F + 460 \tag{6.7}$$

6.3.2 Pressure

Molecules possess a high degree of translational kinetic energy in the gaseous state, which means they are able to move quite freely throughout the body of a gas. If the gas is in a container of some type, the molecules are constantly bombarding the walls of the container. The macroscopic effect of this bombardment by a tremendous number of molecules – enough to make the effect measurable – is called pressure. The natural units of pressure are force per unit area. In the example of the gas in a container, the unit area is a portion of the inside solid surface of the container wall, and the force, measured perpendicularly to the unit area, is the result of the molecules hitting the unit area and giving up momentum during the sudden change of direction.

There are a number of different ways to express a pressure measurement. Some of them are natural units, for example, based on a force per unit area, that is, pound (force) per square inch (abbreviated lb/in^2 or psi) or dyne per square centimeter (dyn/cm^2). Others are based on a fluid height, such as inches of water (in H_2O) or millimeters of mercury (mm Hg); units such as these are convenient when the pressure is indicated by a difference between two levels of a liquid as in a manometer or barometer. Barometric pressure and atmospheric pressure are synonymous and measure the ambient air pressure. Standard barometric pressure is the average atmospheric pressure at sea level, 45° north latitude at 32°F. It is used to define another unit of pressure called the atmosphere (atm). Standard barometric pressure is 1 atm and is equivalent to 14.696 psi or 29.921 in Hg. As one might expect, barometric pressure varies with weather and altitude.

Since gauge pressure is the pressure relative to the prevailing atmospheric pressure, the sum of the two gives the absolute pressure, indicated by the letter a after the unit (e.g., psia (pounds per square inch absolute)):

$$P = P_a + P_g \tag{6.8}$$

where P = absolute pressure, psia; P_a = atmospheric pressure, psia; and P_g = gauge pressure, psig.

The absolute pressure scale is absolute in the same sense that the absolute temperature scale is absolute, for example, a pressure of zero psia is the lowest possible pressure theoretically achievable – a perfect vacuum.

6.3.3 Moles and Molecular Weight

An atom consists of protons and neutrons in a nucleus surrounded by electrons. An electron has such a small mass relative to that of the proton and neutron that the weight of the atom (called the *atomic weight*) is approximately equal to the sum of the weights of the particles in its nucleus. Atomic weight may be expressed in *atomic mass units* (amu) per atom or in g per g-atom. One g-atom contains 6.023×10^{23} molecules (Avogadro's number). The atomic weights of the elements can be easily found on line at https://pubchem.ncbi.nlm.nih.gov/periodic-table/.

The *molecular weight* (MW) of a compound is the sum of the atomic weights of the atoms that make up the molecule. Units of atomic mass per molecule (amu/molecule) or g per g-mole (g/gmol) are used for molecular weight. One gmol contains an Avogadro's number of molecules. For the English system, a pound-mole (lbmol) contains $454 \times 6.023 \times 10^{23}$ molecules.

Molal units are used extensively in engineering calculations as they greatly simplify material balances where chemical (including combustion) reactions are occurring. For mixtures of substances (gases, liquids, or solids), it is also convenient to express compositions in mole fractions or mole percentages instead of mass fractions. The mole fraction is the ratio of the number of moles of one component to the total number of moles in the mixture.

6.3.4 Mass and Volume

The density (ρ) of a substance is the ratio of its mass to its volume and may be expressed in units of pounds per cubic foot (lb/ft^3), kilograms per cubic meter (kg/m^3), etc. For solids, density can be easily determined by placing a known mass of a substance in a liquid and determining the displaced volume. The density of a liquid can be measured by weighing a known volume of the liquid in a volumetric flask. For gases, the IGL, to be discussed in Section 6.4 can be used to calculate the density from the pressure, temperature, and molecular weight of the gas.

Densities of pure solids and liquids are relatively independent of temperature and pressure and can be found in standard reference books. The *specific volume* (v) of a substance is its volume per unit mass (ft^3/lb, m^3/kg, etc.) and is, therefore, the inverse of its density.

The specific gravity (SG) is the ratio of the density of a substance to the density of a reference substance at a specific condition:

$$SG = \rho/\rho_{ref} \tag{6.9}$$

The reference most commonly used for solids and liquids is water at its maximum density, which occurs at 4°C; this reference density is 1.000 g/cm^3, 1000 kg/m^3, or 62.43 lb/ft^3. Note that, since the specific gravity is a ratio of two densities, it is *dimensionless*. Therefore, any set of units may be employed for the two densities as long as they are consistent. The specific gravity of gases is used only rarely; when it is, air at the same conditions of temperature and pressure as the gas is usually employed as the reference substance.

6.3.5 Viscosity

Viscosity is a property associated with a fluid's resistance to flow; more precisely, this property accounts for energy losses which result from shear stresses that occur between different portions of the fluid which are moving at different velocities. The absolute viscosity (μ) has units of mass per length-time; the fundamental unit is the *poise*, which is defined as 1 g/cm-s. This unit is inconveniently large for many practical purposes and viscosities are frequently given in *centipoise* (0.01 poise), which is abbreviated cP. The viscosity of pure water at 68.6°F is 1.00 cP. In English units, absolute viscosity is expressed either as pounds (mass) per foot-second (lb/ft-s) or pounds per foot-hour (lb/ft-h). The absolute viscosity depends primarily on temperature and to a lesser degree on pressure. The kinematic viscosity (ν) is the absolute viscosity divided by the density of the fluid and is useful in certain fluid flow problems in the geothermal energy field; the units for this quantity are length squared per time, for example, square foot per second (ft^2/s) or square meters per hour (m^2/h). A kinematic viscosity of 1 cm^2/s is called a *stoke*. For pure water at 70°F, $\nu = 0.983$ cS (*centistokes*). Because fluid viscosity changes rapidly with temperature, a numerical value of viscosity has no significance unless the temperature is specified.

Liquid viscosity is usually measured by the amount of time it takes for a given volume of liquid to flow through an orifice. The *Saybolt universal viscometer* is the most widely used device in the United States for the determination of the viscosity of fuel oils and liquid wastes. It should be stressed that Saybolt viscosities, which are expressed in *Saybolt second universal* (SSU), are not even approximately proportional to absolute viscosities except in the range above 200 SSU; hence, converting units from Saybolt seconds to other units requires the use of special conversion tables. As the time of flow decreases, the deviation becomes more marked. In any event, viscosity is an important property because of potential flow problems in the geothermal energy field. Viscosities of many substances are available online at The Engineering Toolbox.com (2003a).

6.3.6 Heat Capacity

The *heat capacity* of a substance is defined as the quantity of heat required to raise the temperature of that substance by 1°; the *specific heat capacity* is the heat capacity on a unit mass basis. The term *specific heat* is frequently used in place of specific heat capacity. This is not strictly correct because traditionally specific heat has been defined as the ratio of the heat capacity of a substance to the heat capacity of water. However, since the specific heat of water is approximately 1 cal/g-°C or 1 Btu/lb-°F, the term specific heat has come to imply heat capacity per unit mass. For gases, the addition of heat to cause the 1° temperature rise may be accomplished either at constant pressure or constant volume. Since the amounts of heat necessary are different for the two cases, subscripts are used to identify which heat capacity is being used $-c_P$ for constant pressure and c_V for constant volume. For liquids and solids, this distinction does not have to be made since there is little difference between the two. Values of heat capacities are available online at The Engineering Toolbox.com (2003b).

Heat capacities are often expressed on a molar basis instead of a mass basis, in which case the units become cal/gmol-°C or Btu/lbmol-°F. To distinguish between the two bases, uppercase letters (C_P, C_V) are typically used to represent the molar-based heat capacities, and lowercase letters (c_P, c_V) are used for the mass-based heat capacities.

6.3.7 Thermal Conductivity

Experience has shown that when a temperature difference exists across a solid body, energy in the form of heat will transfer from the high-temperature region to the low-temperature region until thermal equilibrium is reached. This mode of heat transfer in which vibrating molecules pass along kinetic energy through the solid is called *conduction*. Liquids and gases may also transport heat in this fashion. (For a more comprehensive discussion of heat transfer mechanisms, the reader is directed to Chapter 10.) The property of thermal conductivity provides a measure of how fast (or how easily) heat flows through a substance. It is defined as the amount of heat that flows per unit time through a unit surface area of unit thickness as a result of a unit difference in temperature. Typical English units for conductivity are Btu-ft/h-ft^2-°F or Btu/h-ft-°F.

Thermal conductivities for liquids, gases and for some insulating solids are available online at The Engineering Toolbox.com (2003c).

6.3.8 Reynolds Number

The Reynolds number, *Re*, is a dimensionless number that indicates whether a fluid flowing is in the *laminar* or *turbulent* mode. Laminar flow is characteristic of fluids flowing slowly enough so that there are no eddies (whirlpools) or macroscopic mixing of different portions of the fluid. (Note: In any fluid, there is always molecular mixing due to the thermal activity of the molecules; this is distinct from macroscopic mixing due to the swirling motion of different portions of the fluid.) In laminar flow, a fluid can be imagined to flow like a deck of cards, with adjacent layers sliding past one another. Turbulent flow is characterized by eddies and macroscopic currents. In practice, moving gases generally flow in the turbulent region. For flow in a pipe, a Reynolds number above 2,100 is an indication of turbulent flow (Abulencia and Theodore 2009).

The Reynolds number is dependent on the fluid velocity, density, viscosity, and some length characteristic of the system or conduit; for pipes, this characteristic length is the inside diameter:

$$Re = \mathrm{v}D\rho/\mu = \mathrm{v}D/\nu \tag{6.10}$$

where *Re* = Reynolds number, unitless; D = inside diameter of the pipe, ft; v = fluid velocity, ft/s; ρ = fluid density, lb/ft^3; μ = fluid viscosity, lb/ft-s; and ν = fluid kinematic viscosity, ft^2/s. Any consistent set of units may be used with Equation 6.10.

6.3.9 pH

An important chemical property of an aqueous solution is its pH. The pH measures the acidity or basicity of the solution. In a neutral solution, such as pure water, the hydrogen (H^+) and hydroxyl (OH^-) ion concentrations are equal. At ordinary temperatures, this concentration is:

$$C_{H^+} = C_{OH^-} = 10^{-7} \text{ g-ion/L} \tag{6.11}$$

where C_{H^+}= hydrogen ion concentration, g-ion/L; and C_{OH^-} = hydroxyl ion concentration, g-ion/L.

The unit g-ion stands for gram-ion, which represents an Avogadro's number of ions. In all aqueous solutions, whether neutral, basic, or acidic, a chemical equilibrium or balance is established between these two concentrations, so that:

$$K_{eq} = C_{H^+}(C_{OH^-}) = 10^{-14} \tag{6.12}$$

where K_{eq} = equilibrium constant.

The numerical value for K_{eq}, given in Equation 6.12 holds for room temperature and only when the concentrations are expressed in g-ion/L. In acid solutions, C_{H^+} is $>C_{OH^-}$; in basic solutions, C_{OH^-} predominates.

The pH is a direct measure of the hydrogen ion concentration and is defined by:

$$\text{pH} = -\log C_{H^+} \tag{6.13}$$

Thus, an acidic solution is characterized by a pH below 7 (the lower the pH, the greater the acidity); a basic solution, by a pH above 7; and a neutral solution by a pH of 7. It should be pointed out that Equation 6.13 is not the exact definition of pH but is a close approximation to it. Strictly speaking, the activity of the hydrogen ion, a_{H^+}, and not the ion concentration belongs in Equation 6.13. For a discussion of chemical activities, the reader is directed to the literature (Theodore and Reynolds 1991; Theodore, Ricci, and VanVliet 2009).

6.3.10 Vapor Pressure

Vapor pressure is an important property of liquids, and, to a much lesser extent, of solids. If a liquid is allowed to evaporate in a confined space, the pressure in the vapor space increases as the amount of evaporated material increases. If there is sufficient liquid present, a point is eventually reached at which the pressure in the vapor space is exactly equal to the pressure exerted by the liquid at its own surface. At this point, a dynamic equilibrium exists in which vaporization and condensation take place at equal rates and the pressure in the vapor space remains constant. The pressure exerted at equilibrium is called the vapor pressure of the material. The magnitude of this pressure for a given material depends on the temperature, but not on the amount of material present. Solids, like liquids, also exert a vapor pressure. Evaporation of solids (called *sublimation*) is noticeable only for those with appreciable vapor pressures, that is, *dry ice*.

6.4 THE IDEAL GAS LAW

Observations based on physical experimentation often can be synthesized into simple mathematical equations called laws. These laws are never perfect and hence are only an approximate representation of reality. The IGL was derived from experiments in which the effects of pressure and temperature on gaseous volumes were measured over moderate temperature and pressure ranges. This law works well in the pressure and temperature ranges that were used in taking the data; extrapolations outside of the ranges have been found to work well in some cases and poorly in others. As a general rule, this law works best when the molecules of the gas are far apart, that is, when the pressure is low and the temperature is high. Under these conditions, the gas is said to behave *ideally*, that is, its behavior is a close approximation to the so-called *perfect* or *ideal gas*, a hypothetical entity that obeys the IGL perfectly. For energy engineering calculations the IGL is almost always assumed to be valid, since it generally works well (usually within a few percent of the correct result) up to the highest pressures and down to the lowest temperatures used in most engineering applications.

The two precursors of the IGL were *Boyle's* and *Charles'* laws. Boyle found that the volume of a given mass of gas is inversely proportional to the absolute pressure if the temperature is kept constant:

$$P_1V_1 = P_2V_2 \tag{6.14}$$

where V_1 = volume of gas at absolute pressure $P_1$and temperature T; and V_2 = volume of gas at absolute pressure P_2 and temperature T.

Charles found that the volume of a given mass of gas varies directly with the absolute temperature at constant pressure:

$$\frac{V_1}{T_1} = \frac{V_2}{T_2} \tag{6.15}$$

where V_1 = volume of gas at pressure P and absolute temperature T_1; and V_2 = volume of gas at pressure P and absolute temperature T_2.

Boyle's and Charles' laws may be combined into a single equation in which neither temperature nor pressure need be held constant:

$$\frac{P_1V_1}{T_1} = \frac{P_2V_2}{T_2} \tag{6.16}$$

For Equation 6.16 to hold, the mass of gas must be constant as the conditions change from (P_1, T_1) to (P_2, T_2). This equation indicates that for a given mass of a specific gas,

PV/T has a constant value. Since, at the same temperature and pressure, volume and mass must be directly proportional, this statement may be extended to:

$$\frac{PV}{mT} = c \tag{6.17}$$

where m = mass of a specific gas; and c = a constant that depends on the gas.

Moreover, experiments with different gases showed that Equation 6.17 could be expressed in a far more generalized form. If the number of moles (n) is used in place of the mass (m), the constant is the same for all gases:

$$\frac{PV}{nT} = R \tag{6.18}$$

where R = the universal gas constant with numerous units depending on the units for all other variables in the equation.

Equation 6.18 is defined as the IGL. Numerically, the values of R are shown in Table 6.5 based on the units used for P, V, T, and n. In this text, all gases are assumed

TABLE 6.5
Values of *R* in Various Units

R	Temperature Scale	Units of V	Units of n	Units of P	Unit of PV (energy)
10.73	°R	ft^3	lbmol	psia	–
0.7302	°R	ft^3	lbmol	atm	–
21.85	°R	ft3	lbmol	in Hg	–
555.0	°R	ft^3	lbmol	mm Hg	–
297.0	°R	ft^3	lbmol	in H_2O	–
0.7398	°R	ft^3	lbmol	bar	–
1,545.0	°R	ft^3	lbmol	psfa	–
24.75	°R	ft^3	lbmol	ft H_2O	–
1.9872	°R	–	lbmol	–	Btu
0.0007805	°R	–	lbmol	–	hp-h
0.0005819	°R	–	lbmol	–	kW-h
500.7	°R	–	lbmol	–	cal
1.314	K	ft^3	lbmol	atm	–
998.9	K	ft^3	lbmol	mm Hg	–
19.32	K	ft^3	lbmol	psia	–
62.361	K	L	gmol	mm Hg	–
0.08205	K	L	gmol	atm	–
0.08314	K	L	gmol	bar	–
8314	K	L	gmol	Pa	
8.314	K	m^3	gmol	Pa	
82.057	K	cm^3	gmol	atm	
1.9872	K	–	gmol	–	cal
8.3144	K	–	gmol	–	J

to approximate ideal gas behavior. As is generally the case in engineering practice, the IGL is assumed to be valid for all illustrative and assigned problems. If a case is encountered in practice where the gas behaves in very nonideal fashion, for example, a high-molecular-weight gas (such as a chlorinated organic) under high pressures, one of the many real gas correlations found in the literature (Theodore, Ricci, and VanVliet 2009; Society of Petroleum Engineers 2025) may be employed.

Other useful forms of the IGL are shown in Equations 6.19 and 6.20. Equation 6.19 applies to gas flow rather than to gas confined in a container:

$$PQ = \dot{n}RT \tag{6.19}$$

where Q = gas volumetric flow rate, ft^3/h; P = absolute pressure, psia; $\dot{n}$ = molar flow rate, lbmol/h; T = absolute temperature, °R; and R = 10.73 psia-ft^3/lbmol-°R.

Equation 6.20 combines n and V from Equation 6.18 to express the law in terms of density:

$$PM = \rho RT \tag{6.20}$$

where M = molecular weight of gas, lb/lbmol; and ρ = density of gas, lb/ft^3.

Volumetric flow rates are often given not at the actual conditions of pressure and temperature but at arbitrarily chosen standard conditions (*STP, standard temperature and pressure*). To distinguish between flow rates based on the two conditions, the letters *a* and *s* are often used as part of the unit. The units acfm and scfm stand for actual cubic feet per minute and standard cubic feet per minute, respectively. The IGL can be used to convert from *standard* to *actual* conditions, but, since there are many standard conditions in use, the STP being used must be known. Standard conditions most often used are shown in Table 6.6. The reader is cautioned on the incorrect use of acfm and/or scfm. The use of standard conditions is a convenience; when predicting the performance of or designing equipment, the actual conditions *must* be employed. Designs based on standard conditions can lead to disastrous results, with the unit usually under designed. For example, for an air or steam gas stream at 2,140°F, the ratio of acfm to scfm (standard temperature = 60°F) is 5.0. Equation 6.21, which is a form of Charle's law, can be used to correct flow rates from standard to actual conditions:

TABLE 6.6
Common Standard Conditions

System	Temperature	Pressure	Molar Volume
SI	273 K	101.3 kPa	22.4 m^3/kgmol
Universal scientific	0°C	760 mm Hg	22.4 L/gmol
Natural gas industry	60°F	14.7 psia	379 ft^3/lbmol
American engineering	32°F	1 atm	359 ft^3/lbmol
Hazardous waste	60°F	1 atm	379 ft^3/lbmol
Incinerator industry	70°F	1 atm	387 ft^3/lbmol

$$\underline{Qa} = Q_s(T_a/T_s) \tag{6.21}$$

where Q_a = volumetric flow rate at actual conditions, ft³/h; Q_s = volumetric flow rate at standard conditions, ft³/h; T_a = actual absolute temperature, °R; and T_s = standard absolute temperature, °R.

The reader is again reminded that absolute temperatures and pressures *must* be employed in all IGL calculations.

In engineering practice, mixtures of gases are more often encountered than single or pure gases. The IGL is based on the *number* of molecules present in the gas volume; the *kind* of molecules is not a significant factor, only the number. This law applies equally well to mixtures and pure gases alike. Dalton and Amagat both applied the IGL to mixtures of gases. Since pressure is caused by gas molecules colliding with the walls of the container, it seems reasonable that the total pressure of a gas mixture is made up of pressure contributions due to each of the component gases. These pressure contributions are called *partial pressures*. Dalton defined the partial pressure of a component as the pressure that would be exerted if the same mass of the component gas occupied the same total volume alone at the same temperature as the mixture. The sum of these partial pressures equals the total pressure:

$$P = P_a + P_b + P_c + \ldots + P_n = \sum_{i=1}^{n} P_i \tag{6.22}$$

where P = total pressure; n = number of components; and P_i = partial pressure of component i.

Equation 6.22 is known as *Dalton's law*. Applying the IGL to one component (a) only,

$$P_a V = n_a RT \tag{6.23}$$

where n_a = number of moles of component a.

Eliminating R, T, and V between Equations 6.18 and 6.23 yields:

$$\frac{P_a}{P} = \frac{n_a}{n} = y_a; \quad P_a = y_a P \tag{6.24}$$

where y_a = mole fraction of component I.

Amagat's law is similar to Dalton's. Instead of considering the total pressure to be made up of partial pressures where each component occupies the total container volume, Amagat considered the total volume to be made up of partial volumes in which each component is exerting the total pressure. The definition of the partial volume is therefore the volume occupied by a component gas alone at the same temperature and pressure as the mixture:

$$V = V_a + V_b + V_c + \ldots + V_n = \sum_{i=1}^{n} V_i \tag{6.25}$$

where V = total volume; n = number of components; and V_i = partial volume of component i. Amagat's law can also be expressed in a form equivalent to Equation 6.24 as shown in Equation 6.26:

$$\frac{V_a}{V} = \frac{n_a}{n} = y_a; \quad V_a = y_a V \tag{6.26}$$

6.5 ENERGY TERMS

Energy is sometimes classified into various forms as expressed by the terms *potential, kinetic, mechanical, chemical, electrical, atomic*, etc. These terms are in common use and convenient for some purposes and hence deserve some mention.

Potential energy is the energy that a system possesses by virtue of its particular position. The simplest case would be any object in relation to its position above the Earth's surface. Work must be done to elevate any mass above the surface of the Earth, and the energy represented by the work done is said to be stored in the system (i.e., the combination of the mass and the Earth) as potential energy – *potential* because it is capable of yielding work if the body is allowed to fall to the Earth's surface. Like all energy it has no absolute amount but is purely relative, depending on the choice of a datum plane.

T potential energy of a given system may also be *internal* owing to the particular configuration of the molecules, atoms, electrons, etc., that it is composed of. There are forces of attraction between the molecules of substances, and as the distance between them changes, as in the expansion of a gas or the vaporization of a liquid, changes in potential energy occur analogous to that which occurs when a body changes position in the Earth's gravitational field.

Kinetic energy is the energy associated with the motion of a body or particle relative to some reference body arbitrarily chosen. It is quantitatively defined for the simple case of a constant velocity by the well-known equation $KE = \frac{1}{2}mv^2$, where m is mass and v is the relative velocity. Kinetic energy may be associated with the motion of large or tangible masses or with the motion of the ultimate particles such as molecules or atoms.

Energy, whether potential or kinetic, that is associated with relatively large masses (as compared with the ultimate particles of which matter is composed) is generally referred to as mechanical energy. Energy associated with the motion or relative positions of molecules is referred to as *thermal* or *heat energy*.

Energy associated with electromotive forces is called *electrical energy* and generally results from movements of the relatively mobile electrons that are present in varying degree in all substances. When chemical reactions take place, potential energy is stored or released. The exact form in which this energy is stored is not known in detail; it must, of course, be related to the configuration of the atoms and electrons in the molecules. Since it is associated with chemical change, it is conveniently referred to as *chemical energy*.

Note that the internal energy of a substance does not include any energy that it may possess as a result of its position or movement as a whole. Rather it refers to the energy of the molecules making up the substance. The molecules of any substance are believed to be in ceaseless motion and therefore to possess kinetic energy not only of translation but also in many cases of rotation and vibration. The addition of heat to a substance increases this molecular activity, and thus causes an increase in its internal energy. In addition to kinetic energy, the molecules of any substance possess potential energy resulting from the forces of attraction existing among them. On a submolecular scale there is also energy associated with the electrons and nuclei of atoms and bond energy resulting from the forces which hold atoms together as molecules. It has not been possible to determine the total internal energy of a substance, and hence absolute values are unknown. However, this is not a disadvantage in engineering calculations since only *changes* in internal energy are required to quantify. The designation of this form of energy of a substance as *internal* distinguishes it from potential and kinetic energy which the substance may possess because of its position or motion *as a whole*, and which can be thought as *external* forms of energy.

Nuclear energy/reactions are *not* included in the term *ordinary energy*. Such processes require the use of Einstein's equation relating energy and mass:

$$E = \frac{c^2 m}{g_c} \tag{6.27}$$

where c = the velocity of light; m = mass; and g_c = gravitation acceleration. Since the term c^2 is very large, a small quantity of mass can be transformed into an enormous amount of energy. For example, in the usual engineering units c is approximately 10^9ft/s, and 1 lb mass is equivalent to:

$$E = \frac{10^{18}(1)}{32.2} = 3.1\text{x}10^{16}\,\text{ft - lb}_\text{f}$$

Thus the laws of conservation of energy and mass cannot be thought of as existing separately where nuclear reactions are concerned, but must be combined to state that energy and mass together are conserved.

6.6 HEAT, WORK, AND POWER

Energy is commonly used as a general term that embraces not only heat and work but also power.

Heat may be defined in terms of its effect in changing the temperature of some substance to which it is transferred; the substance commonly chosen is water. The units of heat in common use are the calorie and the British thermal unit (Btu).

Work is accomplished whenever a system acts against an external force. It is expressed as the product of an intensity factor and a capacity, or amplitude, factor. For ordinary mechanical work, the intensity factor is force (F) and the capacity factor is the distance through which the force acts (D). Therefore:

$$W = F\,(D) \tag{6.28}$$

or if the force is a variable one, Equation 6.28 becomes Equation 6.29:

$$W = \int_{D_1}^{D_2} F dD \tag{6.29}$$

which can be evaluated if F can be expressed as a function of D.

In addition to mechanical work, which involves the action of mechanical forces, one may have other kinds of work, depending on the nature of the force acting. Work may be accomplished by the action of an electrical force or, more strictly, an electromotive force, represented by E. In this case the capacity, or amplitude, factor is Q, the quantity, or charge, of electricity, usually expressed in coulombs. Hence for electrical work:

$$W = \int_{Q_1}^{Q_2} E dQ \tag{6.30}$$

Power is the time rate of doing work, or, mathematically,

$$P = \frac{dW}{dt} \tag{6.31}$$

For the case of electrical work, this becomes,

$$P = \frac{EdQ}{dt} = EI \tag{6.32}$$

since the current I is, by definition, equal to dQ/dt.

6.7 ELECTRICITY

Electricity is a phenomenon resulting from both the existence of charge and the interaction of charges. When a charge is stationary or static, it produces forces on objects in regions where it is present, and when it is in motion, it produces magnetic effects. Electricity and magnetic effects are caused by the relative position and movement of positively and negatively charged particles of matter. So far as electrical effects are concerned, these particles are either neutral, positive, or negative. Electricity is concerned with the positively charged particles, such as protons, and the negative charged particles, such as electrons, that repel one another of the same charge but attract each other of opposite charge.

Electricity is an extremely useful form of energy. It can be converted to heat, light, and mechanical energy. However, it is best to view electricity as a carrier of energy. Any form of energy, including energy present in a fossil fuel or a nuclear fuel or from a geothermal energy source, can be converted to energy in the form of electricity in order to transport and readily distribute it to customers. Electric power is transmitted and distributed by means of transmission lines to essentially all the residences, industries, and commercial buildings in the United States (Dorf 1978), a topic to be discussed later in this book.

Electricity is a relatively useful energy carrier since it is readily transported with low attendant losses, and improved methods for safe handling of electricity have been developed over the past 80 years. Furthermore, methods of converting fossil fuels to electric power are well developed, economical, and reliably safe. Means of converting solar, wind, and nuclear energy to electric energy are also well developed. Geothermal energy, hydropower, and tidal energy may also be converted to electric energy, and the geothermal energy component represents the main focus of Part III of this text.

REFERENCES

Abulencia, P., and Theodore, L. 2009. *Fluid Flow for the Practicing Engineer.* Hoboken, NJ: Wiley.

Dorf, R. 1978. *Energy Resources and Policy.* Reading, MA: Addison-Wesley.

Society of Petroleum Engineers. 2025. *Real Gases.* Petrowicki content. https://doi.org/10.2118/PW0813. https://onepetro.org/spe/general-information/2031/Real-gases. (Accessed June 15, 2025).

The Engineering ToolBox. 2003a. *Dynamic, Absolute, and Kinematic Viscosity – Definitions & Conversions.* www.engineeringtoolbox.com/dynamic-absolute-kinematic-viscosity-d_412.html. (Accessed June 15, 2025).

The Engineering ToolBox.2003b. *Specific Heat of Common Materials – Engineering Reference.* www.engineeringtoolbox.com/specific-heat-capacity-d_391.html. (Accessed June 15, 2025).

The Engineering ToolBox.2003c. *Thermal Conductivity of Common Materials – Solids, Liquids and Gases.* www.engineeringtoolbox.com/thermal-conductivity-d_429.html. (Accessed June 15, 2025).

Theodore, L., and Reynolds, J. 1991. *Thermodynamics, a Theodore Tutorial.* East Williston, NY: Theodore Tutorials. Originally published by the USEPA/APTI, RTP, NC. Roanoke, VA: ETS Professional Training Institute. https://etsi-inc.com/training/tutorials-and-textbooks/.

Theodore, L., Ricci, F., and VanVliet, T. 2009. *Thermodynamics for the Practicing Engineer.* Hoboken, NJ: Wiley.

7 Conservation Laws

7.1 INTRODUCTION

In order to better understand engineering, it is necessary to first understand the theory underlying it. How can one predict what products will be emitted from effluent streams? At what temperature must a thermal device be operated? How much energy in the form of heat is given off during a combustion process? Is it economically feasible to recover this heat? Is the feed high enough in heating value, or must additional fuel be added to assist in the thermal process? If so, how much fuel must be added? The answers to these questions are rooted in various engineering subject areas.

One of the keys necessary to answering the above questions is often obtained via the application of one or more of the conservation laws, and the contents of this chapter deal with these laws. Topics covered include:

7.2 Introduction to the Conservation Laws
7.3 The Conservation Law for Momentum
7.4 The Conservation Law for Mass
7.5 The Conservation Law for Energy
7.6 Net Energy Analysis
References

Obviously, the hearts of the chapter are Sections 7.4 and 7.5. Four important terms are defined as follows before proceeding to the conservation laws.

1. A *system* is any portion of the universe that is set aside for study.
2. Once a system has been chosen, the rest of the universe is referred to as the *surroundings*.
3. A system is described by specifying that it is in a certain *state*.
4. The *path*, or series of values certain variables assume in passing from one state to another, defines a *process*.

DOI: 10.1201/9781003667759-9

7.2 INTRODUCTION TO THE CONSERVATION LAWS

Momentum, energy, and mass are all conserved. As such, each quantity obeys the conservation law below as applied within a system.

$$\left\{\begin{matrix} quantity \\ into \\ system \end{matrix}\right\} - \left\{\begin{matrix} quantity \\ out\,of \\ system \end{matrix}\right\} + \left\{\begin{matrix} quantity \\ generated\,in \\ system \end{matrix}\right\} = \left\{\begin{matrix} quantity \\ accumulated \\ in\,system \end{matrix}\right\} \quad (7.1)$$

Equation 7.1 may also be written on a time rate basis:

$$\left\{\begin{matrix} rate \\ into \\ system \end{matrix}\right\} - \left\{\begin{matrix} rate \\ out\,of \\ system \end{matrix}\right\} + \left\{\begin{matrix} rate \\ generated\,in \\ system \end{matrix}\right\} = \left\{\begin{matrix} rate \\ accumulated \\ in\,system \end{matrix}\right\} \quad (7.2)$$

The conservation law may be applied at the macroscopic, microscopic, or molecular level. One can best illustrate the differences in these methods with an example. Consider a system in which a fluid is flowing through a cylindrical tube (see Figure 7.1). Define the system as the fluid contained within the tube between points 1 and 2 at any time.

If one is interested in determining changes occurring at the inlet and outlet of the system, the conservation law is applied on a "macroscopic" level to the entire system. The resultant equation describes the overall changes occurring to the system without regard for internal variations within the system. This approach is usually applied by the practicing engineer.

The microscopic approach is employed when detailed information concerning the behavior *within* the system is required, and this is often requested of and by the engineer or scientist working in the energy field. The conservation law is then applied to a *differential* element within the system which is large compared to an individual molecule, but small compared to the entire system. The resultant equation

FIGURE 7.1 Conservation law example.

is then expanded, via an integration, to describe the behavior of the entire system. This is defined by some as the transport *phenomena approach* (Theodore 1971; Bird, Stewart, and Lightfoot 2002).

The molecular approach involves the application of the conservation law to individual molecules. This leads to the study of statistical and quantum mechanics – both of which are beyond the scope of this text. In any case, the description of individual molecules at the molecular level is of little value to the practicing engineer. However, the statistical averaging of molecular quantities in either a differential or finite element within a system can lead to a more meaningful description of the behavior of a system.

It should be noted that applied mathematicians have developed differential equations describing the detailed behavior of systems by applying the appropriate conservation law to a differential element or shell within a system. Equations were derived with each new application. The engineer later removed the need for these tedious and error-prone derivations by developing a general set of equations that could be used to describe systems. These came to be defined as the *transport equations* (Theodore 1971; Bird, Stewart, and Lightfoot 2002). Needless to say, these transport equations have proven an asset in describing the behavior of some systems, operations, and processes.

The macroscopic approach is adopted and applied in this text, and no further reference to microscopic or molecular analyses will be made. This chapter's aim, then, is to express the laws of conservation for momentum, energy, and mass in algebraic or finite difference form.

This chapter departs from the approach of developing these differential equations even though this method has a great deal to commend it. Experience has indicated that the practicing engineer possessing a working knowledge of the conservation laws is likely to obtain a more integrated and unified picture of engineering principles and fundamentals by developing the equations in algebraic form.

7.3 THE CONSERVATION LAW FOR MOMENTUM

The general conservation law for momentum on a rate basis numerates the rate of momentum and forces acting on a moving fluid in a volume element of concern at any time t. Each rate of momentum or force term in Equation 7.3 can be expressed in terms of lb_f in order to maintain dimensional consistency. It is suggested that the reader refer to the literature for more information (Theodore 1971; Bird, Stewart, and Lightfoot 2002). The application of this conservation law finds extensive use in the field of fluid mechanics or fluid flow (Abulencia and Theodore 2009). A sample application of this law follows.

Consider a 10-cm diameter horizontal line carrying saturated steam at a velocity of 420 m/s. Water is entrained by the steam at a rate of 0.15 kg/s. The line has a 90° bend (Figure 7.2). Employing momentum principles, one can calculate the force components in the horizontal and vertical directions (and the magnitude and direction of the resultant force) required to hold the bend in place due to the entrained water.

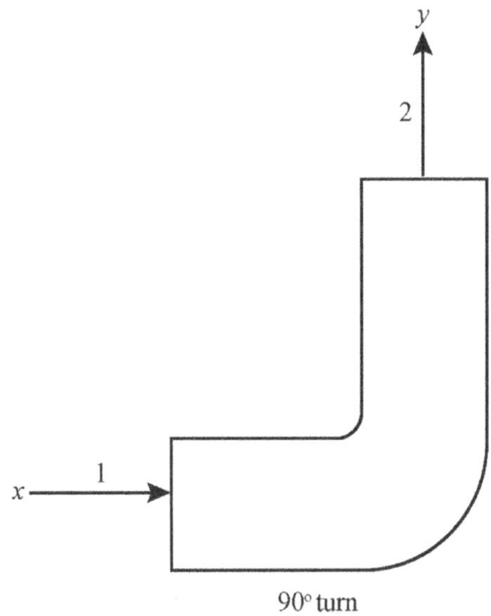

FIGURE 7.2 Diagram to calculate the magnitude and direction of resultant force.

$$\underbrace{\underset{(1)}{\left\{\begin{matrix}\textit{rate of}\\ \textit{momentum}\\ \textit{in by}\\ \textit{convection}\end{matrix}\right\}}-\underset{(2)}{\left\{\begin{matrix}\textit{rate of}\\ \textit{momentum}\\ \textit{out by}\\ \textit{convection}\end{matrix}\right\}}}_{\text{bulk flow}}+\underbrace{\underset{(3)}{\left\{\begin{matrix}\textit{rate of}\\ \textit{momentum}\\ \textit{in by molecular}\\ \textit{diffusion}\end{matrix}\right\}}-\underset{(4)}{\left\{\begin{matrix}\textit{rate of}\\ \textit{momentum}\\ \textit{out by molecular}\\ \textit{diffusion}\end{matrix}\right\}}}_{\text{velocity gradients}}$$

$$\underbrace{\underset{(5)}{\left\{\begin{matrix}\textit{external}\\ \textit{forces}\\ \textit{exerted on}\\ \textit{fluids}\end{matrix}\right\}}}_{\substack{\text{surface forces}\\ \text{body forces}}}=\underbrace{\underset{(6)}{\left\{\begin{matrix}\textit{rate of}\\ \textit{momentum}\\ \textit{accumulation}\end{matrix}\right\}}}_{\text{inventory}} \tag{7.3}$$

A line diagram of the system is provided in Figure 7.2. Select the fluid in the bend as the system and apply the conservation law for mass:

$$\dot{m}_1 = \dot{m}_2 \tag{7.4}$$

Since the density and cross-sectional area are constant,

$$\mathrm{v}_1 = \mathrm{v}_2 \tag{7.5}$$

where $\dot{m}_1$, $\dot{m}_2$ = mass flowrate at 1 and 2, respectively; and v_1, v_2 = velocity at 1 and 2, respectively

A linear momentum ($\dot{M}$) balance in the horizontal direction provides the force applied by the channel wall on the fluid in the x-direction, F_x:

$$F_x g_c = \dot{M}_{x,out} - \dot{M}_{x,in}$$

$$= \frac{d}{dt}(mv)_{x,out} - \frac{d}{dt}(mv)_{x,in} \tag{7.6}$$

Note that the equation assumes that the pressure drop across the bend is negligible.

Since $\mathrm{v}_{x,out} = 0$ and $dm/dt = \dot{m}$,

$$F_x g_c = 0 - \dot{m}_{x,in} = -\frac{(0.15)(420)}{(1)} = -63\text{ N} = -14.1\text{ lb}_f \tag{7.7}$$

The x-direction supporting force acting on the 90° elbow is 14.1 lb_f acting toward the left. A linear momentum balance in the vertical direction results in

$$F_y g_c = \dot{M}_{y,out} - \dot{M}_{y,in} = \dot{m}_{y,out} - \dot{m}_{y,in}$$

$$= \dot{m}\mathrm{v}_2 - 0 = (0.15)(420) = 63\text{ N} = 14.1\text{ lb}_f \tag{7.8}$$

The y-direction supporting free force on the 90° elbow is 14.1 lb_f acting upwards. The resultant supporting force is given by

$$F_{res} = \sqrt{F_x^2 + F_y^2} \tag{7.9}$$

Substituting,

$$F_{res} = \sqrt{(-63)^2 + (63)^2} = 89.1\text{N} = 19.1\text{N lb}_f \tag{7.10}$$

The direction is given by

$$\tan\theta = \frac{F_y}{F_x} = \frac{63}{-63} = -1;\ \theta = 135° \tag{7.11}$$

where θ is the angle between the positive x-axis and the direction of the force. The counterclockwise rotation of the direction from the x-axis is defined as positive. The supporting force is therefore 19.1 lb_f acting in the "northwest" direction.

7.4 THE CONSERVATION LAW FOR MASS

The conservation law for mass can be applied to any process or system. The general form of this law is given as

$$\begin{Bmatrix} mass \\ in \end{Bmatrix} - \begin{Bmatrix} mass \\ out \end{Bmatrix} + \begin{Bmatrix} mass \\ generated \end{Bmatrix} - \begin{Bmatrix} mass \\ consumed \end{Bmatrix} = \begin{Bmatrix} mass \\ accumulated \end{Bmatrix} \quad (7.12)$$

or, on a time rate basis, by

$$\begin{Bmatrix} rate\ of \\ mass \\ in \end{Bmatrix} - \begin{Bmatrix} rate\ of \\ mass \\ out \end{Bmatrix} + \begin{Bmatrix} rate\ of \\ mass \\ generated \end{Bmatrix} - \begin{Bmatrix} rate\ of \\ mass \\ consumed \end{Bmatrix} = \begin{Bmatrix} rate\ of \\ mass \\ accumulated \end{Bmatrix} \quad (7.13)$$

In most energy-related processes, it is often necessary to obtain quantitative relationships by writing mass balances on the various elements in the system. This equation may be applied either to the total mass involved or to a particular species on either a mole or mass basis. This law can be applied to steady-state or unsteady-state (transient) processes and to batch or continuous systems. In order to isolate a system for study, it may be separated from the surroundings by a boundary or envelope. This boundary may be real (e.g., the walls of a thermal device) or imaginary. Mass crossing the boundary and entering the system is part of the *mass in* term in Equation 7.13, while that crossing the boundary and leaving the system is part of the *mass out* term. Equation 7.13 may be written for any compound whose quantity is not changed by chemical reaction and for any chemical element whether or not it has participated in a chemical reaction. It may be written for one piece of equipment, around several pieces of equipment, or around an entire process. It may be used to calculate an unknown quantity directly, to check the validity of experimental data, or to express one or more of the independent relationships among the unknown quantities in a particular application.

A *steady-state* process is one in which there is no change in conditions (pressure, temperature, composition, etc.) or rates of flow with time at any given point in the system. The accumulation term in Equation 7.13 is then zero. (If there is no chemical or nuclear reaction, the generation term is also zero.) All other processes are *unsteady state*.

In a *batch* process, a given quantity of reactants is placed in a container, and by chemical and/or physical means, a change is made to occur. At the end of the process, the container (or containers) to which material may have been transferred holds the product or products.

In a *continuous* process, reactants are continuously fed to a piece of equipment or to several pieces in series, and products are continuously removed from one or more points. A continuous process may or may not be steady state. A coal-fired power plant, for example, operates continuously. However, because of the wide variation in power demand between peak and slack periods, there is an equally wide variation in the rate at which the coal is fired. For this reason, power plant problems may require the use of average data over long periods of time. However, most thermal operations are assumed to be steady-state and continuous.

As indicated previously, Equation 7.13 may be applied to the total mass of each stream (referred to as an *overall* or *total material balance*) or to the individual component(s) of the stream (referred to as a *componential* or *component material balance*).

The primary task in preparing a material balance in engineering calculations is often to develop the quantitative relationships among the streams. The primary factors, therefore, are those that *tie* the streams together. An element, compound, or unreactive mass (e.g., ash) that enters or exits in a single stream or passes through a process unchanged is so convenient for this purpose that it may be considered a *key* to the calculations. If sufficient data are given about this component, it can be used in a component balance to determine the total masses of the entering and exiting streams. Such a component is sometimes referred to as the *key component*. Since a key component does not react in a process, it must retain its identity as it passes through the process. Obviously, except for nuclear reactions, elements may always be used as key components because they do not change identity even though they may undergo a chemical reaction. Thus, CO (carbon monoxide) may be used as a key component only when it does not react, but C (carbon) may always be used as a key component. A component that enters the system in only one stream and leaves in only one stream is usually the most convenient choice for a key component.

Four important processing concepts are *bypass*, *recycle*, *purge*, and *makeup*. With bypass, part of the inlet stream is diverted around the equipment to rejoin the (main) stream after the unit (see Figure 7.3). This bypass stream effectively moves in parallel with the stream passing through the equipment. In recycle, part of the product stream is sent back to mix with the feed. If a small quantity of nonreactive material is present in the feed to a process that includes recycle, it may be necessary to remove the nonreactive material in a purge stream to prevent its building up above a maximum

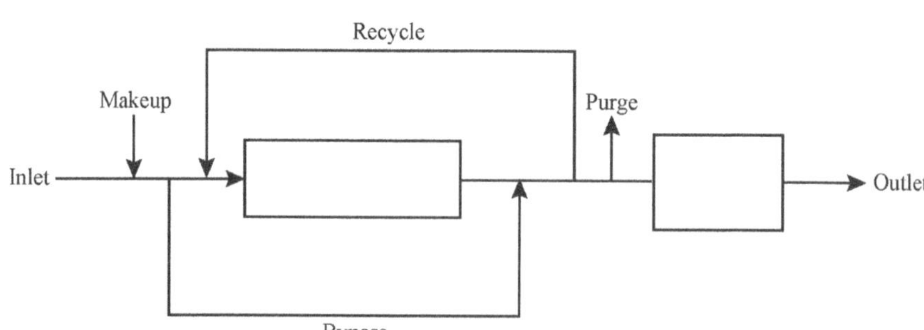

FIGURE 7.3 Bypass, recycle, purge, and makeup flows identified in a process flow diagram.

tolerable value. This can also occur in a process without recycle; if a nonreactive material is added in the feed and not totally removed in the products, it will accumulate until purged. The purging process is sometimes referred to as *blowdown*. Makeup, as its name implies, involves adding or making up part of a stream that has been removed from a process. Makeup may be thought of as the opposite of purge and/or blowdown.

7.5 THE CONSERVATION LAW FOR ENERGY

From the early recognition of energy, man has studied its effects upon objects, its transfer from object to object, and its transformation from one form to another. This field of study is an integral part of engineering. Before proceeding to the first law of thermodynamics, often referred to as the conservation law for energy, certain important terms must be defined.

1. *Isothermal* means constant temperature.
2. *Isobaric* means constant pressure.
3. *Isochoric* means constant volume.
4. *Adiabatic* specifies no transfer or heat to or from systems.

The first law of thermodynamics specifies that energy is conserved. In effect, this law states that energy is neither created nor destroyed. Thus, the change in energy of a system is exactly equal to the negative of the change in energy of the surroundings. For a system of constant mass (a closed system), the only way the system and surroundings may interchange energy is by work and heat. Work and heat are defined as "energy in transit." They are not properties and cannot be stored in a system. Two common forms of work are expansion and electrical. Much of this material will be revisited in more detail in the next chapter on thermodynamics.

Heat is energy in transit because it is transferred due to temperature differences. This heat transfer may take place by conduction, convection, or radiation. Additional details on this topic are provided later in a later chapter concerned with heat transfer.

The energy balance makes use of the conservation law to account for all the energy in a chemical process, or in any other processes for that matter. After a system is defined, the energy balance considers the energy entering the system across a boundary, the energy leaving the system across the boundary, and the accumulation of energy within the system. This may be written in a simplified equation form as:

$$\text{energy in} - \text{energy out} = \text{energy accumulated} \tag{7.14}$$

This expression has the same form as the general law of conservation of mass as well as the conservation law for momentum. It may also be written on a rate basis.

All forms of energy must be included in an energy balance. In many processes, certain energy forms remain constant; therefore, any changes in them may be neglected. However, these forms should be recognized and understood before their magnitude and constancy can be determined. Some forms of energy are easily recognized in everyday life: the energy of a moving object, the energy given off by a fire, and the energy content of a container of hot water. Other forms of energy are less easily recognized. However, the five key energy terms introduced earlier are *kinetic*, *potential*, *internal*, *heat*, and *work*. These are once again briefly reviewed as follows.

1. *Kinetic energy*. The energy of a moving object is called "kinetic energy." A baseball thrown by a pitcher possesses a definite energy as it travels toward the catcher. A pound of flowing fluid possesses kinetic energy as it travels through a duct or pipe.
2. *Potential energy*. The energy possessed by a mass by virtue of its position in the Earth's gravitational field is called "potential energy." A boulder lying at the top of a cliff possesses potential energy with reference to the bottom of the cliff. If the boulder is pushed off the cliff, its potential energy is transformed into kinetic energy as it falls. Similarly, a mass of fluid in a flowing system possesses a potential energy because of its height above an arbitrary reference level.
3. *Internal energy*. The component molecules of a substance are constantly moving within a substance. This motion imparts internal energy to the material. The molecules may rotate, vibrate, or migrate within the substance. The addition of heat to a material increases its molecular activity and hence, its internal energy. The temperature of a material is a measure of its internal energy.
4. *Heat*. As noted earlier, when energy is transferred between a system and it surroundings, it is transferred either as work or as heat. Thus, heat is energy in transit. This type of energy transfer (heat) occurs whenever a hot body is brought into contact with a cold body. Energy flows as heat from the hot body to the cold body until the temperature difference is dissipated, i.e., until thermal equilibrium is established. For this reason, heat may be considered as energy being transferred due to a temperature difference.
5. *Work*. Work is also energy in transit. Work is done whenever a force acts through a distance.

The first law of thermodynamics may be stated formally – as opposed to equation form – in many ways. One of these is as follows: although energy assumes many forms, the total quantity of energy is constant, and when energy disappears in one form, it must appear simultaneously in other forms. This law, in steady-state equation form for batch and flow processes is presented here.

For *batch* processes:

$$\Delta U = Q + W \tag{7.15}$$

For *continuous* processes:

$$\Delta H = Q + W_s \tag{7.16}$$

where potential, kinetic, and other energy effects have been neglected and U = internal energy of the system; Q = energy in the form of heat transferred across the boundaries of the system; W = energy in the form of work transferred across the boundaries of the system; H = enthalpy of the system; W_s = energy in the form of mechanical work transferred across the boundaries of the system; and ΔU, ΔH = changes in the internal energy and enthalpy, respectively, during the process. Contrary to an earlier convention, both Q and W (or W_s) are considered positive if added/transferred to the system. Also note that for a flow process

$$\Delta H = Q;\ \text{if}\ W_s = 0 \tag{7.17}$$

The internal energy and enthalpy in Equations 7.15 and 7.16, as well as the other equations in this section, may be on a *mass* basis (i.e., for 1 gal or 1 lb of material), on a mole basis (i.e., for 1gmol or 1 lbmol of material), or represent the total internal energy and enthalpy of the entire system. As long as these equations are dimensionally consistent, it makes no difference. For the sake of clarity, the same convention used for heat capacities, is employed throughout this text – uppercase letters (e.g., H, U, C_p) represent properties on a mole basis, while lowercase letters (e.g., h, u, c_p) represent properties on a mass basis. Properties for the entire system will rarely be used and therefore require no special designation.

Perhaps the most important thermodynamic function the engineer works with is *enthalpy*. Enthalpy is defined by:

$$H = U + PV \tag{7.18}$$

where P = pressure of the system, and V = volume of the system.

7.6 NET ENERGY ANALYSIS

How much energy does it take to produce useable energy or materials? The term *energy analysis* represents a broad field of study dealing with the development and use of all aspects of energy in human society and the environment. *Net energy analysis*, a more limited field of study, deals with the analysis of the energy made available to society by energy production processes after accounting for energy lost to society/the environment as a result of the processes. This subject can also include the energy analysis of materials production (i.e., how much energy must be invested in the total system needed for the production of a unit of material). Net energy analysis is a topic that will be addressed several times in this book.

Net energy analysis differs significantly from traditional engineering energy efficiency studies. First, net energy analysis is concerned with the total system of production, starting with resources in the ground. Second, it is concerned with the total quantity of energy throughout society that must be input into construction and operation of an energy or material production system up to the point where the produced energy is actually utilized.

The objectives of a net energy analysis are the following:

1. Provide reliable, objective, credible information to government and industry on the net energy inputs and outputs of energy systems.

2. Provide a workable methodology that could be used in subsequent expanded net energy studies.
3. Provide the best possible documentation of data related to net energy.
4. Discuss and describe the usefulness and limitation of net energy studies and their potential values in decision-making.
5. Discuss philosophy and issues pertaining to net energy studies.

Three major concerns or issues to which the general title of net energy analysis might apply include the following:

1. How much energy is required from the industrial component of society to drive or establish and operate an energy production process relative to the energy yield of the process?
2. In extracting, processing, and moving a resource (if applicable) to provide energy to end users, what final yields are obtained relative to losses of the total energy of the recovered (fuel) resources and of the industrial energy needed to establish and operate the energy production systems?
3. For a given output of energy for end use, what total amounts of the gross energy resources and industrial energies are necessary to establish and operate the system?

The issues of the finiteness of energy resources and the rate of depletion are also of concern to society.

Energy must be expended when a material is extracted from its source, is processed, or is transported. As a material moves downstream through a series of processing steps, it represents (or has necessitated) an accumulation of energy expenditures. This energy embodied in the material as a result of processing is called *sequestered* or *imbedded energy*. A petroleum-derived chemical usually has such an energy value. Thus, the energy requirements of finished products include fuel values in some cases, and expended processing energy in all cases, to represent the total imbedded energy of the product.

Net energy analysis should not be used as the sole decision factor in this type of energy analysis. Other factors may generally carry more weight; they include the following:

1. Economics
2. Environment
3. National security
4. Energy mix
5. Lead times
6. Transportation capacities
7. Institutional restraints, such as governmental regulations and incentives
8. Availability of needed materials
9. Local attitudes
10. Socioeconomic impacts

11. Employment needs
12. Needs for energy
13. Safety concerns

Net energy analysis is not a panacea for energy planning, but is worth utilizing as a framework for energy analyses in examining a variety of issues being faced by engineers and planners working in the geothermal energy field.

REFERENCES

Abulencia, P., and Theodore, L. 2009. *Fluid Flow for the Practicing Chemical Engineer*. Hoboken, NJ: Wiley.

Bird, R., Stewart, W., and Lightfoot, E. 2002. *Transport Phenomena, 2nd Edition*. Hoboken. NJ: Wiley.

Theodore, L. 1971. *Transport Phenomena for Engineers, a Theodore Tutorial*. East Williston, NY: Theodore Tutorials. Originally published by the International Textbook Company, Scranton, PA. Roanoke, VA: ETS Professional Training Institute. https://etsi-inc.com/training/tutorials-and-textbooks/.

8 Thermodynamics

8.1 INTRODUCTION

A successful understanding of geothermal energy applications requires a thorough and complete understanding of the fundamental principles of thermodynamics, heat transfer, and fluid flow. These thermodynamic principles provide the framework for analyzing how heat is stored underground and how thermal energy can be efficiently converted to useful forms not only for heating and cooling purposes, but also for electricity production.

Engineers and scientists involved in geothermal energy activities must analyze energy transformations of heat to work and heat transfer between different media, and must understand the transfer efficiency limitations imposed by physical laws. This includes understanding the thermodynamic properties of various working fluids, such as water and steam, plus phase changes such as boiling and condensation that are an integral part of power generation cycles.

Thermodynamics can be defined as the science which deals with the inter-transformation of heat and work. The fundamental principles of thermodynamics are contained in the *first*, *second*, and *third laws of thermodynamics*. These principles have been defined as "pure" thermodynamics. These laws were developed and extensively tested in the latter half of the 19th century and are essentially based on experience. (The third law was developed during the 20th century.)

Historically, the study of the second law of thermodynamics was developed in the middle of the 19th century by individuals such as Carnot (a French engineer), Clausius, and Kelvin. This development was made purely on a macroscopic scale and is referred to as the "classical approach to the second law," which does not require the existence of an atomic or molecular theory.

This chapter lays the theoretical groundwork for many geothermal thermodynamic concepts and principles that include the principle of energy conservation (the first law of thermodynamics), as well as a discussion of entropy generation (the second law of thermodynamics) and an associated concept of the inherent "quality" of energy (i.e., *exergy*). While an effort was made to keep the technical complexity to an absolute minimum, and also avoid explicit mention of the detailed mathematical framework which underlies these topics, one cannot have a meaningful discussion of thermodynamics without at least alluding to some of its most basic tenets.

DOI: 10.1201/9781003667759-10

Although some of the concepts discussed in this chapter may seem completely foreign to a nontechnical individual, they are largely considered elementary to practicing engineers or applied scientists. Therefore, it is assumed that the reader has at least a cursory understanding of college-level general chemistry, wherein some of these topics should have been introduced. Where appropriate suitable literature references are provided for those interested in delving deeper into the topics discussed.

Certain important terms have been discussed in previous chapters but are defined here again to reinforce their concepts before proceeding to the laws of thermodynamics. These critical terms and concepts include the following (Theodore and Reynolds 1991; Theodore, Ricci, and Van Vliet 2009):

1. A *system* is any portion of the universe that is set aside for study.
2. Once a system has been chosen, the rest of the universe is referred to as the *surroundings*.
3. A system is described by specifying that it is in a certain *state*.
4. The *path* or series of values certain variables assume in passing from one state to another defines a *process*.
5. *Isothermal* means constant temperature.
6. *Isobaric* means constant pressure.
7. *Isochoric* means constant volume.
8. *Adiabatic* specifies no transfer or heat to or from systems.
9. A *reversible* process is one where changes occur due to driving forces (e.g., temperature differences that are *differentially* small); the system may therefore be considered to be in a state of equilibrium during the change.

Section numbers and their accompanying titles for this chapter covering thermodynamics are listed as follows:

The topics selected for presentation in this abbreviated chapter are those that find application in geothermal energy analysis and application. The bulk of the material to follow has been drawn from the work of Theodore and Reynolds (1991) and Theodore, Ricci, and VanVliet (2009).

8.2 THE FIRST LAW OF THERMODYNAMICS

Practically all thermodynamics, in the ordinary meaning of the term, is applied thermodynamics in that it is essentially the application of the three fundamental laws of thermodynamics, coupled with certain facts and principles of mathematics, physics, and chemistry, to problems in engineering. The fundamental laws are of such generality that it is not surprising that these laws find application in other disciplines, including physics and chemistry, energy studies, and specifically to chemical, mechanical, and environmental engineering.

With respect to the first law, this section reviews enthalpy effects, including those associated with a chemical reaction. To introduce this subject, the reader is reminded that the practicing engineer and applied scientist are rarely concerned with the *magnitude* or *amount* of the energy in a system; the primary concern is with *changes* in the amount of energy. In measuring energy changes for systems, the enthalpy (H) has been found to be the most convenient term to work with.

The term enthalpy was defined earlier as

$$H = U + PV \tag{8.1}$$

where U is internal energy of the system, P is the pressure of the system, and V is the volume of the system. The terms H and U are *state* or *point* functions. By fixing a certain number of variables upon which the function depends, the numerical value of the function is automatically fixed; that is, it is single-valued. For example, fixing the temperature and pressure of a one-component single-phase system immediately specifies its enthalpy and internal energy.

The change in enthalpy as it undergoes a change in state from (T_1, P_1) to (T_2, P_2) is given by

$$\Delta H = H_2 - H_1 \tag{8.2}$$

Most industrial facilities operate in a steady-state flow mode with no significant mechanical or shaft work added (or withdrawn) from the system. For this condition, the First Law reduces to:

$$Q = \Delta H \tag{8.3}$$

This equation is routinely used in many engineering and energy calculations. If a unit or system is operated *adiabatically*, $Q = 0$ and the above equation becomes:

$$\Delta H = 0 \tag{8.4}$$

Enthalpy values and enthalpy changes may be calculated directly from enthalpy data. These are available in the literature (Theodore and Reynolds 1991; Theodore, Ricci, and VanVliet 2009).

There are many different types of enthalpy effects; these include:

1. Sensible (temperature)
2. Latent (phase)
3. Dilution (with water) and Solution (nonaqueous)
4. Reaction (chemical)

Although details of these enthalpy effects abound in the literature, each is briefly described in the following sections.

8.2.1 Sensible Enthalpy Effects

Sensible enthalpy effects are associated with temperature. There are methods that can be employed to calculate these changes. These methods (Theodore and Reynolds 1991; Theodore, Ricci, and VanVliet 2009) include the use of:

1. Enthalpy values
2. Average heat capacity values, $\bar{C}_p$
3. Heat capacity as a function of temperature, e.g., $C_p = \alpha + \beta T + \gamma T^2$

8.2.2 Latent Enthalpy Effects

It has been observed that there is absorption of heat at constant temperature and pressure that accompanies the transition or equilibrium phase change from solid to liquid and liquid to gas. In terms of molecular theory, this latent enthalpy represents the energy required to overcome inter-molecular forces of attraction and to permit molecules to pass from a more highly restrained condensed phase to a more mobile phase. This relationship indicates that the latent heat is the sum of a change in internal energy and energy of expansion (or contraction).

Three common types of transition (and the terminology applied to them) are:

1. vapor ↔ liquid; enthalpy of condensation or vaporization
2. liquid ↔ solid; enthalpy of fusion or melting
3. solid → vapor; enthalpy of sublimation

Since there is a very substantial increase in volume in passing from the liquid to the vapor state, usually from a few hundred to over a thousand-fold, the heat of vaporization is large and important in many energy engineering applications involving water, that is, steam production/condensation. Since there is a relatively small difference in volume between a solid and the liquid to which it melts, the heat of fusion represents mainly an increase in internal energy (Theodore, Ricci, and VanVliet 2009).

8.2.3 Dilution and Solution Enthalpy Effects

When two or more pure substances are mixed to form a solution, a heat effect usually results. When the mixing occurs with water (H_2O), the process is normally defined as a *dilution* effect. If mixing occurs with a solvent other than H_2O, it is defined as a *solution* effect. The dilution and solution enthalpy effects can be significant in some industrial absorber calculations, but may safely be neglected in most energy-related calculations (Theodore, Ricci, and VanVliet 2009).

8.2.4 Reaction Enthalpy Effects

The enthalpy of reaction is defined as the enthalpy change of a system undergoing chemical reaction; this effect normally cannot be neglected. This effect receives extensive treatment in undergraduate thermodynamics courses. The standard enthalpy (heat) of reaction can be calculated from standard enthalpy of formation data. The (standard) enthalpy of a reaction is obtained by taking the difference between the (standard) enthalpy of formation of products and reactants multiplied by their respective stoichiometric coefficients. If the (standard) enthalpy of reaction or formation is negative (exothermic), as is the case with most combustion reactions, then energy is liberated due to the chemical reaction.

Tables of enthalpies of formation and reaction are available in the literature (particularly thermodynamics text/reference books) for a wide variety of compounds (Theodore, Ricci, and VanVliet 2009). It is important to note that these data are useless unless *both* the stoichiometric equation and the state of the reactants and products are known so proper enthalpy calculations can be carried out.

8.3 THE SECOND LAW OF THERMODYNAMICS

The law of conservation of energy was defined earlier as the first law of thermodynamics. Its applications allow calculations of energy relationships associated with all kinds of processes. The "limiting" law is the "second law of thermodynamics." Applications addressed later involve calculations for maximum power outputs from a power plant and equilibrium yields in chemical reactions. In principle, this law states that water cannot flow uphill and heat cannot flow from a cold to a hot body on its own accord. Other defining statements for this law that have appeared in the literature are:

1. Any process whose sole net result is the transfer of heat from a lower temperature level to higher one is impossible.
2. No apparatus, equipment, or process can operate in such a way that its only effect (in system and surroundings) is to convert heat taken in completely into work.
3. It is impossible to convert the heat taken into a system completely into work in a cyclical process.

As described earlier, the first law of thermodynamics is a conservation law for energy transformations. Regardless of the types of energy involved in processes – thermal, mechanical, electrical, elastic, magnetic, etc. – the change in the energy of a system is equal to the difference between energy input and energy output. The first law also allows free convertibility from one form of energy to another, as long as the overall quantity is conserved. Thus, this law places no restriction on the conversion of work into heat, or on its counterpart – the conversion of heat into work.

The unrestricted conversion of work into heat is well known to most technical individuals. Frictional effects are frequently associated with mechanical forms of work that result in a temperature rise of the bodies in contact. However, the transformation of heat into work is of greater concern. In nations with a partially developed or developing technological society, the ability to produce energy in the form of work takes on prime importance. Work transformations are necessary to transport people and goods, drive machinery, pump liquids, compress gases, and provide energy input to so many other processes that are taken for granted in highly developed societies. Much of the work input in such societies is available in the form of electrical energy, which is then converted to rotational mechanical work. Although some of this electrical energy (work) is produced by hydroelectric power plants, by far the greatest part of it is obtained from the combustion of fossil fuels or nuclear fuels. These fuels act as a thermal (heat) source for the production of work. Hence, the study of the conversion of heat to work is extremely important – especially in the light of developing shortages of fossil and nuclear fuels, along with the accompanying potential environmental problems, particularly with global warming.

The brief discussion of energy conversion devices above leads to an important second-law consideration – that is, that energy has "quality" as well as quantity. Because work is 100% convertible to heat whereas the reverse situation is not true, work is a more valuable form of energy than heat. Although it is not as obvious, it can also be shown through second-law arguments that heat has "quality" in terms of the temperature at which it is discharged from a system. The higher the temperature at which heat transfer occurs, the greater the possible energy transformation into work. Thus, thermal energy stored at high temperatures is generally more useful to society that that available at lower temperatures. While there is an immense quantity of energy stored in the oceans, for example, its present availability to society for performing useful tasks is quite low.

This implies, in turn, that thermal energy loses some of its "quality" or is degraded when it is transferred by means of heat transfer from one temperature to a lower one. Other forms of energy degradation include energy transformations due to frictional effects and electrical resistance. Such effects are highly undesirable if the use of energy for practical purposes is to be maximized. The second law provides some means of measuring this energy degradation through a term referred to as *entropy* and it is the second law (of thermodynamics) that serves to define this important thermodynamic property. Entropy is normally designated as S with units of energy/absolute temperature, for example, Btu/°R or cal/K, and like enthalpy and internal energy is a point function. At the heart of the "quality energy" discussion is the relatively recent concept of *exergy*, which is a direct measure of a system's potential to do useful work (a measure of its quality energy).

Entropy calculations (Theodore and Reynolds 1991; Theodore, Ricci, and VanVliet 2009; Theodore 2014) can provide quantitative information on the "quality" of energy and energy degradation. Detailed second law equations and calculations follow.

It can be shown that the change in entropy for a reversible, adiabatic process is always zero, that is,

$$\Delta S = 0 \tag{8.5}$$

For liquids and solids the entropy change for a system undergoing a temperature change from T_1 to T_2 is given by Equation 8.6:

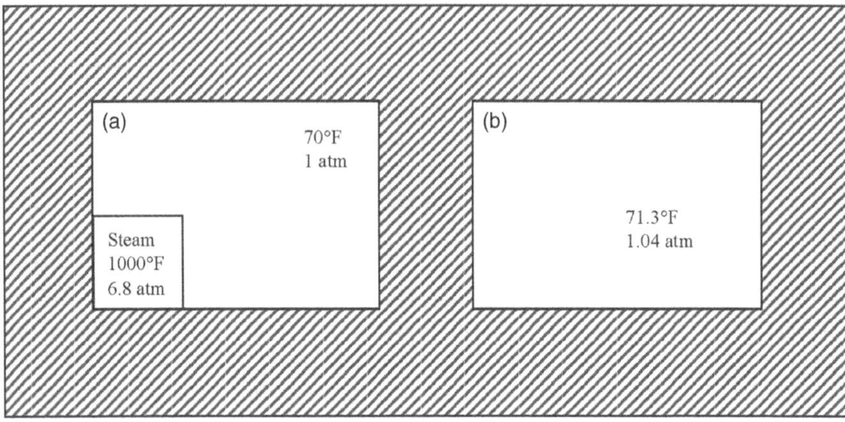

FIGURE 8.1 Entropy application.

$$\Delta S = C_P \ln (T_2/T_1);\ \text{Btu/lbmol-°R} \tag{8.6}$$

Regarding ideal work and lost work, consider the insulated space pictured in Figure 8.1(a) and (b). Since the space is insulated, and is a closed system with no work term, one can conclude from the first law that

$$U_A = U_B \tag{8.7}$$

Although the internal energy levels are the same, note that System A has the capability of doing useful work (because of the high-temperature, high-pressure steam) while System B does not. If an entropy analysis is performed on both A and B, one would deduce that

$$S_A < S_B \tag{8.8}$$

In effect, the entropy level has increased for the system that has lost its ability to do useful work. It is in this manner that the concept of entropy can be used to determine as system's ability to either do useful work or lose its ability to do useful work.

This development can now be extended to introduce the concept of ideal work, W_{ideal}, and lost work, W_{lost}. As noted in the literature (Theodore, Ricci and VanVliet 2009), "in any steady-state flow process requiring work, there is an absolute minimum amount which must be expended to accomplish the desired change of state of the fluid flowing through the system. In a process producing work, there is an absolute maximum amount which may be accomplished as the result of a given change of state of the fluid flowing through the system. For either case, the limiting value for work is obtained when the change of state associated with the process is accomplished *completely reversibly*; for such a process, the entropy generation is zero," which is in line with Equation 8.5.

For real-world, or actual, or irreversible processes, the actual work W_s (or $\dot{W}_s$ on a rate basis) can be compared with the ideal work. If W_{ideal} (or $\dot{W}_{ideal}$ on a rate basis) is positive, it is the minimum work required and is smaller than W_s. The thermodynamic efficiency E_t can then be defined as the ratio of the ideal work to the actual work:

$$E_t(\text{work required}) = \frac{W_{\text{ideal}}}{W_s} = \frac{\dot{W}_{\text{ideal}}}{\dot{W}_s} \tag{8.9}$$

When W_{ideal} (or $\dot{W}_{\text{ideal}}$) is negative, the ideal work is the maximum work obtainable and is larger than the actual work. The thermodynamic efficiency is then defined as:

$$E_t(\text{work produced}) = \frac{W_s}{W_{\text{ideal}}} = \frac{\dot{W}_s}{\dot{W}_{\text{ideal}}} \tag{8.10}$$

Work that is lost because the process is irreversible is defined as the lost work, W_{lost}, and represents the difference between the actual work and the ideal work,

$$W_{\text{lost}} = W_s - W_{\text{ideal}} \tag{8.11}$$

Equation 8.11 may also be written on a time rate basis as

$$\dot{W}_{\text{lost}} = \dot{W}_s - \dot{W}_{\text{ideal}} \tag{8.12}$$

The lost work may be viewed as the decrease in the aforementioned "quality" energy of the system. In any event, the second law leads to the conclusion that the greater the irreversibility of a process, the greater the (rate of) entropy increase and the greater the amount of energy that becomes unavailable for work, i.e., the lost quality work.

8.4 THE THIRD LAW OF THERMODYNAMICS

Based on extensive experimental data, Walter Hermann Nernst (1864–1941), a German physicist, postulated a "heat theorem" that the entropy change for any chemical reaction at absolute zero temperature, that is, 0°R or 0K, was zero. Thus,

$$\Delta S = 0 \text{ at } T = 0 \tag{8.13}$$

Planck, another German physicist, further proposed that at absolute zero temperature

$$\Delta S = 0 \tag{8.5}$$

Experimental data have verified these two conclusions.

The third law of thermodynamics is based on Equation 8.5, that is, the third law of thermodynamics is concerned with the absolute values of entropy. Thus, by definition, the entropy of all pure crystalline materials at absolute zero temperature is exactly zero and in line with Equation 8.5

$$S_{-273°C} = S_{0K} = 0 \tag{8.14}$$

Note, however, that practicing engineers and applied scientists are usually concerned with changes in thermodynamic properties. The third law allows one to calculate the entropy at another temperature via the following equation:

$$\Delta S = \int_0^{T_f} \frac{(C_P)_s}{T} dT + \frac{\Delta H_f}{T_f} + \int_{T_f}^{T_v} \frac{(C_P)_l}{T} dT + \frac{\Delta H_v}{T_v} + \int_{T_v}^{T} \frac{(C_P)_g}{T} dT \tag{8.15}$$

where the first, third, and fifth terms on the right-hand-side (RHS) represent the entropy change associated with temperature changes for the solid, liquid, and gas phases, respectively. The second and fourth RHS terms represent the entropy change associated with fusion and vaporization, respectively. Since the entropy is zero at absolute zero,

$$\Delta S = S_T - S_0 = S_T \tag{8.16}$$

The above equation provides the entropy at the temperature in question, T.

8.5 ENERGY TRANSFORMATIONS

As noted earlier, the first law is concerned with the conservation of energy during energy transfers and transformations. Irrespective of whether the process involves mechanical, thermal, electrical, elastic, magnetic, or even nuclear energy, the energy terms must balance. For example, in a common household boiler which burns natural gas as fuel for indoor heating, the energy that is "generated" by combusting the fuel is actually energy which was already stored within the molecules of fuel and oxygen in the form of chemical bonds between the atoms which compose the molecules. The combustion process results in the breaking of certain chemical bonds, and the formation of new ones. In the case of complete combustion of hydrocarbons such as pure methane, this would result in the conversion of the fuel + oxygen → carbon dioxide + water, resulting in a net liberation of energy in the form of heat. Therefore, in a literal sense, no energy was truly "generated" since the fuel and oxygen mixture already contained the energy within it. Hence, the combustion operation was simply the process which facilitated the transformation of energy from the chemical bonds into heat (Theodore and Theodore 2021). The previous analysis also holds true for any of the aforementioned modes of energy generation, which could be more aptly referred to as modes of "energy transformation."

It is readily apparent to any practicing engineer and applied scientist that not all forms of energy are equivalent in terms of their capacity to be put to use. Within the context of this discussion it is important to note that two main forms of energy transfer are heat (i.e., the flow of energy due to temperature differentials) and work (i.e., energy transfer due to a macroscopic motion resulting in the application of a force over some distance, such as to move a piston and rotate a shaft).

Although dimensionally consistent, work is inherently a more "useful" form of energy than heat. For a system producing a certain quantity of work, this work can conceivably be transformed entirely into heat. However, if a system produces an equivalent quantity of heat, only a fraction of that energy may be converted into work, even in the hypothetical case of a reversible process (Dincer and Cengel 2001). Hence, from a practical perspective, these two forms of energy transfer are *not* equivalent.

Consider the following thought experiment. Supposing that one begins with high-pressure water vapor (steam) contained within the boiler tank of a steam engine, such as that found on a steam locomotive train. The steam within the tank contains a finite amount of internal energy, and for the purposes of the present discussion, simply pick a round number with no concern as to whether it is realistic for a steam engine, say 1,000 kilojoules (kJ). If steam from the tank is allowed to expand into one of the engine's cylinders (a piston-cylinder assembly), the random thermal motion of the water molecules results in a unified macroscopic motion of the rigid piston. In this case, some of the steam's internal energy is transformed into useful work (pushing the piston in order to turn the locomotive wheels, thereby moving the train), and as a result of this process the steam's internal energy decreases. If one instead took a larger mass of steam at a much lower pressure, but with the same overall energy content of 1,000 kJ, it is clear that this steam would not be able to push the piston (produce work) as effectively.

Indeed, if the pressure of the steam is too low, it may not be able to overcome the opposing forces at all and the train will not move. Hence, despite identical quantities of energy contained within the steam in each of these scenarios, it is clear that the high-pressure steam is much better suited to perform useful work. In other words, the 1,000 kJ of energy in the high-pressure steam is of higher quality than that contained in the low-pressure steam. Hence, it is apparent that energy can be described in terms of its quality as well as its quantity. The concept of the quality of energy is central to the concept of *exergy*.

8.6 EXERGY

Exergy is defined as the maximum amount of work that can be produced by a system as it comes into equilibrium with a reference environment. Inversely, exergy may also be equivalently defined as the *minimum* amount of *work required* to bring a system out of equilibrium with a reference environment.

Exergy has units of energy (e.g., J and Btu) and is also sometimes referred to *quality energy*, *availability*, or *availability energy*, because exergy is a direct measure of the maximum amount of work available, which is inherently a measure of the quality of a system's energy. However, unlike energy, exergy is *not* a conserved quantity. This makes sense when one considers that exergy is related to entropy, which is also not conserved, but rather is being constantly generated. Indeed exergy destruction is directly proportional to entropy generation. Because the second law dictates that the overall entropy change for any real process is always positive (entropy is being generated), the exergy change for any real process is always negative (exergy is being destroyed). In other words, because all real processes are irreversible and thereby result in a net generation of entropy, all real processes must result in a net consumption of exergy.

Note that implicit in the definition of exergy is the notion of a reference environment, which indicates that exergy cannot be quantified in an absolute sense, but must rather always be measured with respect to the chosen reference state. When defining a reference environment, key variables which should generally be determined are the reference's temperature, pressure, and chemical composition. Reference environments are generally chosen as the local surroundings of the system. Two of the main criteria for ensuring an adequate reference environment are that it must be:

1. Assumed to be at a state of equilibrium.
2. Large enough that it is virtually unaffected by the system of interest as the process proceeds.

Because the great majority of human-made processes occur on or near the surface of the Earth, the local atmosphere is often taken as the reference environment in exergy analyses. However, it is important to note that Earth's atmosphere does not strictly satisfy both of the aforementioned criteria. In particular, Earth's atmosphere is not in a state of equilibrium, as evidenced by the spatial heterogeneity of its temperature, pressure, and chemical composition (e.g., moisture content) around the globe at any point in time. These gradients results in significant fluctuations of temperature, barometric pressure, and other meteorological phenomena (wind, precipitation, etc.) observed over time at any specific location. Hence, strictly speaking, the exergy of Earth's atmosphere is nonzero. In order to account for realistic deviations from the theoretical behavior of an ideal reference environment, numerous reference environment models are available in the literature (Dincer and Cengel 2001), with each model having its own protocols, assumptions, and range of applications.

In addition to computing the exergy of a system at a given state with respect to a reference environment, it is often of interest to calculate the change in exergy of a system as it changes from an initial state "1" to a final state "2," which is different than that of the reference environment. This difference in exergy is then equivalent to the maximum work obtainable (i.e., the work which would be produced by a reversible process) in transitioning from State 1 to State 2. Note that, while the discussion of exergy changes is being framed with respect to the energy change of a *system*, which is usually of interest for closed-system processes (or transient open-system processes), this discussion is equally applicable to a steady state open-system process (e.g., as in a steam turbine), where X_2–X_1 is not the exergy change of the *system* (system properties do not vary at steady state), but rather the exergy change between the outlet and inlet *flow streams*.

Denoting the exergy of the system as X, one may visualize this change in exergy, and its relationship to the maximum work obtainable as shown in Figure 8.2. Note the difference between the theoretical maximum work obtainable (W_{rev}), and the actual work obtained (W) represents the "lost work" due to the irreversibilities (i.e., entropy generation/exergy destruction) associated with the real process. Magnitudes or absolute values of work ("$|W|$") are used to avoid any confusion regarding the sign conventions associated with work.

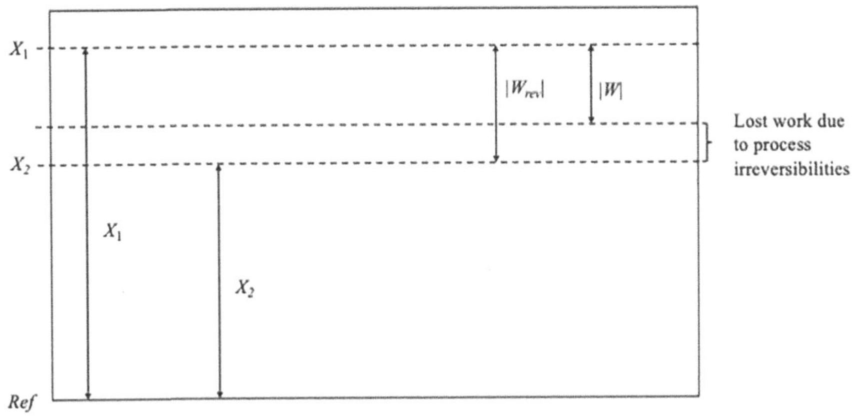

FIGURE 8.2 Illustration of the change in exergy of a system between two states, and the associated amount of work produced, versus the theoretical maximum amount of work possible. The state of the reference environment is by definition = 0.

A concept central to engineering and any conservation study is that of process efficiency. In particular, energy efficiency is a major consideration when analyzing any energy-producing (or consuming) process. In the simplest terms, energy efficiency is a quantification of how ideally the process operates. In direct analogy to energy efficiency, one may also define exergy efficiency. However, unlike a typical energy efficiency which accounts for energy losses, an exergy efficiency must account for both exergy losses and exergy destruction due to the irreversible nature of any real process. Hence, exergy efficiency analysis is more comprehensive than energy efficiency analysis, because it incorporates the "usefulness" of the energy into the analysis. By contrast, energy efficiencies assign the same weight to energy, irrespective of whether it is in the form of shaft work (high in quality energy), or a stream of low temperature fluid (which likely contains little quality energy, depending on the reference environment of interest) (Dincer and Cengel 2001). Theodore et al. (2009) provide examples of steady-state energy and exergy balances, as well as definitions of energy and exergy efficiencies.

In summary, exergy serves as a measure of the maximum work potential that a system can afford due to a separation of the system from equilibrium with the chosen reference environment.

8.7 HEAT ENGINES

There are two major categories of heat engines that employ steam to create mechanical energy: steam engines and steam turbines. Steam engines use reciprocating motion to push a piston back and forth in a cylinder, are bulky, noisy, and are less efficient than steam turbines. Steam turbines use a series of stages to capture the momentum of expanding steam to create mechanical energy. These turbines are generally more compact, quieter, and more efficient than steam engines. Steam turbines are used in thermal power stations and district cooling schemes. They are also a key technology for meeting the world's growing energy needs, and will play a major role in the future if a geothermal energy economy is to become a reality.

The maximum work that can be obtained as a result of any change in state of a fluid is obtained when the change takes place reversibly without friction or turbulence. The Carnot engine is a hypothetical device that represents this maximum thermodynamic achievement in a cyclic process and has been used as a standard in evaluating the efficiency and performance of all actual work producing cycles for transforming heat energy into mechanical energy.

A machine for producing work from heat in a cyclic process is termed a *heat engine*. The key example is the steam power plant in which the medium (steam) periodically returns to its original state. In such a power plant the cycle (in simple form) is shown in Figure 8.3.

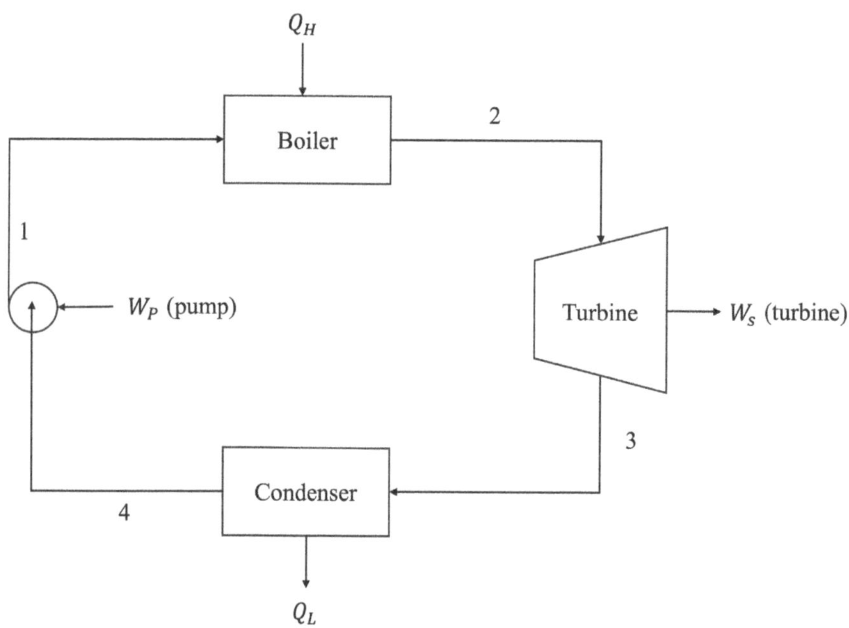

FIGURE 8.3 Simple steam power plant.

Notes:

1. Part of the heat from the fuel (heat of combustion of coal, oil, or natural gas or heat from a nuclear reaction or geothermal source) is transferred to liquid water, converting it into steam at a high pressure and temperature, T_H.

2. Energy is transferred from this steam as shaft work by a device defined as a turbine.

3. The exhaust steam from the turbine is condensed to liquid water at a low temperature, T_C, by transferring heat to cooling water.

4. The liquid water is pumped back to the boiler, thus completing the cycle.

In this cyclic energy process, energy in the form of heat is absorbed by the steam, Q_H. In Step 2, which is adiabatic, energy as shaft work is extracted from the steam, W_s. In Step 3, an amount of heat, Q_L, is transferred to cooling water. The final step (4) is also adiabatic and requires the addition of a relatively small amount of pump work to the water, $W_p \approx 0$. The accompanying temperature – entropy diagram is shown in Figure 8.4.

Figure 8.4 portrays a Carnot-engine cycle that operates reversibly and consists of two isothermal steps connected by two adiabatic steps. In the isothermal step at higher (hot) temperature T_H, heat $|Q_H|$ is absorbed by the working medium of the heat engine, and in the isothermal step at lower (cold) temperature T_L, heat $|Q_L|$ is discarded by the fluid. The work produced is

$$|W| = |Q_H| - |Q_L| \tag{8.17}$$

and the thermal efficiency of the Carnot engine is:

$$E_C \equiv \frac{|W|}{|Q_H|} = 1 - \frac{T_L}{T_H} = \frac{T_H - T_L}{T_H} \tag{8.18}$$

In the cycles of real thermal power plants there is practically only one source of heat (T_H) – the fuel and one sink (T_L) – the atmosphere, to which the residual heat must be discarded. As an example, these temperatures can be taken as 2,000°R and 250°R, respectively. Any ideal heat engine working between these temperatures will then have a limiting thermal efficiency as set by the Carnot cycle. From Equation 8.18 this is:

$$E_C = \frac{2{,}000 - 250}{2{,}000} = 0.825 - 82.5\%$$

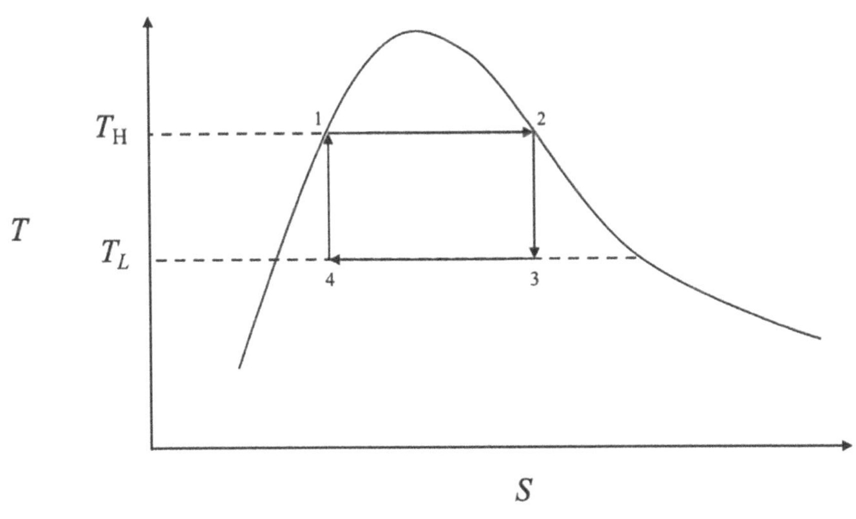

FIGURE 8.4 Carnot cycle on a *T S* diagram.

No power plant can actually even approach this value. The "best" steam plants and internal combustion engines give thermal efficiencies between 0.35 and 0.40 overall. These are the best values obtained in practice, and the average is generally lower.

Regarding refrigeration, the maintenance of any temperature below that of the surroundings requires the continuous abstraction of heat at this temperature level and its rejection at some higher level. Thus, refrigeration may be viewed as an operation involving the pumping of heat from one temperature level to a higher one. The second law of thermodynamics allows one to view a reversible heat engine as a heat pump. A heat engine operating on the Carnot cycle taking in heat at absolute temperature T_{H} and rejecting it at T_{L} yields the maximum efficiency that could be attained by any heat engine operating between the given temperature limits (Figure 8.5). The efficiency, which is the ratio of work done to heat absorbed at T_{H} is given by:

$$\frac{W}{Q_{\mathrm{H}}} = \frac{T_{\mathrm{H}} - T_{\mathrm{L}}}{T_{\mathrm{H}}} \tag{8.19}$$

(The Carnot engine may be imagined to operate in reverse, absorbing heat at T_{L} and rejecting Q_{L} at T_{H} in a refrigeration cycle). Equation 8.20 results for the refrigeration efficiency, E_{R}.

$$E_R = \frac{W}{Q_L} = \frac{T_{\mathrm{H}} - T_{\mathrm{L}}}{T_{\mathrm{L}}} \tag{8.20}$$

The term Q_{L}/W, which is a measure of the refrigerating effect per unit of work done, is known as the "coefficient of performance", *COP*, (Equation 8.21) and establishes

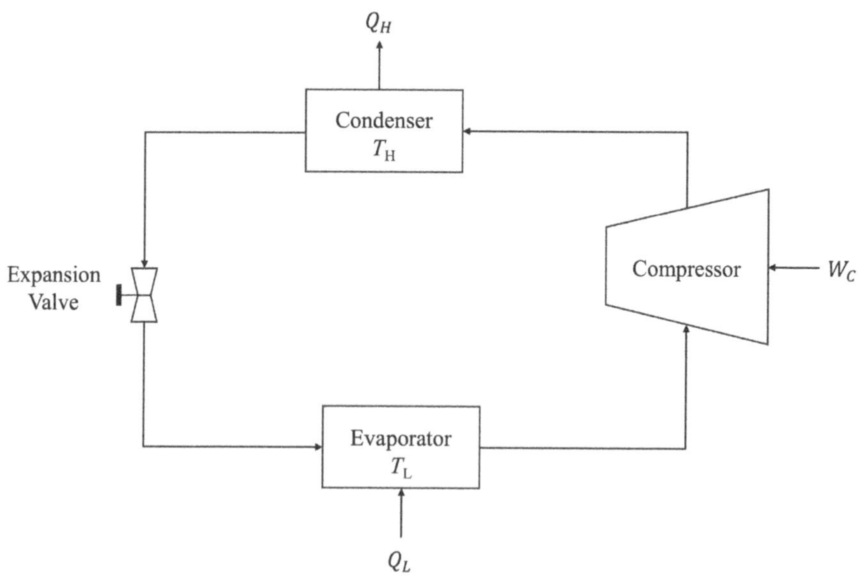

FIGURE 8.5 The refrigeration cycle.

the optimum performance that can be expected of any refrigerating machine operating between given temperature limits. No machine can be devised that will remove heat with less expenditure of work.

$$COP = \frac{Q_L}{W} = \frac{T_L}{T_H - T_L} \tag{8.21}$$

In a broad sense, refrigeration is the science and art of producing and maintaining temperatures below that of the surrounding atmosphere. Low temperatures can be obtained in various ways, as follows:

1. Allowing a phase change to take place in such a way as to extract heat, for example, the evaporation of a liquid such as water or ammonia or the melting of ice or the solution of a salt.
2. Expanding a compressed gas or vapor so that it does external work.
3. Expanding a gas in the Joule-Thomson way.
4. Desorption of a gas.

In fact, any reversible change involving the expenditure of work can be utilized to absorb heat and produce low temperatures. This is the basis on which Ground Source Heat Pumps operate for space cooling in the summer. Note that both heat engines and refrigeration will be revisited along with heat pumps later in Part III of this text.

REFERENCES

Dincer, I., and Cengel, Y.A. 2001. Energy, Entropy, and Exergy Concepts and Their Roles in Thermal Engineering. *Entropy* **3**:116–149.

Theodore, L. 2014. *Chemical Engineering: The Essential Reference*. New York City, NY: McGraw-Hill.

Theodore, L., and Reynolds, J. 1991. *Thermodynamics, A Theodore Tutorial.* East Williston, NY: Theodore Tutorials. Originally published by the USEPA/APTI, RTP, NC. Roanoke, VA: ETS Professional Training Institute. https://etsi-inc.com/training/tutorials-and-textbooks/.

Theodore, L., Ricci, F., and VanVliet, T. 2009. *Thermodynamics for the Practicing Engineer.* Hoboken, NJ: Wiley.

Theodore, M.K., and Theodore, L. 2021. *Introduction to Environmental Management, 2nd Edition*. Boca Raton, FL: CRC Press/Taylor & Francis Group.

9 Heat Exchangers

9.1 INTRODUCTION

Heat transfer is often only one of several unit operations involved in a process, and the interrelations among the different unit operations involved must be understood since the transfer of heat to and from process fluids is an essential part of most energy processes including those involving geothermal energy sources. Heat transfer requires that there be a driving force present for heat to flow. This driving force is the temperature difference between a warmer and colder body. The subject of heat transfer and heat exchangers also plays an important role in many energy conservation studies. Most energy conservation measures in industry, involving energy recovery in the form of heat, utilize any one of a variety of heat exchangers.

Chemical process industries use three principal types of heat exchangers:

1. *Double-pipe exchanger*– the simplest type, used for cooling and heating
2. *Shell and tube exchangers*– used for most applications
3. *Finned (extended surfaces) exchangers*– used for some specialty applications

The mentioned exchangers will be reviewed in detail later in this chapter. However, note that the word *exchanger* applies to all types of equipment in which heat is exchanged but is often used specifically to denote equipment in which heat is transferred between two process streams. An exchanger in which a process fluid is heated or cooled by a plant service stream is referred to as a heater or cooler.

To summarize, a temperature difference between two bodies in close proximity or between two parts of the same body results in heat flow from higher to lower temperatures. There are the three mechanisms by which this heat transfer can occur: *conduction*, *convection*, and *radiation*. In most industrial applications, multiple mechanisms are usually involved in the transmission of heat. However, since each mechanism is governed by its own set of physical laws, it is beneficial to discuss them independently of each other.

Section numbers and their accompanying titles for this chapter covering heat transfer are listed as follows:

DOI: 10.1201/9781003667759-11

The bulk of the material to follow has been drawn from the work of Farag and Reynolds (1996).

9.2 HEAT TRANSFER MECHANISMS

Practically all of the operations that are carried out in industry involve the production or absorption of energy in the form of heat. The laws governing the transfer of heat and the types of equipment that consider heat flow as one of their main objectives are therefore of great importance. Heat may flow by one or more of the following three basic mechanisms:

1. *Conduction*. When heat flows through a body by the transference of the momentum of individual atoms or molecules without bulk mixing, it is said to flow by conduction. For example, the flow of heat through the wall of a furnace or the metal shell of a boiler takes place by conduction.
2. *Convection*. When heat flows by actual mixing of warmer portions with cooler portions of the same material, the heat transfer mechanism is known as convection. Convection is restricted to the flow of heat in fluids. Rarely does heat flow through fluids by pure conduction without some convection resulting from the eddies caused by the changes of density of the fluid with changes in temperature. For that reason, the terms conduction and convection are often used together, although in many cases, heat transfer is primarily by convection. For example, the heating of water flowing by a hot surface is an example of heat transfer due primarily to convection. Finally, heat transfer by convection due to density differences is defined as natural convection.
3. *Radiation*. A body emits radiant energy in all directions. If this energy strikes a receiver, a portion of it may be transmitted, a portion may be reflected, and a portion may be absorbed. It is this absorbed portion that represents heat transfer in the form of radiation that is the transfer of energy through space by means of electromagnetic waves. Radiation passing through empty space is not transformed to heat or any other form of energy. If, however, matter appears in its path, the radiation will be transmitted, reflected, or absorbed by that matter. For example, fused quartz transmits practically all the radiation that strikes it. A polish opaque surface or mirror will reflect most of the radiation impinging on it, while a black, dull, or matted surface will absorb most of the radiation it receives. Any absorbed energy will be transformed quantitatively into heat. Note that it is only the absorbed energy that appears as heat.

A significant amount of material on conduction, convection, and radiation is available in the literature (Kern 1950; Farag and Reynolds 1996; Theodore 2011; Flynn, Akashige, and Theodore 2019). Numerous illustrative examples complement these presentations.

9.3 INTRODUCTION TO HEAT TRANSFER EQUIPMENT

Heat exchangers are defined as equipment that affect the transfer of thermal energy in the form of heat from one fluid to another. The simplest exchangers involve the direct mixing of hot and cold fluids. In most industrial exchangers, the fluids are *separated by a wall*. The latter type, referred to by some as a *recuperator*, can range from a simple plane wall between two flowing fluids to more complex configurations involving multiple passes, fins, and/or baffles. Conductive and convective heat transfer principles are required to describe and design these units; radiation effects are generally neglected.

Heat exchangers used in the chemical, environmental, petrochemical, petroleum, paper, and power industries encompass a wide variety of designs that are available from many manufacturers. Equipment design practice first requires the selection of safe operable equipment. The selection-design process must also seek a cost-effective balance between initial (capital) installation costs, operating costs, and maintenance costs.

There is a near-infinite variety of heat exchange equipment. These can vary from a simple electric heater in the home to a giant boiler in a utility power plant. A limited number of heat transfer devices likely to be encountered by the practicing engineer have been selected for description in this section. Most of the units transfer heat from one fluid to another fluid, with the heat passing through a solid interface such as a tube wall that separates the two (or more) fluids. The size, shape, and material employed that separates the two fluids is, of course, important. Another potential problem or concern in a heat exchanger is the method of confining one or both of the two.

Heat exchangers are classified by their functions. In a general sense, heat exchangers are classified into three broad types:

1. Recuperators or through-the-wall non-storing exchangers (e.g., double-pipe heat exchangers, shell-and-tube heat exchangers).
2. Direct contact nonstoring exchangers.
3. Regenerators, accumulators, or heat storage exchangers.

To design a piece of heat transfer equipment, it is necessary to know the required heat transfer rate calculated by the enthalpy balances described earlier. This is the rate at which heat travels from the hot fluid, through the tube walls, and into the cold fluid and is employed in the calculation of the required contact area. The slower this rate is, for given hot and cold fluid flow rates, the more contact area is required. The rate of heat transfer through a unit of contact area is referred to as the *heat flux* or *heat flux density* and, at any point along the tube length, is given by:

$$\frac{dq}{dA} = U\left(T_H - T_L\right) \tag{9.1}$$

where dq/dA = local heat flux density, Btu/(h-ft^2); U = local overall heat transfer coefficient, Btu/(h-ft^2-°F).

The use of the overall heat transfer coefficient (U) in Equation 9.1 is a simple yet powerful concept. In most applications it combines both conduction and convection effects, although transfer by radiation can also be included. Methods for calculating the overall heat transfer coefficient are presented later in this section. In actual practice it is not uncommon for vendors to provide a numerical value for U for their heat transfer equipment.

With reference to Equation 9.1, the temperatures T_H and T_L are actually average values. When a fluid is being heated or cooled, the temperature will vary throughout the cross-section of the stream. If the fluid is being heated, its temperature will be highest at the tube wall and decrease with increasing distance from the tube wall. The average temperature across the flowing stream's cross-section is therefore, T_H, which is the temperature that would be achieved if the fluid at this cross-section were suddenly mixed to a uniform temperature. If the fluid is being cooled, on the other hand, its temperature will be lowest at the tube wall and increase with increasing distance from the wall.

Equation 9.1 must be integrated to apply the equation to an entire heat exchanger. This cannot be accomplished unless the geometry of the exchanger is first defined. For simplicity, one of the simplest geometries will be assumed here: the double-pipe heat exchanger. This device consists of two concentric pipes. The outer surface of the outer pipe is normally well insulated so that no heat exchange with the surroundings may be assumed. One fluid flows through the center pipe, while the other one flows through the annular channel between the pipes. The fluid flows may be either concurrent, when the two fluids flow in the same direction, or countercurrent, where the flows are in opposite directions. The countercurrent arrangement is more efficient and is more commonly employed in practice. For this configuration of a heat exchanger, integration of Equation 9.1 along the exchanger length, and applying several simplifying assumptions, yields:

$$q = UA\Delta T \tag{9.2}$$

This has come to be defined by some as the *heat transfer equation*. The simplifying assumptions are that U and ΔT do not vary along the length. Since this is not actually the case, both U and ΔT must be regarded as averages of some type. A more careful integration of Equation 9.1, assuming that only U is constant, would show that the appropriate average for ΔT is the log-mean average:

$$\Delta T_{\text{lm}} = \frac{\Delta T_2 - \Delta T_1}{\ln\left(\dfrac{\Delta T_2}{\Delta T_1}\right)} \tag{9.3}$$

where ΔT_1 and ΔT_2 are the temperature differences between the two fluids at the ends of the exchanger. For practical design purposes, when the value of the ratio $\Delta T_2/\Delta T_1$, in Equation 9.3 is not over 1.5, the arithmetic mean of the temperature differences may be used as a reasonable approximation of the true value.

The area term, A, in Equation 9.2 denotes the cylindrical contact area between the fluids. However, since a pipe of finite thickness separates the fluids, the cylindrical area, A, must first be defined. Any one of the infinite number of areas between and including the inside and outside surface areas of the pipe may be arbitrarily chosen for this purpose. The usual practice is to use either the inside, A_i, or outside, A_o, surface areas; the outside area is the more commonly used of the two. Since the value of the overall heat transfer coefficient depends on the area chosen, it should be subscripted to correspond to the area on which it is based. Equation 9.2, based on the outside surface area, now becomes:

$$q = U_o A_o \Delta T_{\text{lm}} \tag{9.4}$$

9.4 DOUBLE PIPE HEAT EXCHANGERS

Of the various types of heat exchangers that are employed in industry, perhaps the two most fundamental are the *double pipe* and the *shell-and-tube* units. Despite the fact that shell-and-tube heat exchangers generally provide greater surface area for heat transfer with a more compact design, greater ease of cleaning, and less possibility of leakage, the double pipe heat exchanger still finds use in practice.

The double pipe unit consists of two concentric pipes. Each of the two fluids – hot and cold – flow either through the inside of the inner pipe or through the annulus formed between the outside of the inner pipe and the inside of the outer pipe. Generally, it is more economical (from a heat efficiency perspective) for the hot fluid to flow through the inner pipe and the cold fluid through the annulus, thereby reducing heat losses to the surroundings. In order to ensure sufficient contact time, pipes longer than approximately 20 ft are extended by connecting them to return bends. The length of pipe is generally kept to a maximum of 20 ft because the weight of the piping may cause the pipe(s) to sag. Sagging may allow the inner pipe to touch the outer pipe, distorting the annulus flow region and disturbing proper operation. When two pipes are connected in a "U" configuration by a return bend, the bend is referred to as a *hairpin*. In some instances, several hairpins may be connected in series to provide the necessary contact area for a given application.

Double pipe heat exchangers have been used in the chemical process industry for over 100 years.

The first patent on this unit appeared in 1923. The unique (at that time) design provided "the fluid to be heated or cooled and flow longitudinally and transversely around a tube containing the cooling or heating liquid, etc." Interestingly, the original design has not changed significantly since that time.

Although this unit is not extensively employed in industry (the heat transfer area is small relative to other heat exchangers), it serves as an excellent starting point from an academic and/or training perspective in the treatment of all the various heat exchangers reviewed in this chapter.

Thus, the simplest kind of heat exchanger, which confines both the hot and cold fluids, consists of two concentric tubes, or pipes. When conditions are such that only a few tubes per pass are required, the simplest construction is the double pipe heat interchanger. This consists of special fittings that are attached to standard iron (typically) pipe so that one fluid (usually liquid) flows through the inside pipe and the second fluid (also usually a liquid) flows through the annular space between the two pipes. Such a heat interchanger will usually consist of a number of passes which are almost invariably arranged in a vertical stack. If more than one pipe per pass is required, the proper number of such stacks is connected in parallel. Although they do not provide a large surface area for heat transfer, double pipe heat exchangers are at times used in industrial settings. A schematic of this unit and flow classifications are provided in Figure 9.1

Recommended standard fittings for double pipe interchangers are provided in the literatures (Kern 1950; Farag and Reynolds 1996; Theodore 2011; Flynn, Akashige, and Theodore 2019). Other types of tubes or pipes available include:

1. Plain tubes
2. Duplex tubes
3. Finned tubes

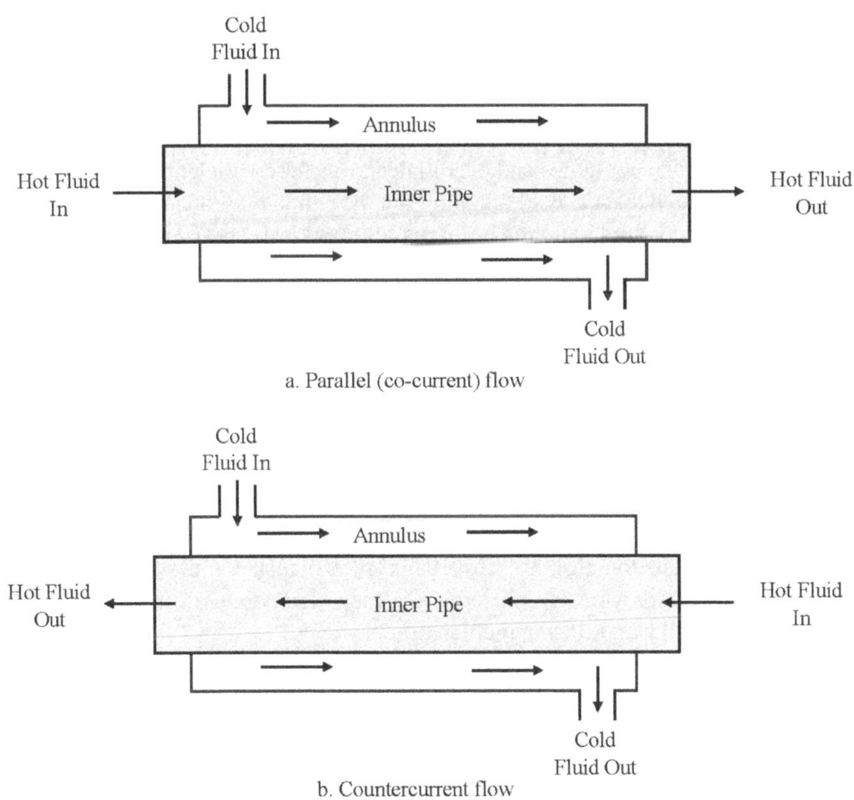

FIGURE 9.1 Double pipe heat exchanger schematics. (a) Parallel (co-current) flow and (b) Countercurrent flow.

Plain tubes are used when the corrosion potential on either the tube or shell side are approximately the same. Duplex tubes have walls of two different metals and are used when it is impossible to find a single metal that can adequately handle corrosion on both sides of the unit. Finned tubes are usually employed for heat transfer to gases where film coefficients are very low. Tube materials include: carbon steel, carbon alloy steels, stainless steels, brass and alloys, cupro-nickel, nickel, monel, glass, reinforced fiberglass plastic, etc.

In a co-current flow heat exchanger, both hot and cold fluids enter on the same side and flow through the exchanger in the same direction. The *temperature approach* is defined as the temperature difference driving force at the heat exchanger entrance, ΔT_1, or $(T_1 - t_1)$. This driving force drops as the streams approach the exit of the exchanger. At the exit, the temperature difference driving force is ΔT_2 or $(T_2 - t_2)$. Thus, the heat exchanger is more effective at the entrance than at the exit. In a countercurrent flow exchanger, the two fluids exchange heat while flowing in opposite directions. The temperature approach at the tube entrance end ΔT_1, or $(T_1 - t_2)$ and at the annular entrance end, ΔT_2 or $(T_2 - t_1)$, usually are roughly the same across the length of the exchange device. The temperature profiles for both parallel and countercurrent systems are presented in Figure 9.2.

Summarizing, the double pipe heat exchanger employed in practice consists of two pipes: an inner and an outer pipe. Hot fluid normally flows in the inner pipe and the cold fluid flows in the annulus between the outer diameter of the inner pipe and inner diameter of the outer pipe. Heat transfer occurs from the inner pipe to the outer pipe as the fluids flow through the piping system. By recording both inner and outer fluid temperatures at various points along the length of the exchanger, it is possible to calculate heat exchanger duties and overall heat transfer coefficients.

It should be noted that the cold fluid flows in the annular region in nearly all double pipe heat exchanger applications. As discussed previously, heat losses to the environment are lower (except in cryogenic applications) with the hot fluid flowing in the inside pipe. Any losses are thus transferred to the cold fluid and not the surroundings.

In the design of a double pipe heat exchanger, mass balances, heat balances, and the applicable heat transfer equation(s) are used. The steady-state heat balance equation is:

$$q = \Delta \dot{h}_H = \Delta \dot{h}_C = \dot{m}_H c_{pH} \left(T_1 - T_2\right) = \dot{m}_C c_{pC} \left(t_1 - t_2\right) \tag{9.5}$$

This equation assumes steady state, no heat loss, no viscous dissipation, and no heat generation. The rate equation used to design an exchanger is the design equation, which includes the previously developed log mean temperature difference (ΔT_{lm} or LMTD) and overall heat transfer coefficient:

$$q = UA\Delta T \tag{9.6}$$

9.5 SHELL AND TUBE HEAT EXCHANGERS

Shell and tube (also referred to as tube and bundle) heat exchangers provide a large heat transfer area economically and practically. The tubes are placed in a bundle

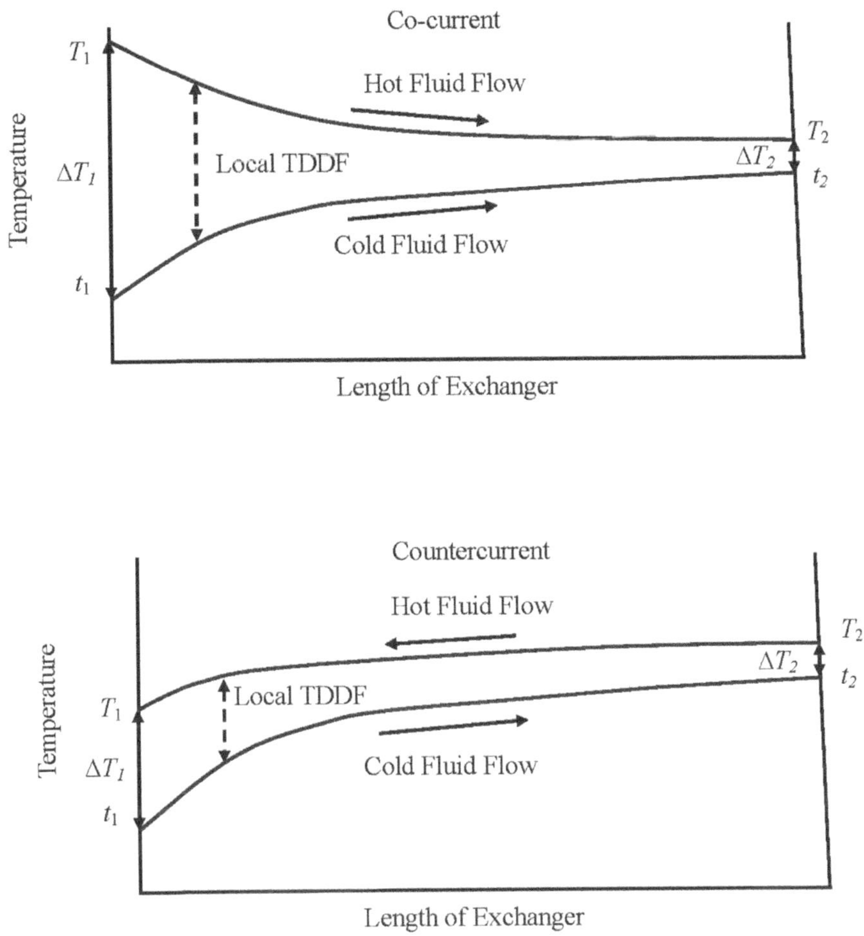

FIGURE 9.2 Co-current and countercurrent flow in a double pipe heat exchanger.

and the ends of the tubes are mounted in tube sheets. The tube bundles are enclosed in a cylindrical shell, through which the second fluid flows. Most shell and tube exchangers used in practice are of welded construction. The shells are built as a piece of pipe with flanged ends and necessary branch connections. The shells are made of seamless pipe up to 24 in diameter; they are made of bent and welded steel plates if above 24 in. Channel sections are usually built-up construction, with welding-neck forged-steel flanges, rolled-steel barrels and welded-in pass partitions. Shell covers are either welded directly to the shell, or are built-up constructions of flanged and dished heads and welding-neck forged-steel flanges. The tube sheets are usually nonferrous castings in which the holes for inserting the tubes have been drilled and reamed prior to assembly. Baffles can be employed to both control the flow of the fluid outside the tubes and provide turbulence.

There are vast industrial and utility uses of shell and tube heat exchangers. These units are used to heat or cool process fluids, either through a single-phase or through a two-phase heat exchanger. In single-phase exchangers, both the tube side and shell side fluids remain in the same phase that they enter. In two-phase exchangers (examples include condensers and boilers), the shell side fluid is usually condensed to a liquid or heated to a gas, while the tube-side fluid remains in the same phase.

Generally, shell and tube exchangers are employed when double pipe exchangers do not provide sufficient area for heat transfer. Shell and tube exchangers usually require less materials of construction and are consequently more economical when compared to double pipe and/or multiple double pipe heat exchangers in parallel.

There are more than 280 different types of shell and tube heat exchangers that have been defined by the Tubular Exchanger Manufacturers Association. The simplest shell and tube heat exchanger has a single pass through the shell and a single pass through the tubes. This is termed a 1-1 shell and tube heat exchanger. A schematic of a 1-2 unit can be found in Figure 9.3. A side view of the tubes in a typical exchanger is shown in Figure 9.4. Fluids that flow through tubes at low velocity result in a low heat transfer coefficients and low pressure drops. To increase the heat transfer rates, multi-pass operations may be used. As noted earlier, baffles are used to divert the fluid within a distribution header.

An exchanger with one pass on the shell side and four tube passes is termed a 1-4 shell and tube heat exchanger. It is also possible to increase the number of passes on the shell side by using dividers. A 2-8 shell and tube heat exchanger has two passes on the shell side and eight passes on the tube side.

Consider the following example. In order to determine the number of passes on the shell side and the number of passes on a the tube side for a 4-6 shell and tube heat exchanger, one must realize that there are four passes on the shell side and six passes on the tube side. The number of shell passes depends on the inner design of the shell while the number of tube passes depends on how many hairpins are formed by the bundle of tubes within the shell.

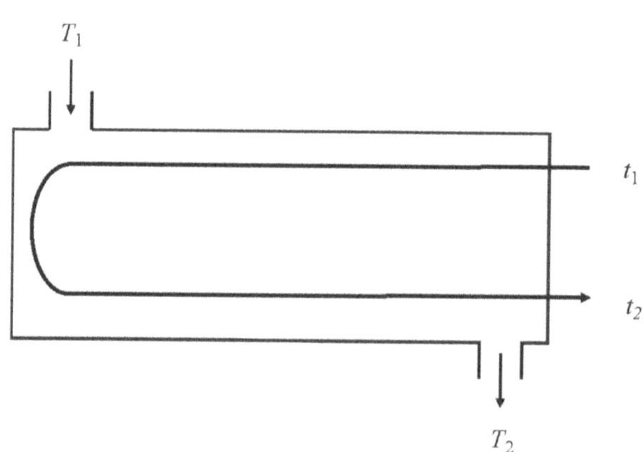

FIGURE 9.3 One shell pass, two tube pass heat exchanger.

FIGURE 9.4 Side view of tube layout in a shell and tube heat exchanger.

Also consider the example line diagram for a 3-6 shell and tube heat exchanger, provided in Figure 9.5.

When designing a shell and tube heat exchanger, the layout of the inner tube bundle is the main physical design criterion and includes:

1. Tube diameter
2. Tube wall thickness
3. Tube length
4. Tube layout
5. Tube corrugation
6. Baffle design

Details on each of the above are provided by Theodore (2011).

Finally, the choice of which fluid should flow in the tubes is an important design decision. If one of the fluids is particularly corrosive, it is typically introduced on the

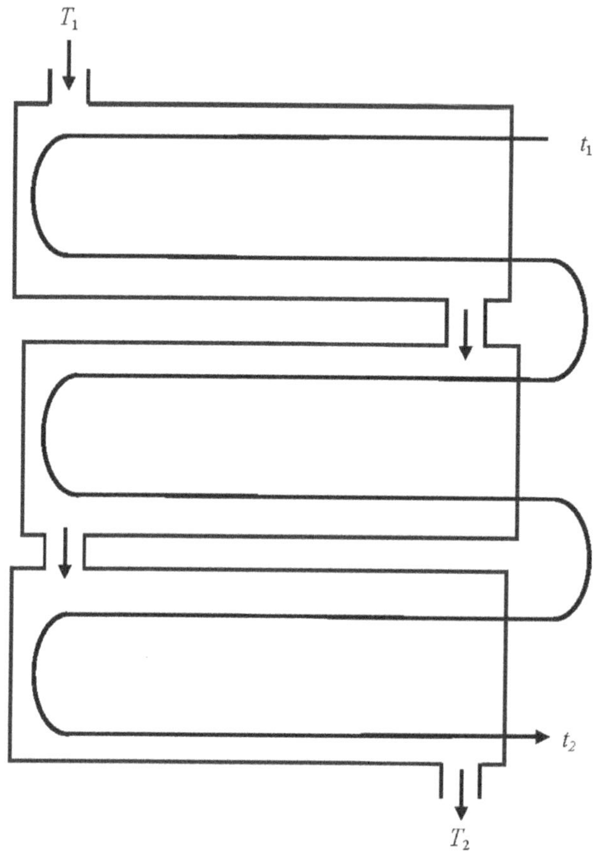

FIGURE 9.5 Stacked three shell pass, six tube pass heat exchanger.

tube side since more expensive resistant tubes can be purchased. If one of the fluids is more likely to form scale and/or deposits, it should be the tube side fluid since the inside of the tubes is much easier to clean than the outside of the tubes. In addition, viscous fluids normally flow on the shell side in order to help induce turbulence. The impact on the overall heat transfer coefficient and pressure drop should also be considered when determining the flow locations of the fluids.

In summary, shell and tube heat exchangers are among the most widely used exchangers due to the large surface area available for heat transfer. Countercurrent operation is the norm in industry but co-current operation is sometimes desired.

When two process fluids at different temperatures pass one another as in a shell and tube exchanger, heat transfer occurs due to the temperature difference between the two streams. The energy required to accomplish this heat transfer, that is, the *heat exchanger duty* can be determined for both process streams. For a well-insulated exchanger, the two duties should be equal to one another. The describing equations for the two duties are similar to those presented for the double pipe heat exchanger.

However, the following equation gives the heat exchanger duty of the hot process stream when saturated steam is introduced on the shell side (enthalpy balances between steam and condensate may also be used):

$$q_h = \dot{m}_s \lambda + \dot{m}_s c_{p,hw} \left(T_{s,i} - T_{s,o}\right) = \dot{m}_s \lambda + \dot{m}_s c_{p,hw} \left(T_1 - T_2\right) \tag{9.7}$$

where q_h = rate of energy lost by steam, Btu/h; $\dot{m}_s$ = total mass flow rate of steam condensate, lb/h; $\lambda = h_{vap}$ = heat of vaporization, Btu/lb; $c_{p,hw}$ = average heat capacity of condensate (hot water), Btu/(lb-°F); $T_{s,i} = T_1$= stream inlet temperature, °F; and $T_{s,o}$ = T_2 = condensate outlet temperature, °F.

The cold process stream duty is calculated similarly

$$q_c = \dot{m}_c c_{p,cw} \left(t_{c,o} - t_{c,i}\right) = \dot{m}_c c_{p,cw} \left(t_2 - t_1\right) \tag{9.8}$$

where q_c = rate of energy gained, Btu/h; $\dot{m}_c$ = total mass flow rate of cold water (all tubes); $c_{p,cw}$ = average heat capacity of cold process stream, Btu/(lb-°F); $t_{c,i} = t_1$ = cold water inlet temperature, °F; and $t_{c,o} = t_2$ = cold water outlet temperature, °F.

If the heat exchanger duty is known, the overall heat transfer coefficient is calculated from either of the two equations provided as follows:

$$U_o = \frac{q_h}{A_o \Delta T_{lm}} \tag{9.9}$$

$$U_i = \frac{q_c}{A_i \Delta T_{lm}} \tag{9.10}$$

where A = surface area for heat transfer (outside or inside), ft^2; q = average heat duty (cold or hot), Btu/h; U_o = overall heat transfer coefficient based on the outside area, Btu/(h-ft^2-°F); and U_i = overall heat transfer coefficient based on the inside area, Btu/(h-ft^2-°F).

Typical values for the overall heat transfer coefficient in steam condensers with water in the tubes ranging from 1,000 to 6,000 W/(m^2-K) (Incropera and DeWitt 1996). *Perry's Chemical Engineers' Handbook* (Perry, Green, and Maloney 1997) reports coefficients ranging from 400 to 1,000 Btu/(h-ft^2-°R). These values are approximately the same after the conversion of units.

9.6 FINNED AND EXTENDED SURFACES HEAT EXCHANGERS

Consider the case of a heat exchanger that is heating air outside the tubes by means of steam inside the tubes. The steam side coefficient will be very high and the air-side coefficient will be extremely low. Therefore, the overall heat transfer coefficient will approximate that of the air side (i.e., the air is the controlling resistance). One method of increasing the heat transfer rate is to increase the surface area of the heat exchanger. Therefore, if the surface of the metal on the air side could be increased, it would increase the area term without putting more tubes in the exchanger. This can be accomplished by mounting metal fins on a tube in such a way that there is good metallic contact between the base of the fin and the wall of the tube. If this contact is secured, the temperature throughout the fins will approximate that of the temperature of the heating (cooling) medium because of the high thermal conductivity of most metals used in practice. Consequently, the surface can be increased without more tubes (Kern 1950; Farag and Reynolds 1996; Theodore 2011; Flynn, Akashige, and Theodore 2019).

Additional metal is often added to the outside of ordinary heat transfer surfaces such as pipes, tubes, walls, etc. These extended surfaces, usually referred to as fins, increase the surface available for heat flow and result in an increase of the total transmission of heat. Examples of fin usage include auto radiators, air conditioners, cooling of electronic components, and heat exchangers. They are primarily employed for heat transfer to gases where film coefficients are very low.

Extended surfaces, or fins, are classified into *longitudinal fins*, *transverse fins*, and *spine fins*. *Longitudinal fins* (also termed straight fins) are attached continuously along the length of the surface. They are employed in cases involving gases or viscous liquids. As one might suppose, they are primarily employed with double pipe heat exchangers. *Transverse* or *circumferential fins* are positioned approximately perpendicular to the pipe or tube axis and are usually used in the cooling of gases. These fins find they major application with shell and tube exchangers. Transverse fins may be continuous or discontinuous (segmented). *Annular fins* are examples of continuous transverse fins. *Spine* or *peg fins* employ cones or cylinders, which extend from the heat transfer surface, and are used for either longitudinal flow or cross flow.

Fins are constructed of highly conductive materials. The optimum fin design is one that gives the highest heat transfer for the minimum amount of metal. The metal used in their manufacture has a strong influence on fin efficiency. Table 9.1 compares the volume and mass of three different metals required to give the same amount of heat transfer for fins with identical shapes. The values in the volume and mass columns are relative to the volume and mass of a copper fin.

9.7 OTHER HEAT EXCHANGERS

The following heat exchangers are briefly addressed: evaporators, waste-heat boilers, condensers, and quenchers. Cooling towers are not discussed; however, descriptive material is available in the literature (Theodore 2011).

TABLE 9.1
Fin Metal Data

Metal	Thermal Conductivity	Specific Gravity	Relative Volume	Relative Mass
Copper	400	8.9	1.00	1.00
Aluminum	210	2.7	1.83	0.556
Steel	55	7.8	7.33	0.43

9.7.1 Evaporators

The vaporization of a liquid for the purpose of concentrating a solution is a common mass transfer-heat transfer operation in the chemical process industry. The simplest device is an open pan or kettle that receives heat from a coil or jacket or by direct firing underneath the pan. In practice, the traditional unit is the horizontal-tube evaporator, in which a liquid (to be concentrated) in the shell side of a closed, vertical cylindrical vessel is evaporated by passing steam or another hot gas through a bundle of horizontal tubes contained in the *lower* part of the vessel. The liquid level in the evaporator is usually less than half the height of the vessel, thus permitting the disengagement of entrained liquid from the overhead vapor (Theodore, Ricci, and VanVliet 2009).

9.7.2 Waste Heat Boilers

Energy has become too valuable to discard. As a result, waste-heat and/or heat recovery boilers are now common in many process plants. As the chemical processing industries are becoming more competitive, few companies can afford to waste or dump thermal energy. This increased awareness of economic considerations has made waste-heat boilers one of the more important heat exchanger units of the boiler industry. The term *waste-heat boiler* includes units in which steam is generated primarily from the sensible heat of an available hot-flue or hot-gas stream rather than by solely firing fuel. The main purpose of a boiler is to convert a liquid (usually water) into a vapor. In most industrial boilers, the energy required to vaporize the liquid is provided by the direct firing of a fuel in the combustion chamber. The energy is transferred from the burning fuel in the combustion chamber by convection and radiation to the metal wall separating the liquid from the combustion chamber. Conduction then takes place through the metal wall and conduction–convection into the body of the vaporizing liquid. In a waste-heat boiler, no combustion occurs in the boiler itself; the energy for vaporizing the liquids is provided by the sensible heat of hot gases, which are usually product (flue) gases generated by a combustion process occurring either upstream or elsewhere in the system. The waste-heat boilers found at many facilities utilize the flue gases for this purpose. Additional details are available in the literature (Ganapathy 1984; Santoleri, Reynolds, and Theodore 2010; Theodore 2011).

9.7.3 Condensers (Connery 1982; Theodore 2008)

Condensation can be accomplished by increasing the pressure, decreasing the temperature (removing heat), or both. In practice, condensers generally operate through extraction of heat. Condensers differ in the means of removing heat and the type of device used. The two different means of condensation are *direct contact*, where the cooling media is intimately mixed and combined with vapors and condensate; and *indirect contact* (or surface), where the cooling media and vapors or condensate are separated by a surface area of some type. Contact condensers are simpler, are less expensive to install, and require less ancillary equipment and maintenance than surface condensers. The condensate or coolant from a contact condenser has a volume flow 10–20 times that of a surface condenser. This condensate usually cannot be reused and may pose a waste disposal problem.

Surface condensers form the bulk of the condensers used in industry. Some of the applicable types of surface condensers similar to those described previously are shell-and-tube, double-pipe, spiral-plate, flat-plate, air-cooled, and various extended-surface tubular units. The choice of coolant, for this as well as other heat exchange equipment will usually depend on the particular plant and the efficiency required of the unit. The most common coolant is usually cooling tower or river water. It has been shown that the vapor outlet temperature is critical to the operating efficiency of a condenser. In some instances, use of chilled brine or a boiling refrigerant can achieve a collection efficiency that will be sufficient without additional devices.

It is not unusual to specify multi-stage condensation in practice, where condensers, usually two, are connected in series and use cooling media with successively lower temperatures. For example, a condenser using cooling tower water can be used prior to a unit using chilled water or brine, thereby achieving higher thermal efficiency while minimalizing the cost of the use of chilled water.

9.7.4 Quenchers (Montaz, Trappi, and Seuvert 1997)

Hot gases must often be cooled before being discharged to the atmosphere or entering another device which normally is not designed for very high temperature operation (<500°F). These gases are usually cooled either by recovering the energy in a (waste heat) boiler, as discussed in the previous subsection, or by quenching. Both methods may be used in tandem. For example, a waste heat boiler can reduce an exit gas temperature down to about ~500°F; a water quench can then be used to further reduce the gas temperature to ~200°F, as well as saturate the gas with water. This secondary cooling and saturation can later eliminate the problem of water evaporation and can also alleviate other potential problems.

Although quenching and the use of a waste heat boiler are the most commonly used methods for gas cooling applications, several other techniques are used. All methods may be divided into two categories: *direct-contact cooling* and *indirect-contact cooling*. Direct-contact cooling methods include:

1. Dilution with ambient air
2. Quenching with water
3. Contact with high heat capacity solids

Among the indirect-contact methods are:

1. Natural convection and radiation from ductwork
2. Forced-draft heat exchangers
3. Waste-heat boilers.

With the *dilution* method, the hot gas effluent is cooled by adding sufficient ambient air that results in a mixture of gases at the desired temperature. The *water quench* method uses the heat of vaporization of water to cool the gases. When water is sprayed into the hot gases under conditions conducive to evaporation, the energy contained in the gases evaporates the water, and this results in a cooling of the gases. The hot exhaust gases may also be quenched using *submerged exhaust quenching*. This is another technique employed in some applications. In the *solids contact method*, the hot gases are cooled by giving up heat to a bed of ceramic elements. The bed, in turn, is cooled by incoming air to be used elsewhere in the process. As discussed earlier, natural convection and radiation occur whenever there is a temperature difference between the gases inside a duct and the atmosphere surrounding it. Cooling hot gases by this method requires the provision of enough heat transfer area to obtain the desired amount of cooling. In *forced draft heat exchangers*, the hot gases are cooled by forcing cooling fluid past the barrier separating the fluid from the hot gases.

9.7.5 Heat Pipes

This last subsection introduces the reader to heat pipes (Farber 2016). A heat pipe is a relatively new heat-transfer device that combines the principles of both thermal conductivity and phase transition to efficiently manage the transfer of heat between two interfaces. Refer to Figure 9.6. When heat is applied to the heat pipe, the liquid in the wick (chosen for compatibility with both the heat pipe material of construction and the temperatures of the hot and cold surfaces) heats and evaporates. As the evaporating fluid fills the heat pipes hollow center, it diffuses axially along its length. Condensation of the downstream vapor occurs wherever the temperature is even slightly below that of the evaporation area. As it condenses, the vapor gives up the heat it acquired during evaporation and establishes a reverse vapor flow pattern in the pipe. The condensed liquid flows through capillary action within the wick, back to the heat source, completing the heat transfer cycle.

Calculational details regarding heat pipes are provided by Hagen (1998). Information on the pipe thermal resistances, the heat source, the cooling (sink) source, evaporator and condenser pipes, and for the wick on both the evaporator and

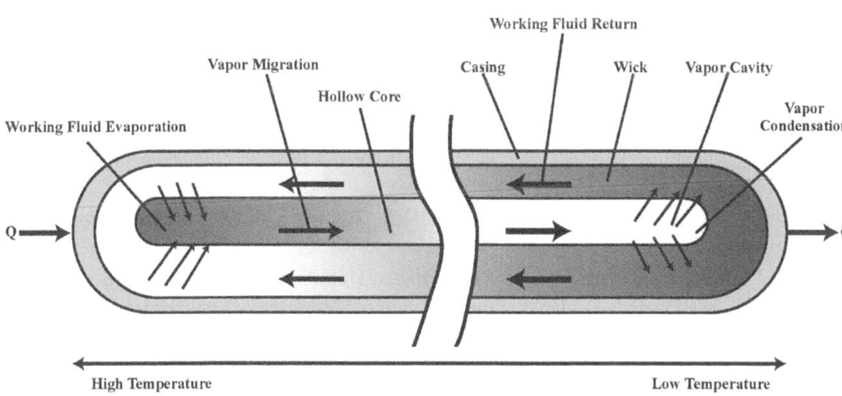

FIGURE 9.6 Heat pipe principles.

condenser sections are also provided. Heat pipes have been employed within computer systems to transfer heat from CPU's to heat sinks so the heat can be rejected to the environment. Heat pipes are also used in HVAC systems to recover heat by transfer from an exit stream to a fresh air stream in the winter and to cool an entering air stream in the summer. Its use in geothermal energy applications has yet to be realized.

9.8 THE HEAT EXCHANGER DILEMMA

One area where meaningful energy conservation measures can be realized is in the design and specification of process (operating) conditions for heat exchangers. This can be best accomplished by including second-law principles in the analysis. The quantity of heat recovered in an exchanger is not alone in influencing size and cost. As the log-mean temperature difference (LMTD) driving force in the exchanger approaches zero, the "quality" of heat recovered increases.

Most heat exchangers are designed with the requirement or specification that the temperature difference between hot and cold fluids be positive at all times and be at least 20°F. This temperature difference or driving force is referred to by some as the *approach temperature*. However, the corresponding entropy change is also related to the driving force, with large temperature difference driving forces resulting in large irreversibilities and the associated large entropy changes.

Heat exchanger designers must decide whether to employ a large LMTD driving force, which will result in a both more compact (smaller area design and a large entropy increase that is accompanied by the loss of quality energy), or a small LMTD driving force, which will result in both a larger heat exchanger and a smaller entropy change and greater recovery of quality energy. This topic receives extensive treatment by Theodore, Ricci, and Van Vliet (2009) and Theodore (2011).

REFERENCES

Connery, W. 1982. Condensers. In *Air Pollution Control Equipment,* ed. L. Theodore and A. J. Buonicore, Chapter 6.Upper Saddle River, NJ: Prentice-Hall.

Farag, I., and Reynolds, J. 1996. *Heat Transfer, a Theodore Tutorial.* East Williston, NY: Theodore Tutorials. Originally published by the USEPA/APTI, RTP, NC. Roanoke, VA: ETS Professional Training Institute. https://etsi-inc.com/training/tutorials-and-textbooks/.

Farber, P. 2016. *Personal Communication to L. Theodore.* Willowbrook, IL.

Flynn, A. M., Akashige, T., and Theodore, L. 2019. *Kern's Process Heat Transfer, 2nd Edition.* Beverly, MA: Scrivener-Wiley.

Ganapathy, V. 1984. Size or Check Waste Heat Boilers Quickly. *Hydrocarbon Process* 9: 169–170.

Hagen, K.D. 1998. *Heat Transfer with Applications.* Hoboken, NJ: Prentice Hall.

Incropera, F., and DeWitt, D. 1996. *Fundamentals of Heat and Mass Transfer, 4th Edition,* Chapters 8, 11. Hoboken, NJ: Wiley.

Kern, D. 1950. *Process Heat Transfer.* New York, NY: McGraw-Hill.

Montaz, W., Trappi, T., and Seuvert, J. 1997. Sizing Up RTO and RCO Heat Transfer Media. *Pollution Engineering* 12: 34–38.

Perry, R., Green, D., and Maloney, J.O., ed. 1997. *Perry's Chemical Engineers' Handbook, 7th Edition.* 10–67 to 10–137. New York, NY: McGraw-Hill.

Santoleri, J., Reynolds, J., and Theodore, L. 2010. *Introduction to Hazardous Waste Incineration, 2nd Edition.* Hoboken, NJ: Wiley.

Theodore, L. 2008. *Air Pollution Control Equipment Calculations.* Hoboken, NJ: Wiley.

Theodore, L. 2011. *Heat Transfer Applications for the Practicing Engineer.* Hoboken, NJ: Wiley.

Theodore, L., Ricci, F., and VanVliet, T. 2009. *Thermodynamics for the Practicing Engineer.* Hoboken, NJ: Wiley.

10 Prime Movers

10.1 INTRODUCTION

If a pressure difference is required between two points in a system, a prime mover such as a fan, pump, or compressor is usually used to provide the necessary pressure and/or flow impetus. Engineers are often called on to specify prime movers more frequently than any other piece of processing equipment, particularly in the chemical and energy industries. In a general sense, these prime movers are to a chemical process or power plant what the engine is to one's automobile. Whether one is processing a waste stream, moving petrochemicals or other products, moving heat transfer fluids, and so on, the material must usually be transferred from one point to another location somewhere in the process. For example, chemicals must be loaded or unloaded, sent to heat exchangers or cooling coils, transferred from one processing unit to another, or packaged for shipment at a chemical plant.

To move material (either a fluid or slurry) through the various pieces of equipment at a facility such as a power plant (including piping and duct work) requires not only conveyance systems but also mechanical energy to impart an initial velocity to the material to overcome pressure and frictional losses that occur throughout the flow path. This energy may be imparted to the moving stream in one or more of three modes: an increase in the stream's velocity, an increase in the stream's pressure, or an increase in stream's height. In the first case, the additional energy takes the form of increased external kinetic energy as the bulk stream velocity increases. In the second, the internal energy (mainly potential energy, but usually some kinetic as well) of the stream increases. This pressure increase may also cause a stream temperature rise, which represents an internal energy increase. In the third case, which may be relatively small for some power plant operations, the bulk fluid experiences an increase in external potential energy in the Earth's gravitational field.

Six topics complement this presentation related to prime movers. Section numbers and their accompanying titles are listed as follows:

10.2 Hydrostatics
10.3 Conveyance Systems
10.4 Fans and Blowers

DOI: 10.1201/9781003667759-12

The bulk of the material to follow has been drawn from the work of Farag (1994).

10.2 HYDROSTATICS

When a fluid is at rest, there is no shear stress and the pressure at any given point in the fluid is the same in all directions. The pressure is also the same across any longitudinal section parallel with the Earth's surface; it varies only in the vertical direction, that is, with depth of the fluid. This phenomenon gives rise to the study of hydrostatics, the subject title for this section. Following this introduction, this section addresses pressure principles, buoyancy effects (including Archimedes' Law), and manometry principles.

Consider a differential element of fluid of height, dz, and uniform cross-section area, S. The pressure, P, is assumed to increase with height, z. The pressure at the bottom surface of the differential fluid element is P; at the top surface, it is $P + \mathrm{d}P$. Thus, the net pressure difference, dP, on the element is acting downward. A force balance on this element in the vertical direction yields:

$$\text{downward pressure force} - \text{upward pressure force} + \text{gravitational force} = 0 \quad (10.1)$$

Therefore,

$$(P+\mathrm{d}P)S - PS + \rho S\left(\frac{g}{g_c}\right)\mathrm{d}z = 0;\ (\mathrm{d}P)S + \rho\left(\frac{g}{g_c}\right)S(\mathrm{d}z) = 0 \quad (10.2)$$

One has a choice as to whether to include g_c in the describing equations. The term g_c is a conversion constant with a given magnitude and units, for example, 32.2 (1b/1bf)(ft/s^2) or dimensionless with a value of unity, for example, $g_c = 1.0$. In this development, g_c is retained. Rearrangement of Equation 10.2 yields:

$$\frac{\mathrm{d}P}{\mathrm{d}z} = -\rho\left(\frac{g}{g_c}\right) = -\gamma \quad (10.3)$$

The term γ is the specific weight of fluid with units of lb/ft^3 or N/m^3. This equation is the hydrostatic or barometric differential equation. The term dP/dz is often referred to as the pressure gradient. Equation 10.3 is a first-order ordinary differential equation. It may be integrated by separation of variables:

$$\int \mathrm{d}P = -\int \rho\left(\frac{g}{g_c}\right)\mathrm{d}z = -\left(\frac{g}{g_c}\right)\int \rho\,\mathrm{d}z \quad (10.4)$$

For most engineering applications involving liquids, and many applications involving gases, the density may be considered constant, that is, the fluid is incompressible. Taking ρ outside the integration sign and integrating between any two limits in the fluid (Station 1 is where the pressure equals P_1, and the elevation is z_1, and Station 2 has a pressure of P_2 and elevation z_2, the pressure-height relationship is:

$$P_2 - P_1 = -\rho\left(\frac{g}{g_c}\right)(z_2 - z_1) \tag{10.5}$$

Equation 10.5 may also be written as:

$$\frac{P_1}{\rho(g/g_c)} + z_1 = \frac{P_2}{\rho(g/g_c)} + z_2 \tag{10.6}$$

The term $P/\rho(g/g_c)$ is defined as the *pressure head* of the fluid, with units of m (or ft) of fluid. Equation 10.6 states that the sum of the pressure head and potential head is constant under hydrostatic conditions. Equation 10.6 is sometimes termed Bernoulli's hydrostatic equation. It is useful in calculating the pressure at any liquid depth.

10.3 CONVEYANCE SYSTEMS

Selection of proper equipment in the construction and design phase of a conveyance system in a power plant is important. Some of the more important conveyance-related equipment is discussed in this section. It should be noted that as a fluid flows through various obstacles, that is, pipes, tubes ducts, etc., it experiences friction resulting in a pressure drop. This pressure drop is calculated employing Fanning's equation (Theodore 2008).

$$\Delta P = \left(\frac{4fL}{D}\right)\left(\frac{\rho v^2}{2g_c}\right) \tag{10.7}$$

where f = the Fanning friction factor, L = pipe length, D = pipe or conduit diameter, ρ = fluid density, v = average fluid velocity, and g_c = gravitational constant.

10.3.1 Pipes and Tubing

The most common conduits for fluids are *pipes* and *tubing*. Both usually have circular cross-sections, but pipes tend to have larger diameters and thicker walls than does tubing. Because of their heavier walls, pipes can be threaded, whereas tubing cannot. Energy systems usually handle large flow rates that require the larger diameters associated with pipes.

Tubing and pipes are manufactured from a large variety of materials. The initial selection of the material for piping primarily depends on the compatibility of the fluid with the piping material in terms of corrosion and the normal system operating pressure. If special piping is required to accommodate corrosive liquids or high standards of purity, stainless steel, nickel alloys, or materials of high resistance to heat and mechanical damage are used. Steel pipe can be lined with tin, plastic, rubber, lead, cement, or other coatings for special purposes. If corrosion problems or contamination are controlling factors, the use of a nonmetallic pipe such as glass, porcelain, thermosetting plastic, or hard rubber is often acceptable.

10.3.2 Ducts

While pipes and tubing are used as conduits for the conveyance of liquids or gases, ducts are used only for gases. Pipes, with their thicker walls, can be used for flows at higher pressures whereas ducts are relatively thin walled (1/16 in maximum) and are generally employed for gas flow pressures below 15 psig. Pipes are usually circular in cross section. Ducts come in many shapes (circular, oval, rectangular, etc.). In general, ducts are much larger in cross-section than pipes because the gases typically transported have low densities and are moved at high volumetric flow rates. Ducts are most often constructed of field-fabricated galvanized sheet steel, although other materials such as fibrous glass board, factory-fabricated round fiberglass, spiral sheet metal, and flexible duct materials are becoming increasingly popular. Other duct construction materials include black steel, aluminum, stainless steel, plastic and plastic-coated steel, cement, asbestos, and copper. All duct work can be fitted with jacketed insulation or sprayed with a fibrous insulation to minimize any heat transfer occurring between the process gas flow and the outside environment. The duct can also be fabricated with acoustical insulation to reduce the noise that may be generated from high gas velocities in the duct.

Duct fittings are similar to pipe fittings in that they allow connections to be made between duct sections and enable flow to be diverted where needed. Typical duct fittings include long and short radii elbows of all degrees, transitions, reducers, "Ts", and "Ys." Joining duct work of a relatively small cross-sectional area, typically below 500 in^2 (e.g., 32 in x 16 in nominal), requires only an inert sealing compound and external restraining guides and clamps (commonly referred to as "slips" and "slides"). The preferred sealing compound for low temperature (below 120°F) and/or noncorrosive gas transport applications is often Room-Temperature Vulcanizing silicon. Those gas transport systems that require higher temperatures and/or corrosion resistance should use an elastomer for joint sealer. Joining ducts of larger sizes requires the use of thicker sheet metal to fabricate a "flanged" surface as part of the duct. The flanged surfaces of ducts that are joined are typically sealed with an inert gasket material appropriate for the intended service (e.g., red rubber below 120°F or "Viton" below 350°F) and securely fastened to each other. In order to properly support a system of ducts and prevent seam leaks, all four sides of the duct work should be creased and at least one support should be located at or near every connection (typically every 6–8 ft).

Dampers, economizers (unlike those used in HVAC applications), and *variable air volume (VAV) boxes* are employed to increase the efficiency of duct systems. Dampers act as flow control valves for gas flow. They are typically actuated by air piston positioners controlled by another system variable or a system "mode" selector switch. Economizers act as regenerative heat exchangers for a duct system. Typically, a hot process flow line or a waste gas (e.g., furnace flue discharge) is placed on the "shell" side, and a cooler gas that requires preheating (e.g., furnace combustion air intake) is passed through the "tube" side of the economizer to take advantage of available low-grade (i.e. low-temperature) heat. Finally, VAV boxes allow the system to intake or discharge a constant mass of gas, regardless of its volume. This is very useful if intake gas temperatures vary more than a few degrees and/or if the process that utilizes the gas to produce a product (e.g., steam) changes its production level due to demand.

10.3.3 Fittings

A *fitting* is a piece of equipment that has one or more of the following functions:

1. The joining of two pieces of straight pipe (e.g., couplings and unions)
2. The changing of pipeline diameter (e.g., reducers and bushings)
3. The changing of pipeline direction (e.g., elbows)
4. The changing of pipeline direction and diameter (e.g., reducer elbows and street elbows)
5. The splitting of a stream into multiple streams (e.g., "T"s, "Y"s, and manifolds)
6. The joining of multiple streams (e.g., "T"s, "Y"s, and manifolds)
7. The mixing of multiple streams (e.g., blender)

Several of these terms are described as follows:

A *coupling* is a short piece of pipe whose ends are threaded on the inside (some plastics are not) used to connect straight sections of pipe. A *union* is also used to connect two straight sections but differs from a coupling in that it can be opened conveniently without disturbing the rest of the pipeline, whereas when a coupling is opened, a considerable amount of piping must usually be dismantled. A *reducer* is a coupling for two pipe sections of different diameters. A *bushing* also connects pipes and/or fitting of different diameters, but unlike the reducer, it is threaded on the outside of one end and on the inside of the other. This allows a larger pipe to screw onto the outside and the smaller pipe to screw into the inside of the bushing. An *elbow* is an angled coupling or bushing used to change flow direction, usually by 30°, 45°, or 90°. An elbow that is an angled bushing is referred to as a *"street" elbow*.

A "T" is also used to change flow direction and/or allow future system add-ons but is more often used to split one stream (supply) into two and later recombine two streams into one (return). A "Y" is similar to the T in that it is used to split a stream and/or combine two streams. A *blender* is a fitting that introduces one stream of a liquid into another in order to achieve a homogenous mixture of the two constituents. Multiple T and Y sections can be combined to form *manifolds* of various configurations to split and/or combine multiple process streams as necessary for a given application.

10.3.4 Valves

Valves have one main function in a pipeline, and that is to control and/or redirect the amount of flow passing through various sections of a system piping. There are many different types of valves, but the two most commonly used are the *gate valve* and the *globe valve*.

The gate valve contains a disk that slides at right angles to the flow direction. This type of valve is used primarily for on-off control of liquid flows. Typically, 70% of system flow will occur in the first 30% of valve stem travel because small adjustments in disk travel cause extreme changes in the flow cross-sectional area. Therefore, this type of valve is not suitable for adjusting flow rates.

Unlike the gate valve, the globe valve is designed for flow control. Liquid passes through a globe valve via a somewhat torturous route. In one form, the globe valve seal is a horizontal ring into which a plug with a beveled edge is inserted when the valve is shut. Better control of flow is achieved with this type of valve due to the fact that stem movement results in relatively small changes in cross-sectional flow area. However, the pressure drop across a fully open globe valve is greater than that in a comparable gate valve.

Other available valve types include the following: *check valves*, which permit flow in one direction only; *butterfly valves*, which operate in a damper-like fashion by rotating a flat plate to either a parallel or a perpendicular position relative to the flow; *plug valves*, in which a rotating tapered plug provides on-off service; *needle valves*, which are a variation of the globe valve which give improved flow control; *diaphragm valves*, which are valves specially designed to handle very viscous liquids, slurries, or corrosive liquids that might clog the moving parts of other valves; and *ball valves*, that provide easy on-off service in addition to good flow control with only small pressure drops when the valves are fully open.

For flow through valves and fittings, one may employ Equation 10.8 (Theodore 2008) to calculate the pressure drop across them.

$$\Delta P = K\left(\frac{\rho v^2}{2g_c D}\right) \tag{10.8}$$

where K = loss or resistance coefficient in consistent units with ρ, v, g_c, and D. Details are available in the literature (Theodore 2008) on the application of this equation to specific system configurations, along with information on expansion losses and resistance coefficients.

10.3.5 Stacks

Gases are discharged into the ambient atmosphere by means of stacks (referred to as chimneys in some industries) of several types. *Stub* or *short stacks* are usually fabricated of steel and extend a minimum distance up from the discharge of an induced draft fan. These are constructed of steel plate, either unlined or refractory lined, or entirely of refractory and structural brick. *Tall stacks*, which are constructed of the same materials as short stacks, provide a driving force (draft) of greater pressure difference than that resulting from the shorter stacks. In addition, tall stacks ensure more effective dispersion of gaseous and particulate effluents into the atmosphere. Some environmental, chemical, and utility applications use metal stacks that are made of a double wall with an air space between the metal sheets. The insulating air pocket created by the double wall prevents condensation on the inside of the stack, thus avoiding corrosion of the metal.

10.4 FANS AND BLOWERS

The terms *fans* and *blowers* are often used interchangeably, and no distinction will be made between the two in the following discussion. Whatever is stated about fans applies equally to blowers. Generally speaking, however, fans are used for low pressure (drop) operation, generally below 2 psi. Blowers are generally employed when generating pressure heads in the 2.0–14.7 psi range. Higher pressure operations than atmosphere require *compressors*.

Fans are usually classified as *centrifugal* or *axial-flow* type. In centrifugal fans, gas is introduced into the center of the revolving wheel (the eye) and discharges at right angles to the rotating blades. In axial-flow fans, the gas moves directly (forward) through the axis of rotation of the fan blades. Both types are used in industry, but it is the centrifugal fan that is employed at most utility and energy facilities.

The gas in a centrifugal fan is subjected to centrifugal forces that compress the gas giving it additional static pressure. Centrifugal fans are enclosed in a scroll-shaped housing that helps convert kinetic energy to static pressure. Gas rotating between the fan blades is compressed in the fan scroll, which increases the static pressure. Centrifugal fans are classified by blade configuration as not only *radial*, *forward curved*, and *backward curved,* but also as *air foil* and *radial tip-forward curved heel.*

Radial or straight blade fans physically resemble a paddle wheel with long radial blades attached to the rotor and are the simplest design of all centrifugal fans. This enables most radial blade fans to be built with great mechanical strength and to be easily repaired. These fans can be used in a variety of situations, especially heavy-duty applications. This type of fan can handle erosive and corrosive gases as well as very viscous gases. It is particularly well-suited for high static pressure operations and can generate pressures in excess of 50 in H_20. When operated properly, the horsepower efficiency range is 55%–69%, with 65% as a typical value.

Forward curved fans are the most popular for general ventilation purposes (high flow rates and low static pressures). These fans have both the heel and the tip of the blade curved forward in the direction of rotation. Blades are smaller and spaced much closer together than in other blade designs. They are generally not used with dirty gases when dust or sticky materials are present because contaminants easily accumulate on the blades and cause imbalance. Efficiencies range from 52% to 71%, with 65% being typical.

Backward curved or backward inclined fans have blades inclined in a direction opposite to that of the direction of rotation. This feature causes the gas to leave the tip of the blade at a lower velocity than the wheel-tip speed, a factor that improves the mechanical efficiency. These types of fans are not suitable for a heavily particulate-laden gas, sticky material, or abrasive dust. Centrifugal forces tend to build up particulate matter on the backside of the fan blades.

The airfoil is similar to the backward curved fan, except that the blade has been contoured to increase stability and operating efficiency. These fans are more expensive to construct than backward curved fans, but have lower power requirements. They are used in applications where the gas is clean and noncorrosive. Since the blades are hollow, abrasion and wear could allow dust, water vapor, and so on, to enter the blade and cause imbalance. This is the most efficient of the various fan types, with typical efficiencies at the 85% level.

A modification of the radial fan is the radial tip-forward curved heel. The blades are curved forward with this unit. This fan is reportedly more dependable than the radial or high tip speed applications (i.e., high static pressures) due to better vibrational characteristics and its ability to resist fatigue. It finds application in the processing of large flow rates (>200,000 acfm) with light to medium particulate loadings. However, sticky material and particulates, in general, can accumulate in the slight curvature of the blades and cause imbalance. Efficiencies here typically range from 52% to 74%, with 70% being common.

Generally, centrifugal fans are easier to control, are more robust in construction, and are less noisy than axial units. They have a broader operating range at their highest efficiencies. Centrifugal fans are better suited for operations in which there are flow variations and they can handle dust and fumes better than axial fans.

A rigorous, extensive treatment on fan selection is beyond the scope of this text. It is common practice among fan vendors to publish voluminous data in tabular form providing flow rate, static pressure, speed, and horsepower at a standard temperature and gas density. These are often referred to as *multi-rating tables*. These tables should not be used for fan selection except by those who have experience in this area. For those who do not, the proper course of action to follow is to provide the fan manufacturer with a complete description of the system and allow the manufacturer to select and guarantee the optimum fan choice for this given application.

To help in the actual selection of fan size, a typical fan rating is provided in Table 10.1. The fan size and dimensions are usually listed at the top of the table. Values of static pressure are arranged as columns that contain the fan speed and brake horsepower required to produce various volumetric flowrates. The point of maximum efficiency at each static pressure is usually underlined or printed in special type. In order to select a fan for the exact condition desired, it is sometimes necessary to interpolate between values presented in the multi-rating tables. Linear interpolation can be used with negligible error for multi-rating tables based on a single fan size.

Some multi-rating tables attempt to show ratings for a whole series of geometrically similar (homologous) fans in one table; in this case, interpolation is not advised.

The fan selection procedure involves, in part, an examination of the *fan curve* and the *system curve*. An example fan curve, relating static pressure with flow rate, is provided in Figure 10.1. Note that each type of fan has its own characteristic curve. Also note that fans are usually tested in the factory or laboratory with open inlets and long smooth straight discharge ducts. Since these conditions are seldom duplicated in the field, actual operation often results in lower efficiency and reduced performance than predicted from the manufacturer's data. A system curve is also shown in Figure 10.1. This curve is calculated prior to the purchase of the fan and provides a best estimate of the pressure drop across the system through which the fan must

TABLE 10.1
Typical Fan Rating Table

Static Pressure (in H_2O)

	½		1		1½		2		2½	
q_a	rpm	bhp	rpm	bhp	rpm	bhp	rpm	bhp	rpm	bhp
5,727	216	0.7	278	1.34	330	2.01	378	2.75	421	3.5
6,873	236	0.9	291	1.67	339	2.42	384	3.24	424	4.06
8,018	250	1.2	305	2.05	352	2.87	393	3.76	430	4.69
9,164	273	1.6	320	2.49	366	3.42	405	4.35	441	5.36
10,309	291	2.1	338	3.05	381	4.06	419	5.06	453	6.14
11,455	315	2.7	356	3.65	396	4.76	432	5.88	468	7.03
12,600	337	3.4	377	4.39	413	5.58	448	6.81	482	8.04
13,746	360	4.3	399	5.25	430	6.48	465	7.82	496	9.16
14,891	382	5.4	421	6.25	451	7.52	481	8.93	512	10.35
16,037	405	6.6	442	7.48	473	8.67	501	10.16	529	11.69
17,182	429	8.0	463	8.97	496	10.05	521	11.54	547	13.18
18,328	451	9.6	486	10.61	517	11.61	543	12.99	566	14.78
19,473	474	11.0	510	12.47	539	13.47	565	14.85	587	16.56
20,619	520	13.0	532	14.55	560	15.6	587	16.82	610	18.54
21,764	520	15.0	556	16.82	584	17.98	608	19.17	632	20.77

Notes:
Backward inclined wheel style, 50½ in wheel diameter, max fan speed 1,134 rpm.
Underline indicates conditions at maximum fan efficiency.

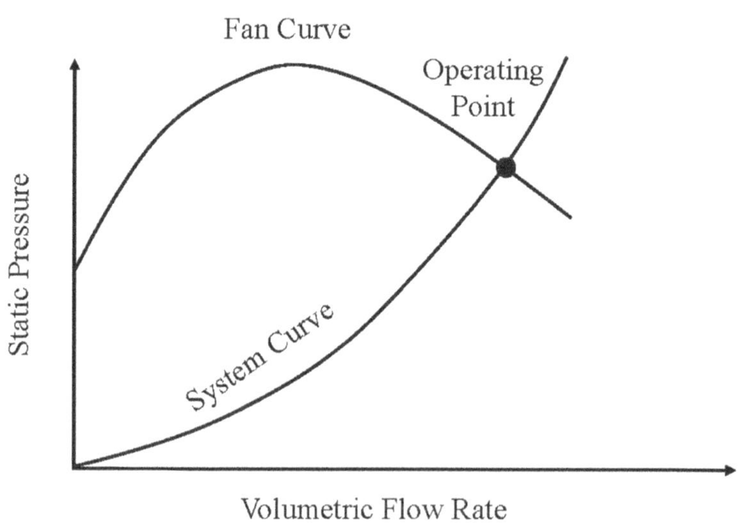

FIGURE 10.1 System and fan characteristic curves and the corresponding operating point.

deliver the gas. (This curve should approach a straight line with an approximate slope of 1.8 on log – log coordinates.) The system pressure (drop) is defined as the resistance through ducts, fittings, equipment, contractions, expansions, etc.

A number of methods are available to estimate the total system pressure change. These vary from very crude approximations to detailed, rigorous calculations. The simplest procedure is to obtain estimates of the pressure change associated with the movement of the gas through all of the resistances just described. The sum of these pressure changes represents the total pressure drop across the system, and represents the total pressure (change) that must be developed by the fan. This calculation becomes more complex if branches (i.e., combining flows) are involved. However, the same stepwise procedure should be employed.

When the two curves are superimposed, the intersection is defined as the *operating point*. The fan should be selected so that it operates just to the right of the peak on the fan curve. The fan operates most efficiently and with maximum stability at this condition. If a fan is selected for operation too close to its peak, it will usually surge and oscillate. Thus, the point of intersection of the two curves determines the actual volumetric flow rate. If the system resistance has been accurately specified and the fan properly selected, the two performance curves will intersect at the design flow rate. If system pressure losses have not been accurately specified or if undesirable inlet and outlet conditions exist, design conditions will not be obtained. Dampers and fan speed changes can provide some variability in operating conditions.

There are a number of process and equipment variables that are classified as part of a fan specification. These include (Theodore 2008):

1. Flow rate (acfm)
2. Temperature
3. Density
4. Gas stream characteristics
5. Static pressure that needs to be developed
6. Motor type
7. Drive type
8. Materials of construction
9. Fan location
10. Noise controls

With respect to drive type, belt drives are usually employed if the power is <200 hp. Direct drives are preferred for large horsepower systems. Direct drive units have lower maintenance costs and lower power transmission losses. However, their speed cannot be varied; if a speed change is required, the cost can be high. For materials of construction, mild carbon steel is commonly employed when treating dry air up to temperatures approaching 1,000°F. Fiberglass may be used for corrosive conditions but the temperature should not exceed 250°F when using fiberglass fan materials (Theodore 2008).

With respect to fan location, the fan may be located downstream or upstream of a particular piece of equipment in the process. Fans located downstream are referred to as *induced draft* or *negative pressure* fans. Leakage occurs into, rather than out of, an induced draft unit. Possible lower flow rates and lower temperatures may lead to a smaller fan and reduced operating costs. Equipment costs could be higher because of the need for heavier construction material when operating under negative (vacuum) pressure. Fans located upstream may have erosion problems due to possible particulate loading; in addition, there may be an accumulation of particulates. A larger volumetric flow rate is possible, which would require a larger fan and high horsepower costs (Theodore 2008).

Gas (or air) horsepower (*ghp*) and brake horsepower (*bhp*) are the two terms of interest. These may be calculated from Equations 10.9 and 10.10 as:

$$ghp = 0.0001575\, q_a \Delta P = q_a \Delta P/6356 \tag{10.9}$$

$$bhp = 0.0001575\, q_a \Delta P/\eta_f \tag{10.10}$$

where q_a = volumetric flow rate, acfm; ΔP = static pressure head developed, in. H_2O; and η_f = fractional fan efficiency.

For cold start-ups, the static pressure can be significantly higher than the normal operating pressure drop, and the fan must be able to handle the increased resistance. The correction for the cold static pressure may be estimated from Equation 10.11 (Theodore 2008):

$$\Delta P_c = \Delta P\, (\rho_c/\rho) \tag{10.11}$$

where c = subscript denoting cold gas; and ρ = gas density, lb/ft^3.

10.5 PUMPS

Pumps are required to transport liquids, liquid-solid mixtures such as slurries and sludges, auxiliary fuel, etc. Pumps are also needed to transport water to and/or from such peripheral devices as boilers, quenchers, scrubbers, and so on. Pumps may be classified as *reciprocating*, *rotary*, or *centrifugal*. The reciprocating and rotary types are referred to as *positive displacement* pumps because, unlike the centrifugal type, the liquid or semiliquid flow is broken up into small portions as it passes through the pump. These three classes of units are described as follows.

Reciprocating pumps operate by the direct action of a piston on the liquid contained in a cylinder. As the liquid is compressed by the piston, the higher pressure forces it through discharge valves to the pump outlet. As the piston retracts, the next batch of low pressure liquid is drawn into the cylinder and the cycle is repeated. The piston may be either directly steam driven or moved by a rotating crankshaft through a crosshead. The rate of liquid delivery is a function of the volume swept out by the piston and the number of strokes per unit time. A fixed volume is delivered for each stroke but the actual delivery may be less because of both leakage past the piston and failure to fill the cylinder when the piston retracts. The volumetric efficiency of the pump is defined as the ratio of the actual volumetric discharge to the pump displacement. For well-maintained pumps, the volumetric efficiency should be at least 95%. Reciprocating pumps can deliver the highest pressure of any type of pump (20,000 psig); however, their capacities are relatively small compared to the centrifugal pump. Also, because of the nature of the operation of the reciprocating pump, the discharge flow rate tends to be somewhat pulsating. Liquids containing abrasive solids can damage the machined surfaces of the piston and cylinder. Because of its positive displacement operation, reciprocating pumps can also be used to meter liquid volumetric flow rates.

The rotary pump combines rotation of the liquid with positive displacement. The rotating elements mesh with elements of the stationary casing in much the same way that two gears mesh. As the rotating elements come together, a pocket is created that first enlarges, drawing in liquid from the inlet or suction line. As rotation continues, the pocket of liquid is trapped, reduced in volume, and then forced into the discharge line at a higher pressure. The flow rate of liquid from a rotary pump is a function of size and speed of rotation and is slightly dependent on the discharge pressure. Unlike reciprocating pumps, rotary pumps deliver nearly constant flow rates. Rotary pumps are used on liquids of almost any viscosity as long as the liquids do not contain abrasive solids. For this reason, they are very effective with many high viscosity mixtures. They operate in moderate pressure ranges (5,000 psig), have small-to-medium capacities, and like the reciprocating pump, can be used for metering liquids.

Centrifugal pumps are the most widely used pumps because of simplicity of design, low initial cost, low maintenance, and flexibility of application. Centrifugal pumps have been built to move as little as a few gallons per minute of liquid against a small head, and as much as several thousand gallons per minute of liquid against a pressure of several hundred psi. In its simplest form, this type of pump consists of an impeller rotating within a casing. Fluid enters the pump near the center of the rotating impeller and is thrown outwards by centrifugal force. The kinetic energy of the fluid increases from the center of the impeller to the tips of the impeller vanes. This high velocity is converted to a high pressure as the fast-moving fluid leaves the impeller and is driven into slower moving fluid in the volute or diffuser (Foust et al. 1950).

In pumping fluids, the operation of a pump may result in either the loss of liquid or the release of air contaminants, or both. The reciprocating and centrifugal pumps can also be sources of hazardous emissions. The opening in the cylinder through which the connecting rod drives the piston is the major source of contaminant release from a reciprocating pump. In centrifugal pumps, the only potential source of leakage normally occurs where the drive shaft passes through the casing.

Several methods have been devised for sealing the clearance between the pump shaft and fluid casing. For most applications, packed seals and mechanical seals are widely used. Packed seals can be used on both positive displacement and centrifugal pumps. A typical packed seal consists of a stuffing box filled with a sealing material or packing that encases the moving shaft. The stuffing box is fitted with a take-up ring that compresses the packing and causes it to tighten around the shaft. Materials used for packing vary with the fluid's temperature, physical and chemical properties, pressure, and pump type. Some commonly used materials are metal, rubber, leather, and plastics. Specialized pumps such as canned-motor, diaphragm, or electromagnetic pumps are used for cases where the use of mechanical seals is not feasible. These specialty pumps are used where no leakage can be tolerated but are available in a limited range of sizes; most are for low flow rates and all are of single- or two-stage construction. They have been used to handle both high temperature and very low temperature liquids. These pumps follow the same hydraulic principles as the traditional centrifugal pump. Because of their small size, they operate with rather low efficiencies, but in dangerous environmental applications, efficiency must be sacrificed for safety (Theodore and Dupont 2012).

Pumps are employed in industry for one and/or a combination of the following reasons:

1. The transportation of liquids from trucks or delivery vehicles to storage tanks.
2. The transportation of liquids from storage tanks to process units.
3. The transportation of slurries and/or sludges to/from storage tanks to/from process units.
4. The transportation of liquids from one process unit to another.
5. The delivery of water to quench units and heat boilers.
6. The delivery and circulation of water and/or caustic solutions in absorbers and scrubbers (where applicable).

Pump discharge pressures for (2) and (3) are roughly 75 and 50 psig, respectively. Actual discharge pressures will vary, however, with fluid viscosity and line pressure drop (which is a function of site-specific logistics and application).

A general relationship that may be used to determine pump hp requirements is (Farag 1994)

$$\text{hp} = \frac{7.27\text{x}10^{-5}\,\dot{m}\Delta P}{\rho\eta_p} \tag{10.12}$$

where $\dot{m}$ is the liquid flow rate, ΔP is the developed pressure drop, ρ is the liquid density, and η_p is the pump-motor efficiency (on a fraction basis).

The pressure increase (head) developed by a centrifugal pump is generally a function of the discharge rate. In an operating pump, as the flow rate becomes large, the total head delivered by the pump decreases. To obtain the discharge rate for a given system, the engineer has to solve, simultaneously, the system curve of head, h_p, versus discharge with the pump curve head, h_c. The system h_p is usually obtained by applying Bernoulli's equation (or the equivalent). The pump characteristic curve is usually supplied by the manufacturer. The point where the two curves intersect (as indicated earlier – see Figure 10.1) yields the operating conditions for the system, as shown in Figure 10.2.

Consider how the fluid flows in Figure 10.3 with respect to the system curve. The liquid is pumped from the large tank (Point 1) to the delivery station (Point 2). Applying a modified form of Bernoulli's equation to the liquid system between Stations 1 and 2 gives:

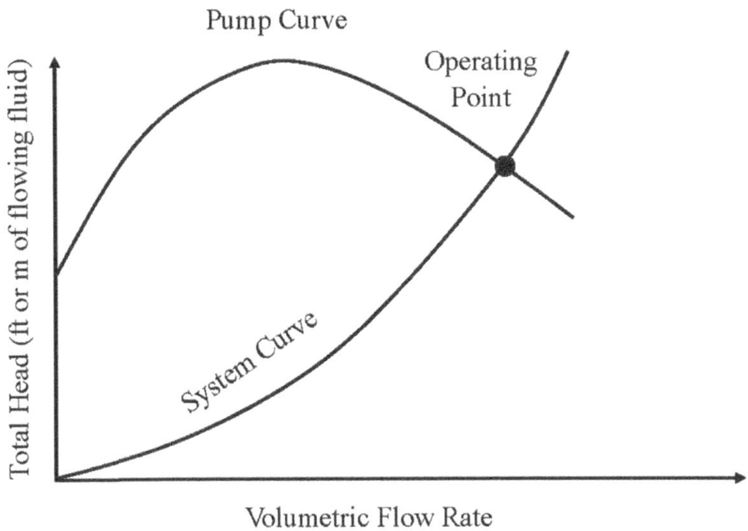

FIGURE 10.2 System and pump characteristic curves and the corresponding operating point.

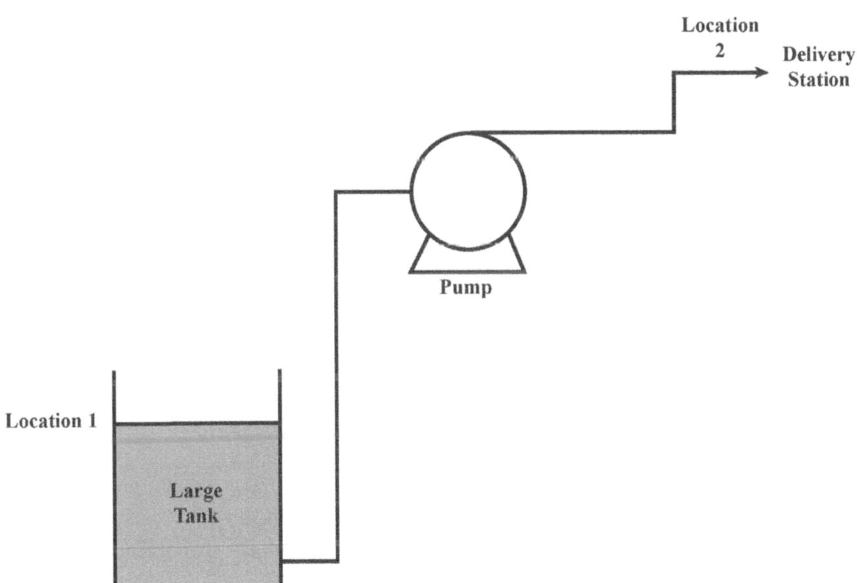

FIGURE 10.3 Pump system for Bernoulli equation application.

$$\frac{P_1}{\rho}+\frac{v_1^2}{2g_c}+\frac{g}{g_c}z_1=\frac{P_2}{\rho}+\frac{v_2^2}{2g_c}+\frac{g}{g_c}z_2-h_s+h_f \tag{10.13}$$

or

$$h_s=\frac{P_2-P_1}{\rho}+\frac{v_2^2-v_1^2}{2g_c}+(z_2-z_1)+h_f \tag{10.14}$$

where h_s is the input work (pump head) and h_f is the head loss due to fluid friction, with both treated as positive numbers. The units of each term as written in the previous equation are energy/mass.

For any flow rate, the pump head required to maintain that flow is calculated from Equation 10.14. In this procedure, one constructs a head-versus-discharge curve. This curve of h_s versus q is termed the system curve, and for a given centrifugal pump, the pump manufacturer supplies a characteristic head-versus-discharge curve. Once again, where the two curves intersect is the operating point of the pump.

When a centrifugal pump is operating at high flow rates, the high velocities occurring at certain points in the eye of the impeller or at the vane tips cause local pressures to fall below the vapor pressure of the liquid. This is predicted by the Bernoulli equation. Vaporization occurs at these points, forming bubbles that collapse violently upon moving to a region of higher pressure or lower velocity. This momentary vaporization and destructive collapse of the bubbles is called *cavitation*. This should be avoided if maximum capacity is to be obtained and damage to the pump prevented. The shock of bubble collapse causes severe pitting of the impeller and creates considerable noise and vibration. Cavitation may be reduced or eliminated by reducing the pumping rate or by slight alterations in the impeller design to give better streamlining. Note, however, that cavitation usually does not occur at low flow rates for any given pump (Brown and Associates 1950).

There are several items of process data that must be specified for pumps (Younger and Ruiter 1981). The first four are basic and can be obtained from the process flowsheet (which should include a heat and material balance) or from tables of physical properties. The basic four process items are:

1. The fluid being pumped
2. Operating temperature
3. Specific gravity, both at 60°F and at operating conditions
4. Viscosity at operating conditions (this is particularly important for positive displacement pumps

The next three items usually require calculation. These are:

5. Capacity
6. Operating head (or pressure)
7. Net Positive Suction Head (NPSH)

Pumps rarely influence plant layout except where a common standby for two services might require the rearrangement of process equipment. Pumps are generally placed close to process vessels and arranged esthetically. Pumps are also selected by pump specialists and the piping designer has little influence on the basic selection. However, the layout designer can request preferred orientation for suction and discharge and the NPSH limitations to meet required equipment elevations (Kern 1977).

10.6 COMPRESSORS

Compressors, unlike fans and pumps, find only limited and specialized application. They are primarily employed to increase the pressure of a fluid in some types of systems, for example, when liquids are broken up into tiny droplets (atomized) before entering a unit. This can be accomplished through the use of a high-pressure stream of air or steam that impinges on the liquid stream and atomizes it. The pressurizing of the air or steam is accomplished through the use of compressors. Compressors are also used for atomization on certain types of air pollution control devices such as venturi scrubbers which require fine water droplets to remove particulates from the flue gas stream (Younger and Ruiter 1981; Theodore 2008).

Gases are compressed for transportation, storage, and liquefaction, or to maintain either high or sub-atmospheric pressures in chemical and metallurgical processes. Although compression is accompanied by an increase in temperature, this temperature rise is rarely the object of the operation, except in special applications such as the ignition of fuel in a diesel engine or in vapor compression evaporation. Where a high temperature of the compressed gas is desired, the temperature rise in compression is useful but must be considered as incidental to the operation, since a temperature increase is produced more economically through direct heating than by the dissipation of valuable shaft work.

Compressors operate in a similar fashion to pumps and have the same classifications, that is, rotary, reciprocating, and centrifugal. An obvious difference between these two prime movers is the large decrease in volume resulting from the compression of a gaseous stream compared to the negligible change in volume caused by the pumping of a liquid stream.

Centrifugal compressors are employed when large volumes of gases are to be handled at low to moderate pressure increases (0.5–50 psi). Rotary compressors have smaller capacities but can achieve discharge pressures up to 100 psi. Reciprocating compressors are the most common type used in industry and are capable of compressing small gas flows to as much as 3,500 psig. With specially designed compressors, discharge pressures as high as 25,000 psig can be reached, but these devices are capable of handling only very small volumes, and do not work well for all gases. Reciprocating compressors are normally used for the applications mentioned earlier, atomizing of liquids for combustion or of venturi scrubber water for gas cleaning.

The following equation may be used to calculate compressor power requirements when the compressor operation is *adiabatic* and the gas (usually air) follows ideal gas behavior (Farag 1994; Theodore 2008):

$$Ws = \left(\frac{kRT}{k-1}\right)\left[\left(\frac{P_2}{P_1}\right)^{(k-1)/k} -1\right] \tag{10.15}$$

where W_s = compressor work required per lbmol of air; k = ratio of the heat capacity at constant pressure to that at constant volume, typically 1.3 for air; R= 1.987 Btu/(lbmol-°R); T = air temperature at compressor inlet conditions, °R; and P_1, P_2= air inlet and discharge pressures, respectively.

The temperature ratio across a compression stage is given as:

$$T_2 / T_1 = \left(P_2 / P_1\right)^{(k-1)/k}, \text{ for adiabatic operation} \tag{10.16}$$

$$T_2 / T_1 = \left(P_2 / P_1\right)^{(N-1)/N}, \text{ for polytropic operation} \tag{10.17}$$

where T_1, T_2 = suction and discharge temperatures, °R, respectively; P_1, P_2 = suction and discharge pressures, psia, respectively; and N = polytropic exponent, where $(N-1)/N = (k-1)/(k\eta_{pc})$, and η_{pc} = polytropic efficiency, fraction.

It is normally assumed that the usual centrifugal compressor is not cooled internally and thus follows a polytropic path. The temperature of the gas is often limited to protect against temperature excursions. Intercooling can be employed to retain temperatures at reasonable levels during high overall compression ratio applications.

Consider the following example. Suppose compressed air is to be employed in a nozzle to assist the atomization of a liquid and the air requirement for the nozzle is 7.5 lb/min at 40 psia. If atmospheric air is available at 60°F and 1.0 atm, one can calculate the power requirement by employing Equation 10.15 above. First note that the Ideal Gas Law applies to these conditions. Set the coefficient k for air equal to 1.3. The compressed energy requirement (delivered to the air) is calculated as follows:

$$W_s = \left(\frac{kRT}{k-1}\right)\left[\left(\frac{P_2}{P_1}\right)^{(k-1)/k} -1\right] = \frac{(1.3)(1.987)(520)}{(1.3-1)}\left[\left(\frac{40}{14.7}\right)^{(1.3-1)/1.3} -1\right]$$

$$W_s = (4{,}477.373)\ (1.26-1) = (4{,}477.373)\ (0.26) = 1{,}164 \text{ Btu/lbmol of air}$$

The corresponding power is hp

$$\text{hp} = (1{,}164)\ (7.5/29)\ (778) = 234{,}205 \text{ ft-lb}_f\text{/min}$$

This result may be divided by 33,000 to yield an hp requirement of 7.1 hp. The power required has the opposite sign. The reader should also note that this represents the minimum power required to accomplish this job.

REFERENCES

Brown, G., and Associates. 1950. *Unit Operations*. Hoboken, NJ: Wiley.

Farag, I. 1994. *Fluid Flow, a Theodore Tutorial.* East Williston, NY: Theodore Tutorials. Originally published by the USEPA/APTI, RTP, NC. Roanoke, VA: ETS Professional Training Institute. https://etsi-inc.com/training/tutorials-and-textbooks/.

Foust, A., Wenzel, L., Chung, C., Maus, L., and Andrews, L. 1950. *Principles of Unit Operations*. Hoboken, NJ: Wiley.

Kern, R. 1977. How to Get the Best Process-Plant Layouts for Pumps and Compressors. *Chemical Engineering* 12:131.

Theodore, L. 2008. *Air Pollution Control Equipment Calculations*. Hoboken, NJ: Wiley.

Theodore, L., and Dupont, R.R. 2012. *Environmental Health and Hazard Risk Assessment: Principles and Calculations*. Boca Raton, FL: CRC Press/Taylor & Francis Group.

Younger, A., and Ruiter, J. 1981. Selecting Pumps and Compressors. *Chemical Engineering* 6:117.

11 Steam
The Quintessential Medium in Energy Applications

11.1 INTRODUCTION

Water, chemical formula H_2O, plays an extremely important role in not only geothermal energy applications but also energy calculations in general. The reader should note that water is one of the most abundant substances on planet Earth and its properties have been determined with considerable accuracy. Since water is used in most chemical operations as a coolant in the liquid form or as a heating medium in the form of steam, the technical individual should be familiar with this unique compound.

Steam is generated in utility boilers for the development of mechanical work, as a source of heat for various processes, and for a combination of these purposes. The public utility and the steamship are examples of its use for the development of mechanical work. The heating system of a building exemplifies its use as process heat.

Steam is particularly favored for many heating operations. This fact and steam's high latent heat, moderate density, and reasonable pressure make it the prevalent, economical process medium for a host of applications. The steam tables presented in Section 11.4 provide some of the physical properties of steam. Steam is also employed beyond its saturation condition to any desired higher value of temperature. Properties for this superheated condition are also widely available.

Section numbers and their accompanying titles for this chapter on steam include:

11.2 Phase Equilibrium Principles
11.3 Steam Details
11.4 Steam Tables
11.5 Steam Uses
11.6 Steam Engines and Steam Turbines
References

11.2 PHASE EQUILIBRIUM PRINCIPLES

No discussion on steam would be complete without a review of phase equilibrium concepts. This is reviewed as follows as it primarily applies to steam production and use.

DOI: 10.1201/9781003667759-13

The term *phase*, for a pure substance indicates a state of matter, that is, solid, liquid, or gas. For mixtures, however, a more stringent definition must be used since a totally liquid or solid system may contain more than one phase (i.e., a mixture of oil and water). A phase is characterized by uniformity or *homogeneity*; the same composition and properties must exist throughout the phase region. At most temperatures and pressures, a pure substance normally exists as a single phase. At certain temperatures and pressures, two or perhaps even three phases can coexist in equilibrium. This is shown on the phase diagram for water (Figure 11.1). Regarding the interpretation of this diagram, the following points should be noted:

1. The line between the gas and liquid phase regions is the *boiling point line* and represents equilibrium between the gas and liquid.
2. The boiling point of a liquid is the temperature at which its vapor pressure is equal to the *external* pressure. The temperature at which the vapor pressure is equal to 1 atm is the *normal* boiling point.
3. The line between the solid- and gas-phase regions is the *sublimation point line* and represents equilibrium between the solid and gas.
4. The line between the solid- and liquid-phase regions is the *melting point* or *freezing point* and represents equilibrium between the liquid and solid.
5. The point at which all three equilibrium lines meet (i.e., the one pressure and temperature where solid, liquid, and gas phases can all coexist) is the *triple point*.

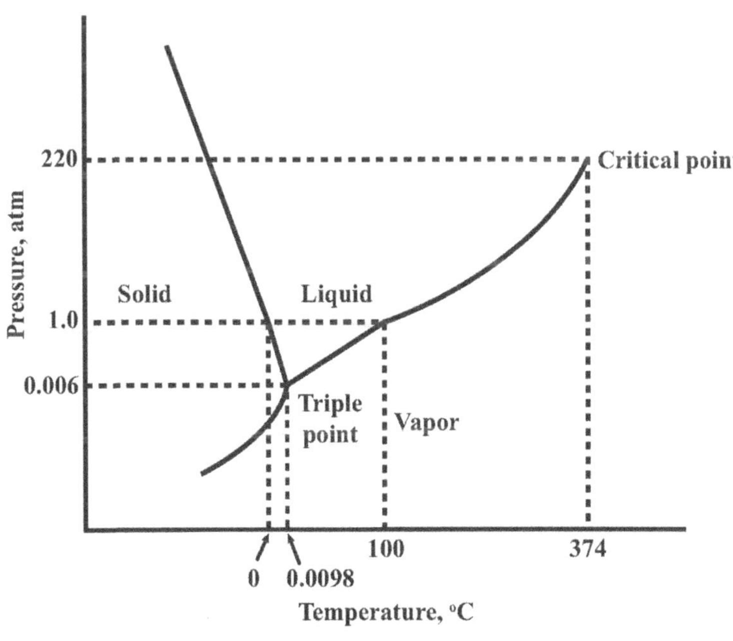

FIGURE 11.1 Phase diagram for water.

6. The liquid–gas equilibrium line is bounded on one end by the triple point and the other end by the *critical point*. The critical temperature (the temperature coordinate of the critical point) is defined as the temperature above which a gas or vapor cannot be liquefied by the application of pressure alone.

The term *vapor*, strictly speaking, is used only for a condensable gas, that is, a gas below its critical temperature, and should not be applied to a noncondensable gas. It should also be pointed out that this phase diagram for water differs from that of other substances in one respect – the freezing point line is negatively sloped; for *all* other materials, the slope of this line is positive. This is a consequence of the fact that liquid water is denser than ice, and isothermal (constant temperature) compression of the liquid can result in a transformation to the solid.

It should be noted that phase equilibria for multicomponent systems are considerably more complex, mainly because of the addition of *composition* variables; for example, in a *ternary* or three-component system, the mole fractions of two of the components are pertinent variables along with temperature and pressure. In a single-component system, dynamic equilibrium between two phases is achieved when the rate of molecular transfer from one phase to the second equals that in the opposite direction. In multicomponent systems, the equilibrium requirement is more stringent – the rate of transfer of each component must be the same in both directions (Theodore and Barden 1995; Theodore and Ricci 2010).

The most important equilibrium-phase relationship is that between liquid and vapor. Raoult's law describes theoretical liquid–vapor behavior and under certain conditions is applicable in practice. It is sometimes useful for mixtures of components of similar structure and states that the partial pressure of any component in the vapor is equal to the product of the vapor pressure of the pure component and the mole fraction of that component in the liquid, that is:

$$P_i = P_i' x_i \tag{11.1}$$

where P_i = partial pressure of component i in the vapor; P_i' = pure component i vapor pressure at the same temperature; and x_i = mole fraction of component i in the liquid. This expression may be applied to all components. If the gas phase is ideal, Equation 11.1 becomes:

$$y_i = (P_i'/P)x_i \tag{11.2}$$

where y_i = mole fraction component i in the vapor; and P = total system pressure.

Thus, the mole fraction of water vapor in a gas that is saturated, that is, in equilibrium contact with pure water (x_i = 1.0), is given simply by the ratio of the vapor pressure of water at the system temperature divided by the system pressure. This finds application in many quencher and heat exchanger calculations (Farag and Reynolds 1994; Theodore 2011).

11.3 STEAM DETAILS

Steam is water in a hot vapor state and is used for the generation of both power and electricity, and for heating purposes. The techniques of generating and using steam are important considerations in geothermal energy engineering technology. The vapor pressure of any liquid is raised by heating because high temperature increases the rapidity of movement of the individual molecules in the liquid. At a certain temperature, which corresponds to the boiling point of the liquid, this molecular activity is great enough to cancel out the forces of attraction between the individual molecules, and the liquid passes into the vapor state. In the case of water, the boiling point at sea-level atmospheric pressure (1.03 kg/cm^2 or 14.7 lb/in^2) is about 100°C (212°F). At this temperature, the addition of 970.3 Btu of heat will convert 0.454 kg (1 lb) of water to 0.454 kg of steam at the same temperature. The boiling point rises with the increase of pressure up to a pressure of 225.8 kg/cm^2 (3,208 lb/in^2) according to Boyle's law. At this pressure, water boils at a temperature of 374.15°C (705.47°F), its critical point.

Pure steam is a dry and invisible vapor. In many cases, however, when water is boiling, a quantity of small droplets or particles of water are taken up with the steam, and the resulting mixture is visible as a white vapor. A similar effect occurs when dry steam is exhausted into the comparatively cool atmosphere. Some of the steam cools and condenses, forming the familiar white vapor seen when a kettle boils on a stove. Such steam is said to be *wet*.

Steam that is heated to the exact boiling point corresponding the existing pressure is called *saturated steam*. So-called *superheated steam* is produced by heating steam beyond this temperature. Superheating also occurs if saturated steam is compressed or if saturated steam is throttled by passing the steam through a valve from a high-pressure vessel to a low-pressure vessel. Throttling causes the temperature of the steam to drop somewhat, but the temperature of the throttled steam is still higher than that of saturated steam at the corresponding pressure.

Steam that is to be used for power or heating purposes is usually generated in boilers. The simplest form of boiler is a closed vessel containing water, which is heated by a flame so that the water turns to saturated steam. The ordinary household-heating system usually has a boiler of this type, but steam-generating plants used for power purposes are more complex in design and are equipped with various auxiliary devices.

In general, a steam-generating plant today consists of a source of feedwater, a feedwater heater that raises the water temperature, the boiler proper in which the water is converted into steam, superheaters that raise the temperature of the vapor, and condensers that cool the steam after uses and reconvert it to the liquid state. The feedwater heater is usually a shell-and-tube heat exchanger in which steam at some intermediate pressure is the heating medium. Just before its reentry into the boiler, the feedwater is often passed through a so-called economizer, which uses the heat of the combustion gases that would otherwise be wasted.

11.4 STEAM TABLES

Extensive steam tables have been compiled, giving the enthalpies and a host of other properties of steam under widely varying conditions for both saturated steam and superheated vapors as well as saturated ice.

Note once again that the term *phase* for a pure substance refers to a state of matter that is gas, liquid, or solid. Latent enthalpy (heat) effects are associated with phase changes. These phase changes involve *no change* in temperature, but there is a transfer of energy to and from the substance. There are three possible latent effects: (1) vapor–liquid, (2) liquid–solid, and (3) vapor–solid. These are detailed as follows.

Vapor–liquid changes are referred to as *condensation* when the vapor is condensing and *vaporization* when liquid is vaporizing. *Liquid–solid changes* are referred to as *melting* when the solid melts to liquid and *freezing* when a liquid solidifies. *Vapor–solid changes* are referred to as *sublimation*. One should also note that there are enthalpy effects associated with a phase change of a solid to another solid form; however, this enthalpy effect is small compared to the other effects mentioned before. Specific volume, enthalpy, and entropy data for saturated steam, superheated steam, and steam-ice are provided in most steam tables. These tables comprise a tabular representation of the thermodynamic properties of water.

Theodore (2014) provides three steam tables employing English units based on the earlier presentation of Keenan and Keyes (1936) and Theodore, Ricci, and VanVliet (2009). These are presented as follows for saturated steam (Table 11.1), superheated steam (Table 11.2) and saturated steam-ice (Table 11.3). The reader should also note that Combustion Engineering (1940) published a 38-page booklet of steam tables that included both supercritical pressure steam and a Mollier diagram. An online version of these steam tables can also be found at The Engineering Toolbox (2003).

In general, two properties must be specified to determine the state of a unit mass of steam. If these properties are pressure, volume, temperature or enthalpy, the state is immediately located in the steam tables or on a Mollier diagram (Theodore, Ricci, and VanVliet 2009).

One property of a *saturated* phase will suffice to identify it since, by the phase rule, there is one less degree of freedom (Theodore, Ricci, and VanVliet 2009). But one additional extensive property, enthalpy, entropy, or specific volume, must be specified to enable determination of the amount of each of the two phases present.

Usually, instead of specifying the net volume of a pound of a two-phase mixture, the quality (weight% dry steam) of the wet steam or the % moisture (100 – % quality) is given. If h_f, h_g, v_f, v_g, are the enthalpy and specific-volume values for 1 lb of saturated liquid and 1 lb of saturated vapor, respectively, and if x is the fraction of moisture in the steam, then the net enthalpy and specific volume of 1 lb of the mixture are given by:

$$h = xh_f + (1 - x)\, h_g \tag{11.3}$$

$$v = xv_f + (1 - x)\, v_g \tag{11.4}$$

The net enthalpy can be read directly from the Mollier chart as the ordinate at the given quality and pressure (or temperature).

When two phases are present, the enthalpy, specific volume, or entropy of a pound of steam are computed to be less than the value for the saturated gas phase at the particular temperature or pressure in question Equations 11.3 or 11.4 will yield the

TABLE 11.1
Properties of Saturated Steam

Temperature, °F	Absolute Pressure, lb_f/in^2	Specific Volume, ft^3/lb			Enthalpy, Btu/lb			Entropy, Btu/lb-°R		
		Saturated Liquid, v_l	Evaporation Difference	Saturated Vapor, v_g	Saturated Liquid, h_l	Evaporation Difference	Saturated Vapor, h_g	Saturated Liquid, s_l	Evaporation Difference	Saturated Vapor, s_g
32	0.08854	0.01602	3306	3306	000	1075.8	1075.8	0.0000	2.1877	2.1877
35	0.09995	0.01602	2947	2947	3.02	1074.1	1077.1	0.0061	2.1709	2.1770
40	0.12170	0.01602	2444	2444	8.05	1071.3	1079.3	0.0162	2.1435	2.1597
45	0.14752	0.01602	2036.4	2036.4	13.06	1068.4	1081.5	0.0262	2.1167	2.1429
50	0.17811	0.01603	1703.2	1703.2	18.07	1065.6	1083.7	0.0361	2.0903	2.1264
60	0.2563	0.01604	1206.7	1206.7	28.06	1059.9	1088.0	0.055	2.0393	2.0948
70	0.3631	0.01606	867.8	867.9	38.04	1054.3	1092.3	0.0745	1.9902	2.0647
80	0.5069	0.01608	633.1	633.1	48.02	1048.6	1096.6	0.0932	1.9428	2.0360
90	0.6982	0.01610	468.0	468.0	57.99	1042.9	1100.9	0.1115	1.8972	2.0087
100	0.9492	0.01613	350.3	350.4	67.97	1037.2	1105.2	0.1295	1.8531	1.9826
110	1.2748	0.01617	265.3	265.4	77.94	1031.6	1109.5	0.1471	1.8106	1.9577
120	1.6924	0.01620	203.25	203.27	87.92	1025.8	1113.7	0.1645	1.7694	1.9339
130	2.2225	0.01625	157.32	157.34	97.90	1020.0	1117.9	0.1816	1.7296	1.9112
140	2.8886	0.01629	122.99	123.01	107..89	1014.1	1122.0	0.1984	1.6910	1.8894
150	3.718	0.01634	97.06	97.07	117.89	1008.2	1126.1	0.2149	1.6537	1.8685
160	4.741	0.01639	77.27	77.29	127.89	1002.3	1130.2	1.2311	1.6174	1.8485
170	5.992	0.01645	62.04	62.06	137.90	996.3	1134.2	0.2472	1.5822	1.8293
180	7.510	0.01651	50.21	50.23	147.92	990.2	1138.1	0.2630	1.5480	1.8109
190	9.339	0.01657	40.94	40.96	157.95	984.1	1142.0	0.2785	1.5147	1.7932
200	11.526	0.01663	33.62	33.64	167.99	977.9	1145.9	0.2938	1.4824	1.7762
210	14.123	0.01670	27.80	27.82	178.05	971.6	1149.7	0.3090	1.4508	1.7598
212	14.696	0.01672	26.78	26.80	180.07	970.3	1150.4	0.3120	1.4446	1.7566
220	17.186	0.01677	23.13	23.15	188.13	965.2	1153.4	0.3239	1.4201	1.7440

(continued)

TABLE 11.1 (Continued)
Properties of Saturated Steam

Temperature, °F	Absolute Pressure, lb_f/in^2	Specific Volume, ft^3/lb			Enthalpy, Btu/lb			Entropy, Btu/lb-°R		
		Saturated Liquid, v_l	Evaporation Difference	Saturated Vapor, v_g	Saturated Liquid, h_l	Evaporation Difference	Saturated Vapor, h_g	Saturated Liquid, s_l	Evaporation Difference	Saturated Vapor, s_g
230	20.780	0.01684	19.365	19.382	198.23	958.8	1157.0	0.3387	1.3901	1.7288
240	24.969	0.01692	16.306	16.323	208.34	952.2	1160.5	0.3531	1.3609	1.7140
250	29.825	0.01700	13.804	13.821	218.48	945.5	1164.0	0.3675	1.3323	1.6998
260	35.429	0.01709	11.746	111.763	228.64	938.7	1167.3	0.3817	1.3043	1.6860
270	41.858	0.01717	10.044	10.061	238.84	931.8	1170.6	0.3958	1.2769	1.6727
280	49.203	0.01726	8.628	8.645	249.06	924.7	1173.8	0.4096	1.2501	1.6597
290	57.556	0.01735	7.44	7.461	259.31	917.5	1176.8	0.4234	1.2238	1.6472
300	67.013	0.01745	6.449	6.466	269.59	910.1	1179.7	0.4369	1.1980	1.6350
310	77.68	0.01755	5.609	5.626	279.92	902.6	1182.5	0.4504	1.1727	1.6231
320	89.66	0.01175	4.896	4.914	290.25	894.9	1185.2	0.4637	1.1478	1.6115
330	103.06	0.01776	4.289	4.307	300.68	887.0	1187.7	0.4769	1.1233	1.6002
340	118.0	0.01787	3.770	3.788	311.13	879.0	1190.1	0.4900	1.0992	1.5891
350	134.63	0.01799	3.324	3.342	321.63	870.7	1192.3	0.5029	1.0754	1.5783
360	153.04	0.01811	2.939	2.957	332.18	862.2	1194.4	0.5158	1.0519	1.5677
370	173.37	0.01823	2.606	2.625	342.79	853.5	1196.3	0.5286	1.0287	1.5573
380	195.77	0.01836	2.317	2.335	353.45	844.6	1198.1	0.5413	1.0059	1.5471
390	220.37	0.01850	2.0651	2.0836	364.17	835.4	1199.6	0.5539	0.9832	1.5371
400	274.31	0.01864	1.8447	1.8633	374.97	826.0	1201.0	0.5664	0.9608	1.5272
410	276.75	0.01878	1.6512	1.6700	385.83	816.3	1202.1	0.5788	0.9386	1.5174
420	308.83	0.01894	1.4811	1.5000	396.77	806.3	1203.1	0.5912	0.9166	1.5078
430	343.72	0.01910	1.3308	1.3499	407.79	796.0	1203.8	0.6035	0.8947	1.4982
440	381.59	0.01926	1.1979	1.2171	418.90	785.4	1204.3	0.6158	0.8730	1.4887
450	422.6	0.0194	1.0799	1.0993	430.1	774.5	1204.6	0.6280	0.8513	1.4793
460	466.9	0.0196	0.9748	0.9944	441.4	763.2	1204.6	0.6402	0.8298	1.4700
470	514.7	0.0198	0.8811	0.9009	452.8	751.5	1204.3	0.6523	0.8083	1.4606
480	566.1	0.0200	0.7972	0.8172	464.4	739.4	1203.7	0.6645	0.7868	1.4513

490	621.4	0.0202	0.7221	0.7423	476.0	726.8	1202.8	0.6766	0.7653	1.4419
500	680.8	0.0204	0.6545	0.6749	487.8	713.9	1201.7	0.6887	0.7438	1.4325
520	812.4	0.0209	0.5385	0.5594	511.9	686.4	1198.2	0.7130	0.7006	1.4136
540	962.5	0.0215	0.4434	0.4639	536.6	656.6	1193.2	0.7374	0.6568	1.3942
560	1133.1	0.0221	0.3647	0.3858	562.2	624.2	1186.4	0.7621	0.6121	1.3742
580	1325.8	0.0228	0.2989	0.3217	588.9	588.4	1177.3	0.7872	0.5659	1.3532
600	1542.9	0.0236	0.2432	0.2668	617.0	648.5	1165.5	0.8131	0.5174	1.3307
620	1786.6	0.0247	0.1955	0.2201	646.7	503.6	1150.3	0.8398	0.4664	1.3062
640	2059.7	0.0260	0.1538	0.1980	678.6	452.0	1130.5	0.8679	0.4110	1.2789
660	2365.4	0.0278	0.1165	0.1442	714.2	390.2	1104.1	0.8987	0.3485	1.2472
680	2708.1	0.0305	0.0810	0.1115	757.3	309.9	1067.2	0.9351	0.2719	1.2071
700	3093.7	0.0369	0.0392	0.0761	823.3	172.1	995.4	0.9905	0.1484	1.1389

* Abridged from *Thermodynamic Properties of Steam*, by Joseph H. Keenan and Frederick G. Keyes. Copyright 1936, by Joseph H. Keenan and Frederick G. Keyes, Published by John Wiley & Sons, Inc., Hoboken, NJ.

TABLE 11.2
Properties of Superheated Steam

Absolute Pressure, lb_f/in^2 (Saturated temperature)		Temperature, °F 200	220	300	350	400	450	500	550	600	700	800	900	1000
1 (101.74)	*v*	392.6	404.5	452.3	482.2	512.0	541.8	571.6	601.4	631.2	690.8	750.4	809.9	869.5
	h	1150.4	1159.5	1195.8	1218.7	1241.7	1264.9	1288.3	1312.0	1335.7	1383.8	1432.8	1482.7	1533.5
	s	2.0512	2.0647	2.1153	2.1444	2.1720	2.1983	2.2233	2.2468	2.2702	2.3137	2.3542	2.3923	2.4283
5 (162.24)	*v*	78.16	80.59	90.25	96.26	102.26	108.24	114.22	120.19	126.16	138.10	150.03	161.95	173.87
	h	1148.8	1158.1	1195.0	1218.1	1241.2	1264.5	1288.0	1311.7	1335.4	1383.6	1432.7	1482.6	1533.4
	s	1.8718	1.8857	1.9370	1.9664	1.9942	2.0205	2.0456	2.0692	2.0927	2.1361	2.1767	2.2148	2.2509
10 (193.21)	*v*	38.85	40.09	45.00	48.03	51.04	54.05	57.05	60.04	63.03	69.01	74.98	80.95	86.92
	h	1146.6	1156.2	1193.9	1217.2	1240.6	1264.0	1287.5	1311.3	1335.1	1383.4	1432.5	1482.4	1533.2
	s	1.7927	1.8071	1.8595	1.8892	1.9172	1.9436	1.9689	1.9924	2.0160	2.0596	2.1002	2.1383	2.1744
14.696 (212.00)	*v*		27.15	30.53	32.62	34.68	36.73	38.78	40.82	42.86	46.94	51.00	55.07	59.13
	h		1154.4	1192.8	1216.4	1239.9	1263.5	1287.1	1310.9	1334.8	1383.2	1432.3	1482.3	1533.1
	s		1.7624	1.8160	1.8460	1.8743	1.9008	1.9261	1.9498	1.9734	2.0170	2.0576	2.0958	2.1319
20 (227.96)	*v*			22.36	23.91	25.43	26.95	28.46	29.97	31.47	34.47	37.46	40.45	43.44
	h			1191.6	1215.6	1239.2	1262.9	1286.6	1310.5	1334.4	1382.9	1432.1	1482.1	1533.0
	s			1.7808	1.8112	1.8396	1.8664	1.8918	1.9160	1.9392	1.9829	2.0235	2.0618	2.0978
40 (267.25)	*v*			11.040	11.843	12.628	13.401	14.168	14.93	15.688	17.198	18.702	20.20	21.70
	h			1186.8	1211.9	1236.5	1260.7	1284.8	1308.9	1333.1	1381.9	1431.3	1481.4	1532.4
	s			1.6994	1.7314	1.7608	1.7881	1.8140	1.8384	1.8619	1.9058	1.9467	1.9850	2.0214
60 (292.71)	*v*			7.259	7.818	8.357	8.884	9.403	9.916	10.427	11.441	12.449	13.452	14.454
	h			1181.6	1208.2	1233.6	1258.5	1283.0	1304.7	1331.8	1380.9	1430.5	1480.8	1531.9
	s			1.6492	1.6830	1.7135	1.7416	1.7678	1.7926	1.8162	1.8605	1.9015	1.9400	1.9762
80 (312.03)	*v*				5.803	6.220	6.624	7.020	7.410	7.797	8.562	9.332	10.077	10.830
	h				1204.3	1230.7	1256.1	1281.1	1305.8	1330.5	1379.9	1429.7	1480.1	1531.3
	s				1.6175	1.6791	1.7078	1.7346	1.7598	1.7836	1.8281	1.8694	1.9079	1.9442
100 (327.81)	*v*				4.592	4.937	5.268	5.589	5.905	6.218	6.835	7.446	8.052	8.656

	h					1200.1	1227.6	1253.7	1279.1	1304.2	1329.1	1378.9	1428.9	1479.5	1530.8
	s					1.6188	1.6518	1.6813	1.7085	1.7339	1.7581	1.8029	1.8443	1.8829	1.9193
120 (341.25)	*v*					3.783	4.081	4.363	4.636	4.902	5.165	5.683	6.195	6.702	7.207
	h					1195.7	1224.4	1251.3	1277.2	1302.5	1327.7	1377.8	1428.1	1478.8	1530.2
	s					1.5944	1.6287	1.6591	1.6869	1.7127	1.7370	1.7822	1.8237	1.8625	1.8990
140 (353.02)	*v*						3.468	3.715	3.954	4.186	4.413	4.861	5.301	5.738	6.172
	h						1221.1	1248.7	1275.2	1300.9	1326.4	1376.8	1427.3	1478.2	1529.7
	s						1.6087	1.6399	1.6683	1.6945	1.7190	1.7645	1.8063	1.8451	1.8817
160 (363.53)	*v*						3.008	3.230	3.443	3.648	3.849	4.244	4.631	5.015	5.396
	h						1217.6	1246.1	1273.1	1299.3	1325.0	1375.7	1426.4	1477.5	1529.1
	s						1.5908	1.6230	1.6519	1.6785	1.7033	1.7491	1.7911	1.8301	1.8667
180 (373.06)	*v*						2.649	2.852	3.044	3.229	3.411	3.764	4.110	4.452	4.792
	h						1214.0	1243.5	1271.0	1297.6	1323.5	1374.7	1425.6	1476.8	1528.6
	s						1.5745	1.6077	1.6373	1.6642	1.6894	1.7355	1.7776	1.8167	1.8534
200 (381.79)	*v*						2.361	2.549	2.726	2.895	3.030	3.380	3.693	4.002	4.309
	h						1210.3	1240.7	1268.9	1295.8	1322.1	1373.6	1424.8	1476.2	1525.0
	s						1.5594	1.5937	1.6240	1.6513	1.6767	1.7232	1.7655	1.8048	1.8415
220 (389.86)	*v*						2.125	2.301	2.465	2.621	2.772	3.033	3.352	3.634	3.913
	h						1206.5	1237.9	1266.7	1294.1	1320.7	1372.6	1424.0	1475.5	1527.5
	s						1.5453	1.5808	1.6117	1.6395	1.6652	1.7120	1.7545	1.7939	1.8308
240 (397.37)	*v*						1.9276	2.094	2.247	2.393	2.533	2.804	3.068	3.327	3.584
	h						1202.5	1234.9	1264.5	1292.4	1319.2	1371.5	1423.2	1474.8	1526.9
	s						1.5319	1.5686	1.6003	1.6286	1.6546	1.7017	1.7444	1.7839	1.8209
260 (404.42)	*v*							1.9183	2.063	2.199	2.330	2.582	2.827	3.067	3.305
	h							1232.0	1262.3	1290.5	1317.7	1370.4	1422.3	1474.2	1526.3
	s							1.5573	1.5897	1.6184	1.6447	1.6922	1.7352	1.7748	1.8118
280 (411.05)	*v*							1.7674	1.9047	2.033	2.156	2.392	2.621	2.845	3.066
	h							1228.9	1260.0	1288.7	1316.2	1369.4	1421.5	1473.5	1525.8
	s							1.5464	1.5796	1.6087	1.6354	1.6834	1.7265	1.7662	1.8033

(*continued*)

TABLE 11.2 (Continued)
Properties of Superheated Steam

Absolute Pressure, lb_f/in^2 (Saturated temperature)		Temperature, °F 200	220	300	350	400	450	500	550	600	700	800	900	1000
300 (417.33)	*v*						1.6364	1.7675	1.8891	2.005	2.227	2.442	2.652	2.859
	h						1225.8	1257.6	1286.8	1314.7	1368.3	1420.6	1472.8	1525.2
	s						1.5360	1.5701	1.5998	1.6268	1.6751	1.7184	1.7582	1.7954
350 (431.72)	*v*						1.3734	1.4923	1.6010	1.7036	1.8980	2.084	2.266	2.445
	h						1217.7	1251.5	1282.1	1310.9	1365.5	1418.5	1471.1	1523.8
	s						1.5199	1.5481	1.5792	1.6070	1.6563	1.7002	1.7403	1.7777
400 (444.59)	*v*						1.1744	1.2851	1.3843	1.4770	1.6508	1.8161	1.9767	2.134
	h						1208.8	1245.1	1277.2	1306.9	1362.7	1416.4	1469.4	1522.4
	s						1.4892	1.5281	1.5607	1.5894	1.6398	1.6842	1.7247	1.7623

TABLE 11.3
Properties of Saturated Steam-Ice

		Specific Volume, ft³/lb		Enthalpy, Btu/lb			Entropy, Btu/lb-°R		
Temperature, °F	Absolute Pressure, lb_f/in^2	Saturated Ice, v_i	Saturated Steam $v_g \times 10^{-3}$	Saturated Ice, h_i	Sublimation Difference, h_{sub}	Saturated Steam, h_g	Saturated Ice, s_i	Sublimation Difference, s_{sub}	Saturated Steam, s_g
32	0.0885	0.01747	3.306	-143.35	1219.1	1075.8	-0.2916	2.4793	2.1877
30	0.0808	0.01747	3.609	-144.35	1219.3	1074.9	-0.2936	2.4897	2.1961
20	0.0505	0.01745	5.658	-149.31	1219.9	1070.6	-0.3038	2.5425	2.2387
10	0.0309	0.01744	9.05	-154.17	1220.4	1066.2	-0.3141	2.5977	2.2836
0	0.0185	0.01742	14.77	-158.93	1220.7	1061.8	-0.3241	2.6546	2.3305
-10	0.0108	0.01741	24.67	-163.59	1221.0	1057.4	-0.3346	2.7143	2.3797
-20	0.0062	0.01739	42.2	-168.16	1221.2	1053.0	-0.3448	2.7764	2.4316
-30	0.0035	0.01738	74.1	-172.63	1221.2	1048.6	-0.3551	2.8411	2.4860

% moisture. For enthalpy, the moisture content can be read by inspection of the Mollier diagram.

Information on the internal energy, u, are not provided in these tables. If any two of the enthalpy terms, pressure, specific volume, and temperature are known, then u can be computed directly from Equation 11.5:

$$h = u + Pv \tag{11.5}$$

Note that h, u, P, and v must be in consistent units. If u and some other property are given to fix the state, a trial and error procedure must be employed to determine the other properties.

11.5 STEAM USES

Steam generated in power plants is used in prime movers for the development of mechanical work, as a source of heat for various processes, and for a combination of these purposes. The heating system of a building exemplifies its use as process heat. A steam plant in a utility, refinery or chemical plant is an example of its use for a combination of the two.

Many often regard steam as being employed solely for the generation of power. However in the chemical process industry, so much heat is obtained by condensation of steam that the dual use of steam for power and for process heat is of major importance. A practicing engineer or applied scientist often endeavors to coordinate and balance the two. In these chemical processes, one often expands the steam through a steam engine or turbine and then employs the exhaust or lower-pressure steam simply for heating purposes. For turbines, where dry steam is necessary to prevent blade erosion, the entering steam is superheated and condensation is prevented.

The turbine is a particularly flexible prime mover for many power plants chemical manufacturing facilities, since it can be designed not only to furnish low-pressure exhaust steam but also by bleeding off, to supply steam at higher pressures. Both the steam engine and the steam turbine are covered in more detail in the next section.

11.6 STEAM ENGINES AND STEAM TURBINES

The *steam engine* is a mechanical device used to transfer the energy of steam into mechanical energy for a variety of applications, including propulsion and generating electricity. The basic principle of the steam engine involves transforming the heat energy of the steam into mechanical energy by permitting the steam to expand and cool in a cylinder equipped with a movable piston.

An important adjunct to all types of reciprocating steam engines is the flywheel, which is driven by the piston crank. Because of its inertia, the flywheel, usually a heavy metal casting, makes continuous the individual surges of power corresponding to the expansive impulses given to the piston and permits the engine to essentially provide a uniform flow of power.

The cylinder of a compound engine, unlike that of a single-cylinder engine of the ordinary type, can be kept at a uniform temperature. A still further improvement in engine efficiency is afforded by the uniflow engine, which uses the piston itself as a valve and in which all portions of the cylinder remain at approximately the same temperature when the engine is operating. In the uniflow engine, steam moves in only one direction while entering the cylinder of the engine, expanding, and then leaving the cylinder. This unidirectional flow is accomplished by employing two sets of inlet ports at either end of the cylinder, together with a single set of outlet ports in the cylinder wall at the center. The flow of steam into the two sets of inlet ports is controlled by separate valves. The inherent advantages of the uniflow system are such that engines of this type are usually specified for large installations, although the initial cost of the engines is considerably higher than that of conventional steam engines. One virtue of the uniflow engine is that it permits the efficient use of high-pressure steam in a single-cylinder engine without the necessity of compounding.

A *turbine* is a rotary engine that converts the energy of a moving stream of water, steam, or gas into mechanical energy. The basic element in a turbine is a wheel or rotor with paddles, propellers, blades, or buckets arranged on its circumference in such a fashion that the moving fluid exerts a tangential force that turns the wheel and imparts energy to it. This mechanical energy is then transferred through a drive shaft to operate a machine, electric generator, compressor, or propeller. Turbines are classified as hydraulic (water) turbines, steam turbines, or gas turbines. Today turbines power generators that produce one-fourth of the world's electrical energy. Interestingly, windmills are technically wind turbines. Only steam turbines are discussed in the following text.

The success of the water turbine inevitably led to consideration of the turbine principle for extracting power from steam where the turbine could achieve higher efficiency by utilizing the kinetic energy of steam flow. The turbine can be made smaller, lighter, and cheaper than a reciprocating steam engine of comparable power. It can also be made in far larger sizes than the conventional steam engine. Mechanically, it has the advantage of producing rotating motion directly without the necessity of using a crankshaft or other means of transforming reciprocal to rotary motion. As a result, the turbine has generally supplanted the reciprocating engine as a prime mover in large steam-power installations, in electricity-generating plants, and on shipboard where space is at a premium.

The essential parts of all steam turbines are similar. They consist of nozzles or jets through which steam flows and expands, dropping in temperature, and gaining kinetic energy, and blades against which the swiftly moving steam exerts pressure. The arrangement of jets and blades, whether fixed or stationary, depends upon the type of turbine. In addition to these two basic components, turbines are equipped with wheels or drums upon which the blades are mounted, a shaft for these wheels or drums, an outer casing that confines the steam to the area of the turbine proper, and various pieces of auxiliary equipment, including lubrication devices and governors.

The simplest form of steam turbine is the so-called *impulse turbine*, in which the turbine jets are fixed in place on the inside of the turbine casing, and the blades are set on the rims of revolving wheels mounted on a central shaft. Steam passing through a nozzle passes over curved blades; these absorb some of the kinetic energy of the expanded steam, turning the wheel and shaft on which they are mounted. A complete turbine consists of a number of such stages set next to one another and designed so that steam entering at one end of the turbine expands through a succession of nozzles until it has lost most of its internal energy. In *reaction turbines*, mechanical energy is obtained to some degree by the impact of steam upon blades, but primarily it is obtained by the acceleration of the steam as it expands past nozzle like pairs of blades.

In order to use the energy available in steam efficiently in a turbine of either type, it is necessary to employ a number of stages, in each of which a small amount of thermal energy is converted to kinetic energy. If the entire conversion of energy took place in a single expansion stage, instead of successive stages, the rotative speed of the turbine wheel would be excessive. In general, reaction turbines have more stages than impulse turbines. A 50-stage reaction turbine is not uncommon, but the average impulse turbine may not exceed 10 stages. Because of the complications of building a turbine with a large number of stages, the two turbine principles are frequently combined in a single impulse-reaction turbine having one or two stages of impulse blading that are followed by 20 or more stages of reaction blading.

Because of the increase in volume as the steam expands through the various stages of a turbine, the size of the openings through which the steam passes must increase from stage to stage. In the practical engineering design of turbines, this increase is accomplished by lengthening the blades from stage to stage and by increasing the diameter of the drum or wheel upon which the blades are mounted. As a result, a complete turbine is more or less conical in shape, with its smallest diameter at the high-pressure, or intake, end, and its largest at the low-pressure or exhaust end.

The staging of an impulse turbine may be accomplished in several ways. In the simplest form, the turbine is pressure-staged, that is, arranged so that the drop in steam pressure is uniform in each stage. This type of staging is commonly called Rateau staging after the French engineer Auguste Rateau (1863–1930). Any stage of a pressure-staged turbine may also be velocity-staged by the use of two sets of movable blades behind each set of nozzles. These blades serve to reduce the velocity of the steam and absorb its energy. Between the moving blades, an additional set of fixed blades is placed. The purpose of these blades is to direct the steam at the proper angle for the second set of moving blades.

Although turbines are comparatively simple machines, having only one moving part, the shaft or rotor, a certain amount of auxiliary equipment is necessary for their operation. In turbines of the reaction type, a large thrust or pressure along the length of the turbine rotor is developed because of the drop in steam pressure at each stage. To compensate for this thrust, such turbines are equipped with devices known as dummy pistons. These are disks mounted on the rotor shaft in such a way that steam pressure can be applied to one side of the disk, creating a thrust in the opposite direction to that of the turbine itself. The speed of rotation of a turbine can be controlled in several ways. Small turbines are usually controlled by a throttle valve at the inlet to the turbine. In turbines having *Curtis* staging (three zone staging with first and third zones moving and middle zone stationary), the speed is varied by varying the number of inlet nozzles in use.

The efficiency of turbines is inherently high because comparatively little of the energy developed by the internal-energy drop in the turbine is lost. Losses due to friction and turbulence of steam flow inside the turbine are not actual losses; the energy given off heats the interior of the turbine, and this heat is passed along to the steam. Thus the actual efficiency of a turbine may be as high as 95% of the theoretical energy made available by the drop in temperature within the turbine. The overall efficiency of a boiler and turbine system is, however, much smaller, because a great deal of energy is lost in the exhaust steam that leaves the turbine.

REFERENCES

Combustion Engineering, Inc. 1940. *Steam Tables. Properties of Saturated and Superheated Steam,3rd Edition*. Windsor, CT: Combustion Engineering, Inc.

Farag, I., and Reynolds, J. 1994. *Heat Transfer, a Theodore Tutorial.* East Williston, NY: Theodore Tutorials. Originally published by the USEPA/APTI, RTP, NC. Roanoke, VA: ETS Professional Training Institute. https://etsi-inc.com/training/tutorials-and-textbooks/.

Keenan, J., and Keyes, F. 1936. *Thermodynamic Properties of Steam*. Hoboken, NJ: Wiley.

The Engineering ToolBox. 2003. *Thermodynamic Properties of Saturated Steam: Data & Charts in Bar*. [online] Available at: www.engineeringtoolbox.com/saturated-steam-properties-d_457.html. (accessed June 16, 2025).

Theodore, L. 2011. *Heat Transfer Applications for the Practicing Engineer*. Hoboken, NJ: Wiley.

Theodore, L. 2014. *Chemical Engineering: The Essential Reference*. New York City, NY: McGraw-Hill.

Theodore, L., and Barden, J. 1995. *Mass Transfer Operations, a Theodore Tutorial.* East Williston, NY: Theodore Tutorials. Originally published by the USEPA/APTI, RTP, NC. Roanoke, VA: ETS Professional Training Institute. https://etsi-inc.com/training/tutorials-and-textbooks/.

Theodore, L., and Ricci, F. 2010. *Mass Transfer Operations for the Practicing Engineer*. Hoboken, NJ: Wiley.

Theodore, L., Ricci, F., and VanVliet, T. 2009. *Thermodynamics for the Practicing Engineer*. Hoboken, NJ: Wiley.

12 Air Conditioning, Refrigeration, and Cryogenics

12.1 INTRODUCTION

The use of air conditioning in plants has become more and more common. Control of the temperature, humidity, and cleanliness of the air is very important in many processes. The comfort of workers is also important to efficient industrial organizations. This latter fact has led to the use of air conditioning in plants and offices where it is not essential for product quality. Air conditioning systems consist of centralized equipment that provides an atmosphere with controlled temperature, humidity, and purity at all times, regardless of weather conditions. A number of manufacturing processes require air conditioning for the control of conditions during manufacture.

Refrigeration is the process of producing cold conditions, particularly referring to cooling below atmospheric temperature. It is a vital factor in many chemical processes where cold or the removal of heat is necessary for optimum reaction control. Examples are the manufacture of azo dyes, the separation of an easily frozen product from liquid isomers or impurities, and products from the food and beverage industries. Emerging AI data centers utilizing vast quantities of heat generating server banks are another example of industries requiring large capacity heat dissipation systems. Refrigeration operations involve a change in phase in a body so that it will be capable of abstracting heat, exemplified by the vaporization of liquid ammonia and the melting of ice.

Mechanical refrigeration can be divided into two general types: *compression systems* and the *absorption systems*. Both systems cause the refrigerant to absorb heat at the low temperature by vaporization and to give up this heat at the higher temperature by condensation. The absorption system is used mainly in household units, but it finds economical industrial applications where exhaust steam is available.

Cryogenics is the study and use of materials at very low temperatures. The upper limit of cryogenic temperatures has not been agreed upon, but the National Bureau of Standards once suggested that the term cryogenics be applied to all temperatures below –150°C (–238°F or 123°K). Some regard the normal boiling point of oxygen (–183°C or – 297°F), as the upper limit. Cryogenic temperatures are achieved either by the rapid evaporation of volatile liquids or by the expansion of gases confined

DOI: 10.1201/9781003667759-14

initially at pressures of 150–200 atm. The expansion may be simple, that is, through a valve to a region of lower pressure, or it may occur in the cylinder of a reciprocating engine, with the gas driving the piston of the engine. The second method is more efficient but is also more difficult to apply.

Section numbers and their accompanying titles for this chapter are listed as follows:

12.2 AIR CONDITIONING

In popular usage, the term *air conditioning* often is applied improperly to air cooling. Many so-called air-conditioning units consist merely of blower-equipped refrigerating units that provide only a flow of cool, filtered air.

A number of manufacturing processes, such as those used in the production of paper, textiles, and printed matter, require air conditioning for the control of conditions during manufacture. Air conditioning of this kind usually is based on adjusting the humidity of the circulated air. When dry air is required, it is usually dehumidified by cooling or by dehydration. In the latter process it is passed through chambers containing adsorptive chemicals such as silica gel. Air is humidified by circulating it through water baths or sprays. When air must be completely free of dust, as is necessary in the manufacture of certain drugs and medical supplies, the air-conditioning system is designed to include some type of filter. The air is passed through water sprays or, in some filters, through a labyrinth of oil-covered plates; in others, dust is removed electrostatically by means of precipitators (Theodore 2008).

The design of an air-conditioning system depends on the type of structure in which the system is to be placed, the amount of space to be cooled, the number of occupants, and the nature of their activity. A room or building with large windows exposed to the Sun, or an indoor office space with many heat-producing lights, requires a system with a larger cooling capacity than an almost windowless room in which cool LED lighting is used. In homes or apartments, most of the cooled or heated air can be recirculated without discomfort to the occupants; but in laboratories or factories employing processes that generate noxious fumes, no air can be recirculated, and a constant supply of cooled or heated fresh air must be supplied.

Small air-conditioning units, which provide only ventilation, filtering, and air cooling, are rated commonly in terms of tons, British thermal units (Btu) per hour, and horsepower. A ton of cooling capacity is the amount of cooling furnished by the melting of 1 metric ton of ice in 1 hour and is equivalent to 12,000 Btu. The only accurate measurement of the cooling power of air conditioning units is a Btu rating. The horsepower rating is misleading because it represents the power of the motor rather than the cooling capacity of the unit. Air conditioners with the same horsepower ratings may differ as much as 4,000 Btu in cooling power based on their energy transfer efficiency.

12.3 REFRIGERATION

Although refrigeration is applied most extensively in the treating, transportation, and preservation of foods and beverages, it is used on a considerable scale in a variety of other industries. Applications to petroleum refining include lubricating oil purification, catalytic processes, and condensation of volatile hydrocarbons. Refrigeration is prominent in other industries concerned with the separation of volatile material by condensation. These include the production of nitrogen and oxygen from air, manufacture of ice, the dehydration of gases, and liquefaction of gases. The growth of air conditioning in houses, larger buildings, and large data centers has, in itself, created a great demand for refrigeration equipment, particularly small, compact units.

Refrigeration is defined as the process of lowering the temperature and maintaining it in a given space for the purpose of chilling foods, preserving certain substances, or providing an atmosphere conducive to bodily comfort. Storing perishable foods, furs, pharmaceuticals, or other items under refrigeration is commonly known as cold storage. Such refrigeration is effective because it checks both bacterial growth and adverse chemical reactions occurring under normal ambient conditions.

The use of natural or manufactured ice for refrigeration was widespread until shortly before World War I, when mechanical or electric refrigerators became available. Ice owes its effectiveness as a cooling agent to the fact that it has a constant melting point of 0°C (32°F). Foodstuffs maintained at this temperature or slightly above have an increased storage life. Solid carbon dioxide, known as dry ice, is also used as a refrigerant. Having no liquid phase at normal atmospheric pressure, it sublimes directly from the solid to a vapor at a temperature of –78.5°C (109.3° F). Because of its low temperature, dry ice is effective for maintaining products at low temperatures during the period of sublimation.

In mechanical refrigeration, constant cooling is achieved by the circulation of a refrigerant in a closed system, in which it evaporates to a gas and then condenses back again to a liquid in a continuous cycle. If no leakage occurs, the refrigerant lasts indefinitely throughout the entire life of the system. All that is required to maintain cooling is a constant supply of energy, or power and a method of dissipating waste heat. The two main types of mechanical refrigeration systems used are the *compression system*, used in domestic units for large cold-storage applications and for most air conditioning; and the *absorption system*, now employed largely for heat-operated air conditioning units.

12.3.1 Compression Systems

Compression systems employ four elements in their refrigeration cycle: a compressor, a condenser, an expansion valve, and an evaporator. In the evaporator the refrigerant is vaporized and heat is absorbed from the material contents or the space being cooled. The vapor next is drawn into a motor-driven compressor and is elevated to high pressure, which raises the temperature. The resulting superheated, high-pressure gas is then condensed to liquid in an air- or water-cooled condenser. The liquid flows from the condenser through an expansion valve, in which its pressure and temperature are reduced to the conditions that are maintained in the evaporator.

12.3.2 Absorption Systems

Household units, called *gas refrigerators*, operate on the absorption principle. In such gas refrigerators, a strong solution of ammonia in water is heated by a gas flame in a container called a generator, and the ammonia is driven off as a vapor, which passes into a condenser. Changed to a liquid state in the condenser, the ammonia flows to the evaporator as in the compression system. Instead of the gas being inducted into a compressor on exit from the evaporator, however, the ammonia gas is re-absorbed in the partially cooled, weak solution returning from the generator, to form the strong ammonia solution. This process of reabsorption occurs in a container called the absorber, from which the enriched liquid flows back to the generator to complete the cycle.

Increasing use of absorption refrigeration now occurs in refrigeration units for comfort space cooling, for which purpose refrigerant temperatures of 45°F–50°F (7.2°C–10°C) are suitable. In this temperature range, water can be used as a refrigerant with an aqueous salt solution, usually lithium bromide, as the absorbent material. The very cold boiling water from the evaporator is absorbed in concentrated salt solution. This solution is then pumped into the generator, where, at elevated temperatures, the surplus water is boiled off to increase the salt concentration of the solution; this solution, after cooling, recirculates back to the absorber to complete the cycle. The system operates at high vacuum at an evaporator pressure of about 7.5 torr (0.0099 atm); the generator and condenser operate at about 75 torr (0.099 atm).

In a broad sense, refrigeration is the science and art of producing and maintaining temperatures below that of the surrounding atmosphere. Low temperatures can be obtained in various ways, as follows (Theodore, Ricci and VanVliet 2009; Theodore 2014):

1. Allowing a phase change to take place in such a way as to extract heat; for example, the evaporation of a liquid such as water or ammonia or the melting of ice or the solution of a salt.
2. Expanding a compressed gas or vapor so that it does external work.
3. Expanding a gas via the Joule-Thomson effect.
4. Desorption of a gas.
5. Demagnetization of a solid.
6. Passage of an electric current through a bimetallic junction.

In fact, any reversible change involving the expenditure of work can be utilized to absorb heat and produce low temperatures. Method 1 is most commonly used for commercial refrigeration. It is also the basis for the use of the large number of cooling mixtures, of which salt and ice is one of the most common, that are useful for intermittent cooling on a small scale. Methods 2 and 3 are most commonly used for reaching very low temperatures on an industrial scale.

12.4 CRYOGENICS

Cryogenics is not, in itself, a stand-alone field. It is the extension of many other fields of science that delve into the realm of the thermodynamic property of temperature. When compared to room temperature, the properties of most substances change dramatically at extremely low temperatures. From a molecular perspective, the atoms in any substance at a lower temperature, while still vibrating, are compressed closer and closer together. Depending on the phase of the substance, various phenomena and changes to physical and chemical characteristics occur at these very low temperatures.

There are many accepted definitions of cryogenics as mentioned before based on temperature ranges and limits. A more elaborate explanation defines it as "the unusual and unexpected property variation appearing at low temperatures and which makes extrapolation from ambient to low temperature reliable" (Green and Perry 1998). Webster's dictionary defines cryogenics as "a branch of physics that deals with the production and effects of very low temperatures" (Merriam-Webster 2025). Cryogenics has also been referred to as "all phenomena, process, techniques, or apparatus occurring or using temperatures below 120K" (Green and Perry 1998). Combining all of these definitions, an all-purpose definition can be stated as "cryogenics is the study of the production and effects of materials at low temperatures" (Santoleri, Reynolds and Theodore 2000).

There have been uses for cryogenic technologies as far back as the latter part of the 19th century. It became common knowledge in the 1840s that in order to store food at low temperatures for long periods of time, it needed to be frozen, a technology that is still utilized today.

At the beginning of the 20th century, the scientist Carl von Linde produced a double distillation column process that separated air into pure streams of its basic components of 78% nitrogen, 21% oxygen, and 1% argon. By 1912, it was discovered that minor modifications to the double distillation column process could separate many other gases from an input stream. Since there are trace amounts of neon, krypton, and xenon in air, this distillation process, with minor modifications, was found capable of separating these gases into relatively pure streams. In the 1930s, the development of the sieve tray brought changes in cryogenic technology. (A sieve tray is a plate utilized in distillation columns with perforated holes about 5–6 mm in diameter, which enhances mass transfer (Theodore, Ricci, and VanVliet 2009)). These trays were highly popular in cryogenics due to their simplicity, versatility, capacity, and cost effectiveness.

When the "Space Race" hit the United States in the 1960s, cryogenic technologies were utilized to develop a process known as *cryopumping*, which is based on the freezing of gases on a cold surface. This process helped produce an ultrahigh vacuum here on the Earth that would be similar to what was to be experienced in outer space. This led to many other discoveries, including rocket propulsion technologies which enabled astronauts to better prepare for their voyages into space.

The *cryopreservation* process is based on the same principles as food storage (i.e., using extremely low temperatures to preserve a perishable item). Cryopreservation has become increasingly popular because of its appeal in preserving living cells. Whole cells and tissue can be preserved by this technique by stopping biological activity at extremely low temperatures. This preservation of organs by cryogenics has been a stepping stone for cryosurgery which relies on the cold temperatures to insure clean and precise incisions. More recently, cryobiology has been applied well below freezing temperatures to living organisms to observe how they "react." Most recently, incorporating electronic systems with cryo-technologies has provided valuable information on superconductivity. Extremely low temperatures have also led to systems that contain near-zero resistance throughout wires.

12.5 REFRIGERATION FUNDAMENTALS AND PRINCIPLES

Refrigeration and cryogenics have aroused considerable interest among those in engineering and science. All refrigeration processes involve work, specifically, the extraction of heat from a body of low temperature and the rejection of this heat to a body willing to accept it. Refrigeration generally refers to operations in the temperature range of 120K–273K.

In addition to being employed for domestic purposes (when a small "portable" refrigerator is required), refrigeration and cryogenic units have been used for the storage of materials such as antibiotics, other medical supplies, specialty foods, etc. Much larger cooling capacities than this are required in air-conditioning equipment. Some of these units, both small and large, are especially useful in applications that require the accurate control of temperature. Most temperature-controlled enclosures are provided with a unit that can maintain a space below ambient temperature (or at precisely ambient temperature) as required. The implementation of such devices led to the recognition that cooling units would be well suited to the refrigeration of electronic components and to applications in the field of instrumentation. Such applications usually require small compact refrigerators, with a relatively low amount of cooling power, where economy of operation is often unimportant.

One of the main cost considerations when dealing with refrigeration and cryogenics is the cost of building and powering the equipment. This is a costly element in the process, so it is important to efficiently transfer heat so that money is not wasted in lost heat in these units. Since the cost of equipment can be expensive, there are a number of factors to be considered when choosing equipment.

Insulation must certainly be considered as an integral part of any refrigeration or cryogenic unit. The extent of the problem of keeping heat out of a storage vessel containing a liquid refrigerant or a cryogenic liquid varies widely. Generally, one must decide on the permissible and/or allowable heat losses (leaks) since insulation costs money, and an economic analysis must be performed. Thus, the main purpose of insulation is to minimize radiative and convective heat transfer and to use as little material as possible in providing the optimal insulation. When choosing appropriate insulation, the following factors are normally taken into consideration (Kern 1950; Farag and Reynolds 1996; Theodore 2011):

1. Ruggedness
2. Convenience
3. Volume and weight
4. Ease of handling
5. Thermal effectiveness
6. Cost

The thermal conductivity (k) of a material is a major consideration in determining the thermal effectiveness of the insulation material. Different types of insulation obviously have different k values and there are five key categories of insulation. These include:

1. *Vacuum insulation* which employs an evacuated space that reduces radiant heat transfer.
2. *Multilayer insulation*, referred to by some as superinsulation, which consists of alternating layers of highly reflective material and low-conductivity insulation in a high vacuum.
3. *Powder insulation* which utilizes finely divided particulate material packed between surfaces.
4. *Foam insulation* which employs nonhomogeneous foam whose thermal conductivity depends on the amount of insulation.
5. *Special insulation* which includes composite insulation that incorporates many of the advantageous qualities of the other types of insulation.

It should also be noted that multilayer insulation has revolutionized the design of cryogenic refrigerant vessels.

The equipment necessary in refrigeration is also dependent upon many factors, including the working substances and fluids in the system. One very important part of refrigeration is the choice of refrigerant being employed and the refrigerant choice obviously depends on the system in which it will be used. The following criteria are usually considered in refrigerant selection:

1. Practical evaporation and condensation pressures
2. High critical and low freezing temperatures
3. Low liquid and vapor densities
4. Low liquid heat capacities
5. High latent heat of evaporation (enthalpy)
6. High vapor heat capacities

Ideally, a refrigerant should also have a low viscosity and a high coefficient of performance (COP) (to be defined later in this chapter). Practically, a refrigerant should have:

1 A low cost
2 Chemical and physical inertness at operating conditions
3 No corrosiveness toward materials of construction

4 Low explosion hazard
5 Be nonpoisonous and nonirritating

Solid refrigerants are not impossible to use but liquid refrigerants are most often used in practice. These liquid refrigerants include hydrocarbon and nonhydrocarbon refrigerants. The most commonly used hydrocarbon refrigerants include:

1. Propane
2. Ethane
3. Propylene
4. Ethylene

Nonhydrocarbon liquid refrigerants include:

1. Nitrogen
2. Ammonia
3. CO_2
4. Water

As noted previously, the maintenance of any temperature below that of the surroundings requires the continuous abstraction of heat at this temperature level and its rejection at some higher level. Thus, refrigeration is essentially an operation involving the pumping of heat from one temperature level to a higher one. This simple concept is of the greatest value in enabling one to visualize the essential elements of any process of refrigeration. In discussing the second law of thermodynamics it was shown that any reversible heat engine can also be regarded as a heat pump. An engine operating on the Carnot cycle taking in heat at absolute temperature T_L and rejecting it at T_H was shown to have the maximum efficiency that could be attained by any heat engine operating between the given temperature limits. The efficiency, which is the ratio of work done to heat absorbed at T_L, is given by:

$$\frac{W}{Q_L} = \frac{T_L - T_H}{T_L} \tag{12.1}$$

For a completely reversible cycle, the Carnot engine may be imagined to operate in the reverse direction, taking in heat at T_H and rejecting the amount Q_L at T_L. The same equation still applies; but since in heat pumping the interest is in knowing how much heat can be removed at the lower temperature with the expenditure of a given amount of work, Equation 12.1 is generally put in the form:

$$\frac{W}{Q_H} = \frac{T_L - T_H}{T_H} \tag{12.2}$$

Here T_H refers to a heat reservoir or refrigerator at a temperature below that of the surroundings and T_L to a heat reservoir or heat receiver at a temperature equal to or greater than that of the lowest temperature of any large mass in the surroundings (e.g., the atmosphere or any large body of water). Furthermore

$$W = Q_L - Q_H \tag{12.3}$$

which follows from the first law of thermodynamics.

Q_H/W, which is a measure of the refrigerating effect per unit of work done, is known as the *COP*, β. Equation 12.2 establishes the optimum performance that can be expected of any refrigerating machine operating between given temperature limits. No machine can be devised that will pump heat with a lower expenditure of work. Furthermore, Equation 12.2 implies that all reversible refrigerating machines operating between given temperature limits will have the same COP irrespective of fluid used or of mechanism employed. In other words, the coefficient depends only on the temperatures. These statements are, in essence, nothing but statements of the second law of thermodynamics.

12.6 CALCULATION DETAILS: THE REFRIGERATION CYCLE

The basic refrigeration cycle is shown in Figure 12.1. The cycle begins when a refrigerant enters the compressor as a low pressure gas [1]. Once compressed, it leaves as a hot, high pressure gas. Upon entering the condenser [2], the gas condenses to a liquid and releases heat to the outside environment, which may be air or water. The cool liquid then enters the expansion valve at a high pressure [3]; the flow is

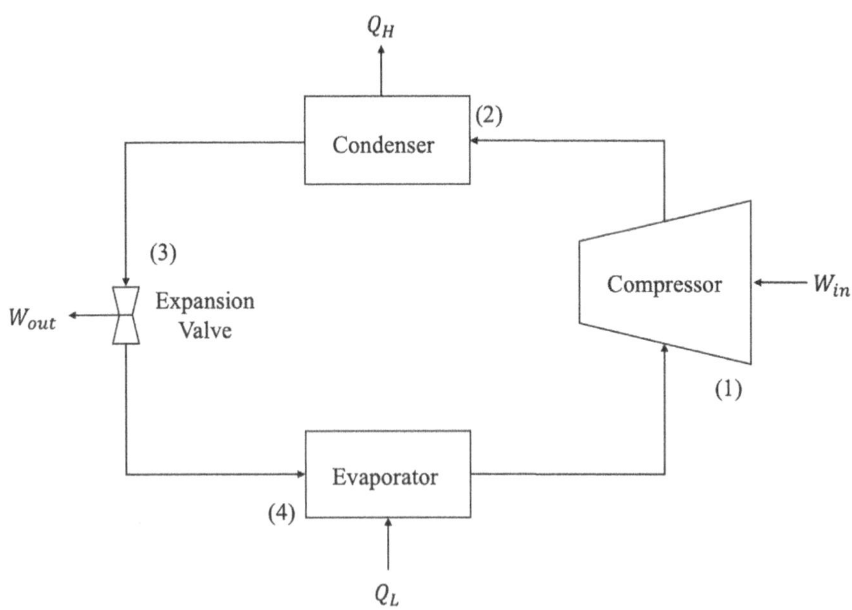

FIGURE 12.1 Basic components of a refrigeration system.

restricted and the pressure lowered. In the evaporator [4], heat from the source to be cooled is absorbed and the liquid becomes a gas. The refrigerant then repeats the process (i.e., the cycle continues).

In the refrigerator, the working fluid enters the evaporator in a wet condition and leaves dry and saturated (or slightly superheated). The heat absorbed, Q_L, by the evaporator can therefore be estimated by multiplying the change in the fluid's entropy, Δs, as it passes through the evaporator by the fluid's saturation temperature, T_s, at the evaporator pressure since the fluid's temperature will be constant while it is in a wet condition at constant pressure. Thus,

$$Q_L = T_s \Delta s \tag{12.4}$$

where Q_L = heat absorbed by the evaporator (cooling effect), kJ/kg; T_s = fluid saturation temperature at evaporator temperature, K; and Δs = fluid entropy change, kJ/kg-K.

The performance and energy saving ability of a refrigerator is measured in terms of the system's COP. This is defined as the heat removed at a low temperature, that is, the cooling effect, Q_L, divided by the work input, W_{in}, into the system:

$$\text{COP} = \beta = \frac{Q_L}{W_{in}} \tag{12.5}$$

where W_{in} = work input, kJ/kg.

The traditional system of units used in refrigeration is English units (i.e., Btu, etc.) but SI units are also acceptable. The cooling effect, Q_L, is equal to the change in enthalpy of the working fluid as it passes through the evaporator, and the work input, W_{in}, is equal to the increase in the working fluid's enthalpy as it passes through the compressor.

The performance of a steam-powered plant process can be measured in a manner somewhat analogous to the COP for the refrigeration system. The thermal efficiency, η_{th} of a work-producing cycle is defined as the ratio of work produced to the heat added. Thus

$$\eta_{th} = \frac{W_{net}}{Q_H} \tag{12.6}$$

where η_{th} = thermal efficiency, dimensionless; W_{net} = net work produced by the cycle, J/kg; and Q_H = heat added to the cycle, J/kg.

This can be rewritten as

$$\eta_{th} = \frac{W_{out} - W_{in}}{Q_H} \tag{12.7}$$

where W_{out} = work produced by the cycle, J/kg; and W_{in} = work consumed by the cycle, J/kg.

Note that for the power plant cycle, the compressor, evaporator, and expansion valve in Figure 12.1 are replaced by a turbine, boiler, and pump, respectively (see Figure 8.3), with both Q_L and Q_H as well as W_{in} and W_{out} reversed. When no velocity information is provided, velocity effects can be *neglected*. In addition, the change in enthalpy across the pump is often *neglected* since it is close to zero relative to the other enthalpy changes in the power plant system.

REFERENCES

Farag, I., and Reynolds, J. 1996. *Heat Transfer, a Theodore Tutorial.* East Williston, NY: Theodore Tutorials. Originally published by the USEPA/APTI, RTP, NC. Roanoke, VA: ETS Professional Training Institute. https://etsi-inc.com/training/tutorials-and-textbooks/.

Green, D., and Perry, R., ed. 1998. *Perry's Chemical Engineers' Handbook, 7th Edition.* New York City, NY: McGraw-Hill.

Kern, D. 1950. *Process Heat Transfer.* New York, NY: McGraw-Hill.

Merriam-Webster. 2025. Cryogenics. In *Merriam-Webster.com Dictionary.* www.merriam-webster.com/dictionary/cryogenics. (accessed July 2, 2025).

Santoleri, J ., Reynolds, J., and Thedore, L. 2000. *Introduction to Hazardous Waste Incineration, 2nd Edition.* Hoboken, NJ: Wiley.

Theodore, L. 2008. *Air Pollution Control Equipment Calculations.* Hoboken, NJ: Wiley.

Theodore, L. 2011. *Heat Transfer Applications for the Practicing Engineer.* Hoboken, NJ: Wiley.

Theodore, L. 2014. *Chemical Engineering: The Essential Reference.* New York, NY: McGraw-Hill.

Theodore, L., Ricci, F., and VanVliet, T. 2009. *Thermodynamics for the Practicing Engineer.* Hoboken, NJ: Wiley.

13 Power Plant Thermodynamic Cycle Analysis

13.1 INTRODUCTION

Thermodynamics is concerned with the mechanics of heat and especially with the conversion of heat into work. There are two basic principles involved which have become firmly established in the scientific field as the first and second laws of thermodynamics. These are once again briefly discussed in the following text.

According to the law of the conservation of energy, or the first law of thermodynamics, energy can be neither created nor destroyed. Thus, in any device or system into which energy is put and out of which some energy is taken, this principle means that an exact balance must prevail between the input and output quantities. There can be no excess or deficiency. In addition, energy may be converted within the device from one form to another, such as heat to work, but there must be complete conservation of the aggregate.

The general energy equation, however, has one very serious shortcoming – it tells practically nothing of the possible extent of conversion of one form of energy into another in many kinds of apparatus found in practice (Theodore, Ricci and Van Vliet 2009; Theodore 2014). Thus, if 100 units of heat were added to a system, how much could be converted into work? According to the first law; not more than the 100 units could be converted, but nothing is said about how much less than the 100 units would actually be converted. One purpose of the second law of thermodynamics is to rigorously define the extent of possible conversion of heat into work. This concept is very broad and is most useful in not only fuel-burning power plants but also geothermal power plants. Therefore, to define the extent of the possible conversion of heat energy into work, the first law of thermodynamics and the general energy equation must be supplemented by the second law. By using the concept of entropy, it is possible to demonstrate the extent of this possible conversion. For ideal processes this property (entropy) is defined mathematically (assuming unit mass of the substance) as

$$ds = dQ / T \tag{13.1}$$

where ds= entropy change, kJ/kg-K; dQ = heat added, kJ/kg; and T = absolute temperature at which heat is added, K.

DOI: 10.1201/9781003667759-15

Entropy can be more generally defined as the property which measures that portion of the heat added which cannot be converted into work no matter how nearly perfect the operation may be. Or, expressed in another way, it is the measure of energy *unavailability*.

The reader is directed to Chapters 7 and 8 to revisit conservation law and thermodynamic concepts presented there. Section numbers and their accompanying titles for this chapter concerned with power plant thermodynamic cycle applications are listed as follows:

13.2 THE FIRST LAW OF THERMODYNAMICS

The first law can be expressed in equation form for a power plant system in Equation 13.2, and pictorially in Figure 13.1. To account for the energy flow, in at Point 1 and

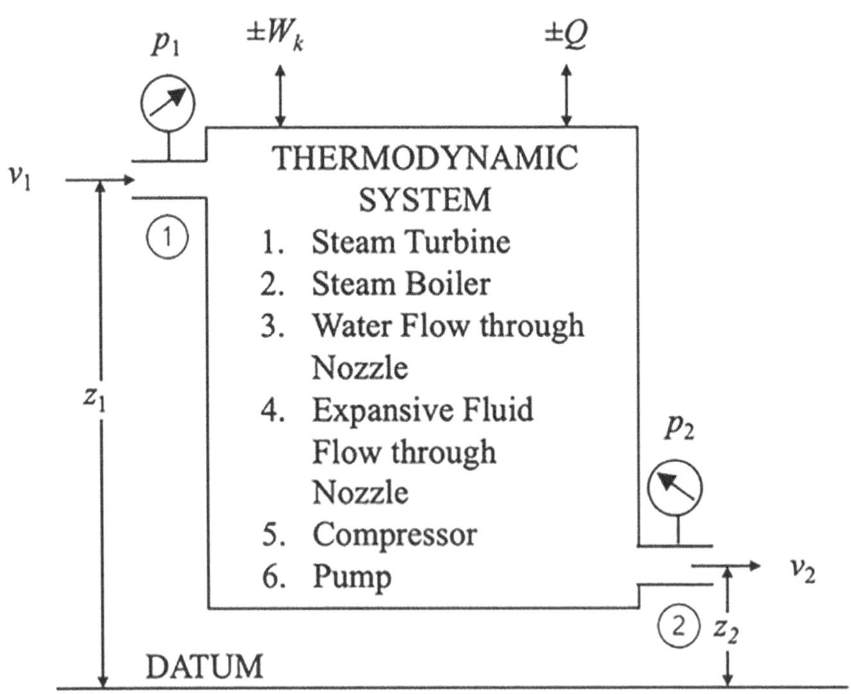

FIGURE 13.1 Diagram to illustrate the application of the general energy equation to thermodynamic systems.

out at Point 2 in Figure 13.1 – which may represent a boiler, a turbine a compressor, or any thermodynamic system – Equation 13.2 can be written assuming for convenience that 1 lb of working fluid moves through the device (Reynolds, Jeris and Theodore 2002; Abulencia and Theodore 2014).

$$\frac{v_1^2}{2g_c} + p_1v_1 + E_1 + z_1 \pm W_k \pm Q = \frac{v_2^2}{2g_c} + p_2v_2 + E_2 + z_2 \qquad (13.2)$$

This is the general energy equation, in which: v = the flow velocity through the intake and discharge pipes, ft/s; g_c = gravitational constant, (lb/lb$_f$)(ft/s^2); $V^2/2g$ = the kinetic energy, ft-lb$_f$/lb; p = the static pressure under which the fluid exists at the inlet and outlet pipes, (lb$_f$/ft^2); v = the specific volume of the fluid, ft^3/lb; pv = the flow work, ft-lb$_f$/lb; E = the internal, or intrinsic energy of the fluid, Btu/lb; z = the elevational head (potential energy), ft-lb$_f$/1b; W_k = work done on or by the fluid while it is in the device, ft-lb$_f$/1b; and Q = heat added to or abstracted from the fluid while it, is in the device, Btu/lb. Note that there is an equivalence in heat and work units i.e., 1 Btu = 778 ft-lb$_f$.

In using the general energy equation it is essential that all terms be expressed in a consistent system of units. It is not customary to measure all the energy items in the same units. The heat unit in use today is the Btu, and the conversion to ft-lb$_f$ given above serves to give the necessary consistency in the use of Equation 13.2, which may now be rewritten as:

$$\frac{v_1^2}{2g_c(778)} + \frac{p_1v_1}{778} + E_1 + \frac{z_1}{778} \pm \frac{W_k}{778} \pm Q = \frac{v_2^2}{2g_c(778)} + \frac{p_2v_2}{778} + E_2 + \frac{z_2}{778} \qquad (13.3)$$

The terms pv and E, representing, respectively, the flow work and the intrinsic energy of a fluid, can be treated conveniently as a group. They are then usually expressed as a single symbol, H, called enthalpy, thus:

$$H = E + pv/778,\ \text{Btu/lb} \qquad (13.4)$$

Enthalpy values for water and steam are available in the steam tables provided earlier in Chapter 11.

The general energy equation can be applied to specific equipment (Figure 13.1) and examples are detailed in the next section.

13.3 SPECIFIC EQUIPMENT CALCULATIONS

Specific equipment calculations, as they relate to geothermal energy equipment shown in Figure 13.1, are presented as follows.

13.3.1 Steam Turbine

For this system, $\Delta v = 0$, $\Delta z = 0$, and $Q = 0$. This leads to the following result when these values are substituted into Equation 13.3.

$$\frac{p_1 v_1}{778} + E_1 = \frac{p_2 v_2}{778} + E_2 + \frac{W_k}{778} \tag{13.5}$$

which may be rewritten as

$$\Delta H = \frac{W_k}{778} \tag{13.6}$$

13.3.2 Steam Boiler

For this system, Δv = 0, $\Delta z = 0$, and W_k =0. This leads to the following result when these values are substituted into Equation 13.3.

$$Q = \Delta H \tag{13.7}$$

13.3.3 Water Flow through Nozzle

For this system, $\Delta z = 0$, $\Delta v = 0$ as water is considered noncompressible, ΔE = 0, W_k = 0, and $Q = 0$. This leads to the following result when these values are substituted into Equation 13.3.

$$\frac{\left(\mathrm{v}_2^2 - \mathrm{v}_1^2\right)}{2g_c} = (p_1 - p_2)v \tag{13.8}$$

13.3.4 Expansive Fluid Flow through Nozzle

Here, for steam or air that would be considered expansive fluids, Δv and ΔE are no longer considered equal to 0. For this system, $\Delta z = 0$, $W_k = 0$, and $Q = 0$. This leads to the following result when these values are substituted into Equation 13.3.

$$\frac{\left(\mathrm{v}_1^2 - \mathrm{v}_2^2\right)}{2g_c} = \Delta H \tag{13.9}$$

13.3.5 Compressor

For this system as with the steam turbine, Δv = 0, $\Delta z = 0$, and $Q = 0$. This leads to the following result when these values are substituted into Equation 13.3.

$$\Delta H = \frac{W_k}{778} \tag{13.6}$$

13.3.6 Pump

For this system pumping a noncompressible liquid such as water, $\Delta v = 0$, $\Delta E = 0$, $\Delta v = 0$, $\Delta z = 0$, and $Q = 0$. This leads to the following result when these values are substituted into Equation 13.3.

$$W_k = (p_2 - p_1)v \tag{13.10}$$

13.4 THE SECOND LAW OF THERMODYNAMICS

The Cases 1–6 noted previously in Section 13.3 represent some of the basic manipulations which can be made in applying the general energy equation to specific equipment. The great significance of the property of enthalpy is amply demonstrated by the examples cited. All of these manipulations are based on the first law of thermodynamics. They account for all energy added or abstracted, and the conclusions satisfy the requirement for an exact balance (thermal) between the energy supplied to a system and the energy rejected from it.

The limitations imposed by the second law of thermodynamics are perhaps best illustrated by considering the temperature–entropy (*T-S*) diagram, Figure 13.2, where absolute temperature is plotted vertically, entropy horizontally, and areas represent heats added (+) or rejected (–). Consider any two temperatures such as T_H, which is the temperature of a heat source, and T_L, the temperature of a heat sink, which for

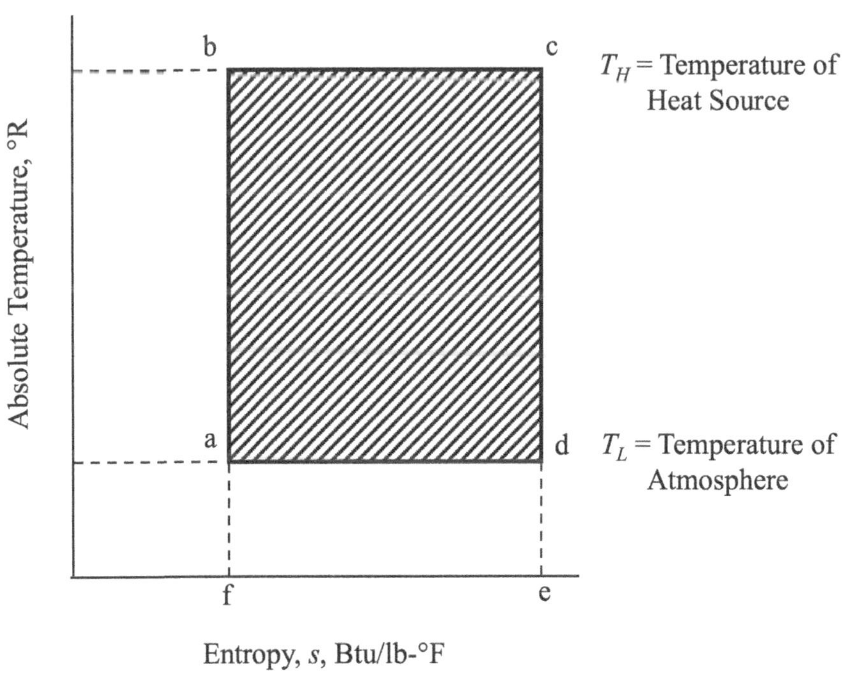

FIGURE 13.2 Carnot cycle temperature–entropy diagram.

practical purposes is limited to a minimum value set by atmospheric temperature. Then by Carnot's principle, the maximum area which can be constructed on the *T–S* diagram between these two temperatures is the rectangle shown. The crosshatched area represents the *ideal* work of this cycle. The area under T_H represents the heat added from the source, and the area under T_L represents the heat rejected to the sink. The thermal efficiency of any ideal cycle is therefore defined as:

$$\text{Thermal Efficiency}, \eta = \frac{\text{Work Done}}{\text{Heat Added}} \tag{13.11}$$

By the first law this can be written as:

$$\text{Thermal Efficiency}, \eta = \frac{\text{Heat Added} - \text{Heat Rejected}}{\text{Heat Added}} \tag{13.12}$$

For the Carnot cycle this is resolved to form:

$$\text{Thermal Efficiency}, \eta = \frac{T_H - T_L}{T_H} \tag{13.13}$$

The value of efficiency given by Equation 13.13 is a rigorous maximum limit imposed by the second law of thermodynamics. No cycle can have a better efficiency between specified temperature limits than the value given for the Carnot cycle. Other cycles may equal it, but none can exceed it.

Because the atmosphere is the usual thermodynamic sink and its temperature is of the order of 500°R, it is evident from Equation 13.13 that the conversion efficiency of heat into work is substantially less than 100%. By what extent it is less is measured by the energy which must be discarded as heat. This is the unavailable portion which cannot be converted into work.

In the cycles of real thermal plants there is practically only one source of heat (T_H) – the heat source and one sink (T_L) – the atmosphere, to which the heat residue must be discarded. As an example, these temperatures can be taken as 3,000°R and 500°R, respectively. Any ideal heat engine working between these temperatures will then have a limiting thermal efficiency as set by the Carnot cycle. From Equation 13.13 this theoretical efficiency limit is

$$\text{Thermal Efficiency}, \eta(\text{fractional basis}) = \frac{3,000 - 500}{3,000} = 0.83 = 83\%$$

No power plant actually even approaches this value. The best steam plants and internal combustion engines give thermal efficiencies between 0.35 and 0.40 over-all. These are the best values obtained in practice, and the average is often lower.

13.5 THERMODYNAMIC ANALYSIS OF A POWER PLANT CYCLE

The power plant cycle received treatment earlier in 8. The line diagram of a typical cycle was given in Figure 8.3 and is show once again in Figure 13.3 for convenience of the reader. Each step in the process can be analyzed by using the equations in the previous section and the thermodynamic properties of steam. This analysis will illustrate how thermodynamics may be used to calculate work and heat effects in terms of changes in the properties of the system.

13.5.1 The Heat Absorbed in the Boiler

The primary purpose of the steam boiler is to transfer heat to the water; energy effects accompanying changes in elevation and velocity are small. Also, there is no mechanical energy exchange with the surroundings so that the work term disappears. The total energy balance equation then reduces to the following form:

$$Q = \Delta H \tag{13.7}$$

In other words, the heat absorbed by the water is equal to its increase in enthalpy. From the state of the steam entering and leaving the boiler this change in enthalpy can be obtained directly from the steam tables provided in Chapter 11.

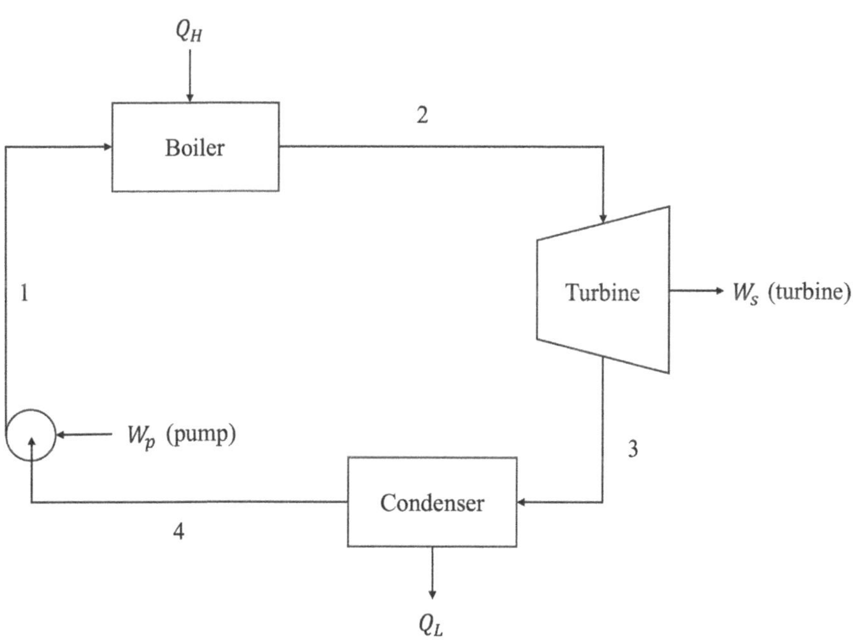

FIGURE 13.3 Steam power plant diagram.

That the kinetic and potential energy effects are negligible in comparison with the heat may be seen by considering a numerical example. Suppose the liquid water velocity entering the boiler is 5 ft/s and the steam velocity leaving is 200 ft/s. The kinetic-energy term $\Delta v^2/2g_c$ would be approximately 0.8 Btu/lb of steam. Similarly, a change in elevation of as much as 100 ft would amount to only about 0.1 Btu/lb. Both these quantities are negligible in comparison with the heat absorbed in the boiler, which would be of the order of 1,000 Btu/lb of steam.

13.5.2 The Work-Producing Step

By applying the total energy balance to the turbine process and assuming adiabatic operation, the following result is obtained:

$$W_s = -\Delta H \tag{13.14}$$

This again neglects changes in kinetic energy and potential energy. Even in a turbine, which depends upon the conversion of kinetic energy into mechanical energy of a rotating shaft, the overall changes in kinetic energy are relatively small, that is, the differences in velocity of the steam in the inlet line to the turbine and in the exhaust line from the turbine are such that the kinetic energy effect is *small* in comparison to the work done. Equation 13.14 indicates that the work obtained is entirely at the expense of the change in enthalpy of the expanding steam. It may be evaluated per pound of steam directly from the initial and final states by using the steam tables. The work is equal to the change in enthalpy (except for sign) even though the process is not reversible. The only requirement is that the path be adiabatic. However, if the steam in a turbine expands in an irreversible manner, the change in enthalpy, and therefore the work, is less than if the process had been reversible to the same final pressure.

13.5.3 The Condenser

In the condenser the heat transferred from the condensing steam to the cooling water is equal to the change in enthalpy of the steam and may be evaluated from enthalpy values corresponding to conditions at the inlet and discharge of the condenser.

The heat transferred in the condenser as well as other heat and work effects can be conveniently represented on a temperature–entropy diagram. Referring to Figure 13.2, heat rejected to the condenser can be designated by the area a–d–e–f. Likewise, the heat absorbed in the boiler can be indicated by the area b–c–e–f. By the total energy balance, the work produced must equal the difference between the two areas, or a–b–c–d. In the case of irreversible operation in the expansion process, the heat rejected to the condenser is increased. Since the heat absorbed in the boiler is the same as in the reversible case, there is a loss in work. It will be noted that this loss is also given by the following expression:

$$W_{\text{lost}} = \int T\, dS \tag{13.15}$$

in which T is the variable temperature at which the heat is transferred. This is in agreement with the general principle developed in connection with the discussion of the second law, that the loss in work accompanying an irreversible process is equal to the total change in entropy accompanying that process times the lowest temperature at which heat can be discarded.

13.5.4 The Pumping Process

If the pump returning the liquid water to the boiler operates adiabatically, the total energy balance becomes

$$-W_p = \Delta H \tag{13.16}$$

From a mechanical energy balance the work of the pump can be expressed in the following manner:

$$-W_p = \int v\,dp + I(\text{Irriversibilities}) \tag{13.17}$$

Since the specific volumes of liquids are small, the pump work will also be small even for relatively large changes in pressure. In general this work can be *neglected* in making an overall analysis of the cycle. If necessary, W_p can be estimated for a reversible process by considering the specific volume to be constant and multiplying it by the change in pressure. However, for a pressure change of 1,000 psia, W_p amounts to about 3 Btu/lb.

13.6 POWER PLANT MONITORING NEEDS

In the last 100 years, progress in instrumentation has resulted in a more accurate and thorough presentation of information to the operating personnel. Early in the 20th century, boilers were operated almost without benefit of instruments. Units were small, pressures and temperatures were low, and firing was mainly by hand or by stokers. The development and availability of more and better instrumentation not only paralleled the needs of the boiler industry but also stimulated the development of larger units. The rising cost of steam production has caused both user and supplier to redouble their efforts to improve the efficiency of the conversion of thermal energy to power.

It is recommended that all steam boilers should be provided with instruments to indicate or record the following variables:

Steam pressure
Steam temperature (if other than saturated)
Water level
Feedwater pressure
Furnace draft or pressure

The advantages of good instrumentation are:

1. Adequate presentation of all key factors indicates readily any malfunctioning of the unit. Operators can then make necessary adjustments before damage to equipment or injury to personnel occurs.
2. Accurate presentation of information allows engineers to detect deterioration of unit performance and to determine the maintenance or operation procedures necessary to rectify the conditions. Records of load and fuel consumption facilitate the efficient distribution of steam load and monitor the extent of capacity utilization.
3. Combustion guides enable personnel to maintain optimum fuel–air ratios and minimize fuel consumption.

While it is possible to operate a steam boiler with indicators and recorders alone, few units can be run efficiently in this manner. Many of the dampers and valves that control the flow of air, fuel, and water require substantial forces for their rapid manipulation. Controls relieve the boiler attendant of these tasks and permit them to devote their time to more "rewarding" problems, such as burner pattern. Manual operation is seldom relied upon because of the extremely flexible control components and computerized systems that are now available.

While each steam boiler installation should be analyzed for specific requirements, regulation is normally provided for:

Fuel flow
Combustion-air flow
Feedwater flow
Furnace draft or pressure
Steam temperature

The advantages obtained by proper application of controls are:

1. The ability to position remotely any of the regulating devices in the system. It is difficult to maintain an adequate operating force to manipulate satisfactorily all of the variables when changing load demands are to be met or when malfunctioning occurs.
2. The reduction of production costs through lower maintenance.
3. Better supervision of operation.
4. The continuous maintenance of optimum plant efficiency.
5. The detection, through system sensitivity, of abnormal conditions that are difficult or impossible to detect with visual observation alone. Automatic correction refines the accuracy of adjustment and improves both efficiency and safety.
6. The release of operating personnel for observations and decisions rather than for routine adjustments.

Instrumentation components as applied to the boiler must be designed for maximum dependability and repeatability. The equipments generally used in the industry are listed as follows:

1. Temperature measurement
2. Pressure measurement
3. Flow measurement
4. Fluid-level measurement

REFERENCES

Abulencia, P., and Theodore, L. 2014. *Fluid Flow for the Practicing Chemical Engineer.* Hoboken, NJ: Wiley.

Reynolds, J., Jeris, J., and Theodore, L. 2002. *Handbook of Chemical and Environmental Engineering Calculations.* Hoboken, NJ: Wiley.

Theodore, L. 2014. *Chemical Engineering: The Essential Reference.* New York, NY: McGraw-Hill.

Theodore, L., Ricci, F., and VanVliet, T. 2009. *Thermodynamics for the Practicing Engineer.* Hoboken, NJ: Wiley.

Part 3

Geothermal Energy Applications

The objective of the third part of this book is concerned with geothermal energy engineering and related subject matter. Chapter numbers and their accompanying titles are listed as follows:

Chapter 14. Deep Well Construction Technologies
Chapter 15. Geothermal Energy: The Quintessential Energy Resource
Chapter 16. Geothermal Energy Conversion Processes
Chapter 17. Geothermal Energy Environmental Impacts and Considerations

DOI: 10.1201/9781003667759-16

14 Deep Well Construction Technologies

14.1 INTRODUCTION

Perhaps the major challenge in the geothermal energy field is the construction of deep wells, their accompanying environmental concerns, and their associated costs in order to reach the energy deep in the Earth's surface that will allow the advanced development of this abundant energy source. The cost of construction has long been an impediment to widespread acceptance of geothermal energy as a viable option in the energy field. Recent developments in the deep well construction field have, however, resulted in an exponential decrease in their costs. This has paved the way for the serious consideration of geothermal energy as a viable part of the solution to future growing energy needs.

The Earth's core is hot enough to satisfy the planet's energy needs indefinitely, and the Earth's core is always conducting its thermal energy from its center to the Earth's surface. Geothermal wells work by tapping into the heat below the Earth's surface. In simple terms, holes are drilled, and then the heat is extracted. A small well for a home or business would see holes drilled and pipes placed around 100–300 feet below the ground, where it is warmer than above ground. Water or a mixed solution passes through the pipes, heating up underground. A heat pump system then brings the heated liquid into a building and pumps the cooler liquids back underground to be reheated. On an industrial scale, holes are drilled up to several kilometers deep. Geothermal power plants are looking to tap into the heat and bring that heat to the surface as hot water or steam to drive turbines and create electricity.

Accessing the vast, high-temperature geothermal resources located deep within the Earth is crucial for expanding geothermal energy production beyond conventional hydrothermal sites. Deep wells, often exceeding depths of 3 km (approx. 10,000 ft or 2 miles) and encountering temperatures above 200°C (392°F), are required for advanced systems like Enhanced Geothermal Systems (EGS), Advanced Geothermal Systems, and Superhot Rock Geothermal Systems (IEA 2024). Drilling these wells presents significant technological challenges compared to conventional oil and gas or shallower geothermal drilling, primarily due to extreme temperatures, hard and abrasive crystalline rock formations, and the associated high costs (U.S. DOE 2025a). Overcoming these challenges through technological innovation is a key focus of

DOI: 10.1201/9781003667759-17

government research and industry efforts aiming to make deep geothermal resources economically viable contributors to a clean energy transition (U.S. DOE 2022).

Despite recent progress, deep geothermal drilling remains expensive, often accounting for over 50% of a project's total cost (U.S. DOE 2025a). Key challenges driving current research are being addressed by the U.S. Department of Energy's (DOE's) Geothermal Technologies Office through a variety of programs including the Frontier Observatory for Research in Geothermal Energy (FORGE) site in southwest Utah, Drilling Demonstrations Initiative, and the Wells of Opportunity programs (US DOE 2025b).

It is fair to say that advancing well drilling technology is paramount to unlocking the full potential of deep geothermal energy resources. While adaptations to conventional rotary drilling have enabled progress, significant cost reductions and performance improvements, particularly in extreme temperature and hard rock conditions, are needed. Emerging and novel drilling methods like thermal spallation and directed energy (microwave or laser high-energy beams) offer promise for fundamentally changing how deep wells are constructed. Continued investment in research, development, and demonstration, strongly supported by government initiatives, is critical to overcoming the existing technological barriers and making deep geothermal energy a widespread, cost-effective clean energy source.

Section numbers and their accompanying titles for this chapter on geothermal deep well construction technologies are listed as follows:

- 14.2 The Planet Earth
- 14.3 Introduction to Drilling Techniques
- 14.4 Geothermal Well Drilling: Challenges and Technological Advances
- 14.5 References

14.2 THE PLANET EARTH

The Earth is the third planet in distance from the Sun and the fifth largest of the planets in the solar system in terms of its diameter. The mean distance of the Earth from the Sun is 149,503,000 km (92,897,000 mi). Its circumference is approximately 40,000 km (25,000 mi) with a surface area of approximately 500,000,000 km^2 (200,000,000 mi^2). Its volume, specific gravity, and mass are 1×10^{12} km^3 (260×10^9 mi^3), 5.5, and 6×10^{21} metric tons, respectively.

The Earth is considered to consist of four parts: the atmosphere, lithosphere, hydrosphere, and barysphere. The *atmosphere* is the gaseous envelope that surrounds the solid body of the planet. Although it has a thickness of more than 1,100 km (more than 700 mi), about half its mass is concentrated in the lower 5.6 km (3.5 mi). The *lithosphere* is the solid rocky crust of the Earth, extending to a depth or perhaps 40 km (25 mi). The *hydrosphere* is the layer of water that, in the form of the oceans, covers approximately 70% of the surface of the Earth. The *barysphere*, sometimes called the *centrosphere*, is below the lithosphere. It is the heavy interior of the Earth, constituting more than 99.6% of the Earth's mass.

The rocks of the lithosphere have an average specific gravity of 2.7 and are almost entirely made up of 11 elements, which together account for about 99.5% of its mass. The most abundant element is oxygen (about 46.60% of the total), followed by silicon (about

27.72%), aluminum (8.13%), iron (5.0%), calcium (3.63%), sodium (2.83%), potassium (2.59%), and magnesium (2.09%). Titanium, hydrogen, and phosphorus make up the balance of these major elements (totaling less than 1%). In addition, 11 other elements are present in trace amounts of from 0.1% to 0.02%. These elements, in order of abundance, are carbon, manganese, sulfur, barium, chlorine, chromium, fluorine, zirconium, nickel, strontium, and vanadium. The elements are present in the lithosphere almost entirely in the form of compounds rather than in their free state. The most common compounds of the Earth's crust are silicates and aluminosilicates of various metals.

The lithosphere is divided into two shells, each of which has a density that increases with depth. The innermost shell contains such rock types as basalt and gabbro and has an average specific gravity of approximately 3. The upper shell is made up of rocks of the granite type with a specific gravity of about 2.7. This structure of higher density basalt lying under a lighter, granitic mass may be compared to a continent floating on a basaltic sea. The surface of the Earth is largely covered with sedimentary rocks and soil.

The hydrosphere consists chiefly of the oceans but technically includes all water surfaces in the world, including inland seas, lakes, rivers, and underground water. The average depth of the oceans is 3,800 m (12,500 ft), more than five times the average height of the continents. The mass of the oceans is approximately 1.35×10^{18} metric tons, or about 1/4,400 of the total mass of the Earth.

The barysphere is divided into an outer zone, called the *mantle*, and an inner zone, called the *core*. The mantle extends from the base of the lithosphere down to approximately 3,000 km (approximately 1,800 mi). It is solid and has a specific gravity that increases with depth, varying from 3.0 to 6.0.

According to current theory, the outer part of the mantle is composed of iron and magnesium silicates. The lower part probably consists of a mixture of magnesium, silicon, and iron oxides. Seismological research has shown that the core which has a radius of some 3,500 km (about 2,200 mi) lacks rigidity and probably is liquid, that is, it consists of molecules having free movement among themselves but tending not to become widely separated. Some geophysicists believe it to be highly compressed molten iron.

Modern knowledge of radioactivity makes possible the absolute dating of certain rocks with considerable accuracy, as the disintegration of a radioactive element occurs at a rate independent of the chemical or physical environment. For example, uranium-238 decays to lead-206 at a rate that results in the reduction of one-half of the uranium to lead in 7.6 billion years. Thus, by careful comparison of the amounts of these isotopes of uranium and lead present in a rock, it is possible to determine the age of the rock. The oldest rocks so far studied are more than 3.5 billion years old. The maximum time required for the formation of all the elements in the crust of the Earth has been determined to be about 5.5 billion years. The age of the Earth, therefore, is probably between 3.5 and 5.5 billion years.

14.3 INTRODUCTION TO DRILLING TECHNIQUES

The idea of using a rotary drill bit is not new, as archeological records show that as early as 3000 BC, the Egyptians may have been using a similar technique. Leonardo

Di Vinci, as early as 1500, developed a design for a rotary drilling mechanism that bears much resemblance to technology used today. Despite these precursors, rotary drilling did not rise in use or popularity until the early 1900s (ISWD 2011). Although rotary drilling techniques had been patented as early as 1833, most of these early attempts at rotary drilling consisted of little more than a mule, attached to a drilling device, walking in a circle. It was the success of the efforts of Anthony Lucas and Patillo Higgins in drilling their 1901 Spindletop well in Texas that solidified rotary drilling as the premier petroleum drilling technology (ISWD 2011).

Traditional deep well drilling technologies include *rotary drilling*, *percussion drilling*, and *rotary–percussion drilling*. These methods can be used to drill for not only geothermal energy but also oil and gas and water. Each of these techniques is briefly described as follows.

14.3.1 Rotary Drilling

A rotary drill consists of an auger like bit fastened to a series of connected steel pipes (drill string). The rotating drill bit cuts through the earth while a drilling fluid is circulated to cool the bit and remove the cuttings. It is used in drilling to great depths, as in drilling water wells and oil wells both on land and at the bottom of the sea.

14.3.2 Percussion Drilling

Percussion drilling, also known as cable drilling, is a method of creating boreholes by repeatedly dropping a heavy, sharp tool (like a chisel or bit) onto the material to be drilled. This action fractures and breaks up the material, which is then removed from the borehole using a bailer or other tools, often after adding water to create a slurry. It is particularly effective for drilling in hard rock materials and can be used both manually and with powered equipment such as winches, compressed air, or other power sources to lift and drop the drilling tool to produce deeper and deeper holes.

14.3.3 Rotary–Percussion Drilling

Rotary–percussion drilling uses a combination of a rotating drill string and a hammering action to create boreholes in a variety of ground conditions, including consolidated and broken formations, as well as transitional hard and soft formations. Either a top hammer outside the borehole or a down-the-hole hammer inside the borehole can be used to drive the drill bit. Compressed air or water is used as a drilling fluid to flush the cuttings and cool the bit.

While these techniques are generally as applicable for geothermal wells as they are for oil and gas production wells for which they were developed, the nature of geothermal formations that represent the highest geothermal energy production potential have unique characteristics that set them apart from oil and gas reservoirs, making drilling in them particularly challenging. Most oils and gas reservoirs are in sedimentary formations while geothermal reservoirs are found in granite, quartzite, basalt, rhyolite, and volcanic tuff (Finger and Blankenship 2010) that are hot, hard, abrasive, highly fractured, and under pressured. They often contain corrosive fluids and high dissolved solids making drilling difficult, slow, and producing conditions highly aggressive on bit, casing, and other equipment wear. These challenges, along with the loss of drilling fluid to pores or fractures in the rock formations being drilled, have encouraged the pursuit of innovative drilling techniques adapted to the rigors of geothermal well placement to increase the speed and reliability of drilling in these harsh conditions to reduce the overall cost of well development for geothermal energy systems.

14.4 GEOTHERMAL WELL DRILLING: CHALLENGES AND TECHNOLOGICAL ADVANCES

14.4.1 Challenges

Drilling for geothermal energy is different from traditional oil and gas drilling due to a number of very specific challenges. These include:

1. *Extreme Temperature and Pressure Conditions.* At depths of 3,000 to greater than 5,000 m temperatures can exceed 300°C, and pressures can reach hundreds of bars. To manage these extreme conditions two basic items are needed: (1) heat-resistant equipment, especially for drill motors and drilling sensors; and (2) *special drilling fluids* that can handle high temperatures without becoming unstable. Water-based muds are common but require specialized high-temperature additives (polymers, viscosifiers, fluid loss control agents) to maintain desired properties (viscosity, density, cooling capacity) above 150°C (300°F) (Finger and Blakenship 2010). Synthetic-based fluids offer higher thermal stability but are more expensive and raise environmental concerns. Preventing thermal degradation of fluid components and ensuring compatibility with reservoir chemistry are critical.
2. *Geological Formation Challenges.* The Earth's crust can contain hard rocks (like granite or basalt), which make drilling harder and more expensive. Key problems to address here are: (1) standard oil and gas drill bits often wear out rapidly in the hard, abrasive igneous and metamorphic rocks common at depth, requiring ultra-tough materials such as Polycrystalline Diamond Compact (PDC) bits. Balancing penetration rate, durability, and cost in high-temperature, hard-rock conditions remains difficult; and (2) the need for thermal or chemical fracturing techniques to improve the permeability of geothermal reservoirs.
3. *Fluid Management and Losses.* Sometimes, natural fractures in the ground cause drilling fluids to leak out, reducing efficiency and significantly increasing exploration costs. This loss of drilling fluid to pores or fractures in the rock formations being drilled represents an average of 10% of total well costs in mature geothermal areas and often accounts for more than 20% of the costs in exploratory wells and developing fields. Well costs, in turn, represent 35%–50% of the total capital costs of a typical geothermal project, making lost circulation roughly 3.5%–10% of the total costs of a geothermal project

(Finger and Blakenship 2010). Solutions like thixotropic fluids or custom cement plugs are necessary to minimize these losses and improve borehole integrity.

14.4.2 Advances in Drilling

As previously noted, EGS leverages drilling methods developed by the oil and gas industry. Applied to geothermal energy development, these innovations are making EGS more feasible across diverse geological areas and more time- and cost-effective. Drilling innovations applied to EGS include:

- *Directional or horizontal drilling* which allows for the creation of controlled fractures to create new pathways for fluid flow and allows for those pathways to extend horizontally across greater distances to acquire heat. Multilateral drilling enables multiple reservoir sections to be accessed from one surface well, potentially reducing the number of wells needed to develop a reservoir and lowering drilling costs.
- *Hydraulic fracturing,* a technique adapted from shale gas recovery, fractures rock with the injection of high-pressure fluids to improve permeability. High-pressure water jets can cut rock, sometimes assisted by abrasives or cavitation (imploding bubbles). This technique can be used alone or in conjunction with mechanical bits.
- The use of *PDC drill bits* in site-specific designs depending on subsurface formation improves drilling productivity in the hard rock found in geothermal

TABLE 14.1
Summary of Emerging Drilling Technologies and Concepts

Emerging Technology	Description
High-pulsed-power drilling	Sends an electrical discharge into the rock, which is less energy-intensive than using compressive force or heat to induce fractures
Thermal-shock drilling	Rapidly heats and cools rocks to weaken its structure and create fractures more easily
Millimeter-wave laser drilling	Combines millimeter-wavelength lasers with traditional rotary drilling to break down rock at the drill bit–rock formation interface, speeding up the drilling process
High-pressure water jet-assisted drilling	Uses high-pressure water jets to cut rock into specific shapes that can then be broken apart more easily by fluid-powered percussive hammers
Directional steel-shot drilling	Incorporates the release of steel-shot particles under high pressure to erode the formation and reduce resistance to subsequent rotary drilling
Plasma-based drilling	Uses electrical discharges to create a plasma torch with high-temperature gas to disintegrate rock

environments. Research focuses on developing bits with improved diamond-metal composites, enhanced thermal stability, better cooling designs, and wear-resistant coatings (US DOE 2025c).

In addition to these now widely used technologies, research is ongoing into advanced hybrid and no-contact drilling techniques to facilitate faster and less energy-intensive well and reservoir creation. Some of these innovations are summarized in Table 14.1.

Drilling costs have typically accounted for half the total cost of EGS projects (Tester 2006). Advances in drilling methods and materials have helped significantly increase drilling efficiency, leading to significant time and cost savings. As shown in Figure 14.1, drilling rates at the DOE's FORGE demonstration site in central Utah have improved by more than 500% over 5 years, bringing them closer to standard drilling rates for the oil and gas industry, and promising a competitive drilling environment for implementation of deep geothermal energy production wells in the very near future.

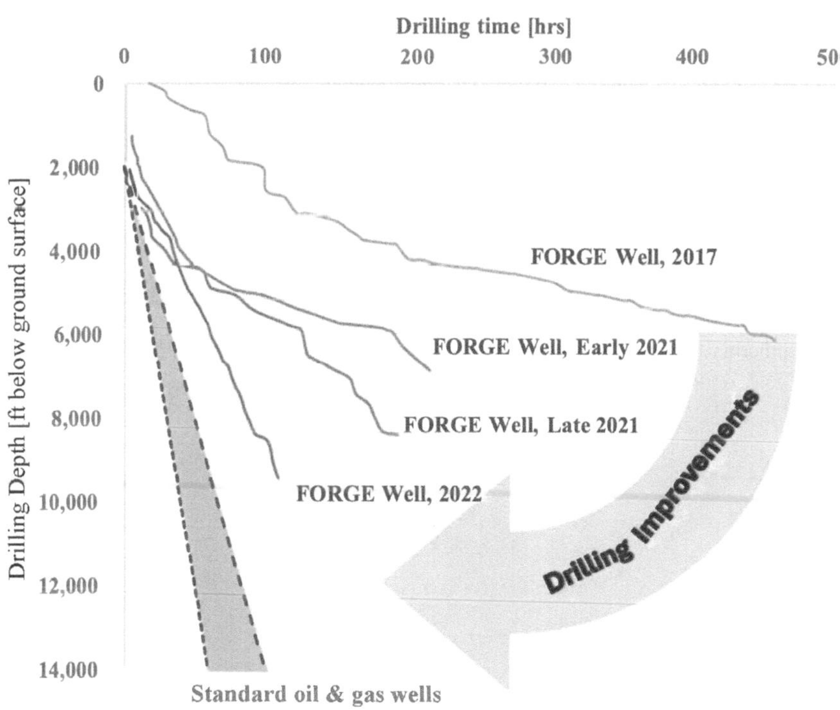

FIGURE 14.1 Drilling timelines at DOE FORGE demonstration site.

Source: Blankenship et al. (2024).

REFERENCES

Blankenship, D., Gertler, C., Kamaludeen, M., O'Connor, M., and Porse, S. 2024. *Pathways to Commercial Liftoff: Next-Generation Geothermal Power*. Washington, D.C.: U.S. Department of Energy. https://climateprogramportal.org/wp-content/uploads/2025/02/LIFTOFF_DOE_NextGen_Geothermal_v14.pdf.

Finger, J., and Blankenship, D. 2010. *Handbook of Best Practices for Geothermal Drilling*. SAND2010-6048.Albuquerque, NM: Sandia National Laboratories. www1.eere.energy.gov/geothermal/pdfs/drillinghandbook.pdf.

International Energy Agency. 2024. *The Future of Geothermal Energy*. Paris, France: International Energy Agency. www.iea.org/reports/the-future-of-geothermal-energy.

International School of Well Dilling (ISWD). 2011. Rotary Drilling: An Overview. *The Driller*, November 1. www.thedriller.com/articles/88669-rotary-drilling-an-overview. (accessed July 6, 2025).

Tester, J.W., Anderson, B.J., Batchelor, A.S., Blackwell, D.D., DiPippo, R., Drake, E.M., Garnish, J., Livesay, B., Moore, M.C., Nichols, K., Petty, S., Toksöz, M.N., and Veatch, R.W. Jr. 2006. *The Future of Geothermal Energy, Impact of Enhanced Geothermal Systems (EGS) on the United States in the 21st Century*. Cambridge, MA: Massachusetts Institute of Technology. www1.eere.energy.gov/geothermal/pdfs/future_geo_energy.pdf.

U.S. Department of Energy. 2022. Enhanced Geothermal Shot™: Unlocking the Power of Geothermal Energy. Washington, D.C.: Geothermal Technologies Office (GTO). U.S. DOE. www.energy.gov/eere/geothermal/articles/enhanced-geothermal-shottm-unlocking-power-geothermal-energy. (accessed July 6, 2025).

U.S. Department of Energy. 2025a. *Drilling Demonstrations Initiatives*. Washington, D.C.: Office of Energy Efficiency and Renewable Energy, U.S. DOE.www.google.com/search?q=https://www.energy.gov/eere/geothermal/geothermal-drilling. (accessed July 6, 2025).

U.S. Department of Energy. 2025b. *Geothermal Everywhere*. Washington, D.C.: Office of Energy Efficiency and Renewable Energy, U.S. DOE. www.energy.gov/eere/geothermal/geothermal-everywhere.(accessed July 6, 2025).

U.S. Department of Energy. 2025c. *Drilling Demonstrations Initiative*. Washington, D.C.: Office of Energy Efficiency and Renewable Energy, U.S. DOE. www.energy.gov/eere/geothermal/drilling-demonstrations-initiative. (accessed October 28, 2025).

15 Geothermal Energy
The Quintessential Energy Resource

15.1 INTRODUCTION

Over the life of planet Earth, thermal energy has been stored within its core. This energy is derived from the natural heat contained within the Earth's mantle and core. Of this energy stored within the Earth, some is transmitted from the interior by means of conduction through the Earth. The average rate of flow of heat to the surface has been estimated to be approximately 0.06 W/m^2. For the Earth's surface of 510 x 10^{12} m^2, the total heat flow amounts to approximately 30 x 10^{12} W. Of this, only 1% of the total rate, or 0.3 x 10^{12} W, is due to heat convection by hot springs and volcanoes.

One of the lesser known of Earth's renewable energy relatives, geothermal energy is now the hot ticket in the energy field. Geothermal energy refers primarily to the generation of electricity from the stored heat in the Earth's crust. Other uses of geothermal energy for process or space heating are termed *direct heat*. Geothermal power may be natural (i.e., from emissions of steam and/or hot water). In addition, cold water may be injected into deep wells where it is heated and returned to the surface for use.

The word geothermal comes from the Greek words *geo* (earth) and *therme* (heat). Thus, geothermal energy is heat from within the Earth that is generated in the Earth's core that can be recovered and used to heat buildings or generate electricity. Temperatures hotter than the Sun's surface are continuously produced inside the Earth by the slow decay of radioactive particles, a process that occurs in all rocks.

Geothermal energy (also referred to by some as *geothermics*) is virtually limitless and is based on the fact that the Earth is hotter the deeper one drills below the surface. Geothermal energy is derived from:

1. Steam trapped deep in the Earth that is brought to the surface and used to drive a turbine to produce electricity.
2. Water heated by pumping it through deep hot rocks.

The Earth's core lies almost 4,000 miles beneath the Earth's surface. The double-layered core is made up of very hot molten iron surrounding a solid iron center. Estimates of the temperature of the core range from 5,000°F to 11,000°F due to radioactive particle decay.

DOI: 10.1201/9781003667759-18

Surrounding the Earth's core is the mantle, which is partly rock and partly magma. The mantle is about 1,800 miles thick. The outermost layer of the Earth is the insulating crust; it is not one continuous sheet of rock but is broken into pieces called *plates*. These plates of continents and ocean floor drift apart and push against each other at the rate of about 1 in per year in a process called *plate tectonics*. This process can cause the crust to become *faulted* (cracked), *fractured*, or *thinned*, allowing plumes of magma to rise up to the crust.

The magma can reach the surface and form volcanoes, but most of it remains underground where it can underlie regions as large as spacious mountain ranges. The magma can take from 1,000 to 1,000,000 years to cool as its heat is transferred to surrounding rocks by conduction. In areas where there is underground water, the magma can fill rock fractures and porous rocks. The water becomes heated and can circulate back to the surface to create hot springs and mud pots, or it can become trapped underground, forming deep geothermal reservoirs. The Earth and its interior are pictured in Figure 15.1.

Section numbers and their accompanying titles for this chapter on geothermal energy are listed as follows:

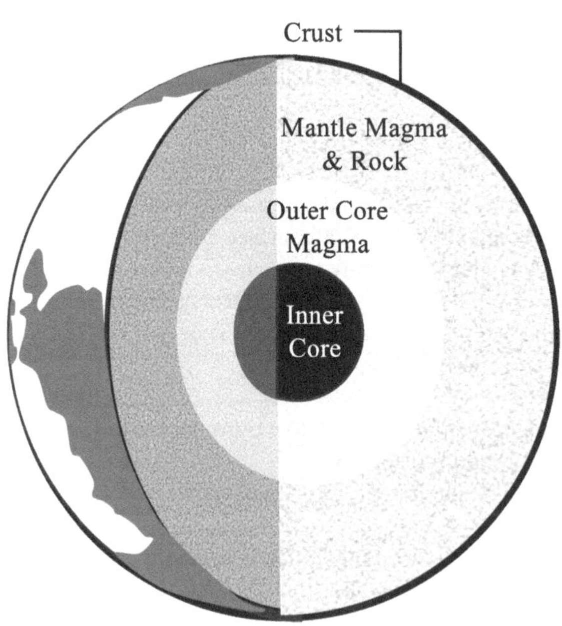

FIGURE 15.1 The Earth's makeup.

15.2 EARLY HISTORY

Humans have been using geothermal energy for millennia – from bathing in hot springs to using it to heat homes. Many ancient peoples including the Romans, Chinese, and Native Americans used hot mineral springs for bathing, cooking, and heating. (Many of these people believed hot mineral springs had natural healing powers.) The oldest known spa is a stone pool on China's Lishan Mountain, which was built in the Qin dynasty in the 3rd century BCE at the same site where the Huaging Chi palace was later built. The Romans conquered Aquae Sulis, now Bath, England, in the 1st century CE and used the hot springs there to feed public baths and underfloor heating. The world's oldest geothermal district heating system in Chaudes-Aigues, France, has been operating since the 14th century (Lund 2007). Roman documents 2,400 years old tell of steam escaping at what is now Larderello, south of Florence, Italy. The earliest industrial exploitation of geothermal energy began in 1827 with the use of geyser steam to extract boric acid from volcanic mud in Larderello. During the 19th century the natural steam was used for heating and limited mechanical power. In 1904, electricity was first produced at Larderello using the steam to drive a turbine and generator. Commercial operation at the Larderello plant began in 1912 when power was supplied to local customers, by using a 250-kW generator. The current generating capacity at Larderello is 406 MW, and this steam source has served effectively for over 60 years.

In 1892, America's first district heating system in Boise, Idaho, was powered directly by geothermal energy and was copied in Klamath Falls, Oregon, in 1900. A deep geothermal well was used to heat greenhouses in Boise in 1926, and geysers were used to heat greenhouses in Iceland and Tuscany at about the same time (Bloomquist 1999). Steam and hot water from geysers began heating homes in Iceland starting in 1943. This was the world's only industrial producer of geothermal electricity until New Zealand built a plant in 1958 at their Wairakei station; the unit was the first to employ flash steam technology. The plant, located in the volcanic belt of the North Island, reached full power of 145 MW in 1964. The plant operates at an annual load factor of 0.9 and provides base-load electricity to the power network (Lund 2007). Further development of geothermal energy in New Zealand has been temporarily halted due to the recent discovery of natural gas. Pacific Gas and Electric (PG&E) began operation of the first successful geothermal electric power plant in the United States at the Geysers in California in 1960. The plant's life was approximately 30 years and generated over 10 MW of power. Mexico began operation of a 75-MW plant in the Cerro Prieto region of northern Mexico in 1973.

In Hungary, the land area covered by greenhouses heated by geothermal water was 1.5×10^6 m^2 in 1973. Any small industrial process that requires hot water or space heat can effectively use naturally escaping steam or heat. However, in order to effectively tap geothermal energy, wells have to be driven to the geothermal reservoir.

The binary cycle power plant was first demonstrated in 1967 in Russia and later introduced to the United States in 1981 (Bloomquist 1999). A binary cycle plant in Chena Hot Springs, Alaska, came online in 2006, generating electricity from a record low fluid temperature of 135°F (Lund 2007). In 2016 there was more than 3,592 MW installed geothermal capacity, and more than 1,217 MW under development throughout the United States. (NREL 2016).

15.3 AVAILABILITY/DISTRIBUTION AND CHARACTERIZATION

While air temperatures can vary widely through the seasons, the temperatures of the shallow ground only range from 50°F to 70°F depending on latitude. Geothermal heat pumps (GHPs) draw on this relatively stable temperature as a source for heating buildings in the winter and keeping them cool in the summer. Through underground piping, a GHP discharges heat from inside a building into the ground in the summer, much like a refrigerator uses electricity to keep its interior cool while releasing heat into a kitchen. In the winter, this process is reversed; the GHP extracts heat from the ground and releases it into a building. Geothermal power plants access underground steam or hot water from wells drilled a mile or more into the Earth. The steam or hot water is piped up from the well to drive a conventional steam turbine which powers an electric generator. Typically, the water is then returned to the ground to recharge the reservoir and complete the renewable energy cycle.

15.3.1 Availability/Distribution

The areas with the highest underground temperatures are in regions with active or geologically young volcanoes. These "hot spots" occur at plate boundaries or at places where the crust is thin enough to allow the heat to pass. The Pacific Rim, often called the "Ring of Fire" for its many volcanoes, has many hot spots in the United States, including those in Alaska, California, and Oregon. In addition, Nevada has hundreds of hot spots, covering much of the northern part of the state. A similar "ring" extends through southern Europe and across the middle of Asia connecting with some of the South Sea Islands.

These regions also exhibit seismic activity. Earthquakes and magma movement break up the rock covering, allowing water to circulate. As the water rises to the surface, natural hot springs and geysers occur, such as Old Faithful at Yellowstone National Park. The water in these systems can be more than 430°F.

Seismically active hotspots are not the only places where geothermal energy can be found. There is a steady supply of milder heat – useful for direct heating purposes – virtually in any location on Earth at depths of anywhere from 10 to a few hundred feet below the surface. In addition, there is a vast amount of heat energy available from dry rock formations deeper below the surface (4–10 km). Some of the emerging technologies may be able to capture this heat for electricity production on a much larger scale than some of the conventional technologies presently allow.

Geothermal resources may be compartmentalized into three main categories:

1. Hydrothermal resources are underground reservoirs of hot water and/or steam. Hydrothermal energy sources consist of vapor-dominated water and liquid dominated water systems. Vapor-dominated systems consist of water that has been vaporized on reaching the Earth's surface. Its initial temperature and pressure are approximately 400°F and below 100 psig, respectively. Real-world applications of this type system have been limited. Liquid-dominated systems consist of hot water in the 375°F-600°F range and at elevated

pressures. The drop in pressure on reaching the Earth's surface results in its flashing into two phases that is predominately liquid, that is, low quality vapor. This type system represents the main source of geothermal energy employed to date.

2. Hot dry rock resources are typically at about 3-mile depths, with little or no fluid present.
3. Geopressured resources are hot saline fluids in the hydrothermal state, found at very high pressures and at great depths in porous formations. These are believed to contain large amounts of dissolved natural gas, and the production of natural gas and heat may be equally important.

In 2008, there were geothermal power plants in 19 countries, generating 60,435 MW of electricity. Direct uses of geothermal reservoirs amount to over 10,000 MW of thermal energy in 24 countries. An additional 22 countries have new geothermal electricity projects in development.

Recent data from the International Geothermal Association (IGA) indicates that in 2023 a total of 88 countries had geothermal heating and cooling projects on line with a total installed capacity of 173.3 GW (thermal) (IGA 2005a), 88% for building heating and cooling, 9% for health, recreation, and tourism, 2.5% for agriculture and food processing, and the balance for industrial process heat and other uses. In addition IGA indicated that 31 countries have active geothermal power systems in operation that include 197 geothermal fields, 671 power units, and an operating capacity of 16.3 GW of electricity (IGA 2005b).

Table 15.1 provides data on the geothermal capacities for the top 20 countries (IGA 2005c). Geothermal energy comprised 5.2% of the total U.S. domestic electric net generation by the electric power sector in 2024 (IEA 2005a). Most of the geo thermal reservoirs in the United States are located in the western states and Hawaii. California generates the most electricity from geothermal energy. "The Geysers" dry steam reservoir in northern California is the largest known dry steam field in the world and has been producing electricity since 1960. The United States led the world in geothermal electricity production in 2023 with 3,889 MW of installed capacity from 64 power plants (IGA 2005c), with the largest group of geothermal power plants in the world located at the Geysers in California.

The majority of the geothermal resources located in the United States are in the following western states: California, New Mexico, Arizona, Utah, Nevada, Washington, Oregon, and Idaho. Sixty-seven percent of the 3,889 MW is located in California, where more than 30 geothermal plants providing 5% of the state's electricity. Another 23% of the geothermal generating capacity is located in New Mexico.

15.3.2 Characterization

The bulk of thermal energy stored on the Earth is located within its core. Some of this energy is transmitted from the interior by means of conduction through the Earth. The average temperature difference driving force is approximately 10°C/km. It is this gradient that transports this energy from the core of the Earth to its surface. As

TABLE 15.1
International Geothermal Power Capacities in 2023

Country	Capacity, MW
United States of America	3,889
Indonesia	2,335
Philippines	1,952
Turkey	1,717
New Zealand	1,055
Mexico	1,002
Kenya	952
Italy	916
Iceland	755
Japan	546
Costa Rica	252
El Salvador	204
Nicaragua	165
Russia	82
Chile	81
China	52
Papua-New Guinea	50
Germany	47
Guatemala	46
Honduras	35

Source: IGA (2005c).

indicated before, the average rate of flow of heat to the surface has been estimated to be 0.063 W/m^2. Since the Earth's surface is 510 x 10^{12} m^2, the total heat flow amounts to 32 x 101^2 W. Only 1% of this total can be attributed to heat transfer by convection (Theodore 2011).

The total amount of heat stored within the Earth is of the order of 10^{31} J. Most of this energy is not presently available through any technical means. The present energy consumption of the Earth is in the order of 6 x 10^{20} J. The heat flow rate to the surface by conduction is approximately 30–40 terawatts (TW). This energy is replenished by the aforementioned radioactive decay at a rate of approximately 30 TW. This nuclear process has been going on since the Earth's formation and will effectively continue for many millennia.

Harder (1982) has provided pressure–temperature data at various locations within the Earth as well as the boiling temperature of water corresponding to these pressures. This information is provided in Tables 15.2 and 15.3, respectively.

15.4 EXTRACTION

When geothermal reservoirs are located near the surface, they can be reached by drilling wells. Some wells are more than 2 miles deep. Exploratory wells are drilled to

TABLE 15.2
The Earth's Temperature at Various Depths

Depth	Temperature, °F
Surface	50
1,000 ft (305 m)	65
1 mile (1.61 km)	129
1.4 miles (2.25 km)	161
2.05 miles (3.29 km)	212
10 miles (16.1 km)	842

Source: Harder (1982).

TABLE 15.3
Boiling Temperature of Water at Various Depths

Depth	Temperature, °F
Surface	212
1,000 ft (305 m)	455.9
1 mile (1.61 km)	655.9
1.4 miles (2.25 km)	705

Source: Harder (1982).

search for reservoirs. Production wells are drilled once a reservoir has been located. Hot water and steam – at temperatures of 300°F to 700°F – are extracted to the surface and used to generate electricity at power plants near the production wells.

The usual way of extracting the energy from geothermal sources is to tap into naturally occurring hydrothermal convection systems where cooler water seeps into the Earth's crust, is heated up, and then rises to the surface. When heated water and/or steam is forced to the surface, it is a relatively simple matter to capture that steam and use it to drive electric generators.

At present, geothermal wells are rarely more than 2-miles deep. Upper estimates of geothermal resources assume wells as deep as 6 miles. Drilling at this depth is now possible, although it is obviously an expensive process. The deepest research well in the world is presently 7.5 miles deep. The challenges facing engineers are to drill wide holes at minimum cost and to disengage rocks.

It should be noted that the exploitation pattern in geothermal energy is similar to that for fossil fuels. High-quality resources with long-term, low-cost, high-volume production potential are first sought. Geothermal resources with these initial cost-effective attributes are relatively rare but exist as reservoirs of very hot water or steam. Because they are usually in remote places, the emphasis is on conversion of the heat energy to "transportable" electricity. Secondary, but apparently increasing in importance, is the potential use of more plentiful lower temperature resources for direct heating applications. Dry steam resources are rare, and the only U.S power generation facility using dry steam resources is at the Geysers, California. Future development potential is highest for hot-water resources.

15.5 ENVIRONMENTAL ISSUES

Geothermal sources have a reputation for being "clean" reservoirs of energy, but geothermal development has numerous *potential* environmental impacts, including air, water, and noise pollution; land subsidence; and induced seismic activity. A number of noncondensable gases are released to the atmosphere during the drilling, testing, and operations phases of geothermal plant development. The nature and quantity of these emissions are site-specific since the chemical composition of the geothermal fluid can vary considerably from site to site. Carbon dioxide is usually the principal component while ammonia, methane, hydrogen sulfide, and nitrogen are usually present in smaller amounts; and radon, mercury vapor, and argon are typically only present in trace amounts. Small quantities of particulates are also found in the steam released to the atmosphere. Many of the air pollution abatement efforts have gone into the control of the hydrogen sulfide emissions. The conversion of hydrogen sulfide to other compounds of sulfur leads to an increase of the ambient levels of sulfur oxides in the region. Atmospheric dilution is usually sufficient to keep concentrations below critical levels near the plant site, but the concentrations may still create occupational safety and (perhaps) acid rain problems.

The major water pollution problems with geothermal energy occur with actual operation of the plant. Ammonia and boron in the blowdown wastewater from the cooling towers could cause surface water pollution and harm to aquatic life at the Geysers in Sonoma County, California. The wastewater is reinjected into the geothermal reservoir. Reinjection appears to be a very promising technique for minimizing wastewater impacts. As might be expected, the magnitude of water pollution problems is considerably larger at liquid-dominated systems as opposed to vapor-dominated systems. Many geothermal fluids have a high total dissolved solids content, and many are saline, but some are pure enough to actually be used in agriculture and industry. This wide variability has contributed to a broad range of disposal techniques, including direct release to surface waters, evaporation, and reinjection.

Noise pollution occurs at various stages of geothermal development, including the construction phase and the operation phase. The greatest noise problem occurs during the drilling phase. The "blowing" of the well after it has been drilled, the testing of the well, and unanticipated well blowouts all cause steam at great pressures to escape to the atmosphere, generating high-decibel noise levels. Extensive work has been done with various types of *mufflers* to reduce the impact of this noise pollution.

The withdrawal of large quantities of underground fluids can often lead to both horizontal and vertical ground movement. This has been noted in the petroleum industry where oil extraction has often led to land subsidence. While reinjection can be used to control land subsidence, it is not without problems of its own, such as plugging of reinjection pipes and high costs. The subsurface conditions that lead to geothermal phenomena can also lead to earthquakes and other seismic activity. No current data indicate that the extraction of geothermal energy has increased the seismicity of an area, but some research is focusing on the impacts of reinjection techniques.

The release of substantial amounts of water vapor has led to instances of local fogging around some geothermal sources. Waste heat is dissipated to the environment, primarily to the atmosphere through cooling towers (Theodore and Ricci 2010), although the use of once through cooling can have a substantial impact on surface waters. The question of land use also arises, since geothermal operations are industrial in nature, but the sites of some geothermal resources are often recreational.

Technologies are available for controlling geothermal pollution in most cases and for mitigating other environmental effects. Most pollutants are inorganic and can be subjected to physical and/or chemical treatment where necessary (Theodore 2008). The costs of their application to any expected large discharges may be high, however, and in a few cases might be limiting. Underground injection of the spent fluid, considered to be an attractive alternative, would eliminate the need for treatment and would aid in reservoir maintenance and in preventing subsidence. This technology, however, needs to be proven in the long term. Difficulties with plugging due to precipitating salts have been encountered in the past. Finally, regulatory issues do not appear to be a problem (Stander and Theodore 2008).

On the whole, the environmental problems in geothermal development are not complex. The pollutants are not "exotic" and they are contained in the raw geothermal fluid and are natural constituents derived from rock and mineral components dissolved in water. Geothermal energy conversion does not materially change these constituents, except for the removal of heat, which may lead to precipitation of dissolved solids.

A more detailed discussion of environmental issues related to geothermal energy is presented later in Chapter 17.

15.6 FUTURE PROSPECTS AND CONCERNS

The Earth's geothermal resources are theoretically more than adequate to supply humanity's energy needs, but only a very small fraction is currently profitably exploited. Drilling and exploration for deep resources are expensive, but fortunately, costs continue to decrease. Forecasts for the future of geothermal power depend on assumptions about technology, energy prices, existing subsidies, and the value of money (interest rates). The Geothermal Technologies Office of the U.S. Department of Energy has as its mission to increase geothermal energy deployment through research, development, and demonstration of innovative technologies that enhance exploration and production to realize its potential to provide clean energy for electricity generation as well as heating and cooling across the entire United States.

For the near future, energy from near-surface geothermal resources will be the primary source of geothermal energy. The economics of these energy systems are still not fully understood and only those sources/processes that promise the lowest initial costs will be marketable. For example, GHPs can generally extract enough heat from shallow ground anywhere in the world to provide some form of home heating, but industrial applications require the higher temperatures of deep resources since the thermal efficiency and profitability of electricity generation are particularly sensitive to temperature (Theodore, Ricci and VanVliet 2009).

Future geothermal resources include the following:

1. *Magma geothermal energy*. This has been called the ultimate energy source. A magma power plant would use a process similar to hot dry rock. Water would be injected directly into the magma, cooling and hardening the rock around the well. The resulting steam would be pumped out through a pipe in the well.
2. *Geopressurized resources*. These are reservoirs of hot water and natural gas (primarily methane) locked in deep sedimentary rocks, under great pressure from the overlying sediments. The heat, pressure, and natural gas can be used to produce electricity.

In any event, maintenance of the environmental quality of the air, water, and land surrounding a geothermal power plant will be a subject for research and development as new processes/resources for geothermal power arrive on the scene. It appears that geothermal energy may have difficulty in ever contributing more than about 8% of U.S. needs. On the other hand, locally and perhaps regionally, its contribution may be very important, particularly in the western and southwestern United States (IEA 2024).

REFERENCES

Bloomquist, R. 1999. Geothermal Heat Pumps, Four Plus Decades of Experience. *Geo-Heat Centre Quarterly Bulletin* 20(4):13–18.Klamath Falls, OR: Oregon Institute of Technology.

Energy Information Agency. 2025a. Table 7.2b Electricity Net Generation: Electric Power Sector. www.eia.gov/totalenergy/data/browser/?tbl=T07.02B#/?f=A&start=1949&end=2024&charted=1-2-3-5-8-14. (accessed July 1, 2025).

Harder, E. 1982. *Fundamentals of Energy Production*. Hoboken, NJ: Wiley.

International Energy Agency.2024. *2022 United States Country Report*. Geothermal Technologies Office. Taupo, New Zealand: International Energy Agency. https://drive.google.com/file/d/1rQv7U1RIy5M4Gc6C5C0MQRBoPJmP7_n6/view. (accessed July 1, 2025).

International Geothermal Association. 2005a. Geothermal Heating and Cooling. https://worldgeothermal.org/geothermal-data. (accessed July 1, 2025).

International Geothermal Association. 2005b. Geothermal Power. https://worldgeothermal.org/geothermal-data. (accessed July 1, 2025).

International Geothermal Association. 2005c. Geothermal Energy Database. https://worldgeothermal.org/geothermal-data/geothermal-energy-database. (accessed July 1, 2025).

Lund, J. 2007. Characteristics, Development and Utilization of Geothermal Resources. *Geo-Heat Centre Quarterly Bulletin* 28(2):1–9.Klamath Falls, OR: Oregon Institute of Technology.

National Renewable Energy Laboratory. 2016. *Geothermal Power Generation, Current and Planned Nameplate Capacity (MW) by State*. Golden, CO: NREL, U.S. Department of Energy. www.nrel.gov/images/libraries/gis-images/geothermal-capacity.jpg?Status=Master&sfvrsn=2f93660d_2. (accessed July 1, 2025).

Stander, L., and Theodore, L. 2008. *Handbook of Environmental Regulatory Calculations*. Hoboken, NJ: Wiley.
Theodore, L. 2008. *Air Pollution Control Equipment Calculations*. Hoboken, NJ: Wiley.
Theodore, L. 2011. *Heat Transfer for the Practicing Engineer*. Hoboken, NJ: Wiley.
Theodore, L., and Ricci, F. 2010. *Mass Transfer Operations for the Practicing Engineer*. Hoboken, NJ: Wiley.
Theodore, L., Ricci, F., and VanVliet, T. 2009. *Thermodynamics for the Practicing Engineer*. Hoboken, NJ: Wiley.

16 Geothermal Energy Conversion Processes

16.1 INTRODUCTION

Practically all chemical reactions involve energy changes that may be considered as potential sources of power or that require a supply of power for their accomplishment. The two major forms of energy used in the process, chemical manufacturing, and utility industries are heat and electric power. It must be remembered that heat becomes more difficult to produce, and thus more expensive, the higher the temperature at which it is used. Electrical energy is the most valuable as well as normally the most expensive form of energy. It can be converted at high efficiency into heat, mechanical work, light, sound, or used directly in ionic reactions.

Heat, on the other hand, is the cheapest and most generally available form of energy, but its conversion to other forms of energy is limited by complex thermodynamic relations discussed earlier, and by the properties of materials for construction of power generating equipment, boilers, engine cylinders and pistons, or turbine blades.

Geothermal energy has two primary process applications: heating/cooling and electrical generation. Most heating and cooling applications utilize low enthalpy heat stored in the Earth's crust. Ground source heat pumps (GSHPs) are employed for heating and cooling and often use 75% less energy than traditional heating and cooling systems. Geothermal energy production can be carried out by a variety of process alternatives, but all use reservoirs of hot water found a few miles or more below the Earth's surface. Process alternatives include dry steam, flash steam, and binary cycle systems. Each is briefly discussed in the following text.

Dry steam geothermal plants draw from underground resources of steam. The steam is piped directly from underground wells to the power plant where it is directed into a turbine/generator unit. Since Yellowstone is protected from development, the only dry steam plants in the country are at The Geysers.

Flash steam power plants are the most common and use geothermal reservoirs of water with temperatures greater than 360°F (182°C). This very hot water flows up through wells in the ground under its own pressure. As it flows upward, the pressure decreases and some of the hot water boils into steam. The steam is then separated from the water and used to power a turbine/generator. Any leftover water and condensed steam are injected back into the reservoir, making this a sustainable resource.

DOI: 10.1201/9781003667759-19

Binary cycle power plants operate on water at lower temperatures of about 225°F–360°F (107°C–182°C). Binary cycle plants use the heat from the hot water to boil a working fluid, usually an organic compound with a low boiling point. The working fluid is vaporized in a heat exchanger and used to turn a turbine. The water is then injected back into the ground to be reheated. The water and the working fluid are kept separated during the whole process, so there are little or no air emissions nor reservoir contamination. Currently two types of geothermal resources can be used in binary cycle power plants to generate electricity: *enhanced geothermal systems* (*EGS*) and *low-temperature or co-produced resources*. EGS provide geothermal power by tapping into the Earth's deep geothermal resources that are otherwise not economical due to lack of water, location, or rock type. Low-temperature and co-produced geothermal resources are typically found at temperatures of 300°F (150°C) or less. Some low-temperature resources can be harnessed to generate electricity using binary cycle technology. Hot water is a byproduct of some oil and gas wells in the United States. This hot water is being examined for its potential to produce electricity, helping to lower greenhouse gas emissions and extend the life of oil and gas fields.

Section numbers and their accompanying titles for this chapter concerned with geothermal energy conversion processes are listed as follows:

16.2 CYCLIC PROCESSES

A series of operations so conducted that all changes are periodically repeated in the same order is defined as a *cyclic process*. This may comprise a sequence of either nonflow or flow operations or a combination of the two types, either repeated on the same substance or involving a new mass of substance for each cycle. For example, a reciprocating air compressor produces a repeated series of nonflow operations in which the conditions at a selected point of any cycle duplicate those at the same point of any other cycle.

The production of mechanical energy from the great store of chemical energy in fossil fuels is presently accomplished by the highly irreversible process of combustion, whereby heat is released at high temperatures and transferred to a working fluid at a much lower temperature. In a steam-power plant, part of the heating value of the fuel is transferred to water in a boiler; the steam generated expands against the piston of a steam engine, or through a turbine and is then condensed with rejection of heat at a lower temperature to a cold fluid flowing in a condenser; the condensate is pumped back into the boiler. A flow diagram of these operations is shown in Figure 16.1. With reference to the figure, the feedwater is pumped with a corresponding increase in

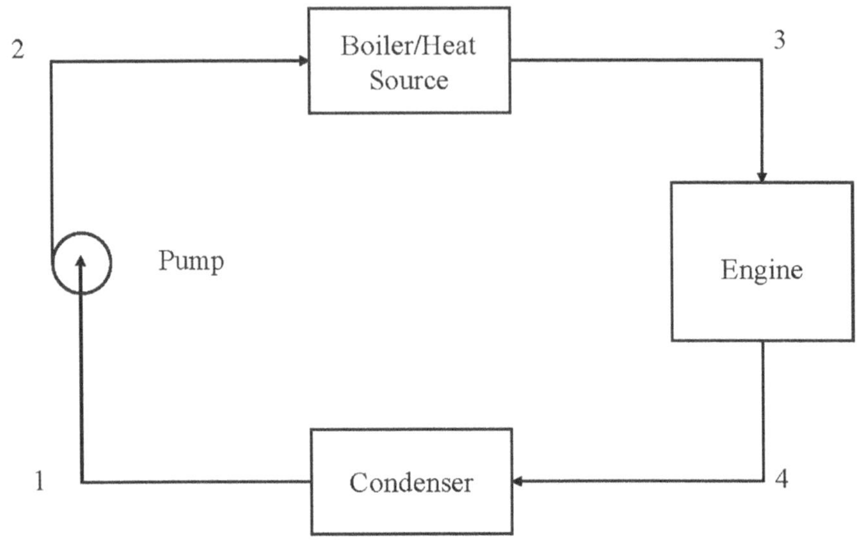

FIGURE 16.1 Power plant cycle.

enthalpy (and with a negligible rise in temperature). The circulating medium (usually water) enters the boiler subcooled, that is, at a temperature below that corresponding to its equilibrium pressure. In the next stage, the vapor expands with negligible changes in kinetic energy, and the work extracted is equal to the loss in enthalpy. In the final (condensation) stage, the enthalpy loss is equal to the heat absorbed by the condenser.

The net work of the cycle represents the engine work (3–4) minus the small work input to the pump (1–2). Since the process is cyclic with no change in any point property, it follows that the net work accomplished is equal to the heat absorbed less the heat rejected if kinetic (and any potential) energy changes are negligible.

The gross work performed results from the flow and expansion of the fluid through the engine. The net work is slightly lower than this because of the work required for pumping. Maximum gross work is obtained by ideal isentropic expansion. However, actual engines (or turbines) – as a result of mechanical friction and deviations from isentropic conditions of the fluid – deliver an amount of energy less than that calculated ideally. The ratio of the shaft work actually delivered to that ideally obtainable is generally termed *engine efficiency*. Engine efficiencies vary widely with the size and design of the engines, ranging from as high as 80% for large well-designed turbines and reciprocating engines down to 30% or less for small units of inferior design.

16.3 HEATING/COOLING AND DIRECT USE

Geothermal resources are reservoirs of hot water that exist or are human made at varying temperatures and depths below the Earth's surface. Wells ranging from a few feet to several miles deep can be drilled into underground reservoirs to tap steam and very hot water that can be brought to the surface for use in a variety of applications, including heating/cooling and direct use.

Geothermal resources such as naturally occurring underground reservoirs of hot water or the stable temperature of the subsurface can be used to heat and cool buildings. Geothermal heat pumps provide heating and cooling using the ground as a heat sink, absorbing excess heat when the aboveground temperatures are warmer, and as a heat source when aboveground temperatures are cooler. District heating and cooling systems use one or more types of geothermal systems, such as a series of geothermal heat pumps, in order to heat and cool groups of buildings, campuses, swimming pools, spas, greenhouses, and even entire communities.

So-called direct use geothermal applications use wells – usually deeper than those for heat pumps – to draw hot water from the subsurface to directly provide hot water to buildings, space heating, or heat for industrial processes ranging from fish farming and greenhouses to drying pulp, paper, lumber, and other materials. These wells work by tapping into the heat below the Earth's surface. In simple terms, holes are drilled, and then the heat is extracted.

For example, a small well, for a home or business premise would see holes drilled and pipes placed around 100–300 feet below the ground, where it is warmer than above ground. Water or a mixed solution passes through a circuit of pipes, heating up underground. A heat pump system then brings the heated water into a building, radiates the heat into the building, and pumps the cooler liquids from the building back underground to be reheated and recycled back into the building again and again as required by the heating demand of the building.

A typical GSHP is pictured in Figure 16.2. Here the flowing medium, e.g., a refrigerant, circulates through piping in the ground (1) to absorb heat (2) in the winter. The heat is then transferred at the GSHP to the home for heating and direct use (3), and the cooled refrigerant is then recirculated (4) to repeat the process.

Steps 1, 2, and 3 in Figure 16.2 involve heat exchange. The heat exchange occurs via the heat exchangers. They were defined earlier in Chapter 9 as equipment that effects the transfer of thermal energy in the form of heat from one fluid to another. The simplest exchangers involve the direct mixing of hot and cold fluids. Most industrial exchangers are those in which the fluids are separated by a wall. The latter type, referred to by some as a *recuperator*, can range from a simple plane wall between two flowing fluids to more complex configurations involving multiple passes, fins, or baffles.

The heat exchanged from any fluid during the heat transfer process results in a minimum entropy increase and hence retains highest availability *if* this transfer is made under the smallest temperature drop, subject to a host of other considerations. Theodore has proposed employing a heat pipe to accomplish the transfer of heat to/from a geothermal source from/to the surroundings.

According to Farber (2016) the heat pipe thermal cycle operates in the following manner (see Figure 9.6):

1. Working fluid evaporates to vapor, absorbing thermal energy.
2. Vapor migrates along a cavity to the lower temperature end.

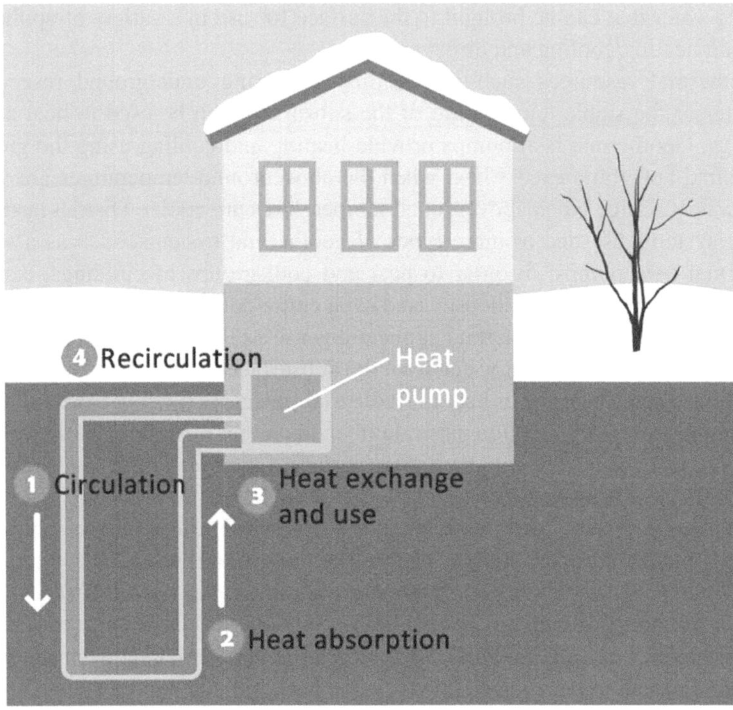

FIGURE 16.2 Typical GSHP in heating mode.

Source: **Belmont Citizens Forum (2019).**

3. Vapor condenses back to the fluid releasing thermal energy in the cooler end and is absorbed by a wick.
4. The working fluid flows back to the higher temperature end.

Thus, a heat pipe (referred to by some as a superthermal conductor) consists of three components:

1. A containing enclosure
2. A wick (or equivalent) unit
3. The circulating fluid

The unit process itself consists of three operating sections:

1. The evaporator section
2. The adiabatic section
3. The condenser

The enthalpy of phase change is employed to achieve the desired heat transfer. The overall heat transfer process is accomplished by evaporating part of the circulating fluid in the evaporator section. A pressure difference driving force is thus created that moves the moving vapor to the condenser section where the enthalpy of condensation converts the vapor to liquid. The resultant flowing liquid is transferred into the wick (or equivalent) and back into the evaporator section. The net effect of the cyclic process is that the heat transferred in the evaporator section is rejected out in the condenser section; in effect, the transfer of heat occurs *without* direct contact of the source and receiver.

Heat pipes are thus capable of transferring process heat to the surroundings. They may also be used to recover thermal energy from process gas streams. It may also be employed to heat up iced surfaces of ground by transferring the heat from below the Earth's surface to cooler external surfaces. A host of other applications are predicted for the future, including a desalination unit developed by one of the authors (Theodore, 2024).

Calculational details regarding heat pipes are provided by Hagen (1999). Information on the pipe thermal resistances, the heat source, the cooling (sink) source, evaporator and condenser pipes, and for the wick on both the evaporator and condenser sections are provided.

Heat pipes have been employed within computer systems to transfer heat from central processing units (CPUs) to heat sinks so the heat can be rejected to the environment. Heat pipes are also used in Heating, Ventilation, and Air Conditioning (HVAC) systems to recover heat by transfer from an exit stream to a fresh air stream in the winter and to cool an entering air stream in the summer.

16.4 DIRECT DRY STEAM GEOTHERMAL POWER PLANTS

As noted earlier, electric power is the most valuable commonly used type of energy. Its value lies in the fact that it is very cheaply transmitted over considerable distances, is easily and accurately controlled, and is convertible at high efficiency to mechanical work, to heat at any required temperature, or to any other form of energy required, for example, in electrochemical processes. Furthermore, large amounts of power may be concentrated in a very small space.

At a geothermal power plant, hot water is pumped, under high pressure, from deep underground. As it reaches the surface, its pressure drops. The water turns to steam, spinning a turbine that, via a generator, creates electricity. The steam eventually cools – most plants use cooling towers for their cooling systems – and the condensed water is injected back into the ground via injection wells, and the cyclic heating process is repeated. Simply stated, dry steam power plants use steam from a geothermal reservoir and route it *directly* through turbines, which drive generators to produce electricity. These geothermal electric plants operate with a thermal efficiency in the 10%–23% range because most geothermal fluids often do not reach the high temperatures of steam in boilers. The laws of thermodynamics limit the efficiency of heat engines in extracting useful energy. Exhaust heat is wasted, unless it can be used directly and/or locally. This efficiency does not materially affect operational costs as it would for plants that use fuel, but it can affect return on the capital used to build the plant. Thus, in order to produce more energy than the pumps consume, electricity generation requires relatively hot fields and specialized heat cycles. Is should also be

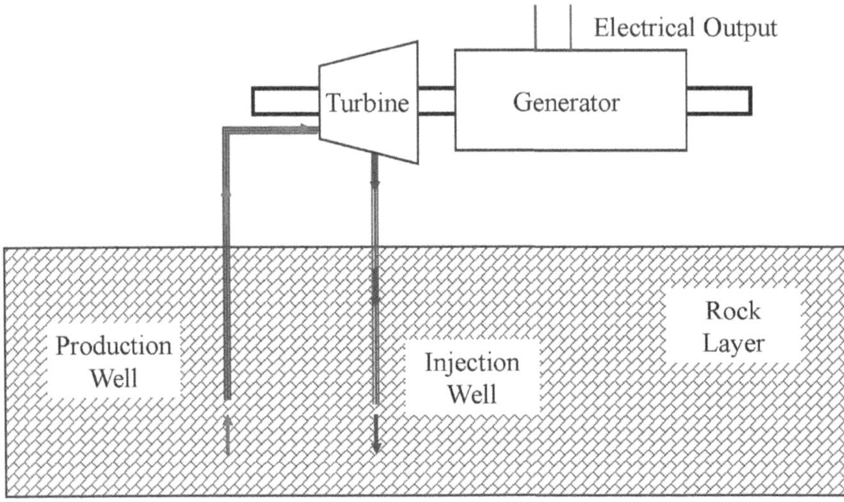

FIGURE 16.3 Dry steam geothermal power plant.

noted that these plants emit only excess steam and very minor amounts of other gases to the environment.

Although the development of geothermal power requires a large capital investment, geothermal has low operating costs and a capacity factor (ratio of actual power production to production potential) of >90%. In 2016, geothermal electricity costs between 7.8 and 22.5 cents/kWh. As of May 2020, geothermal plants qualified for the federal Production Tax Credit (PTC). In 2023, the Inflation Reduction Act renewed and expanded the PTC, which provides up to 2.75 cents/kWh for electricity generated from geothermal energy resources. Starting in 2025, the PTC ranges from 0.3 to 1.5 cents/kWh, with an additional 10% bonus for projects located in energy communities or that meet domestic content requirements.

A few geothermal reservoirs produce mostly steam and very little water. In dry steam plants, the steam from the reservoir shoots directly through a *rock-catcher* (effectively a filtering action) into the turbine generator. The rock-catcher protects the turbine from small rocks that may be carried along with the steam from the reservoir. These plants are the simplest and oldest in design, employing steam at or greater than 300°F (150°C). The Larderello geothermal power plant, first used in Tuscany, Italy, in 1904, is the oldest dry steam power plant in the world. The Geysers dry steam reservoir in northern California is the largest known dry steam field in the world and, after 60 years, still produces enough electricity to meet the needs of a city the size of San Francisco. A line diagram of a dry steam power plant is shown in Figure 16.3.

16.5 HOT WATER/FLASH STEAM POWER PLANTS

Hot water or flash steam power plants are the most common geothermal power plants. Typically, they draw from geothermal reservoirs of water whose temperatures exceed 360°F (182°C). Very hot water flows upward, eventually turning, or flashing, into steam as the pressure that forced the water upwards decreases. The flow of this steam powers turbines to create electricity. Any leftover water and condensed steam is returned to the reservoir. It is reheated and repeats its cyclic journey making hot water (flash steam) as a sustainable resource and renewable energy source.

A line diagram of a flash steam geothermal power plant is provided in Figure 16.4, with an included cooling tower shown in Figure 16.5.

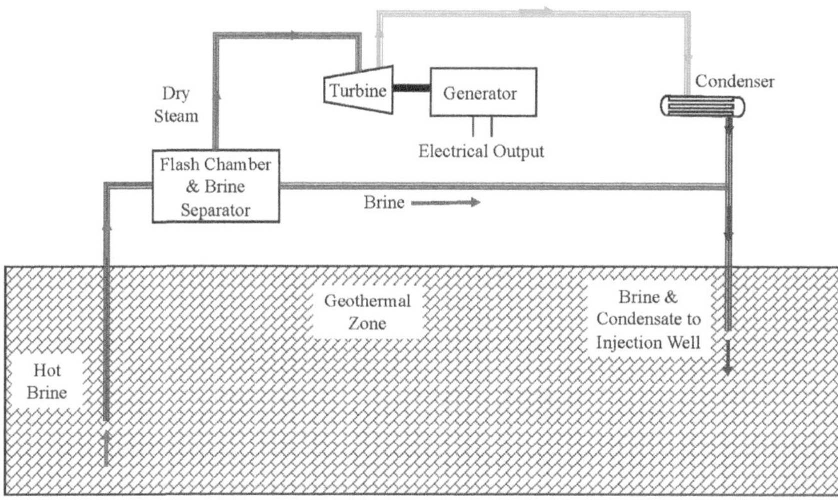

FIGURE 16.4 Flash steam geothermal power plant.

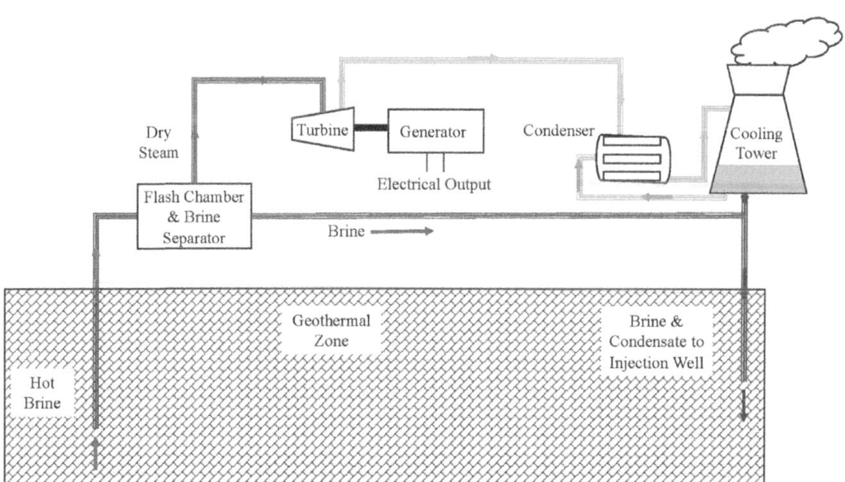

FIGURE 16.5 Flash steam geothermal power plant with cooling tower.

16.6 BINARY CYCLE POWER PLANTS

Binary cycle geothermal power plants can use lower temperature geothermal resources, making them an important technology for deploying geothermal electricity production in more locations. Binary cycle geothermal power plants differ from dry steam and flash steam systems in that the geothermal reservoir fluids never come into contact with the power plant's turbine units. A low-temperature (below 360°F/180°C) geothermal fluid (usually water) passes through a heat exchanger to boil a secondary working or "binary" fluid. This binary fluid (usually isobutane or ammonia) has a much lower boiling point than water, and the modest heat from the geothermal fluid causes it to flash to vapor, which then drives a turbine, which in turn spins a generator and creates electricity.

The water and the working fluid are kept separated during the entire process so there are little or no air emissions from the plant. The water is reinjected underground where it is reheated and used again in the cyclic, closed-loop system. Since the system is closed, there is little heat loss, almost no water loss, and virtually no emissions. These are the most recent plant designs to arrive on the scene and can generally accept fluid temperatures as low as 135°F (55°C). Most geothermal power plants of the future will likely be of the binary cycle type. A line diagram of this class of geothermal power plants is provided in Figure 16.6.

16.7 OTHER GEOTHERMAL ENERGY CONVERSION PROCESSES

Three other geothermal energy production processes have received attention in the last decade. These include:

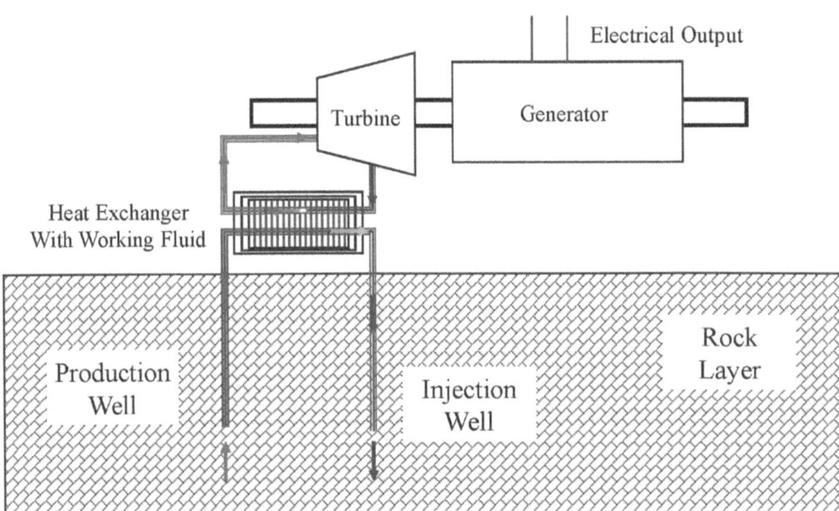

FIGURE 16.6 Binary cycle geothermal power plant.

1. Enhanced Geothermal Systems (EGS)
2. Combined Cycle Power Plants
3. Hybrid Power Plants

Each is briefly discussed in the following sections.

16.7.1 Enhanced Geothermal Systems (EGS)

EGS provide geothermal power by tapping into the Earth's deep geothermal resources that are otherwise not economical due to lack of water, location, or rock type. EGS creates a subsurface fracture system to increase the permeability of rock and allow for the injection of a heat transfer fluid (typically water). The injected fluid is heated by the rock and returned to the surface to generate electricity. This development greatly expands the potential use of geothermal resources to new geographic areas. The U.S. Geological Survey (Burns, Williams, and DeAngelo 2025) recently completed a provisional assessment of the geothermal-electric resources associated with high-temperature, low-permeability rock formations of the Great Basin, Southwestern United States. If sufficient technological advances to commercialize EGS occur, then a current best provisional estimate for electric-power generation capacity of 135 GW electric are available from the upper 6 km of the Earth's crust in this region of the U.S. alone.

16.7.2 Combined Cycle Power Plants

A combined cycle plant works to produce electricity and capture waste heat from a gas turbine to increase efficiency and electrical output. The gas turbine burns fuel; it compresses air and mixes it with fuel that is heated to a very high temperature. The hot air–fuel mixture moves through the gas turbine blades, making them spin. The fast-spinning turbine drives a generator that converts a portion of the spinning energy into electricity. A heat recovery steam generator captures exhaust heat from the gas turbine that would otherwise escape through the exhaust stack and creates steam to drive a steam turbine. The steam turbine sends its energy to a generator drive shaft, where it is converted into additional electricity.

16.7.3 Hybrid Power Plants

In some power plants, flash and binary systems are combined to make use of both the steam and the hot water. The Puna Geothermal Venture Facility on the big island of Hawaii produces 30 MW of power, or 20% of the island's electricity needs.

REFERENCES

Belmont Citizens Forum. 2019. *Ground Source Heat Pumps Make Heating Easy.* Belmont, MA. www.belmontcitizensforum.org/2019/09/16/ground-source-heat-pumps-make-heating-easy/. (accessed September 29, 2025).

Burns, E.R., Williams, C.F., and DeAngelo, J. 2025. *Enhanced Geothermal Systems Electric-Resource Assessment for the Great Basin, Southwestern United States.* U.S. Geological Survey Fact Sheet 2025–3027. Reston, VA: U.S. Geological Survey. https://doi.org/10.3133/fs20253027.

Farber, P. 2016. *Personal Communication to L. Theodore*. Willowbrook, IL.

Hagen, K. 1999. *Heat Transfer with Applications.* Upper Saddle River, NJ: Prentice-Hall.

Theodore, L. 2024. U.S. Patent No. 12,145,862B2. *A System and Process for Obtaining Potable Water Employing Geothermal Energy.* Granted to L. Theodore, November 19, 2024.

17 Geothermal Energy Environmental Impacts and Considerations

17.1 INTRODUCTION

Humans have been safely harnessing geothermal resources for thousands of years, from Roman baths to the first-known district heating in France in the 14th century to the first commercial geothermal power plant in Lardarello, Italy, built in 1913. While every human interaction with nature comes with some inherent risk, using geothermal energy has thus far been a sustainable and low-risk undertaking when compared to other energy sources such as fossil fuels or nuclear power.

As with any power plant development in the United States, a comprehensive scope of laws and regulations apply to ensure minimal risk to public health and the environment. Still, rapid progress in enhanced geothermal systems (EGS) and other next-generation technologies suggests that the widespread deployment of geothermal energy is at hand, and associated effects on surrounding communities and environments should be understood and assessed.

Section numbers and their accompanying titles for this chapter on environmental considerations for geothermal energy systems are listed as follows:

DOI: 10.1201/9781003667759-20

17.2 SEISMIC ACTIVITY AND SUBSIDENCE

17.2.1 Induced Seismicity

Induced seismicity is seismic activity that occurs because of human action and that is not likely to have occurred otherwise. Induced seismic events have been observed for more than a century as a result of an array of industrial activities, including mining, construction, reservoir impoundment, and oil and gas recovery. The first documented induced earthquake stemmed from a mining operation in Johannesburg, South Africa, in 1894. In most cases, diligent study and careful engineering measures have allowed industrial operations to effectively manage seismic risk and to function safely and economically. The vast majority of induced earthquakes are low-intensity and barely perceptible, usually in the range of 1.0–3.5 on the logarithmic Richter Scale.

Induced seismic events are generally caused when mass is either added to or removed from the Earth, either adding stress or reducing stress on a fault, which can cause movement. General useful criteria for determining whether a particular seismic event was induced include:

- Spatial correlation: how close the seismic event was to human activity.
- Temporal correlation: whether the seismic event and the activity happened close in time or after some change in activity.
- Near surface: how close the seismic event is to the surface where most human activity is taking place (Rubinstein 2018).

In the context of geothermal energy, the subsurface conditions that produce geothermal phenomena are inherently tied to an increased risk for seismic activity. Natural geothermal reservoirs are typically located near boundaries of tectonic plates, where there is already high natural seismicity near active fault lines. Hydrothermal (conventional geothermal) operations that recover thermal energy from these natural reservoirs have not been shown to increase the likelihood of significant seismic events.

In EGS, however, a geothermal reservoir is engineered by artificial means in an area where there is ample subsurface heat but low permeability. During reservoir creation, the repeated injection and extraction of water deep into the subsurface rock amounts to the *intentional* triggering of small movements on preexisting fractures as the fluid flow alters the rock formation and its natural stress patterns (Majer et al. 2012). The induced seismic event occurs when the friction between layers of rock is overcome and the rock shifts. During geothermal plant operation, however, the risk for induced seismic activity is mitigated by the continuous withdrawal and reinjection of geothermal fluids in a cycle that helps to maintain pressure and volume in the reservoir.

There have been some notable seismic events related to early EGS projects. A magnitude 3.4 earthquake near Basel, Switzerland, in 2006 occurred near an EGS project that was ultimately cancelled. A 5.5 moment magnitude (M_w) earthquake in Pohang, South Korea, in 2017 was linked to hydraulic stimulation activities at a nearby EGS site. That quake injured 70 people and caused extensive structural damage in the surrounding area (Grigoli et al. 2018). In general, though, most documented earthquakes related to EGS development have been microearthquakes too small to be felt. Observations and data collected from these induced seismic events can be a valuable tool, however. Careful monitoring and imaging allow operators to learn how fractures move at various pressures and volumes of fluid injection, offering a better understanding of the physical processes underway as the reservoir is formed (Rubinstein 2018).

Since EGS methods are based largely on well-established methods historically used in the oil and gas industry, it is useful to note how induced seismic activity differs in oil and gas recovery and in EGS. Increased use of hydraulic fracturing in the oil and gas industry since the 2000s brought with it a marked increase in induced seismicity in regions with drilling operations, including throughout the central United States. While most of these events were low magnitude, a few higher-intensity induced earthquakes include a 5.8 magnitude quake in Pawnee, OK, in 2016 that was one in a series of earthquakes in the region over a decade (USGS 2016).

Studies have found that most of this oil and gas-related induced seismicity is not caused by hydraulic fracturing itself but rather by the injection of large volumes of wastewater and spent fluids into deep disposal wells over the long term. Estimates suggest that more than 2 billion gallons of this wastewater are injected into disposal wells daily (U.S. EPA 2024). Geothermal systems do not involve large quantities of wastewater, and, as noted above, the recirculation of geothermal fluids helps to maintain a consistent volume and pressure in the reservoir, reducing induced seismicity.

In an effort to proactively manage seismic risks, guidance from the DOE recommends the development of mitigation plans on a case-by-case basis that account for the specific geology of the site, history and potential for seismic activity, and the surrounding community and infrastructure (Levine et al. 2017). The DOE developed its *Protocol for Addressing Induced Seismicity Associated with EGS* (Mayer et al. 2012) (based on earlier guidelines from the IEA) to establish a more systematic approach to assessing induced seismicity. It offers guidelines for communication among project stakeholders and for data and knowledge sharing to inform best practices for future EGS projects. The protocol includes the following recommended actions to address induced seismicity during project development:

1. Perform a preliminary screening evaluation.
2. Implement an outreach and communication program.
3. Review and select criteria for ground vibration and noise.
4. Establish seismic monitoring.
5. Quantify the hazard from natural and induced seismic events.
6. Characterize the risk of induced seismic events.
7. Develop a risk-based mitigation plan.

Ongoing efforts to research and proactively mitigate seismic risks are aided by the growth of EGS and lessons learned from each specific project moving forward. Data sharing can help identify the kinds of rock formations and natural faults that are most susceptible to drilling-induced seismic activity. Recent advances in technology are also enabling helpful research, such as sensors to monitor seismicity in real time and at high temperatures, and high-performance computer modeling with predictive capabilities (Blankenship et al. 2024).

17.2.2 Land Subsidence

Land or ground subsidence refers to the sinking of the ground due to compaction of below-surface materials. While subsidence can sometimes occur due to natural compaction of subsurface materials, it is most often caused by human activity that removes large volumes of water from below ground. In geothermal systems, if the produced geothermal fluid is extracted at rates much higher than the working fluid is injected, there could be a risk of surface subsidence, particularly in natural hydrothermal systems with shallow wells. Active, responsible reservoir management significantly diminishes the risk by keeping reservoir pressures consistent. In EGS reservoirs in hard rock formations at much greater depths, the same possibility for subsidence is not a typical concern (Tester et al. 2006). Where needed, subsidence can be measured and monitored using radar, GPS technology, repeated leveling surveys, or via ground or water sensors (NOAA 2024).

17.3 WATER USE

Water plays a critical role in the harvesting of geothermal energy. Geothermal fluid, also referred to as *geothermal water*, serves as the carrier for heat as it circulates through fractured rock and is pumped to the surface. Responsible water management – that safeguards both quantity and quality – is key to geothermal energy's strength as a sustainable resource.

17.3.1 Water Consumption and Withdrawal

Water use impacts of any large-scale facility should be carefully evaluated, especially in regions where the stability of the long-term water supply is a concern. Over the life of a geothermal power plant, water is central to nearly every phase of its operation: during reservoir development and testing, plant construction, and in the regular operation of the plant to maintain geothermal fluid levels and for cooling.

Despite it being critical to the process, however, the total water usage for geothermal systems is actually much lower than for most other forms of power generation. By design, geothermal plants effectively circulate water rather than consume it. Most hot water drawn out of a geothermal reservoir is reinjected back into the same reservoir to regain heat it lost to the power production process. As shown in Figure 17.1, geothermal energy uses less water in terms of both withdrawal and consumption than in fossil fuel energy production and less than most other renewable energy sources. Water withdrawal refers to water diverted from a source for a given use and then returned to the source, whereas water consumption amounts to a net loss from the source because the water is used or lost elsewhere in the process (U.S. DOE 2019).

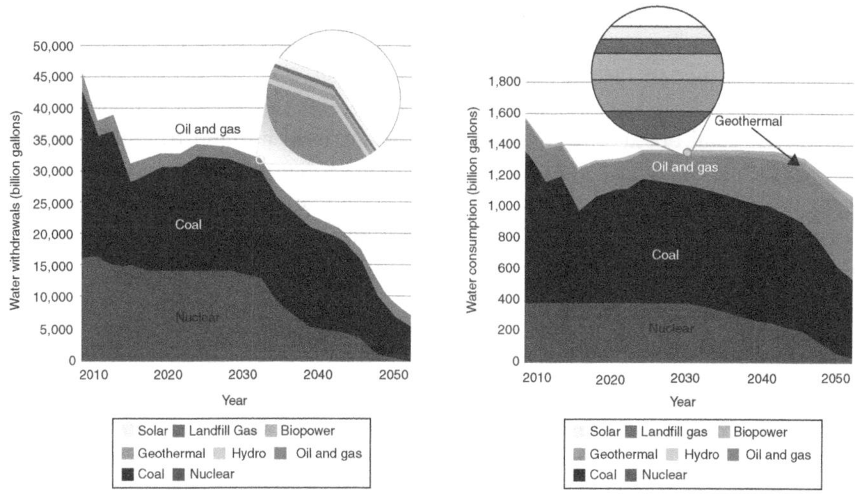

FIGURE 17.1 Power-sector water withdrawals and water consumption in billions of gallons.

Source: U.S. DOE (2019).

Water losses can occur throughout the lifecycle of a geothermal plant and fall into these general categories:

- Drilling and construction of pipelines and the power plant
- EGS reservoir stimulation and testing
- Belowground operational (or reservoir) loss, referring to flow lost to rock surrounding an EGS reservoir
- Cooling-related operational loss for wet or hybrid cooled plants, including makeup water that compensates for evaporation, drift, and blowdown in cooling towers
- Noncooling operational and other maintenance losses

Drilling and pipeline construction account for less than 5% of water losses in hydrothermal plants and less than 0.5% in EGS. In both hydrothermal systems and EGS, the majority of water loss is either belowground reservoir loss or aboveground related to cooling (Schroeder et al. 2014).

Water consumption in geothermal operations can vary widely and is dependent on several factors, including the temperature of the reservoir and the type of power plant. Higher temperature reservoirs generally produce lower water consumption rates since they often utilize flash steam and dry steam plants where the condensed steam is reinjected back into the reservoir. Lower temperature reservoirs in EGS tend to use binary cycle plants that operate a closed-loop system that minimizes water loss. However, higher flow rates are required for lower temperature resources to generate the same amount of energy (Schroeder et al. 2014).

A benefit of geothermal power plants is that they do not require freshwater resources, minimizing impacts on local ecosystems and eliminating competition with drinking water or agricultural needs. Since the water is reinjected into the reservoir to maintain pressure, it does not come into contact with the ambient environment, so power plants can instead make use of brackish water, industrial or municipal wastewater, or stormwater runoff for reservoir pressure maintenance. Water produced *from* geothermal operations could conceivably be used by industrial or agricultural applications such as irrigation, assuming that adequate geothermal fluids are being replaced in the reservoir for sustainable long-term maintenance (Finster et al. 2015). Geothermal waste is treated to water-quality levels required for the desired reuse application.

17.3.2 Water Quality

There is no evidence that geothermal development has ever caused groundwater contamination. Most natural hydrothermal reservoirs are located deep underground, at depths that have little to no effect on the quality or volume of groundwater sources (Kagel, Bates and Gawell 2007). Geothermal production and injection wells are lined with casing made of steel or titanium to prevent geothermal working fluids from ever coming into contact with groundwater sources. Further, those working fluids are continuously reinjected back into the reservoir. This method eliminates the need for waste disposal, conserves water resources, and aids in reservoir stability by keeping pressures and volumes constant.

In the case of EGS, reservoir development takes advantage of the same hydraulic fracturing techniques long used by the oil and gas industry, but the water quality risks associated with oil and gas "fracking" are not the same in the context of geothermal development (Blankenship et al. 2024). These differences are outlined as follows:

- The geothermal fluid chemical composition in EGS is primarily water, with fewer chemical additives than are needed for oil and gas recovery.
- As noted previously, wells used to access EGS reservoirs are fully cased in steel and completely cemented on the outside, and are different from most oil and gas wells that are designed to allow selective flow of hydrocarbons into them.
- EGS reservoirs, reaching depths of 8,000 ft or more, are typically much deeper than average oil and gas reservoir at about 5,000 ft. EGS reaches "basement" bedrock formations, with thousands of feet of low-permeability rock between the EGS reservoir and drinking water supplies.
- Geothermal fluids in EGS are recirculated in a continuous loop to maintain reservoir pressure, whereas oil and gas operations must remove spent working fluids and dispose of them.

17.4 LAND USE

The physical footprint of a facility influences its production capacity, where installations can be sited, public acceptance, and its potential impact on the local environment. Relative to other power generating operations, geothermal power plants typically use less land and have smaller surface footprints since the bulk of the heat recovery process actually takes places underground. While a well field may extend 5–10 km^2 or more on the land above the reservoir's wide network of fractures, the well pads themselves take up just about 2% of that space. Directional drilling allows for multiple wellheads to operate from the same well pad on just hundreds of square meters (Tester et al. 2006).

Geothermal power plants occupy just 404 m^2 of land per GWh of electricity produced (Figure 17.2), compared to 3,642 m^2 for coal and 3,237 m^2 for solar photovoltaic (PV) (DOE 2019). The total footprint for coal power accounts for land occupied by mining operations. Utility-scale solar PV plants typically require large amounts of space for solar panel installation (though continuing improvements in PV technology are driving down land requirements for equivalent electricity produced). For wind power plants, turbines must be spaced widely apart to optimize energy extraction.

As with other industrial activity, the construction of geothermal wells and power plants raises the risk of habitat disruption and other ecological impacts. These impacts are generally temporary, however, and facilities can be designed to minimize land clearing and the long-term loss of trees and other vegetation as part of native ecosystems. Once constructed, the small footprint of a geothermal plant has less of a visual impact and is generally less disruptive to its surroundings than other

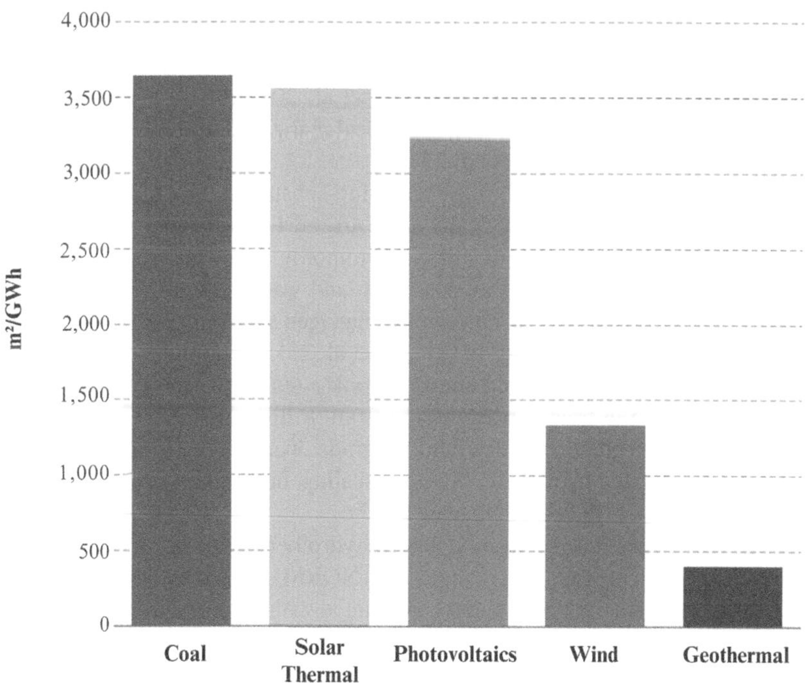

FIGURE 17.2 Land footprint per GWh for different electricity-generating technologies.

Source: U.S. DOE (2019).

large-scale power-generating facilities. Depending on its location, much of the area surrounding a geothermal plant can be used for other purposes, such as agricultural or grazing sites or recreation.

Similarly, a major potential advantage of EGS and its geographic flexibility is the opportunity for co-location of geothermal power plants with complementary end-use applications. Their relatively small size and ability to supply uninterrupted power lends geothermal plants to be sited near energy-intensive or industrial facilities like data centers or direct air capture operations (Stanford University 2024).

17.5 SOLID WASTE AND BYPRODUCTS

The nature of geothermal reservoirs and power plants make the risk for contamination by solid wastes exceedingly low. The primary solid waste from drilling activities is mostly mud and drill cuttings, which are typically stored in lined sumps for disposal. Quantities of additional solid waste generated during drilling and operation at most reservoirs tend to be very small. In many cases, the solids are returned with spent geothermal fluid back into a closed-loop system without coming into contact with ambient air, water, or land. It is important to evaluate materials present at any site, however, to be aware of potential hazards in case of some unforeseen accident or equipment failure.

The makeup of solid waste from a given facility is site-specific and depends on the geological composition of the reservoir and the geothermal fluid extracted from it. Concentrations can vary widely, but some of the most commonly found constituents in geothermal fluid include corrosive salts like sodium chloride (NaCl), bicarbonate (HCO_3^-), and sulfate (SO_4^{2-}) (Finster et al. 2015). Where necessary to prevent corrosion, these salts are typically removed using desalination methods and are disposed of (Theodore and Dupont 2022).

While quantities are low, naturally occurring substances that may be found near geothermal systems include heavy metals like arsenic (As) and lead (Pb), both known to be highly toxic to humans. Arsenic is commonly found in areas with volcanic activity and may be present in soil, sediment, and water sources. These metals can also be corrosive to equipment but can be managed and removed through various methods, such as electrocoagulation (Finster et al. 2015).

Silicon dioxide (silica) and calcium carbonate (calcite) may also be present in brine reservoirs and in treated wastewaters from drilling operations, contributing to scaling on pipes and equipment that could degrade operating performance. Silica and calcite can be mitigated to prevent or remove scaling, but concentrations are generally nonhazardous (Kagel, Bates, and Gawell 2007).

Some geothermal sites, especially natural hydrothermal reservoirs, may also contain naturally occurring radioactive materials (NORM) caused by isotopes of radium. There is evidence that certain processes associated with geothermal heat recovery can lead to higher concentrations of these materials forming in operational waste, such as scale and spent fluids. These "technically enhanced" levels of NORM could pose a risk to workers and lead to possible contamination if not properly abated (Finster et al. 2015). The impacts of NORM and possible inhibitors have been investigated in the oil and gas sector in the past and are now receiving more attention in the context of geothermal energy.

With expanding geothermal operations, there is growing interest in co-located mineral extraction. In some instances, spent geothermal fluid or brine contains useable materials, such as precious metals, zinc, or lithium (Li) that can be recovered and recycled or sold for other uses. These byproducts can be separated from spent fluid before it is reinjected into the reservoir, offering significant process efficiencies and savings in costs, land use, water usage, and time compared with the exploration and operation of separate mineral extraction sites. The viability of these efforts depends on the concentrations of the desirable material, the cost and complexity of separation methods, and economic conditions impacting market prices.

Of particular interest is the recovery of Li, a principal component of rechargeable lithium-ion batteries for electronic devices, electric vehicles, and grid energy storage – all markets that are growing rapidly. Li is considered a critical mineral by the U.S. Department of the Interior, essential to national security and economic stability. Sourcing it from geothermal operations could significantly boost the domestic supply chain. Li may be present at high concentrations in brines at some natural geothermal reservoirs, including the Salton Sea in southern California, where there has been significant focus for research and pilot operations.

17.6 AIR EMISSIONS

In contrast to combustion-based processes that produce air pollutants harmful to both human health and the environment, geothermal energy operations boast exceedingly low direct air pollution emissions and low lifecycle greenhouse gas (GHG) emissions. There is no input fuel required, so there are no emissions associated with fuel transport, and no combustion takes place. Emissions from geothermal facilities are gaseous releases of naturally existing noncondensable gases (NCG) found in the geothermal fluid in subsurface reservoirs. The variability of each geothermal site's geologic conditions lends some uncertainty to estimating all possible emissions, but the presence of serious hazards would likely be discovered early in the site selection process through exploration and monitoring (Millstein et al. 2019).

Though the composition varies widely, common NCG found in geothermal fluids can include carbon dioxide (CO_2) and hydrogen sulfide (H_2S), and often lower concentrations of gases including ammonia (NH_3), methane (CH_4), and nitrogen (N_2). Emissions may also contain trace amounts of radon (Rn), mercury (Hg) vapor, and argon (Ar). In hydrothermal reservoirs, these gases can flow from degassing sources deeper in the Earth's crust, or be produced by chemical reactions of natural geothermal fluids with the subsurface rock. They can be emitted naturally into the atmosphere through steam vents or soil or via groundwater (Fridiksson et al. 2017).

The above NCGs may also be released during the drilling, testing, and operations phases of geothermal plant development. During operation, they are released when the steam is flashed in flash and dry steam power plants. Notably, these releases do not occur in binary cycle plants, which are closed loop systems where the geothermal fluid is recycled into the reservoir, eliminating virtually all air emissions from the generation process. Nearly 90% of geothermal power plants built since 2000 are binary cycle (EIA 2020).

Abatement mechanisms to manage air emissions are built into the system design, as with any other power-producing facility. Processes such as scrubbing are commonly used to manage H_2S emissions and minimize conversion to hazardous sulfur oxides that contribute to air pollution and acid rain (Skipka and Theodore 2014). CO_2 and other GHG can be minimized by compressing and reinjecting them back into the reservoir in the geothermal fluid. Releases of other regulated air pollutants, sulfur dioxide (SO_2), nitrous oxides (NO_x), and fine particulate matter ($PM_{2.5}$), are minimal, especially when compared to other power-generating facilities, as shown in Table 17.1. With wider use of EGS in areas where these gases are not naturally occurring, it is expected that geothermal fluids in those facilities will contain such small concentrations of these air pollutants that their emissions may not require active management (Tester 2006). These emission reductions and further potential benefits of wider deployment of geothermal energy are discussed in Section 17.9.

Because geothermal power plants are relatively small in number compared with other electricity generating technologies, there is a shortage of reliable data on their lifecycle emissions. Estimates include an exhaustive modeling of emissions based on construction materials, including such components as pipes, piping insulation and liners, fuel consumption of drill rigs during well construction, fluids to cool and lubricate drill bits, cement for well casing, and the operation of injection and production

TABLE 17.1
Averages of Four Significant Pollutants, as Emitted from Power-Generating Facilities

Emissions	NOx	SO_2	$PM_{2.5}$	CO_2
Pollutant Impacts	Lung irritation, coughing, smog formation, water quality deterioration	Wheezing, chest tightness, respiratory illness, ecosystem damage	Asthma, bronchitis, cancer, atmospheric deposition, visibility impairment	Global warming increases sea level, flood risk, glacial melting
Coal emissions, lb/MWh	4.31	10.39	2.23	2,191
Natural gas emissions, lb/MWh	2.96	0.22	0.14	1,212
Geothermal emissions (flash plant), lb/MWh	0	0.35	0	60
Geothermal emissions (binary and flash/binary plant), lb/MWh	0	0	negligible	0

Source: Adapted from Kagel, Bates and Gawell (2007).

pumps (Sullivan et al. 2010). As with other renewable energy sources, the majority of GHG emissions tend to occur during the construction phase of the plant. Those emissions are quite low and can be managed with effective abatement methods.

Depending on specific geological conditions of the site, it is possible that GHG emissions shift over time during the operation phase of some power plants. Adding to the complexity of natural subsurface processes, the continuous extraction and reinjection of geothermal fluids, and pressure changes in the reservoir, can impact GHG concentrations in the produced steam and at the surface (Fridriksson et al. 2017). There is generally a shortage of data on these potential changes and further investigation is needed to understand them, but the likelihood of significant variations has been deemed low.

17.7 NOISE POLLUTION

Noise from geothermal power plants falls within the typical range for industrial facilities. The highest noise levels are temporary and occur during well drilling and plant construction. High-decibel noise levels are generated during drilling, by construction equipment, and by high-pressure steam releases after drilling and during testing. Intermittent sound levels associated with drilling and testing operations range from 80 to 115 decibels A-weighted (dBA), which is in the range of a motorcycle (Tester et al. 2006; Theodore and Theodore 2021, NIH 2025).

During normal plant operations, however, typical noise levels are lower, emanating mainly from cooling towers, transformers, and the powerhouse (housing the turbine and generator). These sound levels from 71 to 83 dBA are similar to levels in a congested urban setting (Tester et al. 2006). The impact of noise pollution can be reduced using various noise muffling techniques, such as temporary noise shields around drilling sites and additional insulation or mufflers on operating equipment (Kagel, Bates and Gawell 2007; Theodore and Theodore 2021).

17.8 RISK ASSESSMENT

Risk assessment is a critical tool in the development and operation of any industrial facility, and especially when considering a geothermal energy project with an array of site-specific variables. An advantage of many geothermal energy operations is that they have been utilized and tested over decades, if not centuries, with well-known risks and standard procedures. As the sector evolves, however, established methods and technologies applied in novel ways may present new risks to be managed.

In this context, risk assessment is a decision-making process for characterizing the nature and magnitude of risks to people and the environment, including natural resources and wildlife. It is a continuous process that should engage many stakeholders and be practical to undertake while providing thorough and transparent results. A comprehensive risk analysis should include the following phases (U.S. EPA 2025):

- Planning the assessment process, identifying goals, and communicating with interested parties.
- Problem formulation to refine objectives, identify what entities and resources are at risk and what components are important to protect, and develop an analysis plan.

- Analysis of exposure and effects to evaluate responses to stressors and possible outcomes.
- Risk characterization, using the results of the analysis to estimate the risk posed.
- Communicating results and implementing measures to effectively manage or mitigate the risk.

Ideally, the assessment does not end with risk management but rather makes use of continually gathered data and monitoring to improve the process and reduce risk based on learned information or other new factors (Theodore and Dupont 2012).

When assessing risks associated with a geothermal energy project, this analysis may be applied to any kind of potential hazards where advanced planning is possible and could help to mitigate negative outcomes, such as:

- Seismic and other geological risks, such as induced seismicity or unknown subsurface conditions.
- Environmental risks, such as accidental groundwater contamination or extreme weather event.
- Health and safety risks, such as skin burns or exposure to toxic materials.
- Operational risks, such as equipment failure.

The two subsections to follow offer greater detail about *how* to evaluate risks to health and the environment. Health risk and hazard risk are topics that must be included in any management plan or study, particularly as they apply to the geothermal energy systems. It should be noted that the practicing engineer or scientist must initially determine which category of risk – health or hazard – is to be analyzed or studied (Theodore and Dupont 2012; Theodore and Theodore 2021).

17.8.1 The Health Risk Evaluation Process

For the purposes of this subsection, a few definitions of common terms will apply. *Risk* is the probability that a person or the environment will suffer adverse consequences as a result of exposure to a substance. In this case, the amount of risk is determined by: (a) the concentration of chemical a person is exposed to, (b) the rate of intake or dose of the substance, and (c) the toxicity of the material. *Risk assessment* is the procedure used to attempt to quantify or estimate risk.

Two general types of potential health, safety, and accident exposures exist. These are classified as:

- *Chronic*: Continuous exposure occurs over long periods, generally several months to years. Concentrations of inhaled contaminants are usually relatively low; direct skin contact by immersion, by splash, or by contaminated air involves contact with substances exhibiting low dermal toxicity.
- *Acute*: Exposures occur for relatively short periods, generally minutes to 1– 2 days. The concentration of air contaminants is usually high relative to their concentrations of concern. In addition to inhalation, airborne substances

might directly contact the skin, or liquids and sludges may be splashed on the skin or into the eyes, leading to toxic effects.

In general, acute exposures to chemicals in the air are more typical in either transportation accidents and fires, or releases at chemical manufacturing or storage facilities. High concentrations of contaminants in the air usually do not persist for long periods of time. Acute skin exposure may occur when workers come in close contact with the substances to control a release, such as while offloading a corrosive material or while containing and treating a spilled material. Chronic exposures, on the other hand, are usually associated with longer-term removal and remedial operations. Contaminated soil and debris from emergency operations may be involved, soil and groundwater may be polluted, or temporary impoundment systems may contain diluted chemicals.

In recent years, several guidelines and handbooks have been produced to help explain approaches for health risk assessments. As discussed by a National Academy of Sciences committee convened in 1983, most human or environmental health problems can be evaluated by dissecting the analysis into four parts: health problem identification, dose-response assessment or toxicity assessment, exposure assessment, and risk characterization (Figure 17.3) (Theodore and Theodore 2012; Theodore and Dupont 2022).

For some perceived health problem, the risk assessment might stop with the first step, health problem identification, if no adverse effect is identified or if an agency elects to take regulatory action without further analysis. *Health problem identification* refers to presence of a toxic agent or a set of conditions that has the potential to cause adverse effects to human health. This identification involves an evaluation of various forms of information in order to identify different health risks. *Dose-response* or *toxicity assessment* is required in an overall assessment; responses/effects can vary widely since all chemicals and contaminants vary in their capacity to cause adverse effects. This step frequently requires that assumptions be made to relate experimental data for animals and humans. *Exposure assessment* is the determination of the magnitude, frequency, duration, and routes of exposure of human populations and ecosystems. Finally, in *risk characterization*, toxicology and exposure data/information are combined to obtain a qualitative or quantitative expression of risk.

Health risk assessment involves the integration of the information and analysis associated with the previous four steps to provide a complete characterization of the nature and magnitude of the risk and the degree of confidence associated with this

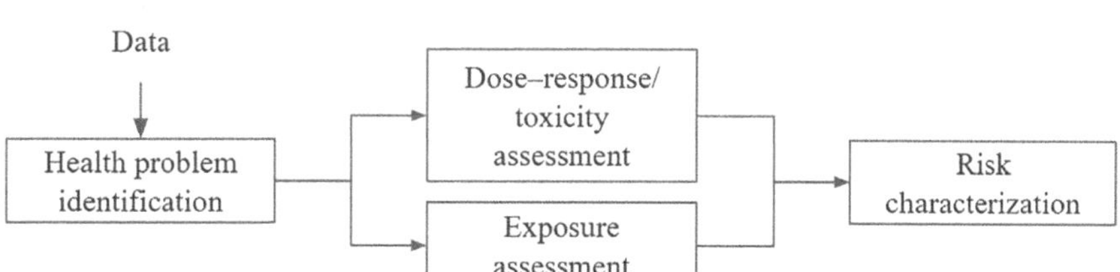

FIGURE 17.3 Health risk assessment flowchart.

characterization. A critical component of the assessment is a full elucidation of the *uncertainties* associated with each of the major steps. Encompassed in this broad concept of risk assessment are all of the essential problems of toxicology. Risk assessment takes into account all available dose–response data. It should treat uncertainty not by the application of arbitrary safety factors but by stating them in quantitatively and qualitatively explicit terms so that they are not hidden from decision-makers or the public. Risk assessment, defined in this broad way, compels an assessor to confront scientific uncertainties and to outline in explicit terms the means used in specific cases to deal with these uncertainties.

17.8.2 The Hazard Risk Assessment Process

Planning for an emergency must begin at the very start, when the facility itself is being planned. As emphasized in *Contingency Planning for Industrial Emergencies*, "The first line of defense against industrial accidents begins at the design stage. It should be obvious that it is much easier to prevent an accident rather than to try to rectify the situation once an accident has occurred" (Armenante 1991). There are several steps in evaluating the risk of an accident. These are detailed as follows (Theodore and Theodore 2012; Theodore and Dupont 2022):

1. A brief description of the equipment and chemicals used in the plant is needed.
2. Any hazard in the system must be identified. Hazards that may occur in a geothermal power plant could include electrical hazards, high-temperature pipe ruptures or leaks, corrosion, toxic vapor release, or slippage.
3. The event or a series of events that will initiate an accident need to be identified. An event could be a failure to follow correct safety procedures, improperly repaired equipment, or failure of a piece of process equipment or a safety device.
4. The probability that the accident will occur is determined. The probability can be ranked from low to high, with low probability meaning that it is unlikely for the event to occur during the operating life of a plant. A medium probability suggests there is a possibility that the event will occur. A high probability means that the event will likely occur at least once during the life of the plant.
5. The severity of the consequences of the failure event leading to an accident must be determined.
6. If the probability of the accident and the severity of its consequences are low, then the risk is usually deemed acceptable, and the plant should be allowed to operate. If the probability of occurrence is too high or the damage to the surroundings is considerable, then the risk is usually unacceptable and the system needs to be modified to minimize either the likelihood of the event and/or the consequences of these effects.

The heart of a hazard risk assessment algorithm provided is enclosed in the dotted box in Figure 17.4. This procedure allows for a reevaluation of the process if the risk is deemed unacceptable. The process is then repeated, starting with either Step 1

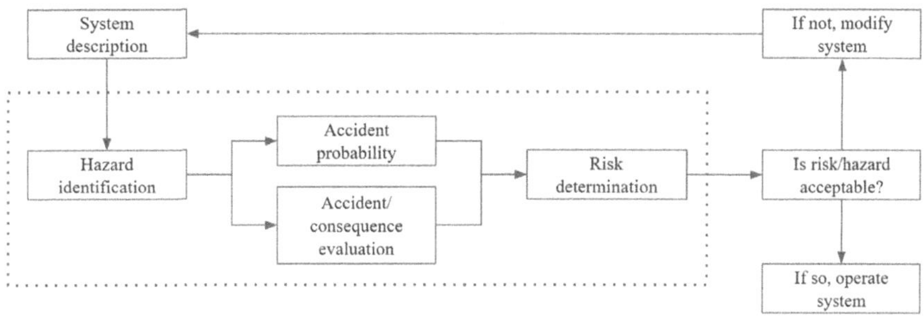

FIGURE 17.4 Hazard risk assessment flowchart.

or 2, and incorporating that completed assessment and new information to improve planning and mitigation efforts (Theodore and Dupont 2012).

17.9 ENVIRONMENTAL ADVANTAGES OF GEOTHERMAL ENERGY

The previous sections have summarized some of the environmental implications and potential hazards associated with geothermal energy. It should be emphasized, however, that geothermal energy is considered a safe and environmentally benign energy source, and geothermal operations to date support that characterization. With the pace of development picking up and the widespread deployment close at hand, it is useful to summarize the advantages that widespread geothermal energy deployment offers from an environmental, public health, and safety perspective.

17.9.1 Improved Air Quality

With continued improvements in technology, adoption of geothermal energy to reach 60 GW in electric generation capacity could reduce SO_2 emissions by 279,000 MT, NOx pollution by 417,000 MT, and $PM_{2.5}$ by 54,000 MT by 2050 (U.S. DOE 2022). For a public health outlook, a conservative estimate suggests that U.S. air quality and health benefits as a result of those emissions reductions would save at least $6 billion, preventing 5,000 premature deaths and more than 450,000 missed work and school days (Millstein et al. 2019).

While this discussion focuses mainly on geothermal power plants for electricity production, air emissions benefits of *geothermal heat pumps* (GHPs) in the heating and cooling sector are also worth noting. Public health benefits from *avoided* air emissions in this sector are even larger than for geothermal electricity generation, because GHPs can be utilized throughout more densely populated urban and suburban settings, as opposed to less-populated areas where a power plant is typically located. GHPs offset the use and storage of natural gas, propane, and oil for combustion-based heating and cooling equipment. Estimates for emissions reductions from widespread GHP adoption vary by location (i.e., are site-specific), depending on the regional climate, the annual need for heating and cooling, and existing state regulations on pollutants. But average estimates suggest avoided emissions equivalent to two to three times the total single-year emissions of SO_2 and NOx and one-fifth of a single-year of emissions of $PM_{2.5}$ from residential combustion sources, avoiding 8,700 premature deaths and yielding up to $61 billion in air pollution-related economic health benefits. Further, GHP systems do not require electricity to operate, and in most locations, their use reduces both on-site fuel and electricity requirements. GHP adoption thus offers further benefit by offsetting air emissions associated with electricity generation, providing an additional $4 billion in public health benefits in the United States by 2050 (Millstein et al. 2019).

17.9.2 Reduced GHG Emissions

As with air pollutant emissions, the fact that geothermal power generation does not require fuel or combustion means that lifecycle GHG emissions are also very low when compared with other energy sources, as shown in Figure 17.5.

Estimates suggest that widespread deployment of geothermal energy for electricity production (60 GW electric generation capacity) and heating and cooling (320 GW_{th} installed) by 2050 could avoid a combined 1,734 MMT of CO_2-equivalent emissions, comparable to taking 26 million cars off the road annually (U.S. DOE 2019). Indeed, technology improvements in EGS development that allow for heat recovery at lower temperatures and shallower depths with less-intensive drilling operations should result in even lower GHG emissions in the future (Sullivan et al. 2010).

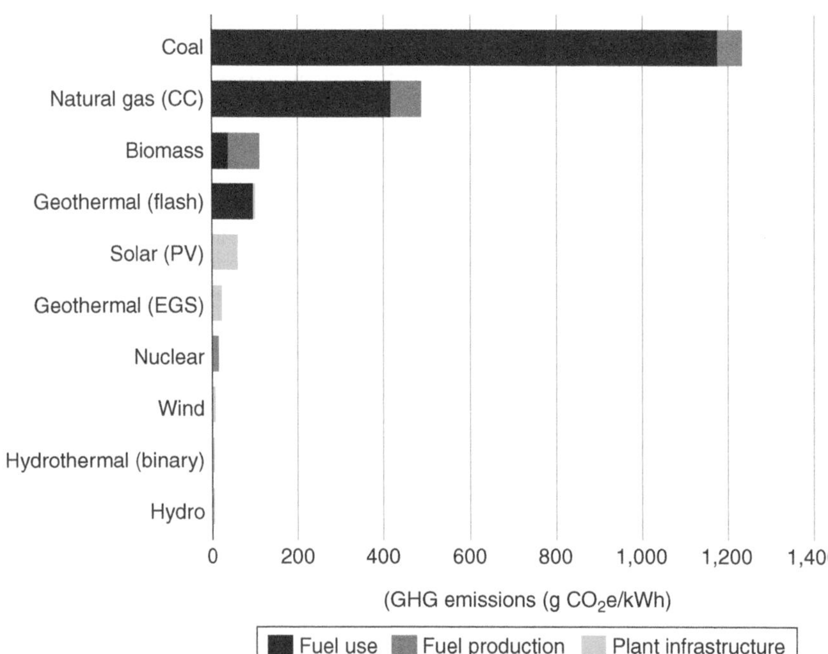

FIGURE 17.5 Lifecycle GHG emissions by power generation.

Source: **Center for Sustainable Systems (2024).**

17.9.3 Reliable and Renewable Energy

Renewable energy typically refers to forms of energy that are sustainable because their source is continuously replenished, such as wind or solar energy. In theory, tapping geothermal energy could extract heat from the Earth at a rate that is unsustainable, since there is, at any given time, a finite amount of heat available to extract. But the heat stored in the Earth since its formation 3 to 4.5 billion years ago is estimated to be equivalent to 44.2 terawatts-thermal, more than double the amount needed to match total global primary energy consumption in 2015 (U.S. DOE 2019). In addition, heat is continuously generated by the decay of naturally occurring radioactive materials and by friction from the shifting of materials in the Earth's crust (Kaven 2023). Thus, this steady supply of geothermal heat ensures a renewable and sustainable energy source for effectively an unlimited period into the future.

Responsible management of individual geothermal reservoirs requires that energy is extracted at a sustainable rate that allows heat in reinjected geothermal fluid to be replenished by surrounding rock. Heat replenishment and fluid reinjection help to extend the working life of the reservoir by maintaining consistent fluid levels and geochemistry of the site, while promoting continuous heat extraction and electricity generation (Smith 2022). The heat in a well-managed geothermal reservoir can last for decades, and production can be controlled by adjusting flow rates and volumes to allow for heat replenishment when necessary.

Unlike many other forms of renewable energy, geothermal energy is considered a stable, baseload power source that can operate 24 hours a day, regardless of ambient weather or other external conditions. It is also dispatchable, since geothermal fluid flow, and therefore production, can be monitored and closely controlled. When demand is lower, flow into the injection well can be slowed or redirected into different zones of the reservoir, essentially storing the heat until it is needed.

As the economy moves away from fossil fuel-based sources, one can safely conclude that geothermal energy holds *enormous* potential as an efficient foundational resource in the transition to secure renewable energy, particularly due to the exponential *decrease* in drilling costs.

17.10 CHAPTER SUMMARY

- Induced seismic events, typically very low in magnitude, can occur as a result of construction and operation of geothermal power plants. Careful monitoring and following established protocols can help mitigate seismic risks.
- Total water usage for geothermal systems is much lower than for most other forms of power generation and geothermal development poses low risk to groundwater sources.
- Geothermal power plants have small surface footprints and can be designed to minimize disruption to surrounding communities and ecosystems, also allowing for co-location with other land uses.
- The likelihood of contamination by solid waste and toxic materials in geothermal operations is low, as closed systems and mitigation methods reduce these risks.

- Geothermal energy operations have very low to zero direct air pollutant emissions, especially when compared with combustion-based power sources.
- Noise pollution associated with geothermal power plants is typical for industrial activity in the short-lived construction phase, and noise during normal operation is manageable, posing no significant hazards.
- Risk assessment is a decision-making process available for characterizing the nature and magnitude of risks to people and the environment, including natural resources and wildlife. Risks from geothermal energy development appear to be minimal.
- Geothermal energy is considered a safe, renewable, environmentally benign energy source. Widespread adoption could substantially reduce air pollution and GHG emissions, leading to considerable environmental and public health benefits. It appears to be the most attractive option for the next-generation energy economy due to the significant decrease in drilling costs.

REFERENCES

Armenante, P.M. 1991. *Contingency Planning for Industrial Emergencies*. Florence, KY: Van Nostrand Reinhold.

Blankenship, D., Gertler, C., Kamaludeen, M., O'Connor, M., and Porse, S. 2024. *Pathways to Commercial Liftoff: Next-Generation Geothermal Power*. Washington, D.C.: U.S. Department of Energy. https://climateprogramportal.org/wp-content/uploads/2025/02/LIFTOFF_DOE_NextGen_Geothermal_v14.pdf.

Center for Sustainable Systems. 2024. *Geothermal Energy Factsheet*. Ann Arbor, MI: University of Michigan. https://css.umich.edu/publications/factsheets/energy/geothermal-energy-factsheet. (accessed April 30, 2025).

Energy Information Administration. 2020. *Most U.S. Utility-Scale Geothermal Power Plants Built since 2000 Are Binary-Cycle Plants*. Today in Energy, July 30. Washington, D.C.: U.S. EIA. www.eia.gov/todayinenergy/detail.php?id=44576. (accessed July 15, 2025).

Finster, M., Clark, C., Schroder, J., and Martino, L. 2015. Geothermal Produced Fluids: Characteristics, Treatment Technologies, and Management Options. *Renewable and Sustainable Energy Reviews* 50:952–966. https://doi.org/10.1016/j.rser.2015.05.059.

Fridriksson, T., Merino, A.M., Orucu, A.Y., and Audinet, P. 2017. Greenhouse Gas Emissions from Geothermal Power Production. *Proceedings, 42nd Workshop on Geothermal Reservoir Engineering*. Stanford, CA: Stanford University. https://documents1.worldbank.org/curated/en/875761592973336676/pdf/Greenhouse-Gas-Emissions-from-Geothermal-Power-Production.pdf.

Grigoli, F., Cesca, S., Rinaldi, A.P., Manconi, A., López-Comino, J.A., Clinton, J.F., Westaway, R., Cauzzi, C., Dahm, T., and Wiemer, S. 2018. The November 2017 Mw 5.5 Pohang Earthquake: A Possible Case of Induced Seismicity in South Korea. *Science* 4:1003–1006. https://doi.org/10.1126/science.aat2010.

Kagel, A., Bates, D., and Gawell, K. 2007. *A Guide to Geothermal Energy and the Environment*. Washington, D.C.: Geothermal Energy Association. www.ourenergypolicy.org/wp-content/uploads/2016/02/Environmental-Guide.pdf.

Kaven, J. 2023. *Geothermal Energy Research Within the USGS: Goldilocks to Electricity With Earthquakes in Between*. USGS Public Lecture Series. Reston, VA: U.S. Geological Survey. www.usgs.gov/media/videos/pubtalk-092023-geothermal-energy-research-within-usgs-goldilocks-electricity. (accessed May 10, 2025).

Levine, A., Cook, J., Beckers, K., and Young, K.R. 2017. *Geothermal Induced Seismicity National Environmental Policy Act Review*. Golden, CO: U.S. Department of Energy National Renewable Energy Laboratory. https://docs.nrel.gov/docs/fy18osti/70203.pdf.

Majer, E., Nelson, J., Robertson-Tait, A., Savy, J., and Wong, I. 2012. *Protocol for Addressing Induced Seismicity Associated with Enhanced Geothermal Systems.* Energy Efficiency and Renewable Energy Geothermal Technologies Program, Washington, D.C.: U.S. Department of Energy. www1.eere.energy.gov/geothermal/pdfs/geothermal_seismicity_protocol_012012.pdf. (accessed May 27, 2025).

Millstein, D., McCall, J., Macknick, J., Nicholson, S., Keyser, D., Jeong, S., and Heath, G. 2019. *GeoVision Analysis Supporting Task Force Report: Impacts—The Employment Opportunities, Water Impacts, Emission Reductions, and Air Quality Improvements of Achieving High Penetrations of Geothermal Power in the United States,* NREL/TP-6A20-71933.Berkeley, CA and Golden, CO: Lawrence Berkeley National Laboratory and National Renewable Energy Laboratory. www.nrel.gov/docs/fy19osti/71933.pdf and https://emp.lbl.gov/publications/geovision-analysis-impacts-task-force.

National Institutes of Health. 2025. Noise-Induced Hearing Loss. www.nidcd.nih.gov/health/noise-induced-hearing-loss. (accessed June 5, 2025).

National Oceanic and Atmospheric Administration. 2024. What Is Subsidence? https://oceanservice.noaa.gov/facts/subsidence.html. (accessed May 27, 2025).

Rubinstein, J. 2018. *Yes, Humans Really Are Causing Induced Earthquakes!* Washington, D.C.: U.S. Geological Survey. www.usgs.gov/media/videos/pubtalk-52018-yes-humans-really-are-causing-induced-earthquakes. (accessed May 21, 2025).

Schroeder, J.N., Harto, C.B., Horner, R.M., and Clark, C.E. 2014. *Geothermal Water Use: Life Cycle Water Consumption, Water Resource Assessment, and water Policy Framework.* Lemont, IL: Environmental Science Division, Argonne National Laboratory, U.S. DOE. https://publications.anl.gov/anlpubs/2014/06/105260.pdf.

Skipka, K.J., and Theodore, L. 2014. *Energy Resources: Availability, Management, and Environmental Impacts*. Boca Raton, FL: CRC Press/Taylor & Francis Group.

Smith, M. 2022. *Enhanced Geothermal Systems: Introduction and Issues for Congress.* R47256. Washington, D.C.: Congressional Research Service. www.congress.gov/crs_external_products/R/PDF/R47256/R47256.6.pdf.

Stanford University. 2024. *Geothermal Energy*. Stanford, CA: Understand Energy Learning Hub, Stanford University. https://understand-energy.stanford.edu/energy-resources/renewable-energy/geothermal-energy. (accessed May 27, 2025).

Sullivan, J.L., Clark, C.E., Han, J., and Wang, M. 2010. *Life-Cycle Analysis Results of Geothermal Systems in Comparison to Other Power Systems*. ANL/ESD/10-5. Lemont, IL: Argonne National Laboratory. www.energy.gov/eere/geothermal/articles/life-cycle-analysis-results-geothermal-systems-comparison-other-power.

Tester, J.W., Anderson, B.J., Batchelor, A.S., Blackwell, D.D., DiPippo, R., Drake, E.M., Garnish, J., Livesay, B., Moore, M.C., Nichols, K., Petty, S., Toksöz, M.N., and Veatch, R.W. Jr. 2006. *The Future of Geothermal Energy, Impact of Enhanced Geothermal Systems (EGS) on the United States in the 21st Century*. Cambridge, MA: Massachusetts Institute of Technology. www1.eere.energy.gov/geothermal/pdfs/future_geo_energy.pdf.

Theodore, L., and Dupont, R.R. 2012. *Environmental Health and Hazard Risk Assessment: Principles and Calculations.* Boca Raton, FL: CRC Press/Taylor & Francis Group.

Theodore, L., and Dupont, R.R. 2022. *Introduction to Desalination: Principles, Processes, and Calculations.* Hoboken, NJ: Wiley.

Theodore, M.K., and Theodore, L. 2021. *Introduction to Environmental Management, 2nd ed.* Boca Raton, FL: CRC Press/Taylor & Francis Group.

U.S. Department of Energy. 2019. *GeoVision: Harnessing the Heat Beneath Our Feet.* Washington, D.C.: U.S. Department of Energy. www.energy.gov/sites/default/files/2019/06/f63/GeoVision-full-report-opt.pdf.

U.S. Department of Energy. 2022. *Enhanced Geothermal Shot Frequently Asked Questions.* Washington, D.C.: U.S. Department of Energy. www.energy.gov/sites/default/files/2022-09/EGSFAQ_Sept22.pdf.

United States Environmental Protection Agency. 2024. *Class II Oil and Gas Related Injection Wells.* Washington, D.C.: U.S. EPA.. www.epa.gov/uic/class-ii-oil-and-gas-related-inject ion-wells . (accessed May 29, 2025).

United States Environmental Protection Agency. 2025. *Conducting an Ecological Risk Assessment.* Washington, D.C.:U.S. EPA. www.epa.gov/risk/conducting-ecological-risk-assessment. (accessed June 9, 2025).

United States Geological Survey. 2016. *M 5.8–14 km NW of Pawnee, Oklahoma.* Washington, D.C.: U.S. Geological Survey. https://earthquake.usgs.gov/earthquakes/eventpage/us10006jxs/ executive#executive. (accessed May 18, 2025).

Part 4

Epilogue

The concluding part of this book contains two chapters. The chapter numbers and accompanying titles are listed as follows:

DOI: 10.1201/9781003667759-21

18 The Future of Geothermal Energy

18.1 INTRODUCTION

The vast amounts of geothermal energy locked in the Earth are now within reach thanks to recent advances in technology across an array of fields. These breakthroughs come at a time when energy needs are skyrocketing. With the right combination of technological progress, policy support, and market validation, the sector is poised for remarkable growth in the near term.

Section numbers and their accompanying titles for this chapter on the future of geothermal energy systems are listed as follows:

18.2 Next-Generation Geothermal Energy
18.3 Recent and Emerging Technological Advancements
18.4 The Role of AI
18.5 Applications and Market Potential
18.6 Policy and Regulatory Tools to Support Geothermal Energy
18.7 Current Barriers to Deployment
18.8 Companies Involved in the Geothermal Sector
18.9 Summary
References

18.2 NEXT-GENERATION GEOTHERMAL ENERGY

As noted throughout this text, the geothermal energy industry has made advances in recent decades that build on centuries old systems to harness the Earth's heat. Until recently, development has been limited to those relatively rare sites where hot rock, fluid, and permeability all converge naturally to enable economically feasible heat recovery from hydrothermal (conventional geothermal) systems. Next-generation geothermal energy technologies use, for example, engineered reservoirs to recover heat from superhot rock (SHR) in locations beyond those scarce natural sources. Estimates suggest that the heat energy accessible via these next-generation systems is capable of meeting global electricity demand 140 times over (IEA 2023).

DOI: 10.1201/9781003667759-22

18.2.1 Enhanced Geothermal Systems

An *enhanced geothermal system*, or EGS, refers to a system where the low permeability of subsurface rock is enhanced by some artificial means in order to allow for heat recovery. EGS technologies leverage advanced drilling methods that have been perfected and long-used by the oil and gas industry, including horizontal drilling, well stimulation, and hydraulic fracturing. These methods create new fractures in, or widen existing fractures in subsurface deposits to increase pathways for fluid flow through the hot rock, enabling EGS to unlock heat energy from previously inaccessible or unfavorable geothermal sources. As shown in Figure 18.1, while hydrothermal resources are geographically limited to areas mainly in the western United States, all 48 contiguous U.S. states are found to have some moderately favorable EGS resources at temperatures above 150°C (300°F) at 7 km (4.3 miles) depth below the surface (Smith 2022).

The first EGS pilot project took place in the early 1970s with the U.S. Department of Energy's (DOE's) Fenton Hill Hot Dry Rock project in Los Alamos, New Mexico, and research on EGS has continued worldwide. Breakthroughs in oil and gas drilling techniques in the 1990s and 2000s that were applied to EGS have moved development forward significantly, with commercialization beginning in the mid-2020s.

18.2.2 Closed-Loop Geothermal Systems

Still in the development phase, *closed-loop geothermal systems* (sometimes referred to as *advanced geothermal systems*, or AGS) circulate working fluid in boreholes that are completely closed to the environment. Unlike EGS, where geothermal fluid flows from the injection well through many small fractures in the rock to the production well, closed-loop systems use just *one* pathway for geothermal fluid to cycle through the hot rock (see Figure 18.2). The loop essentially serves as an underground heat exchanger, heating the geothermal fluid via conductive heat transfer without it ever having direct interaction with the rock (IEA 2023).

While a variety of designs and fluid compositions are being investigated, closed-loop systems in general would offer a few principal advantages:

- They can, in theory, be sited anywhere, with fewer concerns over site-specific geological conditions. Since the geothermal fluid has no contact with the rocks or surrounding materials, there are no chemical interactions or waste streams to manage.
- They have even lower water consumption than EGS, since there is no below ground water loss, no reservoir stimulation, and less long-term reservoir management involved.
- Without the need for reservoir stimulation, there is a much lower risk for increased seismicity.
- The engineered system allows for precise temperature control and production optimization.

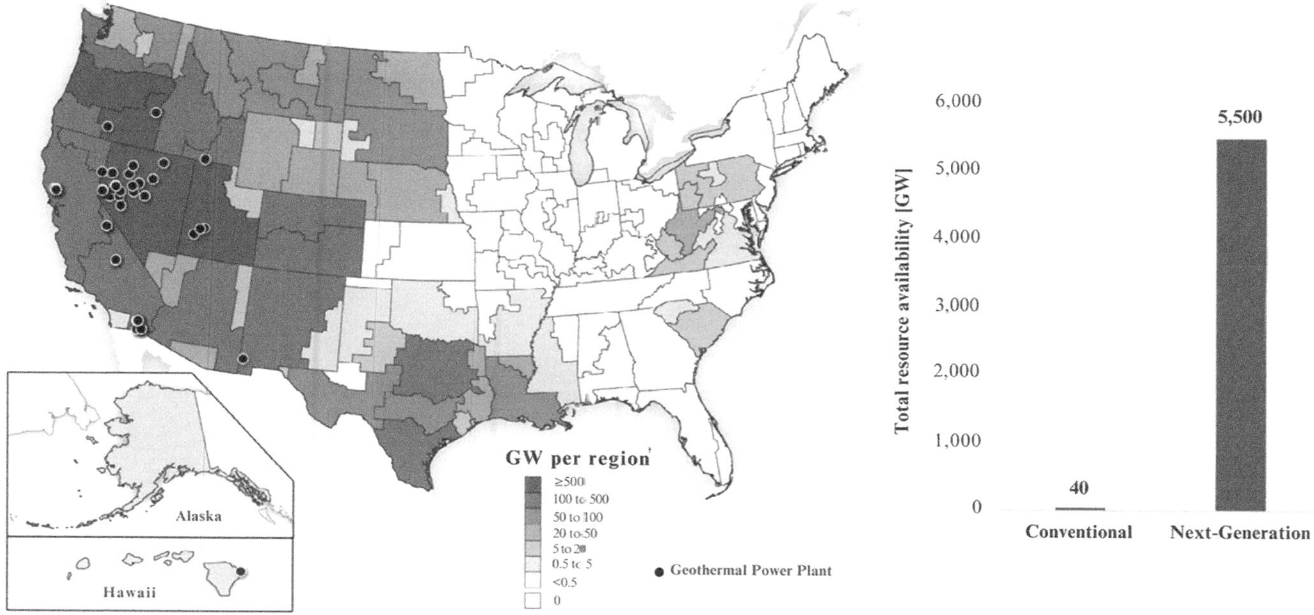

FIGURE 18.1 At left, total next-generation potential across the United States, overlain by locations of current conventional geothermal plants producing 3.7 GW or more of power. At right, comparison between total available resource for conventional geothermal and next-generation geothermal.

Source: Blankenship et al. (2024).

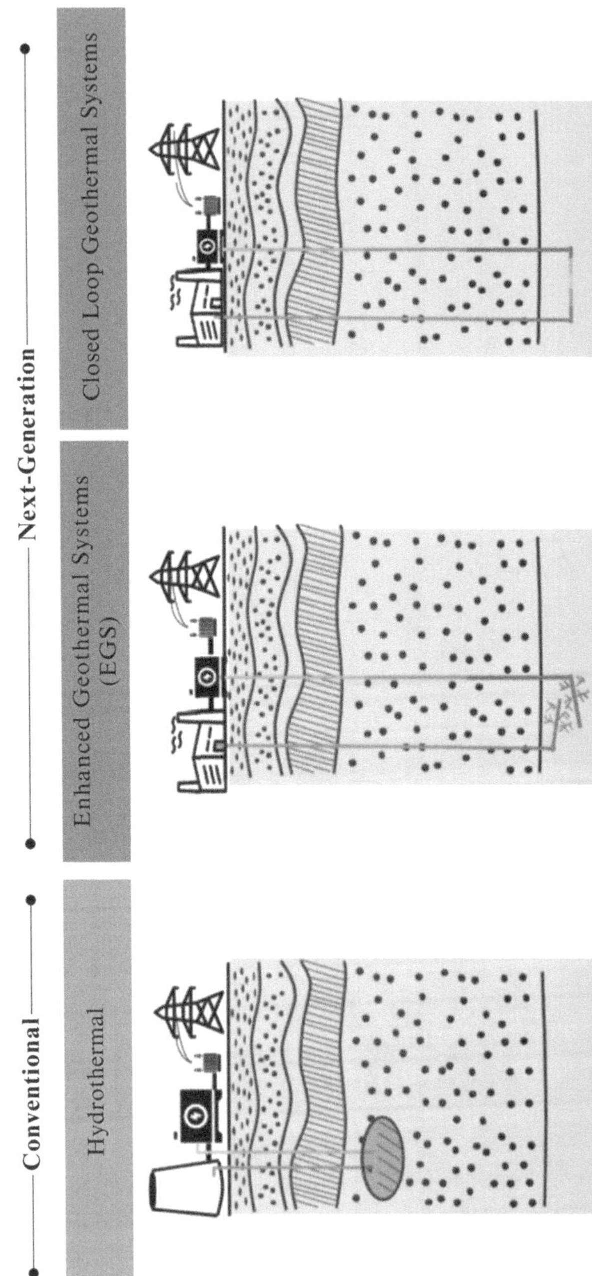

FIGURE 18.2 Conventional and next-generation geothermal technologies.

Source: Blankenship et al. (2024).

Closed-loop geothermal systems have been the subject of research for many years and are in development with support from both government and private interests. Like EGS, advances in drilling methods are driving the technology forward, but technical challenges remain. To achieve high-enough levels of heat transfer through the single fluid pathway, closed-loop systems are used at much greater depths and higher temperatures, with future projects already expected to reach 8 km (5 miles) or more underground. For commercial power production, closed-loop systems will require hundreds of kilometers of *horizontal* length to provide enough surface area for adequate heat recovery, which pushes the limits of current drilling capabilities (Blankenship et al. 2024).

18.2.3 Superhot Rock

Superhot rock (sometimes called superdeep or ultradeep) geological systems are the deep-underground source where high-enthalpy fluids (typically water with high salinity) reach supercritical temperatures that are above 400°C. This supercritical fluid has a very high energy density, and the energy output from these SHR zones is estimated at ten times higher than other geothermal wells (NREL 2025b). EGS and closed-loop geothermal systems refer to the *methods* that can be used to access that heat energy. In most areas of the United States, reaching those resources would require drilling beyond 10 km (6.25 miles) deep, which is not currently economically feasible but may become achievable in the very near future. Research is underway to explore and develop drilling methods, stimulation techniques, and equipment design that can withstand the extreme conditions in SHR systems.

18.3 RECENT AND EMERGING TECHNOLOGICAL ADVANCEMENTS

Recent technological advances in many different areas are enabling breakthroughs in geothermal development, accelerating the pace of progress, and moving next-generation geothermal closer to commercialization.

18.3.1 Advances in Drilling

As previously noted, EGS leverages drilling methods developed by the oil and gas industry. Applied to geothermal energy development, these innovations are making EGS more feasible across diverse geological areas and more time- and cost-effective. Innovations discussed in Chapter 14 include:

- *Directional* or *horizontal drilling*
- *Hydraulic fracturing*
- The use of *polycrystalline diamond compact* drill bits (U.S. DOE 2025a)

Drilling costs have typically accounted for half the total cost of EGS projects (Tester et al. 2006). Advances in drilling methods and materials can help increase drilling efficiency, leading to significant time and cost savings, and have led to 500% improvements in geothermal well drilling rates at the U.S. DOE's Frontier Observatory for Research in Geothermal Energy (FORGE) demonstration site over 5 years, bringing them closer to par with standard drilling rates for the oil and gas industry (Blankenship et al. 2024).

18.3.2 Binary Cycle Power Plants

The advent of binary cycle power plants represents a notable breakthrough in power generation by geothermal energy systems over the last decades and is key to expanding the development of geothermal energy from low-temperature resources. While flash and dry steam plants are used for geothermal resources hotter than 200°C, binary cycle power plants are designed to operate at lower temperatures. They employ a heat exchanger to transfer heat from the geothermal fluid to a secondary (or "binary") fluid that has a lower boiling point than water (Figure 18.3). Coupled with EGS, binary cycle power plants can take advantage of more commonly found moderate temperature reservoirs and enable wider use of once marginal geothermal energy resources.

18.3.3 Data Collection and Processing

An array of technology-enabled data collecting and processing tools is emerging that helps identify, monitor, and evaluate geothermal resources that are not easily observed otherwise. Data collection, analysis, and sharing can help operators optimize performance, better manage risk, and establish and improve standard practices. Artificial intelligence (AI) can play an important role in managing and utilizing these data, applying existing knowledge and machine learning to reduce learning curves and uncertainty. The potential for AI applications in geothermal energy development is discussed in greater detail in Section 18.4.

Remote sensing refers to methods used to observe physical characteristics of an area by measuring natural reflected and emitted radiation from a distance, often from satellites or aircraft. In geothermal resource development, remote sensing can be used for large-scale mapping of surface features to assist in identifying potential geothermal resources. It can also be applied to fault studies near geothermal fields to improve understanding of induced seismicity (Porse et al. 2022; Cochran et al. 2024).

Once wells are drilled, fiber-optic cables can be incorporated into boreholes and reservoirs as *in-situ* measurement tools. Data can be used to map the underground system and to monitor strain, temperatures, and acoustics in real time, providing a view into subsurface conditions and fluid flow changes. Combining data from traditional monitoring techniques with fiber-optic strain meters and acoustic sensors allows for more accurate monitoring of microseismicity.

18.4 THE ROLE OF AI

As previously noted, AI is set to significantly impact the expansion of geothermal energy across various stages of project development, from site exploration to optimized

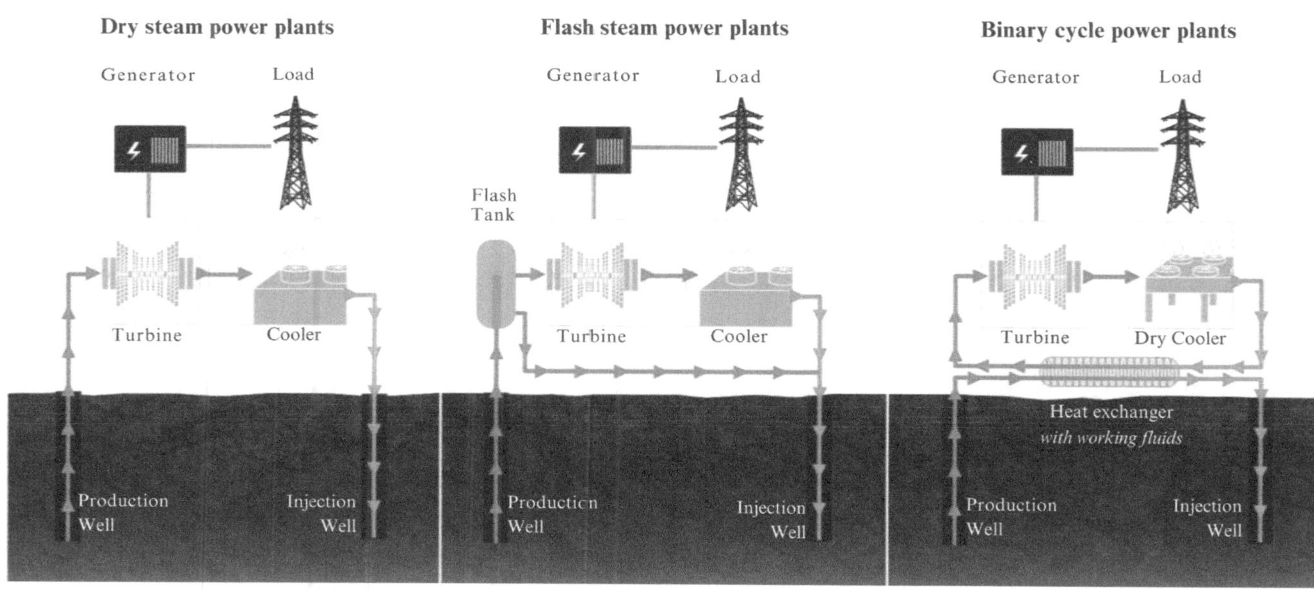

FIGURE 18.3 Geothermal power-plant configurations.

Source: U.S. DOE (2019).

drilling and operations. In each stage, AI has the potential to increase efficiency, reduce costs, and mitigate risks. Additional details are provided in the next five subsections.

18.4.1 Use of AI in Resource Exploration and Site Selection

Exploratory drilling accounts for a substantial portion of the upfront costs associated with geothermal projects. AI algorithms can analyze and interpret vast amounts of data collected from diverse sources like seismic surveys, well logs, satellite imagery, geological maps, and geochemical data. This capability enables the identification of patterns and anomalies to locate potential geothermal reservoirs with much greater accuracy than traditional methods.

The use of AI can assist with more accurately estimating the size and potential of geothermal resources, guiding investment decisions and prioritizing exploration efforts that reduce unsuccessful drilling attempts. AI-enabled three-dimensional modeling of subsurface geological structures also allows for improved evaluation of reservoir characteristics, including fault lines, fluid pathways, and heat distribution. Mapping techniques that utilize remote sensing and GIS data produce precise predictive maps of geothermal potential, allowing for informed decisions on promising drilling locations.

18.4.2 Use of AI in Drilling Operations

Once drilling is underway, AI can analyze real-time monitoring data to reduce drilling time and costs. Continuous measurement and adjustment of parameters such as rotation speed and mud flow rate can maximize penetration rates, while monitoring of pore pressure and fracture gradients can help predict wellbore instability. Algorithms can also be used to monitor the status of drilling equipment to minimize wear and predict potential failures to allow for timely preventive maintenance. *Geosteering* involves making real-time adjustments to a drilling path, and AI algorithms can be used to guide a drill bit in near real-time toward the most productive zones in the geothermal reservoir by analyzing subsurface data as it is collected.

18.4.3 Use of AI in Reservoir Management and Production

In an operating reservoir, AI can be used to continuously monitor reservoir behavior, allowing for real-time adjustments to injection and production rates. Models can simulate how geothermal fluids will flow through the subsurface in order to optimize well placement and maintain long-term reservoir stability. In the context of a power plant as a whole, there are many opportunities to use AI to optimize the entire operation, from heat extraction and fluid flow rates to turbine and generator controls, ensuring consistent power output and maximizing the efficiency of energy extraction.

18.4.4 Use of AI for Risk Mitigation and Enhancing Economic Viability

The use of AI can significantly reduce risks and uncertainties associated with geothermal operations. In resource identification and verification, machine learning models can process and interpret complex data beyond what traditional assessments have allowed. The combined effects of improved exploration, drilling, and operations would lead to substantial reductions in overall project costs, making geothermal development projects more economically viable and attractive to investors.

AI's monitoring and predictive capabilities also have the potential to assist in assessing and mitigating environmental risks (Chapter 17) associated with development projects. Continuous monitoring and processing of real-time data can ensure compliance with applicable regulations and minimize potential hazards or ecological disturbances. AI tools can evaluate real-time strain and pressure data to anticipate and reduce induced seismicity risk, and to inform future seismic modeling to improve prediction capabilities.

18.4.5 Limitations and Challenges of AI in Geothermal Energy Systems

Despite the immense potential, there are currently limitations to AI's capabilities and challenges to its implementation in geothermal development and operations. As it continues to "learn," an AI model is only as good as the data it is trained on, and challenges to incorporating AI effectively into any complex system can include the following:

- Systems across one operation may rely on an array of different and inconsistent data sources. Incomplete, inaccurate, or biased data can lead to flawed predictions.
- Historical reliance on manual data and individual expertise can hinder the collection of consistent data needed for AI training.
- Some complex algorithms can be difficult to interpret, with opaque reasoning or logic that can raise concerns about accountability and accuracy.

The availability of high-quality, standardized data is critical to the successful application of AI models. Data sharing and transparency among various stakeholders should be encouraged to ensure comprehensive data from a variety of sources and its responsible implementation. In addition, diligent attention should be paid to security of data and AI systems to reduce vulnerability to cyberattacks that could disrupt operations or cause physical damage.

Training and operating large AI models can be highly energy-intensive, which can contradict the sustainability goals of a renewable energy source like geothermal. A growing number of technology companies are investing in off-grid solutions to power data centers and AI capabilities. These providers should be urged to pursue efficiency-minded, long-term pathways to managing those power needs, such as co-location with renewable energy sources (discussed further in Section 18.5).

AI offers transformative potential for geothermal drilling by enhancing every stage of the process, making it more efficient, cost-effective, and environmentally responsible. Addressing the challenges related to data, transparency, and cybersecurity will be key to fully realizing these benefits and accelerating the growth of geothermal energy as a vital part of the future global energy mix.

18.5 APPLICATIONS AND MARKET POTENTIAL

Rapidly rising energy demand, strained supply chains, and an increase in extreme weather events all add stress to the U.S. electricity grid. Geothermal energy can play a significant role in meeting demand and providing a stable, secure domestic baseload power source. The DOE projects that, with the right support, geothermal energy could provide 90 GW of capacity and 12% or more of annual U.S. electricity generation by 2050 (Augustine et al. 2023).

With EGS commercialization, utility-scale electricity generation from geothermal energy brings a number of grid benefits, including:

- *Firm, baseload supply*: Geothermal power plants provide consistent baseload power, with the capability to operate continuously at a steady rate to meet consistent demand. This reliable supply is critical to grid stability and to reduce the likelihood or length of power outages during severe weather events.
- *Flexible and dispatchable power*: With the ability to control flow rates and pressures, geothermal power plants can slow, pause, and restart electricity generation relatively easily. They can be responsive to fluctuating demand and access stored power as needed.
- *High capacity factor*: With relatively few operational efficiency losses, geothermal's capacity factor – the ratio of actual electrical energy output to the maximum possible electrical energy output for a given period – is near 90%.

Geothermal power plants occupy compact surface footprints, and flexibility in siting for EGS provides diverse opportunities for co-location with other industrial facilities or even agricultural uses. Co-location can improve the economic feasibility of project development by bringing shared capital investment and offering a secondary revenue stream. A co-located facility could also reduce infrastructure needs and costs, taking advantage of shared equipment or operational services. Depending on the application, additional employment opportunities and cooperation with nearby businesses can help support local economies.

As a firm, flexible power source, geothermal power plants can complement *variable renewable energy* (VRE) sources like wind and solar that rely on favorable ambient conditions or time-of-day. When electricity supplies from VRE are low, a geothermal plant can increase production on demand. Co-location of these facilities could share transmission lines to maximize investment and productivity.

Co-siting geothermal power plants with hyper-scale data centers represents a significant opportunity for growth, particularly with the increasing use of AI and the need for electricity to power it. Geothermal plants can provide this demand reliably and cleanly, and without the need for significant surface infrastructure. Off-grid "behind-the-meter" arrangements, where data center operators purchase power directly from the supplier rather than a utility company, have the added benefit of bypassing the lengthy grid interconnection process. While there is uncertainty about future needs, estimates suggest that EGS power plants could meet 64% of anticipated power center demand by the early 2030s (King et al. 2025).

Geothermal and oil and gas hybrid systems can be used to coproduce energy from active oil and gas reservoirs that are producing large amounts of hot water as part of resource recovery. Heat can be recovered from the spent fluid and used for energy production or for direct use, providing cost and energy efficiencies and reducing greenhouse gas emissions. In another potential synergy with oil and gas infrastructure, well retrofits can make use of abandoned wells to recover geothermal energy and use it for heating, cooling, or electricity generation (NREL 2025a, 2025c).

Outside of direct electricity generation, there is growing interest in reservoir thermal energy storage (*RTES*), which takes advantage of geothermal reservoirs to store heat for 100 hours or more without converting it to electricity (U.S. DOE 2025b). RTES has potential applications in an array of industries, including energy storage from VREs or for industrial heating needs such as pulp and paper drying and food processing.

18.6 POLICY AND REGULATORY TOOLS TO SUPPORT GEOTHERMAL ENERGY

Strong policy and regulatory guidelines will aid in promoting geothermal energy if it is to reach widespread commercialization. While policy and market outlooks are constantly changing, there have been solid demonstrations of geothermal energy's potential as a safe, reliable, and renewable domestic energy source. In the United States, some key policy frameworks are being developed to drive growth in the burgeoning sector.

18.6.1 Expanded Financing and Tax Credits

Along with private investment, financing opportunities with the backing of the federal government have helped spur projects that continue to demonstrate geothermal's capabilities. The Bipartisan Infrastructure Law of 2021 earmarked $84 million for EGS pilot projects through the U.S. DOE. The Clean Electricity Production Tax Credit (45Y) and Investment Tax Credit (48E) provide credits for qualified investments in clean electricity production and energy storage technologies, both relevant for the geothermal sector. But these policies tended to attract investment in energy sources with low capital costs. Future policies should incentivize investments in geothermal energy, specifically, to drive support for long-term, high-capital-cost projects. Funding policies like these are essential for early developers to be able build projects at-scale and show repeatable successes, helping to build confidence in the technology and the industry (Hill and Garth 2024).

18.6.2 Support for Expanded Research and Demonstration Sites

The U.S. government has led research into geothermal energy for decades, including the first EGS demonstration project in 1971 and the recent FORGE (www.energy.gov/eere/geothermal/forge) site near Milford, Utah, established in 2014. These demonstration sites provide critical opportunities for in-field testing and innovation but are somewhat limited by lack of funding and geography. Expanded federal support for multiple demonstration projects in diverse sites across the United States could provide demonstration opportunities for researchers and startups across next-generation technologies. Increased access to in-field testing can further accelerate innovation in both EGS and closed-loop systems, facilitate repeatable and scalable successes, and consequently reduce the risk of investment in early-stage projects. Public funding should also seek to support long-range research goals at early stages where private investment is unlikely (Hill and Garth 2024).

18.6.3 Streamlined Permitting

Most geothermal projects in the United States take 7–10 years to develop, with much of that time spent navigating the complex and unpredictable federal permitting process. The resulting uncertainty before the project even begins can stifle investment and exploration of new technologies. There are many ways the permitting process could be modified to increase efficiency and reduce timelines without eliminating critical environmental assessments. Some suggested reforms are outlined as follows and illustrated in Figure 18.4 (Blankenship et al. 2024):

- Streamline phases that are presently distinct from each other, such as exploration and exploration drilling, so that they can occur concurrently.
- Modify the process to differentiate between the technical siting requirements for conventional and next-generation geothermal projects.
- Centralize permitting expertise for renewable energy and geothermal development to improve inter-agency coordination.

18.6.4 Workforce Development and Public Outreach

A robust workforce is essential to a healthy industry, and policies that promote workforce training for the geothermal energy industry will be important as widespread deployment is reached. While there is overlap with oil and gas sector expertise that will drive many workers to geothermal, academic training and trade programs specific to next-generation technologies can help prevent a shortage of workers in the future. Federal agencies and national laboratories can work in partnership with academia and industry to create large-scale programs to engage potential workers. These programs should incorporate field-based experience and include re-training for workers in declining industries. In addition, educating the public at-large about geothermal energy can help raise awareness about its benefits and challenges while dispelling misperceptions. Such outreach can also foster good working relationships among local stakeholders and potentially serve as an entry point for local workers.

The geothermal energy sector is at an inflection point similar to the rapid expansion of environmental management following the founding of the U.S. Environmental Protection Agency in December 1970, when training and professional development

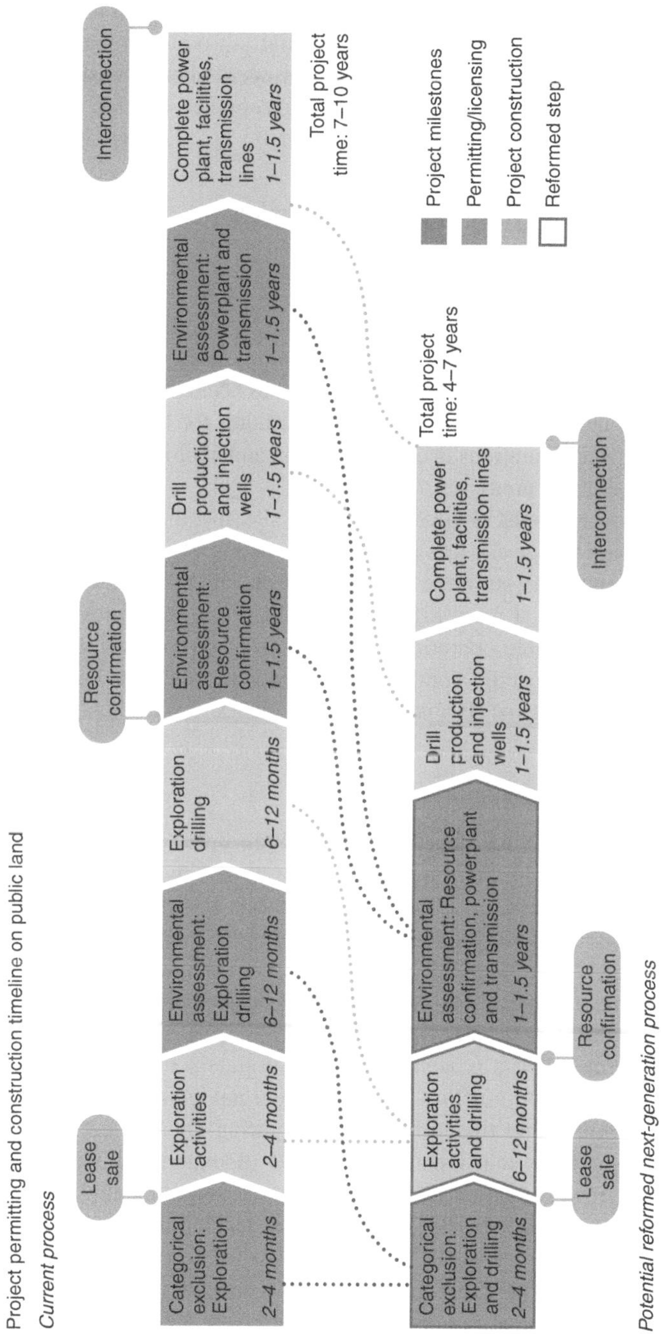

FIGURE 18.4 Permitting timelines for geothermal projects today (top) and in a potential streamlined future scenario (bottom).

Source: Blankenship et al. (2024).

activities emerged in order to help establish effective management strategies and operations for this new discipline. Timely information and technology transfer is essential in disseminating geothermal knowledge, tools, and best practices. Coordinated action across the DOE and other agencies, national lab-oratories, research universities, trade programs, labor unions, and the private sector can accelerate knowledge sharing to allow workforce training and public acceptance to keep pace with the industry's momentum and growing workforce needs.

18.7 CURRENT BARRIERS TO DEPLOYMENT

18.7.1 Technical Barriers

As previously discussed, there are inherent challenges to geothermal development because it is a difficult-to-access subsurface resource. Many of these are technical barriers that must be addressed through ongoing research and demonstration. Locating and characterizing new geothermal resources is difficult for both hydrothermal and EGS development, and limitations in current drilling and exploration technologies are tempering growth in these areas.

Because many facets of EGS development are site-specific, it can be difficult to apply operational knowledge and lessons learned from one site to another. Uncertainty about subsurface conditions, resource characterization, and the longevity of a specific engineered resource make it more difficult to realize iterative improvements and to establish repeatable, scalable operations that lead to economies of scale and wider deployment than if site-specific conditions were not so critically important to the success of geothermal systems (Blankenship et al. 2024).

18.7.2 Nontechnical Barriers

Key challenges for geothermal development are high upfront costs and risk at the start of new projects. Exploration and drilling are costly investments that may not pay off if a potential project site is deemed unsuitable. In-field testing and learning from active geothermal developments can help to identify viable new resources more quickly and accurately. Coupled with continued improvements in drilling, specialized financing arrangements such as cost-sharing and insurance programs can ease initial costs and risks for developers and investors (Blankenship et al. 2024).

As noted in Figure 18.4, a typical current development timeline for geothermal projects is 7–10 years, with much of that time spent navigating the permitting process. It can be complex and time-consuming, requiring engagement with multiple federal and state agencies that have varied levels of experience with geothermal projects. There is growing awareness that cumbersome and often duplicative permitting requirements are a major barrier to geothermal development and to new energy projects in general.

Related to permitting, working through land access and ownership issues for potential geothermal facilities is a complex but necessary step even before environmental studies and site leases can move forward. Important factors to consider include whether the site is a tribal or cultural resource, if it is in or near an environmentally sensitive area (such as a wildlife refuge or national park), or if it could potentially impact animal species and their habitats. Some modeling suggests that permitting and land access challenges are hindering deployment by 50% (Young et al. 2019).

Electricity market conditions pose another challenge for the geothermal sector. The value proposition of geothermal power as both a clean and firm baseload power source is not yet recognized in existing business models. Power purchase agreements with utilities or other end-users most often handle clean electricity sources and baseload electricity sources as two distinct products. This approach makes geothermal power less cost-competitive, undervaluing its advantages and future market potential (Blankenship et al. 2024).

18.8 COMPANIES INVOLVED IN THE GEOTHERMAL SECTOR

The following listings include companies providing drilling services, manufacturing specialized equipment, developing novel drilling technologies, or otherwise significantly involved in enabling deep well drilling for geothermal energy extraction. This partial list is accurate as of the time of publication, but the reader should be aware that market and regulatory conditions change rapidly, and can lead to uncertainty as to the success or longevity of any specific company, project, or investment. The reader is directed to the internet to search for postings in the categories of: *Top Commercial Geothermal Companies in the U.S.*; *Top Geothermal Drilling Companies*; *Geothermal Energy Systems Development Companies*, etc., for up-to-date information on the status of commercial geothermal systems development and the geothermal energy industry.

18.8.1 Major Oil and Gas Exploration Companies with Interests in Geothermal Energy Development

These large companies leverage their extensive oil and gas drilling experience and technology portfolios, adapting them for the challenges of deep, high-temperature geothermal wells. They include the following:

- SLB (formerly Schlumberger). Provides a wide range of services including directional drilling, high-temperature logging/measurement while drilling (LWD/MWD), specialized drill bits, cementing solutions, and project management for geothermal projects.
- NOV Inc. (formerly National Oilwell Varco). Manufactures drilling rigs, specialized geothermal drill pipe, high-temperature power sections, drill bits (e.g., Tektonic bits), tubulars/casing (including connectors like XLF), cementing equipment, and temperature management systems (mud chillers) suitable for geothermal applications.
- Baker Hughes. Offers drilling services, drill bits (like Kymera™ hybrid bits), high-temperature downhole tools and sensors, completion technologies, and integrated project services applicable to deep geothermal.

- Halliburton. Provides drilling fluids, cementing services, drill bits, directional drilling, LWD/MWD, and completion technologies designed to withstand high-temperature geothermal environments.

18.8.2 Geothermal Project Developers and Technology Innovators

These companies are primarily focused on developing geothermal energy projects, often driving innovation in drilling and reservoir creation, especially for EGS or closed-loop/AGS.

- Berkshire Hathaway Energy Renewables. Des Moines, Iowa-based Berkshire Hathaway, has increased its investments in geothermal, focusing on California and Nevada. In February 2024, it acquired two California-based geothermal plants, extending its renewable energy portfolio. Key features include multiple facilities in Western states and a strong focus on research and development activities.
- Calpine Corporation. Calpine, headquartered in Houston, Texas, is a leading player in the U.S. natural gas market and the largest producer of geothermal energy in the United States and the world, producing 725 MW of geothermal power from its California Geysers complex alone. With a sustainability-focused approach, Calpine invested $200 million in geothermal expansion in April 2024, supplying 725,000 homes with geothermal-based power.
- Controlled Thermal Resources. Based in Imperial Valley, California, it is a pioneer in lithium extraction and geothermal power generation, with its Hell's Kitchen Project, launched in 2024, aiming to revolutionize clean energy production and mineral extraction.
- Cyrq Energy. Cyrq Energy, based in Salt Lake City, Utah, is a leading environmentally conscious company that operates sustainable geothermal plants in the western United States. It implemented innovative cooling technology in 2024 to reduce environmental impact and improve power efficiency.
- Energy Development Corporation. EDC, headquartered in Pasadena, California, is a key expert in commercial-scale high-temperature heat pump systems with over 200 installations across the United States. In 2024, it introduced green financing solutions to promote renewable heating and cooling, reducing energy costs and CO_2 emissions.
- EnergySource. EnergySource, a pioneer in geothermal development in the Salton Sea region, partnered with Tesla in March 2024 to implement high-efficiency energy storage systems. This showcased its role in the future of renewable energy development through mineral recovery integration and novel power plant technology.
- Northern Power Systems. A Vermont-based company primarily known for wind power, Northern Power Systems has extended into geothermal and hybrid renewable energy systems, introducing district heating solutions for commercial and residential buildings in 2023.
- Ormat Technologies. Headquartered in Reno, Nevada, it is a leading geothermal power plant developer and operator globally. Also it manufactures

power generation equipment. It has participated in EGS demonstration projects requiring advanced drilling and stimulation techniques.
- Sage Geosystems. Sage Geosystems is developing geopressured geothermal systems and earth-based energy storage, requiring deep drilling and well technologies adapted from the oil and gas industry. It is also involved in U.S. Department of Defense (DoD) demonstration projects.
- Terra-Gen Power. Terra-Gen, headquartered in San Diego, California, is boosting sustainable energy projects with over 300 MW of geothermal capacity, incorporating unique drilling and extraction technologies in Nevada, aiming for a cleaner, more efficient energy future.
- U.S. Geothermal. U.S. Geothermal is a key player in mid-sized geothermal projects, headquartered in Boise, Idaho. It expanded its three operating power plants in Idaho, increasing total capacity by 15 MW in 2024. Key operations include strategic development projects in Idaho.

18.8.3 Novel Drilling Technology Specialists

The following companies are focused on developing potentially disruptive drilling technologies aimed at overcoming the challenges of cost, speed, depth, and hard rock associated with conventional geothermal well drilling technologies.

- AltaRock Energy. AltaRock Energy focuses on developing EGS and SHR geothermal resources. They work on advanced drilling, stimulation techniques, high-temperature tools, and novel drilling concepts (partnering with Quaise on mmWave testing).
- Bedrock Energy. Bedrock Energy is developing autonomous drilling platforms and associated technologies primarily for faster, cheaper installation of ground-source heat pump boreholes, but innovations could have broader relevance.
- Eavor Technologies. It develops a proprietary closed-loop EGS (Eavor-Loop™) requiring advanced directional drilling to create a subsurface radiator-like heat exchanger. It has drilled deep demonstration wells.
- Fervo Energy. It develops EGS projects using techniques adapted from the oil and gas industry, particularly horizontal drilling and well stimulation. It actively drills deep wells for commercial projects and DoD initiatives.
- GA Drilling. GA Drilling is developing plasma-based (PLASMABIT®) and other anchor-bit drilling technologies designed for faster penetration in hard rock and high-temperature environments.
- GreenFire Energy. GreenFire Energy is developing closed-loop EGS technology (GreenLoop™) applicable to various geothermal settings, including deep wells, and is involved in DoD demonstration projects.
- Petra. Petra is developing noncontact thermal drilling technology capable of boring through very hard rock, initially focused on undergrounding utilities but potentially applicable to geothermal well development.
- Quaise Energy. Quaise Energy is the lead developer of millimeter wave (mmWave) directed energy drilling using gyrotrons to vaporize rock, aiming

for ultra-deep (up to 18 km or 12 miles) geothermal access. They are currently conducting field tests of their advanced technology.

18.8.4 Drilling Contractors and Specialized Equipment Providers

These companies provide drilling rigs, crews, and specific equipment, often adapting their fleets and services for geothermal work.

- Geoprobe Systems. Geoprobe Systems manufactures drilling rigs, including models marketed for geothermal applications (though perhaps more focused on shallower ground source heat pump systems than deep power generation wells).
- Helmerich & Payne (H&P). H&P are major land drilling contractors adapting advanced rigs and drilling automation/digital technologies for geothermal applications. They are major investors in Fervo Energy.
- Nabors Industries. Nabors Industries are a major drilling contractor offering automated drilling solutions and advanced rig technology adaptable to geothermal systems.

18.8.5 Major Global Energy Companies Involved in Geothermal Energy Development

The following companies are major global energy companies from around the world that are investing in and developing solutions for geothermal energy systems.

- United States
    - Chevron. Chevron is actively exploring and investing in geothermal energy. It has formed a joint venture with Baseload Capital to develop projects in the United States and is collaborating with Mitsui Oil Exploration Co. (MOECO) in Japan and Pertamina in Indonesia on geothermal projects/studies. Chevron is a member of the Texas Geothermal Energy Alliance (TXGEA) and works with Sonoma Clean Power on geothermal opportunities. They view next-generation geothermal as a fit for their subsurface expertise.
- China
    - PetroChina. PetroChina is actively promoting a green and low-carbon energy transition for China. Reports indicate efforts to secure and convert targets for geothermal energy alongside wind and solar power, suggesting exploration and potential development activities, likely including direct geothermal energy use for heating.
    - Sinopec (China Petroleum & Chemical Corporation). Sinopec is heavily involved in refining and petrochemicals production. Sinopec is a major developer of geothermal district heating projects within China, leveraging low-to-medium temperature resources. It is considered one of the largest geothermal heating providers globally based on this direct use application.

- France
    - Électricité de France (EDF). EDF is actively involved in geothermal energy systems development through subsidiaries like ÉS Géothermie, which develops and operates deep geothermal power and heating plants, particularly in the Rhine Rift Valley region (France/Germany).
    - TotalEnergies. TotalEnergies is actively investing in geothermal energy as part of its multi-energy strategy. They are involved in projects in various regions, including Europe (e.g., France, Germany) and potentially Kenya, exploring both geothermal power generation and heating applications.
- Germany
    - RWE. As part of its transition to renewables, RWE is exploring geothermal potential, particularly within Germany, often in partnership with municipal utilities or technology providers.
- India
    - Oil and Natural Gas Corporation (ONGC). ONGC is actively exploring and developing geothermal potential within India, with pilot projects planned or underway in areas like Puga Valley (Ladakh) and potentially other regions.
- Italy
    - Enel. Enel is a global leader in geothermal power through its Enel Green Power division. Enel operates significant geothermal capacity in Italy (historic Larderello fields), the United States, and Latin America, and actively researches and develops geothermal technology.
    - Eni. Eni is pursuing geothermal energy projects, leveraging its oil and gas expertise, exploring conventional hydrothermal resources and potentially more innovative concepts, sometimes integrating geothermal with other industrial processes or energy sources.
- Japan
    - Tokyo Electric Power Company (TEPCO). As Japan's largest utility in a country with significant geothermal resources, TEPCO utilizes geothermal power sourced through power purchase agreements or potentially via subsidiary operations, although it's not globally highlighted as a primary developer in the same way as companies like Ormat or Enel Green Power.
- Mexico
    - Comisión Federal de Electricidad (CFE). Mexico has substantial geothermal power capacity, and it is primarily developed and operated by the state utility, CFE.
- Netherlands
    - Shell PLC. Shell is actively involved in geothermal energy. It has dedicated geothermal ventures, invests in geothermal technology companies, and participates in projects exploring conventional and next-generation (EGS) geothermal, leveraging its global footprint and subsurface expertise.
- Norway
    - Equinor. Equinor has explored geothermal potential and participated in research, particularly related to EGS, leveraging its offshore drilling expertise. However, its primary renewable focus remains offshore wind.

18.9 SUMMARY

- EGS and closed-loop geothermal systems are next-generation technologies that are expanding access to heat energy beyond natural hydrothermal resources, including in SHR.
- Technological advancements in areas such as drilling, AI, remote sensing, and fiber optics are facilitating the growth of geothermal development.
- Geothermal energy is a consistent, clean baseload power source that can play a significant role in meeting rising demand for utility-scale electricity generation and that offers co-location and energy storage benefits.
- AI is poised to significantly impact the expansion of geothermal energy across various stages of project development, with the potential to increase efficiency, reduce costs, and mitigate risks.
- Strong policies, such as financial incentives, expanded support for research and demonstration, permitting reform, and workforce development, can help promote widespread commercialization of geothermal energy.
- Barriers to the deployment of geothermal energy include challenges in identifying and characterizing potential resources, high upfront development cost and risks, complex permitting processes, land access issues, and existing business models that undervalue the potential for next-generation technologies.
- An array of companies in the United States and globally are playing a role in providing drilling services, manufacturing specialized equipment, or developing novel technologies to enable deep well drilling for geothermal energy extraction.

REFERENCES

Augustine, C., Fisher, S., Ho, J., Warren, I., and Witter, E. 2023. *Enhanced Geothermal Shot Analysis for the Geothermal Technologies Office.* NREL/TP-5700-84822. Golden, CO: National Renewable Energy Laboratory. www.nrel.gov/docs/fy23osti/84822.pdf.

Blankenship, D., Gertler, C., Kamaludeen, M., O'Connor, M., and Porse, S. 2024. *Pathways to Commercial Liftoff: Next-Generation Geothermal Power.* Washington, D.C.: U.S. Department of Energy. https://climateprogramportal.org/wp-content/uploads/2025/02/LIFTOFF_DOE_NextGen_Geothermal_v14.pdf

Cochran, E.S., Rubinstein, J.L., Barbour, A.J., and Kaven, J.O. 2024. *Induced Seismicity Strategic Vision. Circular 1509.* Reston, VA: U.S. Geological Survey. https://pubs.usgs.gov/circ/1509/circ1509.pdf .

Hill, J., and Garth, A. 2024. *Policy Brief: Earth Energy Innovation, Supporting the Research, Development, Demonstration, and Deployment (RDD&D) of Next-Generation Geothermal Energy.* Boston, MA: Clean Air Task Force. https://www.catf.us/resource/policy-brief-earth-energy-innovation/. (accessed June 10, 2025).

International Energy Agency. 2023. *The Future of Geothermal Energy.* Paris, France: International Energy Agency. https://iea.blob.core.windows.net/assets/fe8d11b6-b1a6-43d9-9351-47d86ed1dfbc/TheFutureofGeothermal.pdf.

King, B., Ricks, W., Pastorek, N., and Larsen, J. 2025. *The Potential for Geothermal Energy to Meet Growing Data Center Electricity Demand.* Energy and Climate. New York,

NY: Rhodium Group. https://rhg.com/research/geothermal-data-center-electricity-demand/.

National Renewable Energy Laboratory. 2025a. *Advanced Wells.* Golden, CO: National Renewable Energy Lab, DOE. www.nrel.gov/geothermal/advanced-wells. (accessed June 7, 2025).

National Renewable Energy Laboratory. 2025b. *Next-Generation Geothermal Technologies. Geothermal Research.* Golden, CO: National Renewable Energy Lab, DOE. www.nrel.gov/geothermal/next-generation. (accessed June 8, 2025).

Porse, S., Winick, J., Tasca, C., and Blankenship, D. 2022. *Geothermal Technologies Office Multi-Year Program Plan (FY 2022-2026).* Washington, D.C.: Geothermal Technologies Office, Office of Energy Efficiency and Renewable Energy, U.S. Department of Energy. www.energy.gov/eere/geothermal/articles/geothermal-technologies-office-multi-year-program-plan-fy-2022-2026.

Smith, M. 2022. *Enhanced Geothermal Systems: Introduction and Issues for Congress.* R47526. Washington, D.C.: Congressional Research Service. www.congress.gov/crs_external_products/R/PDF/R47256/R47256.6.pdf.

Tester, J.W., Anderson, B.J., Batchelor, A.S., Blackwell, D.D., DiPippo, R., Drake, E.M., Garnish, J., Livesay, B., Moore, M.C., Nichols, K., Petty, S., Toksöz, M.N., and Veatch, R.W. Jr. 2006. *The Future of Geothermal Energy, Impact of Enhanced Geothermal Systems (EGS) on the United States in the 21st Century* . Cambridge, MA: Massachusetts Institute of Technology. www1.eere.energy.gov/geothermal/pdfs/future_geo_energy.pdf.

U.S. Department of Energy. 2019. *GeoVision: Harnessing the Heat Beneath Our Feet.* Washington, D.C.: U.S. Department of Energy. www.energy.gov/sites/default/files/2019/06/f63/GeoVision-full-report-opt.pdf.

U.S. Department of Energy. 2025a. *Diamonds are a Drill's Best Friend: How PDC Drill Bits are Helping Geothermal Shine.* Washington, D.C.: Geothermal Technology Office, U.S. Department of Energy. www.energy.gov/eere/geothermal/articles/diamonds-are-drills-best-friend-how-pdc drill bits are-helping-geothermal. (accessed June 7, 2025).

U.S. Department of Energy. 2025b. *Reservoir Thermal Energy Storage.* Washington, D.C.: Geothermal Technology Office, U.S. Department of Energy. www.energy.gov/eere/geothermal/reservoir-thermal-energy-storage. (accessed June 8, 2025).

U.S. Department of Energy. 2025c. *Wells of Opportunity.* Washington, D.C.: Geothermal Technology Office, U.S. Department of Energy. https://www.energy.gov/eere/geothermal/wells-opportunity. (accessed June 10, 2025).

Young, K., Levine, A., Cook, J., Heimiller, D., and Ho, J. 2019. *GeoVision Analysis Supporting Task Force Report: Barriers—An Analysis of Non-Technical Barriers.* NREL/TP-6A20-71641. Golden, CO: National Renewable Energy Laboratory, U.S. DOE. https://docs.nrel.gov/docs/fy19osti/71641.pdf.

19 Geothermal Energy Applications and Illustrative Examples

19.1 INTRODUCTION

This last chapter of Part 4 is concerned with applications presented as a series of illustrative examples. There are six categories of examples used to explore concepts related to geothermal energy. These categories and their corresponding section numbers are listed as follows:

There are three illustrative examples in each section (Section 19.5 has eight example problems) and most have been adopted from the literature (Reynolds, Dupont and Theodore 1991; Reynolds 1992; Theodore and Reynolds 1991; Farag 1994; Theodore and Nueser 1996; Reynolds, Jeris and Theodore 2004; Theodore 2014).

19.2 ENGINEERING FUNDAMENTALS

Three illustrative examples complement the presentation for this section concerned with engineering fundamentals.

19.2.1 Illustrative Example 1 – Process Flow Diagram

As part of a geothermal energy study, heated gas from a geothermal source is mixed with recycled gas from a downstream absorber, and the mixture passes through an indirect waste heat boiler (H), which uses water as the cooling medium. It then passes through a water spray quencher (Q) in which the temperature of the mixture is further decreased, and finally through an absorber (A) in which a solvent is employed to

DOI: 10.1201/9781003667759-23

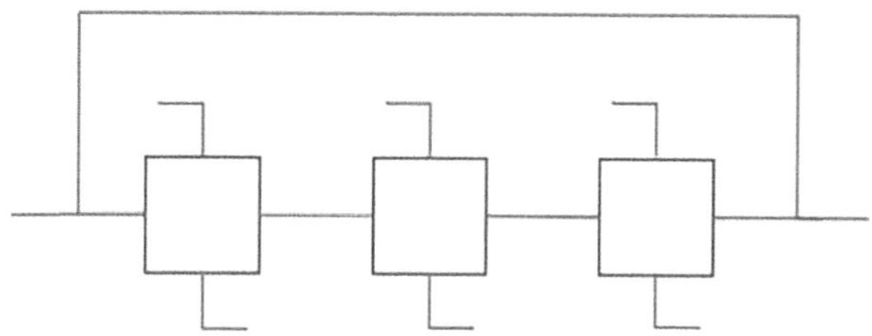

FIGURE 19.1 Line diagram.

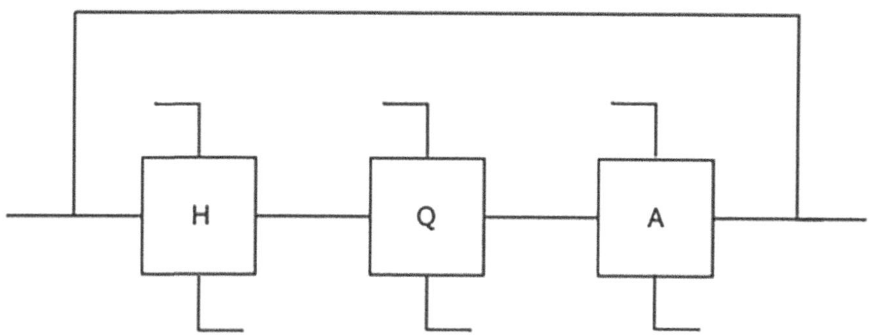

FIGURE 19.2 Line diagram with equipment labels.

absorb one of the contaminant species in the flowing gas stream. Prepare a simplified flow diagram for this process (Reynolds 1992).

Solution: Before attempting to calculate the raw material or energy requirements of a process, it is desirable to attain a clear picture of the process. The best way to do this is to draw a *process flow diagram*. A process flow diagram is a line diagram showing the successive steps of a process by indicating the pieces of equipment in which the steps occur and the material streams entering and leaving each piece of equipment. Process flow diagrams are used to conceptually represent an overall process

Lines are used to represent flow streams and boxes are used to represent equipment. A line diagram of the process described previously is first prepared as a series of lines and boxes as shown in Figure 19.1. The equipment boxes in Figure 19.1 can then be labeled as shown in Figure 19.2 along with the flow streams in Figure 19.3.

19.2.2 Illustrative Example 2 – Velocity Magnitude and Direction

A fluid device has four openings as shown in Figure 19.4. The fluid has a constant density, ρ, of 80.0 kg/m^3. Steady-state information on the system is also provided in Table 19.1.

Determine the magnitude and direction of the velocity, v_4. What is the mass flow rate $\dot{m}_4$?

Solution: Calculate the volumetric flow rate, q, through each pipe as $Ax\ v$:

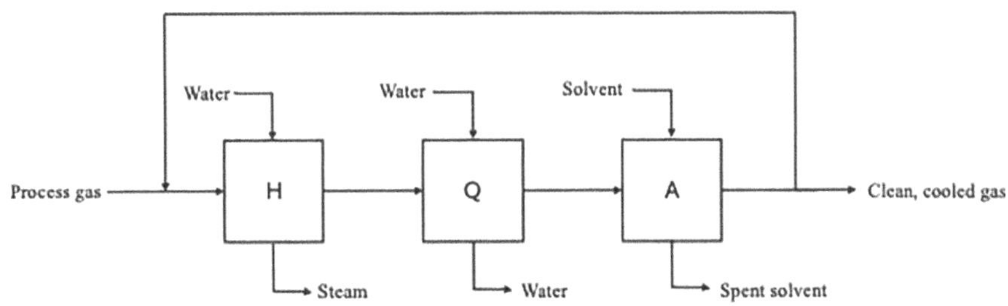

FIGURE 19.3 Line diagram with equipment and stream labels.

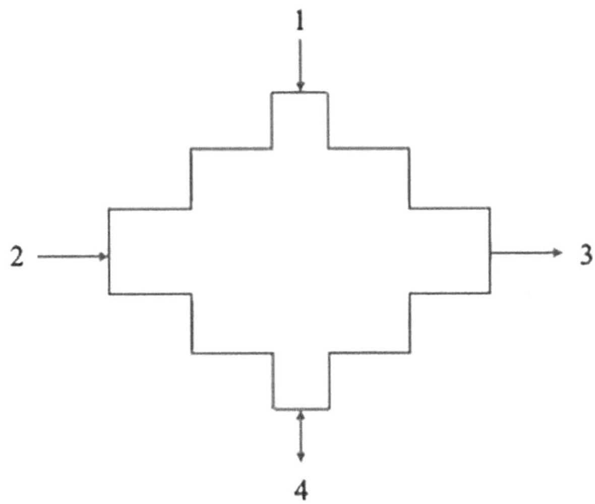

FIGURE 19.4 Diagram for Illustrative Example 19.2.2.

TABLE 19.1
Steady-State Flow Information for Illustrative Example 2

Pipe	Flow area, m²	Velocity, m/s	Direction relative to device
1	0.2	5	In
2	0.3	7	In
3	0.25	12	Out
4	0.15	?	?

$$q_1 = (0.2\ \text{m}^2)\ (5\ \text{m/s}) = 1\ \text{m}^3/\text{s}$$

$$q_2 = (0.3\ \text{m}^2)\ (7\ \text{m/s}) = 2.1\ \text{m}^3/\text{s}$$

$$q_3 = (0.25\ \text{m}^2)\ (12\ \text{m/s}) = 3\ \text{m}^3/\text{s}$$

Since the fluid is of constant density, the continuity equation may be applied as follows:

$$q_1 + q_2 = q_3 + q_4;\ 1\ \text{m}^3/\text{s} + 2.1\ \text{m}^3/\text{s} = 3\ \text{m}^3/\text{s} + q_4$$

Solving for q_4 yields the following:

$$q_4 = 3.1\ \text{m}^3/\text{s} - 3\ \text{m}^3/\text{s} = 0.1\ \text{m}^3/\text{s out}$$

The magnitude of the velocity is calculated to be:

$$0.1\ \text{m}^3/\text{s} = v \times 0.15\ \text{m}^2;\ v = 0.1/0.15 = 0.67\ \text{m/s}$$

The mass flow rate out of Pipe 4 is then:

$$\dot{m}_4 = (\rho)(q) = (80.0\ \text{kg}/\text{m}^3)(0.1\ \text{m}^3/\text{s}) = 8.0\ \text{kg}/\text{s}$$

19.2.3 Illustrative Example 3 – Equivalence of Mass and Energy

Discuss the equivalence of mass and energy.

Solution: The equivalence of mass and energy is generally discussed in the context of nuclear reactions. In a nuclear reaction it is the nucleus of the atom that undergoes rearrangement, releasing a significant quantity of energy, occurring with a miniscule loss of mass. The classic Einstein equation related energy, E, to mass as indicated in Equation 19.1.

$$\Delta E = (\Delta m)\, c^2 \tag{19.1}$$

where Δm = change in mass, kg; and c = the speed of light, 3×10^8 m/s.

When a uranium atom is split apart (fissioned), the mass of the *fragments* is less than the mass of the original atom. The energy corresponding to this loss of mass is defined as *fission* energy. One can show that each kilogram of mass converted liberates approximately 10^{17} J. This thermal energy is enough to run a large turbine generator (a 1 GW unit) for a year at 35% thermal efficiency and is equivalent to the energy possessed by 2.5 million tons of coal (Skipka and Theodore 2014).

Similarly, if two or more light elements (e.g., hydrogen) are combined (*fused*) to form a heavier one, the resulting mass of the atom is less than the sum of the two original atoms. The difference (or loss) in mass is converted to energy, as with the fission process. This energy is defined as *fusion* energy.

19.3 ENGINEERING PRINCIPLES

Three illustrative examples complement the presentation for this section concerned with engineering principles.

19.3.1 Illustrative Example 1 – Potential and Kinetic Energy Calculations

A process plant pumps 1,000 lb of water to an elevation of 1,200 ft above a series of turbogenerators at a geothermal power plant. Determine the change in potential energy in Btu and ft-lb$_f$. If this water has its velocity increased from 8 to 30 ft/s, also calculate the change in kinetic energy of the water in Btu and ft-lb$_f$.

Solution: The potential energy change, ΔPE, is given by:

$$\Delta PE = \frac{mg}{g_c} \Delta Z \tag{19.2}$$

where m = mass of water, lb; g = gravitational acceleration, 32.2 ft/s^2 at sea level; g_c = gravitational constant, 32.2 ft-lb/(lb$_f$-s^2); and ΔZ = change in elevation, ft. Substituting the values given yields:

$$\Delta PE = \frac{(1{,}000 \text{ lb})\left(32.2\dfrac{\text{ft}}{s^2}\right)}{\left(32.2 \text{ ft} - \text{lb} / (\text{lb}_\text{f} - s^2)\right)}(600 \text{ ft}) = 6 \times 10^5 \text{ ft - lb}_\text{f}$$

$$\Delta PE = (6 \text{ x } 10^5 \text{ft-lb}_\text{f})\ (1 \text{ Btu}/778.17 \text{ ft-lb}_\text{f}) = 771 \text{ Btu}$$

Kinetic energy, KE, is given by:

$$KE = \frac{mv^2}{2g_c} \tag{19.3}$$

where v = velocity of water, ft/s. Equation 19.3 allows the evaluation of the kinetic energy of a body of mass m having a velocity v. Substituting the values given for the initial condition of the water yields:

$$KE_i = \frac{(1{,}000 \text{ lb})\left(8\dfrac{\text{ft}}{\text{s}}\right)^2}{2\left(32.2 \text{ ft} - \text{lb} / (\text{lb}_\text{f} - s^2)\right)} = 993.8 \text{ ft - lb}_\text{f}$$

The kinetic energy at the final velocity of 30 ft/s is given as

$$KE_f = \frac{(1{,}000 \text{ lb})\left(30\dfrac{\text{ft}}{\text{s}}\right)^2}{2\left(32.2 \text{ ft} - \text{lb} / (\text{lb}_\text{f} - s^2)\right)} = 13{,}975 \text{ ft - lb}_\text{f}$$

The change in kinetic energy, ΔKE is given as

$$\Delta KE = KE_i - KE_f = (993.8 - 13{,}975) = -12{,}981\,\text{ft-lb}_\text{f}$$

$$\Delta KE = (-12{,}981\,\text{ft-lb}_\text{f})\ (1\ \text{Btu}/778.17\ \text{ft-lb}_\text{f}) = -16.7\ \text{Btu}$$

19.3.2 Double Pipe Heat Exchanger Calculation

The following data are provided for a double pipe heat exchanger located in a geothermal heating/cooling facility.

$$\dot{Q} = 6{,}000\,\text{Btu/h};\ U = 24\,\text{Btu/ft}^2\text{-°F-h};\ \Delta T_{lm} = 50°\text{F}$$

Calculate the area required for the heat exchanger by applying the general heat exchanger equation (Theodore 2014).
Solution: The general heat exchanger equation is given as

$$\dot{Q} = UA\Delta T_{lm} \tag{19.4}$$

where $\dot{Q}$ = heat transfer rate, Btu/h; U = local overall heat transfer coefficient, Btu/ft^2-°F-h; A = total heat transfer area, ft^2; and ΔT_{lm} = the log-mean average temperature, °F. Rearranging Equation 19.4 to solve for A yields:

$$A = \frac{\dot{Q}}{U\Delta T_{lm}} \tag{19.5}$$

Substituting the given input values yields:

$$A = \frac{\left(6{,}000\,\dfrac{\text{Btu}}{\text{h}}\right)}{\left(24\ \text{Btu/ft}^2-°\text{F}-\text{h}\right)\left(50°\text{F}\right)} = 5\ \text{ft}^2$$

As described briefly earlier in Chapter 9, the overall heat transfer coefficient, U, for flow in a tube is related to the individual heat transfer coefficients by the following equation:

$$\frac{1}{UA} = \frac{1}{h_iA_i} + \frac{R_{f,i}}{A_i} + \frac{\ln\left(D_o/D_i\right)}{2\pi kL} + \frac{R_{f,o}}{A_o} + \frac{1}{h_oA_o} \tag{19.6}$$

where h_i = inside heat transfer coefficient, Btu/ft^2-h-°F; A_i = inside area of heat exchanger, ft^2; $R_{f,i}$ = fouling factor based on inner tube surfaces, unitless; D_o, D_i = diameter of pipe, outside, inside, ft; k = pipe thermal conductivity, Btu/ft-h-°F; L = tube length, ft; $R_{f,o}$ = fouling factor based on outer tube surfaces, unitless; A_o = outside area of heat exchanger, ft^2; and h_o = outside heat transfer coefficient, Btu/ft^2-h-°F.

If the fouling factors and the tube wall resistance (middle term on the right hand side of Equation 19.6) are considered negligible, then the relationship between the overall heat transfer coefficient and the individual coefficients simplifies to:

$$\frac{1}{U} = \frac{D_o}{h_i D_i} + \frac{1}{h_o} \tag{19.7}$$

Individual coefficients h_i and h_o can be calculated using empirical equations provided in the literature (Theodore 2014).

19.3.3 River Cooling Water Requirement

Determine the percentage of a river's stream flow available to an industry for cooling such that the river temperature does not increase more than 10°F. Fifty percent of the industrial withdrawal is lost by evaporation and the industrial water returned to the river is 60°F warmer than the river. *Note*: This problem is a modified and edited version (with permission) of an illustrative example prepared in 1980 by Ms. Marie Gillman, a graduate student in Mechanical Engineering at Manhattan College, Bronx, New York.

Solution: Draw a flow diagram representing the process as shown in Figure 19.5. Express the volumetric flow lost by evaporation from the process in terms of that entering the process as

$$q_{\text{lost}} = 0.5q_{\text{in}} \tag{19.8}$$

Express the process outlet temperature and the maximum river temperature in terms of the upstream temperature as

$$T_{\text{out}} = T_{\text{up}} + 60°F \tag{19.9}$$

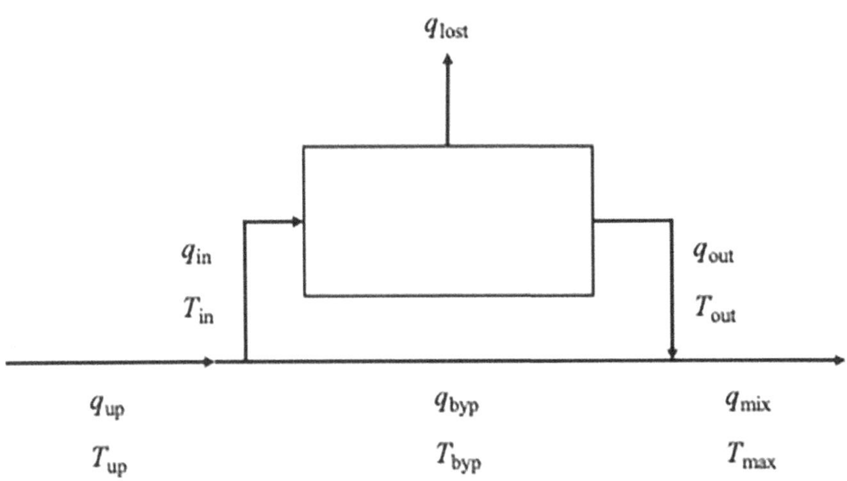

FIGURE 19.5 Flow diagram for Illustrative Example 19.3.3.

$$T_{max} = T_{up} + 10°F \tag{19.10}$$

Using the law of conservation of mass, the process outlet flow can be expressed in terms of the process inlet flow. Also, the process bypass flow can be expressed in terms of the upstream and process inlet flow as follows:

$$q_{out} = 0.5\, q_{in} \tag{19.11}$$

$$q_{byp} = q_{up} - q_{in} \tag{19.12}$$

$$q_{mix} = q_{up} - 0.5q_{in} \tag{19.13}$$

The flow diagram with the mass conservation expressions developed before is shown in Figure 19.6.

Noting that the enthalpy of any fluid stream can be represented by $qc_p\rho\left(T - T_{ref}\right)$, an energy balance around the downstream mixing point leads to:

$$(q_{up} - q_{in})c_p\rho\left(T_{up} - T_{ref}\right) + 0.5q_{in}c_p\rho\left(T_{up} + 60 - T_{ref}\right)$$
$$= \left(q_{up} - 0.5q_{in}\right)c_p\rho\left(T_{up} + 10 - T_{ref}\right) \tag{19.14}$$

Note that T_{ref} is arbitrary and indirectly defines a basis for the enthalpy calculation. Setting T_{ref} to 0 and assuming the density and heat capacity of the fluid is constant yields:

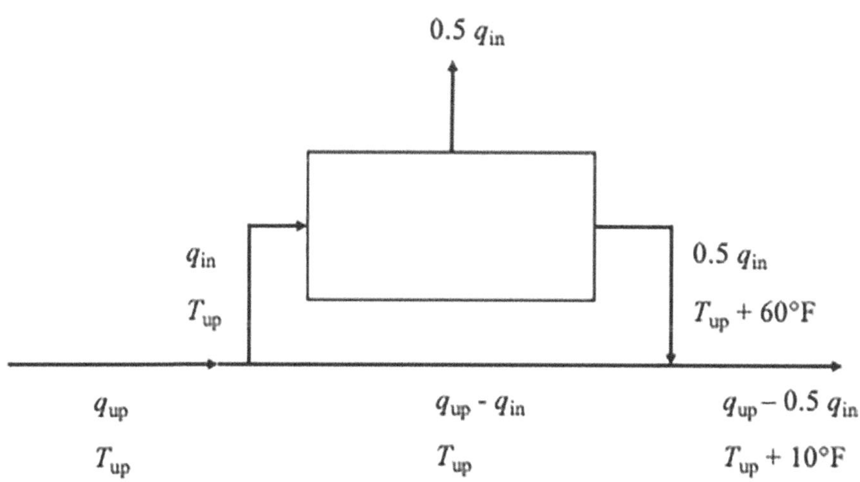

FIGURE 19.6 Flow diagram after applying mass conservation equations for Illustrative Example 19.3.3.

$$(q_{\mathrm{up}} - q_{in})(T_{up}) + 0.5q_{\mathrm{in}}(T_{up} + 60) = (q_{\mathrm{up}} - 0.5q_{in})(T_{up} + 10) \tag{19.15}$$

Equation 19.15 may now be solved for the inlet volumetric flow to the process in terms of the upstream flow as

$$q_{\mathrm{up}}T_{up} - q_{in}T_{up} + 0.5q_{\mathrm{in}}T_{up} + 30q_{\mathrm{in}} = q_{\mathrm{up}}T_{up} + 10q_{\mathrm{up}} - 0.5q_{in}T_{up} - 5q_{in} \tag{19.16}$$

Solving and canceling terms yields the following results:

$$35q_{\mathrm{in}} = 10q_{\mathrm{up}};\ q_{\mathrm{in}} = 0.286q_{\mathrm{up}}$$

Therefore, 28.6% of the original flow, q_{up}, is available for cooling. Note that the problem can also be solved by setting $T_{\mathrm{ref}} = T_{\mathrm{up}}$. Since for this condition, $T_{\mathrm{ref}} - T_{\mathrm{up}} = 0$, the solution to the problem is greatly simplified.

19.4 THERMODYNAMICS

Three illustrative examples complement the presentation for this section concerned with thermodynamics.

19.4.1 Illustrative Example 1 – Entropy Calculations (Theodore and Reynolds 1991)

A copper block with a mass of 30 lb is originally at 100°C (212°F). It is placed in a pail containing 9 gal of water at 25°C (77°F). Neglecting heat losses, calculate the change in entropy of the water and copper block after they achieve equilibrium. The average heat capacity at constant pressure, c_p, for copper and water are approximately 0.092 and 1.0 Btu/lb-°F, respectively. Also calculate the overall entropy change of the system and comment on the results.

Solution: First write the equation describing the energy change of copper in cooling from 100°C to T_{f} as

$$Q_{\mathrm{c}} = m_{\mathrm{c}}c_{p,c}(212 - T_f) \tag{19.17}$$

where Q_{c} = heat lost from the copper block.

Write the equation describing the energy change of the water in being heated from 25°C to T_{f} as

$$Q_{\mathrm{w}} = m_{\mathrm{w}}c_{p,w}(T_f - 77) \tag{19.18}$$

Calculate the final temperature, T_f, in C by setting Equation 19.17 equal to 19.18, and solving for T_f as follows:

$$Q_c = m_c c_{p,c}\left(212 - T_f\right) = Q_w = m_w c_{p,w}\left(T_f - 77\right)$$

$$(30\text{ lb})(0.092\text{ Btu/lb}-°\text{F})\left(212 - T_f\right) = (9\text{ gal})(8.34\text{ lb/gal})(1.0\text{ Btu/lb}-°\text{F})\left(T_f - 77\right)$$

$$\left(585.12 - 2.76T_f\right)\text{Btu/°F} = \left(75.06T_f - 5{,}779.62\right)\text{Btu/°F}$$

$$6{,}364.74\text{ Btu} = \left(72.3T_f\right)\text{Btu/°F}$$

$$T_f = 88°\text{F} = 31.1°\text{C} = 304.3\text{K}$$

Calculate the entropy change of the copper in Btu/°F using Equation 19.19.

$$\Delta S_c = m_c c_{p,c} \ln\left(T_f / 373\right) \tag{19.19}$$

$$\Delta S_c = (30\text{ lb})(0.092\text{ Btu/lb}-°\text{F})\ln(304.3/373)$$
$$= 2.76\text{ Btu/°F }(-0.20) = -0.55\text{ Btu/°F}$$

Calculate the entropy change for the water using Equation 19.20 as

$$\Delta S_w = m_w c_{p,w} \ln\left(T_f / 298\right) \tag{19.20}$$

$$\Delta S_w = (75.06\text{ lb})(1.0\text{ Btu/lb}-°\text{F})\ln(304.3/298)$$
$$= 75.06\text{ Btu/°F }(0.021) = 1.58\text{ Btu/°F}$$

Finally, the overall entropy change of the process is determined from Equation 19.21 as

$$\Delta S_T = \Delta S_c + \Delta S_w \tag{19.21}$$

$$\Delta S_T = -0.55\text{ Btu/°F} + 1.58\text{ Btu/°F} = 1.03\text{ Btu/°F}$$

Note that the overall entropy change is *positive*. If ΔS_{syst} and ΔS_{surr} represent the entropy change of a system and its surroundings, respectively, it can be shown that for a particular process and as a consequence of the second law, the total entropy change, ΔS_{tot}, is given by

$$\Delta S_{tot} = \Delta S_{syst} + \Delta S_{surr} \geq 0 \tag{19.22}$$

In effect, the second law requires that for every process, the total entropy change is positive; the only exception is if the process is reversible and then

$$(\Delta S_{tot})_{rev} = 0 \tag{19.23}$$

Thus no real process can occur for which the total entropy change is zero or negative. The fundamental facts relative to the entropy concept are that the entropy change of a system may be positive, negative, or zero; the entropy change of the surroundings during this process may likewise be positive, negative, or zero. However, the total entropy change, ΔS_{tot}, must be ≥ 0. Once again, the equality sign applies if the change occurs *reversibly* and *adiabatically*.

19.4.2 Illustrative Example 2 – The Heat Pump

The heat pump, a reversed heat engine, is a device for heating houses and commercial buildings during the winter and cooling them during the summer. In the winter it operates to absorb heat from the surroundings and reject heat into the building. Refrigerant evaporates in coils placed underground or in the outside air; vapor compression is followed by condensation, heat being transferred to air or water, which is used to heat the building. Compression must be to a pressure such that the condensation temperature of the refrigerant is higher than the required temperature level of the building. The operating cost of the installation is the cost of electric power to run the compressor. Any economic advantage of the heat pump as a heating device depends on the cost of electricity in comparison with the cost of fuels such as oil and natural gas. The heat pump also serves for air conditioning during the summer. The flow of refrigerant is simply reversed, and heat is absorbed from the building and rejected through underground coils or to the outside air.

A house has a winter heating requirement of 40 kJ/s and a summer cooling requirement of 50 kJ/s. Consider a heat-pump installation to maintain the house temperature at 20°C (68°F) in the winter and 25°C (77°F) in the summer. This requires circulation of the refrigerant through interior exchanger coils at 30°C (86°F) in winter and 5°C (41°F) in summer. Underground coils provide the heat source in winter and the heat sink in summer. For a year-round ground temperature of 15°C (59°F), the heat-transfer characteristics of the coils necessitate refrigerant temperatures of 10°C (50°F) in winter and 25°C (77°F) in summer. What are the minimum power requirements for winter heating and summer cooling? Solve this example employing English Engineering units.

Solution: The minimum power requirements are provided by a Carnot heat pump. For winter heating, the house coils are at the higher temperature T_H, and the heat requirement is Q_H. Employing English Engineering units, $T_L = 50°F$ and $T_H = 86°F$.

$$|Q_L| = |Q_H|\frac{T_L}{T_H} = (40\text{kJ}/\text{s})\left(\frac{50+460}{86+460}\right) = 40(0.934) = 37.36 \text{ kJ}/\text{s} \tag{19.24}$$

This is the heat absorbed in the ground coils. The power requirement for the winter is

$$W = |Q_H| - |Q_L| \tag{19.25}$$

$$W = (40–37.36) = 2.64 \text{ kJ/s } (0.948 \text{ Btu/s})/\text{kJ/s} = 2.50 \text{ Btu/s} = 150 \text{ Btu/min}$$

$$= 3.53 \text{ hp}$$

For summer cooling, the house coils are at the lower temperature T_L, and the heat requirement is Q_L. For summer conditions, T_L = 41°F and T_H = 77°F. The power requirement for summer cooling can be solved by combining Equations 19.24 and 19.25 and simplifying to Equation 19.26 as follows:

$$W = |Q_L|\frac{T_H}{T_L} - |Q_L| = |Q_L|\left(\frac{T_H - T_L}{T_L}\right) \tag{19.26}$$

$$W = (50\text{kJ}/\text{s})\left(\frac{77-41}{41+460}\right) = 50(0.072) = 3.6 \text{ kJ}/\text{s } (0.948 \text{ Btu}/\text{s})/\text{kJ}/\text{s}$$

$$= 3.41 \text{ Btu/s} = 204.6 \text{ Btu/min} = 4.82 \text{ hp}$$

19.4.3 Illustrative Example 3 – Maximum Work Calculation (Theodore and Reynolds 1991)

Calculate the maximum possible work, W_{ideal}, which can be obtained from 15 lb/s of steam in a flow process as it undergoes a change in state from 212°F and 1 atm (saturated) to liquid water at 32°F and 1 atm. Is this amount of work sufficient to operate a reversible Carnot heat pump delivering 12,000 Btu/s as heat at 350°F and taking heat from the unlimited supply of cooling water at 32°F?

Solution: From the steam tables,

$$\Delta h = 0.0–1150.4 = –1150.4 \text{ Btu/lb}$$

$$\Delta s = 0.0–1.7566 = –1.7566 \text{ Btu/lb-°R}$$

Assuming ideal work

$$W_{ideal} = \overset{\text{U}}{m}_s (T_0 \Delta s - \Delta h) \tag{19.27}$$

$$W_{ideal} = (15 \text{ lb/s}) \{(492°\text{R } (–1.7566 \text{ Btu/lb-°R})) – (–1150.4 \text{ Btu/lb})\}$$

W_{ideal} = (15 lb/s) {(–864.25 Btu/lb) + 1150.4 Btu/lb)} = (15 lb/s) (286.15 Btu/lb) = 4,292 Btu/s

If this amount of work, the maximum attainable from the steam, is used to drive a reversible Carnot heat pump operating between the temperatures of 32°F and 350°F, the heat transferred at the higher temperature is

$$Q = W\left(\frac{T_H}{T_H - T_L}\right) \tag{19.28}$$

$$Q = (4,292\ \text{Btu/s})\left(\frac{350+460}{350-32}\right) = (4,292\ \text{Btu/s})(2.55) = 10,945\ \text{Btu/s}$$

This represents the maximum possible heat release at 350°F and is less than the required value of 12,000 Btu/s. Thus it is concluded that the process described *cannot* provide sufficient heat to operate the reversible Carnot heat pump.

19.5 GEOTHERMAL ENERGY CONVERSION PROCESSES

Eight illustrative examples complement the presentation for this section concerned with geothermal energy conversion processes.

19.5.1 Illustrative Example 1 – Selecting a Geothermal Energy Power Plant Site

List several guidelines that should be followed when selecting a "safe" site for a geothermal energy power plant.

Solution: The following guidelines should be followed when selecting a "safe" site for a geothermal energy power plant:

1. *Topography*. A fairly level site is needed to contain spills and prevent spills from migrating and creating more of a hazard. Firm soil above water level is recommended.
2. *Utilities and water supply*. The water supply must be adequate for fire protection and cooling. Reliable access to the electric grid should be available to prevent unplanned shutdowns and service interruptions.
3. *Roadways*. Roadways should allow access in and out of the site by emergency vehicles such as ambulances and fire engines in the event of an emergency.
4. *Neighboring communities and plants*. Population density and proximity to the plant should be considered for the initial site selection and in anticipation of a possible future expansion.

5. *Waste disposal.* Waste disposal systems containing any flammable, corrosive, or toxic materials should be a minimum distance of 250 ft from plant equipment.
6. *Climate and natural hazards.* Lightning arrestors should be installed to reduce/eliminate ignition sources in flammable areas. Storm drainage systems should be maintained. Earthquake history in the potential site location is also a factor.
7. *Emergency services.* Emergency services should be readily available, well trained, and appropriately equipped.

19.5.2 Illustrative Example 2 – Alternate Energy Sources

A lake is located at the top of a mountain. A geothermal power plant has been constructed at the bottom of the mountain. The potential energy of the water traveling downhill can be used to spin turbines and generate electricity. This is the operating mode in the daytime during peak electrical demand. At night, when demand is reduced, the water is pumped back up the mountain. The operation is shown in Figure 19.7.

Using the method of power "production" described before, determine how much power (watts) is generated by the lake located at an elevation of 3,000 ft above the power plant. The flow rate of water is 500,000 gpm. The turbine efficiency is 35%. Neglect friction effects and solve the problem using SI units.

Note: This programmed-instructional problem is a modified and edited version (with permission) of an illustrative example prepared by Marie Gillman in 1981, a graduate student in Mechanical Engineering at Manhattan College.

Solution: First, convert height and flow rate to SI units in order to solve for the power in watts:

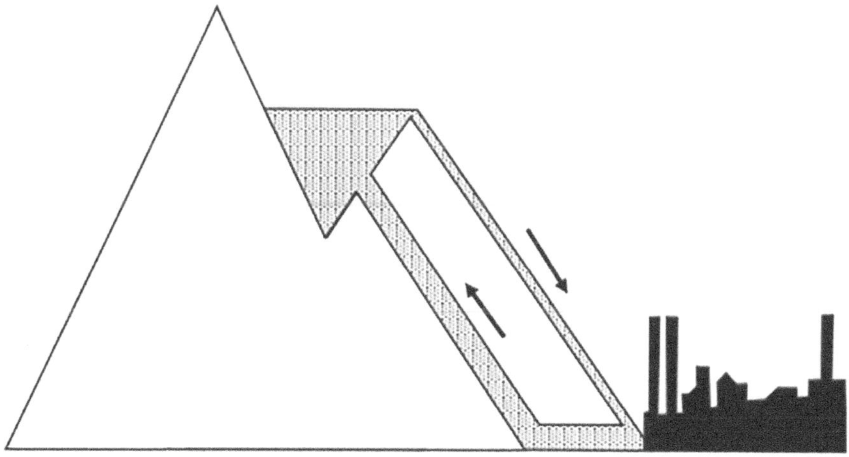

FIGURE 19.7 Schematic for pumped storage energy problem for Illustrative Example 19.5.2.

$$(3000 \text{ ft}) (0.3048 \text{ m/ft}) = 914.4 \text{ m}$$

$$(500{,}000 \text{ gal/min}) (0.003785 \text{ m}^3\text{/gal}) = 1{,}892.5 \text{ m}^3\text{/min}$$

The mass flow rate of the water in kg/s is:

$$(1{,}892.5 \text{ m}^3\text{/min}) (1{,}000 \text{ kg/m}^3) (1 \text{ min/60 s}) = 31{,}542 \text{ kg/s}$$

The drop in potential energy, ΔPE, of the water flow is given by

$$\Delta PE = \frac{mgh}{g_c} \tag{19.29}$$

Substituting yields:

$\Delta PE = (31{,}542 \text{ kg/s}) (9.8 \text{ m/s}^2) (914.4 \text{ m}) = 2.83 \times 10^8 \text{ kg-m}^2\text{/s}^3 = 2.83 \times 10^8 \text{ N-m/s} = 283 \text{ MW}$

Note that $g_c = 1$ in the SI system of units.

Assuming that the potential energy decrease is entirely converted to energy input to the turbine, the actual power output is:

$$P = (0.35) (283 \text{ MW}) = 99 \text{ MW}$$

This is enough power (nearly 100 MW) for a small town. No pollution is generated during power production because no fossil fuel is required. Energy from the geothermal power plant is used during the evening when water is pumped back up into the lake. The initial construction expense would be quite high, but the long-term cost of producing electricity would probably be very economical.

19.5.3 Illustrative Example 3 – Traditional Power Generation

A power plant employs steam to generate power and operates with a steam flow rate of 225,000 lb/h (Table 19.2). For the adiabatic conditions listed, determine the power produced in horsepower, kilowatts, Btu/h, and Btu/lb of steam.

Solution: Apply the general energy equation presented in Chapter 13:

$$\frac{v_1^2}{2g_c} + h_1 + z_1\left(\frac{g}{g_c}\right) + Q = \frac{v_2^2}{2g_c} + h_2 + z_2\left(\frac{g}{g_c}\right) + W \tag{19.30}$$

where v_1, v_2 are steam velocity at inlet/outlet, respectively, ft/s; h_1, h_2 are steam enthalpy at inlet/outlet, respectively, Btu/lb; z_1, z_2 = vertical position at inlet/outlet, respectively, ft; $Q = 0$ under adiabatic conditions; and W = work extracted from the

TABLE 19.2
Power Plant Data

Parameter	Inlet	Outlet
Pressure, psia	100	1
Temperature, °F	1,500	350
Steam velocity, ft/s	120	330
Enthalpy of supersaturated steam, Btu/lb	1,802.2	1,218.7
Steam vertical position, ft	0	-20

system, Btu/lb. Note that 1 Btu = 778 ft-lb$_f$. Substituting given values into Equation 19.30 results in the following:

$$\frac{(120\,\text{ft/s})^2}{2(32.2\,\text{lb}-\text{ft/lb}_f-\text{s}^2)}+1{,}802.2\frac{\text{Btu}}{\text{lb}}+0\,\text{ft}\left(\frac{\text{lb}_f}{\text{lb}}\right)+0$$

$$=\frac{(330\,\text{ft/s})^2}{2(32.2\,\text{lb}-\text{ft/lb}_f-\text{s}^2)}+1{,}218.7\frac{\text{Btu}}{\text{lb}}-20\,\text{ft}\left(\frac{\text{lb}_f}{\text{lb}}\right)+W$$

223.6 ft-lb$_f$/lb (1 Btu/778 ft-lb$_f$) + 1,802.2 Btu/lb = 1,691 ft-lb$_f$/lb (1 Btu/778 ft-lb$_f$) + 1,218.7 Btu/lb – 20 ft-lb$_f$/lb (1 Btu/778 ft-lb$_f$) + W

0.287 Btu/lb + 1,802.2 Btu/lb – 2.17 Btu/lb = 1,218.7 Btu/lb – 0.026 Btu/lb + W

Solving for W yields:

$$W = 581.5\ \text{Btu/lb}$$

The total work of the turbine is equal to:

(225,000 lb/h) (581.5 Btu/lb) = 1.31 x 10^8 Btu/h

Converting this value to horsepower gives:

(1.31 x 10^8 Btu/h) (1 hp/2,545 Btu/h) = 5.14 x 10^4 hp

Converting this value to kW gives:

(1.31 x 10^8 Btu/h) (1 kW/3,412 Btu/h) = 3.83 x 10^4 kW

The reader is left with the exercise showing that both kinetic and potential effects could have been neglected in the solution of the problem.

19.5.4 Illustrative Example 4 – Weibull Distribution (Theodore and Taylor 1994; Shafer and Theodore 2007)

The life in days of a key component in a recently constructed geothermal energy power plant is a random variable having a Weibull distribution with $\alpha = 0.025$ and $\beta = 0.50$. What is the average life of the component? What is the probability that the component will last more than 4,000 days?

Solution: Frequently, the failure rate of equipment exhibits three stages: a break-in stage with a declining failure rate, a useful life stage characterized by a fairly constant failure rate, and a wear out period characterized by an increasing failure rate (Theodore and Dupont 2012). An example of this type of failure rate is the death rate of humans. A failure rate curve exhibiting these three phases is called a "bathtub curve." The Weibull distribution provides a mathematical model of all three stages of the bathtub curve. This probability distribution function (PDF) (Theodore and Taylor 1994; Shafer and Theodore 2007) is given by

$$f(t) = \alpha\beta t^{\beta-1} \exp\left(-\int_0^t \alpha\beta t^{\beta-1} dt\right) = \alpha\beta t^{\beta-1} \exp(-\alpha t^{\beta});\ t > 0;\ \alpha > 0,\ \beta > 0 \quad (19.31)$$

where α and β are constants, referred to as the shape parameters of curve-fitting parameters; and t = time. For $\beta< 1$ the failure rate $Z(t)$ decreases with time. For $\beta= 1$ the failure rate is constant and equal to α. For $\beta> 1$ the failure rate increases with time.

Let T denote the life in days of the key component in the power plant. The PDF of T is obtained by applying the earlier formula and substituting $\alpha = 0.025$ and $\beta = 0.50$ to yield:

$$f(t) = (0.025)(0.50)t^{(0.5-1)}e^{(0.025t^{0.5})} = (0.0125)t^{(-0.5)}e^{(0.025t^{0.5})};\ t > 0$$

By definition, the average or expected value of T is given by

$$E(T) = \int_0^{\infty} tf(t)dt \quad (19.32)$$

$$E(T) = \int_0^{\infty} t(0.0125)t^{(-0.5)}e^{(0.025t^{0.5})}dt = \int_0^{\infty}(0.0125)t^{(0.5)}e^{(0.025t^{0.5})}dt$$

Let $u = 0.025\ t^{0.5}$, then $u^2 = 0.000625\ t$, *and* $t = u^2/0.000625 = 1{,}600\ u^2$. Differentiating both sides yields $dt = 3{,}200\ u\ du$. Also, $t^{0.5} = u/0.025 = 40\ u$. If these values are substituted back into the integral the following results:

$$E(T) = \int_0^{\infty}(0.0125)(40u)e^{-u}(3{,}200u)du = 1{,}600\int_0^{\infty}u^2e^{-u}du \quad (19.33)$$

The term under the integral sign in Equation 19.33 is a form of the Gamma Function:

$$\Gamma(z) = \int_0^{\infty} t^{z-1} e^{-t} dt \tag{19.34}$$

where z–1 = 2, so that z = 3. For positive integers, n, the Gamma Function $\Gamma(n)$ = (n-1)!, so for $z = n = 3$, the value of the Gamma Function is (3–1)! = 2! = 2 x 1 = 2. The final result of the integration then is

$$E(T) = 1{,}600 \text{ x } 2 = 3{,}200 \text{ d}$$

The probability that the component will last more than 4,000 d is given by

$$P(T > 4{,}000\ d) = \int_{4{,}000}^{\infty} f(t)dt = \int_{4{,}000}^{\infty} 0.0125 t^{(-0.5)} e^{(0.025 t^{0.5})} dt \tag{19.35}$$

$$P(T > 4{,}000\ d) = -e^{\infty} + e^{(0.025 x 4000^{0.5})} = 0 + e^{-1.58} = 0.2057 = 21\%$$

19.5.5 Illustrative Example 5 – Refrigeration Application

It was shown earlier that any reversible heat engine can be regarded also as a heat pump. An engine operating on the Carnot cycle taking in heat at absolute temperature T_H and rejecting it at T_L was shown to have the maximum efficiency that could be attained by any heat engine operating between the given temperature limits. The efficiency, which is the ratio of work done to heat absorbed at T_H, is given by

$$E_C = \frac{W}{Q_H} = \frac{T_H - T_L}{T_L} \tag{19.36}$$

With refrigeration, the Carnot engine may be imagined to operate in the reverse direction, taking in heat at T_L and rejecting the amount Q_H at T_H. The same equation still applies but in slightly modified form as

$$E_C = \frac{W}{Q_L} = \frac{T_H - T_L}{T_L} \tag{19.37}$$

where T_L refers to a heat reservoir or refrigerator at a temperature below that of the surroundings and T_H refers to a heat reservoir or heat receiver at a temperature equal to or greater than that of the lowest temperature of any large mass in the surroundings. Furthermore,

$$W = Q_H – Q_L \tag{19.38}$$

which follows from the first law. The ratio Q_L/W, which is a measure of the refrigerating effect per unit of work done, is known as the "coefficient of performance" β.

Equation 19.37 establishes the optimum performance that can be expected of any refrigerating machine operating between given temperature limits. No machine can be devised that will pump heat with less expenditure of work. Furthermore, Equation 19.37 implies that all reversible refrigerating machines operating between given temperature limits will have the same coefficient of performance irrespective of the fluid used or of the mechanism. In other words, the coefficient depends only on the temperatures.

Ten thousand Btu/min are to be continuously removed from an industrial facility which is maintained at a constant temperature of 10°F. The lowest temperature of any large mass in the surroundings is 60°F. What is the least amount of power, expressed as hp, that is necessary to accomplish this?

Solution: The heat removed at T_L must be brought to at least 60°F before it can be transferred out of the system to the surroundings. Therefore $T_H = 460 + 60 = 520°\text{R}$. Applying Equation 19.37 yields:

$$W = (10{,}000\ \text{Btu/min})\frac{60-10}{470} = 1{,}064(\text{Btu/min})$$

$$W = 1{,}064\ (\text{Btu/min})\ (60\ \text{min/hr})\ (1\ \text{hp}/2{,}545\ \text{Btu/h}) = 25.1\ \text{hp}$$

19.5.6 Illustrative Example 6 – Turbine Shaft Work Calculation

Ricci Energy Associates recently provided the following information from one of their geothermal power plants. Steam enters a steam turbine irreversibly at 20,000 kg/hr under steady-state conditions. The inlet temperature and pressure are 500°C and 2500 kPa. As a result of poor insulation, the turbine loses heat to the surroundings at a rate of 0.25 MW. The outlet temperature and pressure of the steam are 175°C and 250 kPa, respectively.

Based on the information given, draw a schematic of the steam turbine system, including all material and energy flows. Also, determine the shaft work produced by the turbine in kJ/(kg inlet steam).

Solution: A schematic of the steam turbine system is provided in Figure 19.8.

The general energy balance is

$$E = \Delta KE + \Delta PE + \Delta h = Q + W_s \tag{19.39}$$

As is customary, potential and kinetic energy effects are ignored, and the resulting energy equation can be solved for W_s as

$$W_s = (h_2 - h_1) - Q \tag{19.40}$$

Converting the mass flow rate to kg/s and Q to kJ/kg inlet steam yields:

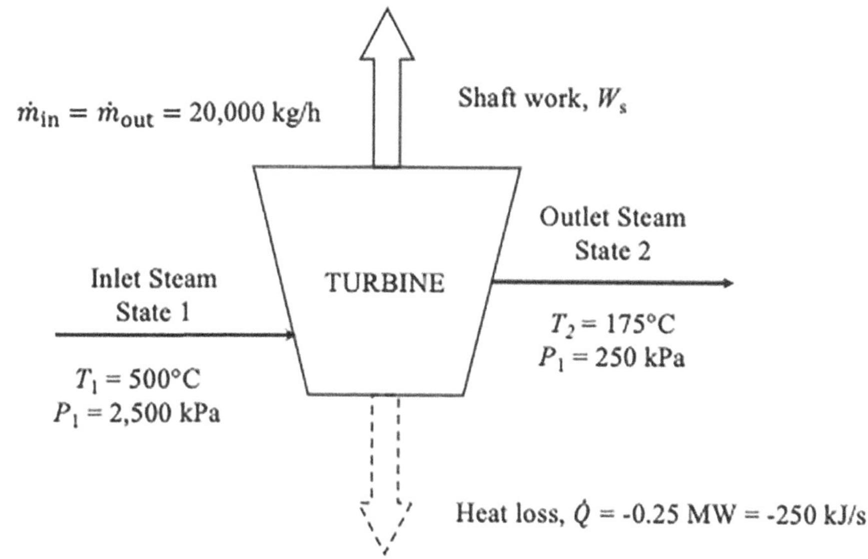

FIGURE 19.8 Schematic of steam turbine system for Illustrative Example 19.5.6.

$$\dot{m} = (20{,}000 \text{ kg/h})\left(\frac{1 \text{ h}}{3{,}600 \text{ s}}\right) = 5.55 \text{ kg/s}$$

And knowing 0.25 MW = 250 kJ/s,

$$Q = \frac{\dot{Q}}{\dot{m}} = \frac{-250 \text{kJ/s}}{5.55 \text{kg/s}} = -45.0 \text{ kJ/kg} \tag{19.41}$$

Employing the steam tables for the supersaturated steam at the inlet and outlet conditions yields, h_1 = 3,462.24 kJ/kg at inlet conditions; h_2 = 2,816.7 kJ/kg at outlet conditions. Using these values and solving Equation 19.40 for the shaft work yields:

$$W_s = (2{,}816.7 \text{ kJ/kg} - 3{,}462.24 \text{ kJ/kg}) - (-45 \text{ kJ/kg}) = -600.5 \text{ kJ/kg}$$

This represents the actual shaft work performed by the steam.

19.5.7 Illustrative Example 7 – Electricity Transmission Process

Describe the process by which electricity generated at a geothermal power plant is delivered to a domestic household.

Solution: In a typical geothermal power plant, electricity is produced with a voltage of around 13,000 V. This voltage is stepped up by means of transformers to values ranging from 138,000 to 500,000 V used on the primary transmission lines. At terminal substations, the voltage is stepped down again by transformers to the voltage of the secondary transmission lines or distribution lines on the order of 13,000 V. The voltage is stepped down once or twice again from the distribution line voltage by transformers near the point of use to levels used by consumers.

19.5.8 Illustrative Example 8 – Electricity Generation Process

Describe the main components of a geothermal power plant generating electricity.

Solution: A geothermal power plant consists of components to extract and utilize heat from the Earth's interior to generate electricity. Key components include production and injection wells, plant piping, a turbine, a generator, a condenser, and a cooling tower. These work together to convert geothermal energy into usable electricity.

The *production* and *injection wells* are the entry and exit points for the geothermal fluid (hot water or steam) circulating between the Earth's interior and the power plant. *Plant piping* is used to transport the geothermal fluid between the wells, power plant, and cooling system. The *turbine* converts the thermal energy of the steam into mechanical energy by rotating its blades as the steam passes through it and operates an *electric generator*. An electric generator is a device that converts this geothermal energy into electricity. There are many different types of electricity generators. These generators use an electromagnet that has a series of insulated coils of wire that form a stationary cylinder – called a stator – surrounding an electromagnetic shaft – called a rotor. Turning the rotor makes an electric current flow in each section of the wire coil, which becomes a separate electric conductor. The currents in the individual sections combine to form one large current. This current is the electricity that moves from generators through power lines to consumers (Figure 19.9).

A *condenser* is used to cool the steam after it has passed through the turbine, converting it back into water to be recirculated. A *cooling tower* can be used to further cool the water from the condenser, releasing excess heat into the atmosphere, and preparing the condensed, cooled water for reinjection into the geothermal reservoir.

Three primary types of geothermal power systems are used as described in Chapter 18. These are: dry steam, flash steam, and binary cycle power plants. More details on these systems are provided in Section 18.3.

19.6 ENVIRONMENTAL ISSUES

Three illustrative examples complement the presentation for this section concerned with environmental issues related to geothermal energy systems.

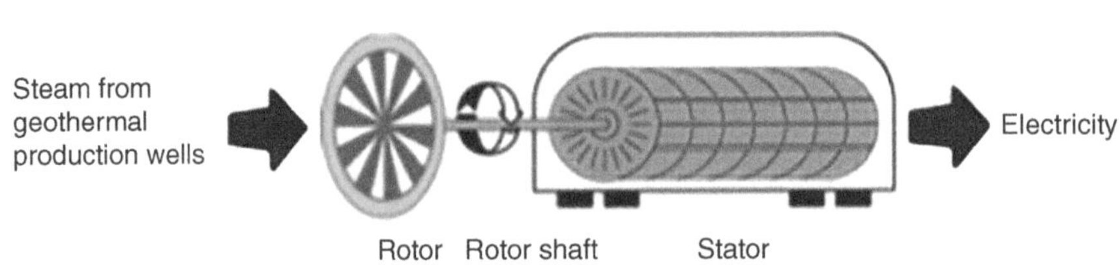

FIGURE 19.9 Electricity generation from an electric turbine for Illustrative Example 19.5.8.

19.6.1 Illustrative Example 1 – Richter Scale Earthquake Measurements

The magnitude of an earthquake occurring on an annual basis, as measured on the Richter scale, is a random variable X having the PDF (Shafer and Theodore 2007; Theodore and Dupont 2012):

$$f(x) = 1.5\, e^{-1.5x};\ x > 0 \tag{19.42}$$

$$= 0;\ \text{elsewhere}$$

Calculate $P(2 < X < 6)$.

Solution: Applying the definition of CDF (Shafer and Theodore 2007; Theodore and Dupont 2012) leads to:

$$P(2 < X < 6) = \int_2^6 f(x)\,dx \tag{19.43}$$

Substituting and integrating yields:

$$P(2 < X < 6) = \int_2^6 1.5\, e^{-1.5x}dx = e^{-3.0} - e^{-9.0} = 0.050 - 0.00012 = 0.0499 = 5\%$$

19.6.2 Illustrative Example 2 – Stack Design Recommendations

List some of the key rules of thumb of traditional power plant stack design (Theodore 2008) that can be applied to geothermal energy power plants. Provide ten such guidelines.

Solution: As experience in designing stacks has accumulated over the years, several guidelines have evolved. These include

1. Stack heights should be at least 2.5 times the height of any surrounding buildings or obstacles so that significant turbulence is not introduced by these factors.
2. The stack gas exit velocity should be greater than 60 ft/s so that stack gases will escape the turbulent wake of the stack. In many cases, it is good practice to have the gas exit velocity on the order of 90 or 100 ft/s.
3. A stack located on a building should be set in a position that will assure that the exhaust escapes the wakes of nearby structures.
4. Gases from the stacks with diameters less than 5 ft and heights less than 200 ft will hit the ground part of the time, and the ground concentration will be excessive. In this case, the plume becomes unpredictable.
5. The maximum ground concentration of stack gases subjected to atmospheric dispersion occurs about 5–10 effective stack heights downwind from the point of emission.

6. When stack gases are subjected to atmospheric diffusion and building turbulence is not a factor, ground-level concentrations on the order of 0.001%–1% of the stack concentration are possible for a properly designed stack.
7. Ground-level concentrations can be reduced by the use of higher stacks; the ground concentration varies inversely as the square of the effective stack height.
8. Average concentrations of a contaminant downwind from a stack are directly proportional to the discharge rate; an increase in the discharge rate by a given factor increases ground-level concentrations at all points by the same factor.
9. In general, increasing the dilution of stack gases by the addition of excess air in the stack does not affect ground-level concentrations appreciably. Practical stack dilutions are usually insignificant in comparison to the much larger atmospheric dilutions resulting from plume diffusion. Addition of dilution gas will increase the effective stack height, however, by increasing the stack exit velocity. This effect may be important at low wind speeds. On the other hand if the stack temperature is decreased appreciably by the diluting gas, the effective stack height may be reduced.
10. Stack dilution has an appreciable effect on the concentration in the plume close to the stack.

These ten guidelines represent the basic design elements of a stack "pollution control" system. An engineering approach suggests that each element be evaluated independently and as part of the whole control system. However, the engineering design and evaluation must be an integrated part of the complete pollution control program (Theodore 2008).

19.6.3 Illustrative Example 3 – HAZOP of a Boiler System

Consider a small industrial boiler system generating low-pressure saturated steam at a geothermal power plant facility. The boiler drum receives preheated feedwater that is required to maintain the water level in the drum between 30% and 40% of its total height. There is a visual level indicator on the drum in addition to a level indicator controller (LIC) that is connected to a level indicator control valve (LICV) in the feedwater line. Elevated temperature geothermal production water is used to heat the boiler drum. This production water line contains a pressure control valve (PCV) that is connected to a pressure controller (PC) located in the steam line in and leaving the drum.

Perform a HAZOP study on the system with the intention of providing the required amount of feedwater to the boiler drum to maintain the prescribed water level and the desired steam production.

Solution: A HAZOP study is a very useful technique that may lead to a more reliable and safer process. Whether it is applied at preliminary design stages or to the detailed layout of an existing plant, its benefits can be invaluable. It reduces the possibility of accidents for the process involved, improves on-line availability of the process, can lead to a better understanding of the process and possible malfunctions, and provides training for the evaluation of any process. Finally, it is also a way of optimizing a process and providing a reliable and cost-effective system.

Generally, a HAZOP focuses on a major piece of equipment, although a lesser piece of equipment, such as a pump or a valve, may be chosen depending upon the nature of the materials being handled and the operating conditions. Once an intended operation is defined, a list of possible deviations from the intended operation is developed. The degrees of deviation from normal operation are conveyed by a set of guide words, some of which are listed in Table 19.3.

The overall HAZOP method is summarized in the following steps:

1. Define objectives.
2. Define plant limits.
3. Appoint and train a team.
4. Obtain complete preparative work.
5. Conduct examination meetings in order to:
   a. Select a manageable portion of the process.
   b. Review the flowsheet and operating instructions.
   c. Agree on how the process is intended to operate.
   d. State and record the intention.
   e. Search for possible ways to deviate from the intention, utilizing the guide words provided in Table 19.3.
   f. Determine possible causes for the deviation.
   g. Determine possible consequences of the deviation.
   h. Recommend action(s) to be taken.
6. Issue meeting report.
7. Follow up on recommendations.

TABLE 19.3
HAZOP Guide Words

Guide Words	Meaning	Examples
NO or NOT	No part of the intention is achieved but nothing else happens.	No flow, no agitation, no reaction.
MORE and LESS	Quantitative increase(s) or decrease(s) to the intended activity	More flow, higher pressure, lower temperature, less time.
AS WELL AS	All of the intention is achieved but some additional activity occurs.	Additional component contaminants, extra phase
PART OF	Only part of the intention is achieved, part is not.	Component omitted, part of multiple destinations omitted.
REVERSE	The opposite of the intention occurs.	Reverse flow, reverse order of addition.
OTHER THAN	No part of the intention is achieved. Something different happens.	Wrong component, startup and shutdown problems, utility failure.

After the serious hazards have been identified with a HAZOP study or some other type of qualitative approach, a quantitative examination should be performed. Hazard quantification or hazard analysis (HAZAN) involves the estimation of the expected frequencies or probabilities of events with adverse or potentially adverse consequences. It logically ties together historical occurrences, experience, and imagination. To analyze the sequence of events that lead to an accident or failure, event and fault trees are used to represent the possible failure sequences (Theodore and Taylor 1994; Shafer and Theodore 2007; Theodore and Dupont 2012).

A line diagram of the boiler system is provided in Figure 19.10.

Apply the first guide word NO to the system and list the deviation(s), cause(s), and consequence(s).

Deviations	Causes	Consequences
NO flow of feed water.	LICV closed.	Loss of level in drum and explosion of drum by heat supplied to dry shell if geothermal production water flow continues
NO level in drum.	Loss of feed water pressure. Massive leak.	Same as above.
NO geothermal production water flow.	Geothermal production water line closed.	No heat supplied. No steam generation.

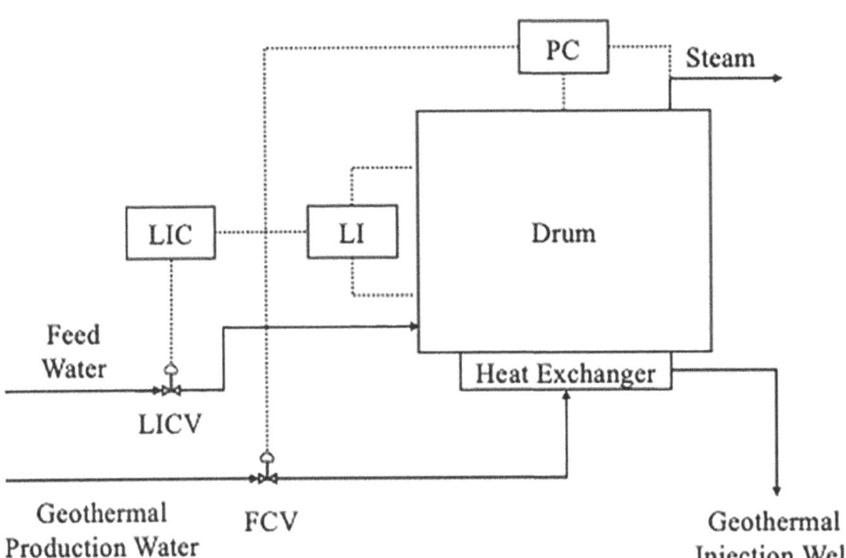

FIGURE 19.10 Line diagram for the boiler system for Illustrative Example 19.6.3.

Apply the second guide word MORE to the system and list the deviation(s), cause(s), and consequence(s).

Deviations	Causes	Consequences
MORE level in drum than 40%.	LIC fault (failure).	Excessive entrainment in steam.

Apply the guide word LESS to the system.

Deviations	Causes	Consequences
LESS level in drum than 30%.	Similar to NO level. LICV fault (failure)	Same as NO level.

Apply the guide word AS WELL

Deviations	Causes	Consequences
Contaminants AS WELL AS feedwater.	Water treatment fault.	Fouling of boiler. Corrosive water. Potential boiler failure due to severe corrosion over time.

Apply the guide word PART OF to the system.

Deviations	Causes	Consequences
PART OF feedwater (treatment chemical) omitted.	Water treatment fault.	Same as above.

Apply the guide word REVERSE to the system.

Deviations	Causes	Consequences
REVERSE flow in feed water line.	Loss of feed water pressure.	Steam in feed water system. Same as NO flow of feed water above.

Apply the guide word OTHER THAN to the system.

Deviations	Causes	Consequences
Unplanned shutdown (OTHER THAN normal operation).	Utility failure.	Does LICV fail closed or open? Same consequences as NO level in drum or MORE level in drum.
OTHER THAN feedwater	Not possible	

19.7 OPTIMIZATION

Three illustrative examples complement the presentation for this section concerned with optimization. One is concerned with optimum pipe diameter calculations, one with optimum geothermal power plant profit, and the third with optimizing the performance of a proposed geothermal energy plant.

19.7.1 Illustrative Example 1 – Optimizing Pipe Diameter

When power plants are designed and developed, designers try to minimize the cost of operations and maximize efficiency. One way this is commonly done in plants is by strategically selecting the piping and pipe fittings to maximize flow and minimize cost. For any given set of flow conditions, as pipe diameter increases, capital cost correspondingly increases. However, with increasing pipe diameter, comes decreasing pumping or fan charges due to lower frictional resistance. Conversely an increase in pressure drop will generally result from a decreased pipe size. Therefore, an optimum economic pipe diameter must exist. This optimum pipe diameter is found at the point when the sum of pumping or fan costs *plus* fixed capital costs are at a minimum. Outline how to determine the optimum economic pipe diameter for a given flow system.

Solution: The usual calculation procedure is as follows:

1. Select a pipe diameter.
2. Obtain the annual operating cost.
3. Obtain the capital equipment cost.
4. Convert the capital cost to an annual basis.
5. Sum the two annual costs in Steps 2 and 4.
6. Return to Step 1, selecting a different pipe diameter.

The only variable that will appear in the resulting total cost expressions is the pipe diameter. The economically optimum pipe diameter can be generated *analytically* by taking the derivative of the total annual cost expressed in terms of pipe diameter, setting the result equal to zero, and solving for the diameter. The derivative operation can be replaced by a trial-and-error procedure that involves calculating the total cost for various diameters and simply selecting the minimum. These later calculations can be used to generate a graph of the form provided below in Figure 19.11.

19.7.2 Illustrative Example 2 – Optimizing Geothermal Energy Power Plant Profit

Plans are underway to construct and operate a small geothermal energy power plant. The company is still undecided as to whether to employ Power Plant A, B, or C. Power Plant A is less expensive to purchase and operate than either a comparable Plant B or Plant C. However, the projected incomes from Plant B and Plant C are higher.

Based on the economic and financial data provided in Table 19.4, select the process that will yield the highest annual profit. Calculations should be based on an

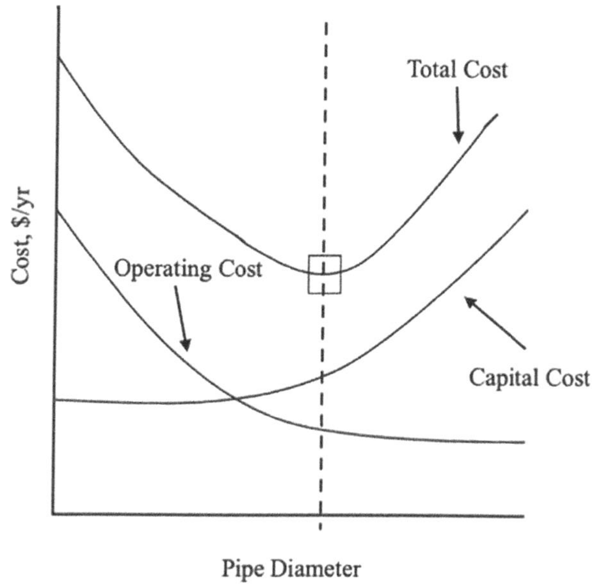

FIGURE 19.11 Graphical solution of optimal pipe diameter selection for Illustrative Example 19.7.1.

TABLE 19.4
Costs Associated with Geothermal Power Plant Systems A, B, and C

Costs/Credits	Plant A	Plant B	Plant C
Capital, $1,000	26,250	29,750	34,350
Construction, $1,000	15,750	17,000	17,550
Operations, $1,000/yr	4,000	5,500	5,900
Maintenance, $/yr	65,000	77,500	84,500
Income, $1,000/yr	20,000	25,000	25,500

interest rate of 5% and a plant lifetime of 20 years for the three geothermal power plant options in question.

Solution: To determine an annual cost for the fixed costs in this problem, the capital recovery factor (CRF) needs to be calculated. The CRF is given as

$$CRF = \frac{i}{\left(1-(1+i)^{-n}\right)} \tag{19.44}$$

where i = interest rate, decimal; and n = number of payment periods of a loan. Substituting for the given input values yields the following:

TABLE 19.5
Annualized Costs Associated with Geothermal Power Plant Systems A, B, and C

Costs/Credits	Plant A	Plant B	Plant C
Capital, \$/yr	2,100,000	2,380,000	2,748,000
Construction, \$/yr	1,260,000	1,360,000	1,404,000
Operations, \$/yr	4,000,000	5,500,000	5,900,000
Maintenance, \$/yr	65,000	77,500	84,500
Income, \$/yr	20,000,000	25,000,000	25,500,000
Total Profit, \$/yr	12,575,000	15,682,500	15,363,500

$$CRF = \frac{0.05}{\left(1-(1+0.05)^{-20}\right)} = \frac{0.05}{(1-0.377)} = \frac{0.05}{0.623} = 0.08$$

Using this CRF, the annualized costs for each plant can be determined for their capital and construction costs given in Table 19.4. These annualized capital costs along with a total annual cost and expected profit for each plant are summarized in Table 19.5. As indicated there, Plant B is the option that should be selected because it provides the best profit, \$15,682,500/yr based on all fixed and operating costs plus expected income.

19.7.3 Illustrative Example 3 – Optimizing Geothermal Energy Power Plant Performance

The cost associated with the operation of a geothermal energy power plant, *CO*, is linearly related to both the operating temperature *T*, °R, and well depth below 10,000 ft, Z. That relationship (in consistent absolute units, \$/h) is given by

$$CO = 0.18\,T + 0.05\,Z \tag{19.45}$$

However, the following system performance and environmental constraints must be satisfied:

$$a_1T + b_1Z \geq S, \text{ for steam quality considerations} \tag{19.46}$$

$$a_2T + b_2Z \geq \mathrm{E}, \text{ for steam quantity considerations} \tag{19.47}$$

If the coefficients for these constraint equations are $a_1 = 0.2$, $a_2 = 0.4$, $b_1 = 0.05$, $b_2 = 0.02$, $S = 600$, $E = 450$, $T \geq 0$, and $Z \geq 0$, calculate the operating temperature and well depth that minimizes the cost of operation of the geothermal power plant and satisfies the constraints associated with the steam quality and quantity constraints (Behan and Theodore 2018).

	A	B	C	D	E	F	G	H
1								
2	**Illustrative Example 19.7.3**							
3								
4		Cost						
5	*CO* =	$ 540.00	/h					
6								
7	Variables	*T*, °R	*Z*, >10,000 ft					
8	Objective Coefficients	0.18	0.05					
9	Values	3000	0		<- Optimal Values appear here after solving			
10								
11	Coefficients	*T*	*Z*		Total		Limit	
12	Constraint Equation 1	0.2	0.05		600	>=	600	
13	Constraint Equation 2	0.4	0.02		1200	>=	450	
14	Constraint Equation 3	1			3000	>=	0	
15	Constraint Equation 4		1		0	>=	0	

FIGURE 19.12 Excel solution for the geothermal power plant performance optimization for Illustrative Example 19.7.3.

Solution: This is a typical linear programming problem (Behan and Theodore 2018). For this example, the objective function is Equation 19.45, subject to the constraints provided in Equation 19.47 as follows:

$$0.2\,T + 0.05\,Z \geq 600 \tag{19.48}$$

$$0.4\,T + 0.02\,Z \geq 450 \tag{19.49}$$

Microsoft Excel can be used to solve this problem with the inputs shown in Figure 19.12, along with the optimal solution that minimizes the objective function subject to the stated constraints. The minimum cost is achieved when the operating temperature is 3,000°R = 2,540°F and the well depth is 10,000 ft (approximately 2 miles).

REFERENCES

Behan, K., and Theodore, L. 2018. *Introduction to Optimization for Environmental and Chemical Engineers*. Boca Raton, FL: CRC Press/Taylor & Francis Group.

Farag, I. 1994. *Fluid Flow, a Theodore Tutorial.* East Williston, NY: Theodore Tutorials. Originally published by the USEPA/APTI, RTP, NC. Roanoke, VA: ETS Professional Training Institute. https://etsi-inc.com/training/tutorials-and-textbooks/.

Reynolds, J. 1992. *Material and Energy Balances. A Theodore Tutorial.* East Williston, NY: Theodore Tutorials. Originally published by the USEPA/APTI, RTP, NC. Roanoke, VA: ETS Professional Training Institute. https://etsi-inc.com/training/tutorials-and-textbooks/.

Reynolds, J., Dupont, R.R., and Theodore, L. 1991. *Hazardous Waste Incineration Calculations: Problems and Software*. Hoboken, NJ: Wiley.

Reynolds, J., Jeris, J., and Theodore, L. 2004. *Handbook of Chemical and Environmental Engineering Calculations*. Hoboken, NJ: Wiley.

Shafer, S., and Theodore, L. 2007. *Probability and Statistics Applications in Environmental Science*. Boca Raton, FL: CRC Press/Taylor & Francis Group.

Skipka, K.J., and Theodore, L. 2014. *Energy Resources: Availability, Management, and Environmental Impacts.*Boca Raton, FL: CRC Press/Taylor & Francis Group.

Theodore, L. 2008. *Air Pollution Control Equipment Calculations*. Hoboken, NJ: Wiley.

Theodore, L. 2014. *Chemical Engineering: The Essential Reference*. New York City, NY: McGraw-Hill.

Theodore, L., and Dupont, R.R. 2012. *Environmental Health and Hazard Risk Assessment: Principles and Calculations*. Boca Raton, FL: Taylor & Francis Group.

Theodore, L., and Nueser, K.M. 1996. *Engineering Economics and Finance, a Theodore Tutorial.* East Williston, NY: Theodore Tutorials. Originally published by the USEPA/APTI, RTP, NC. Roanoke, VA: ETS Professional Training Institute. https://etsi-inc.com/training/tutorials-and-textbooks/.

Theodore, L., and Reynolds, J. 1991. *Thermodynamics, a Theodore Tutorial.* East Williston, NY: Theodore Tutorials. Originally published by the USEPA/APTI, RTP, NC. Roanoke, VA: ETS Professional Training Institute. https://etsi-inc.com/training/tutorials-and-textbooks/.

Theodore, L., and Taylor, F. 1994. *Probability and Statistics*. Theodore Tutorials, originally published by U.S. EPA, RTP, NC. East Williston, NY. Roanoke, VA: ETS Professional Training Institute. https://etsi-inc.com/training/tutorials-and-textbooks/.

Index

F

The Ethics of Neoliberalism

The 21st century is the age of "neo-liberalism"—a time when the free market is spreading to all areas of economic, political and social life. Yet how is this changing our individual and collective ethics? Is capitalism also becoming our new morality? From the growing popular demand for corporate social responsibility to personal desire for "work-life balance," it would appear that nonmarket ideals are not only surviving but also thriving. Why then does it seem that capitalism remains as strong as ever?

The Ethics of Neoliberalism boldly proposes that neoliberalism strategically co-opts traditional ethics to ideologically and structurally strengthen capitalism. It produces "the ethical capitalist subject" who is personally responsible for making their society, workplace and even their lives "more ethical" in the face of an immoral but seemingly permanent free market.

Rather than altering our morality, neoliberalism "individualizes" ethics, making us personally responsible for dealing with and resolving its moral failings. In doing so, individuals end up perpetuating the very market system that they morally oppose and feel powerless to ultimately change.

This analysis reveals the complex and paradoxical way capitalism is currently shaping us as "ethical subjects." People are increasingly asked to ethically "save" capitalism both collectively and personally. This can range from the "moral responsibility" to politically accept austerity following the financial crisis, to the willingness of employees to sacrifice their time and energy to make their neoliberal organizations more "humane," to the efforts by individuals to contribute to their family and communities despite the pressures of a frenetic global business environment. Neoliberalism, thus, uses our ethics against us, relying on our "good nature" and sense of personal responsibility to reduce its human cost. Ironically, in the new millennium, it seems the more ethical we are, the stronger capitalism becomes.

Peter Bloom is a lecturer in the Department of People and Organisations at the Open University, UK.

Routledge Studies in Business Ethics

For a full list of titles in this series, please visit www.routledge.com

Originating from both normative and descriptive philosophical backgrounds, business ethics implicitly regulates areas of behaviour which influence decision making, judgment, behaviour and objectives of the leadership and employees of an organization. This series seeks to analyse current and leading edge issues in business ethics, and the titles within it examine and reflect on the philosophy of business, corporations and organizations pertaining to all aspects of business conduct. They are relevant to the conduct of both individuals and organizations as a whole.

Based in academic theory but relevant to current organizational policy, the series welcomes contributions addressing topics including: ethical strategy; sustainable policies and practices; finance and accountability; CSR; employee relations and workers' rights; law and regulation; economic and taxation systems.

8 **Ethics and Morality in Consumption**
Interdisciplinary Perspectives
Edited by Deirdre Shaw, Andreas Chatzidakis and Michal Carrington

9 **Truth in Marketing**
A Theory of Claim-evidence Relations
Thomas Boysen Anker

10 **Mismanagement, "Jumpers," and Morality**
Covertly Concealed Managerial Ignorance and Immoral Careerism in Industrial Organizations
Reuven Shapira

11 **Restructuring Capitalism**
Materialism and Spiritualism in Business
Rogene A. Buchholz

12 **The Ethics of Neoliberalism**
The Business of Making Capitalism Moral
Peter Bloom

The Ethics of Neoliberalism

The Business of Making Capitalism Moral

Peter Bloom

Routledge
Taylor & Francis Group
LONDON AND NEW YORK

First published 2017 by Routledge

2 Park Square, Milton Park, Abingdon, Oxfordshire OX14 4RN
52 Vanderbilt Avenue, New York, NY 10017

Routledge is an imprint of the Taylor & Francis Group, an informa business

First issued in paperback 2019

Library of Congress Cataloging-in-Publication Data
A catalog record for this book has been requested

ISBN: 978-1-138-66724-2 (hbk)
ISBN: 978-0-367-24304-3 (pbk)

Typeset in Sabon
by Apex CoVantage, LLC

"To Tomas—the world is yours to change"

Contents

Acknowledgments

> I'm no longer accepting the things I cannot change. . . . I am changing the things I cannot accept.
>
> —Angela Davis

This book was written in the hope of discovering a more radical and liberating way of ethically being in the world. I am supremely grateful for all the wonderful people who have intellectually and personally contributed to the completion of this project. Specifically, I would like to thank all those in the Department of People and Organisations at the Open University who have supported me and served as valuable sources of insight. I would also like to extend my gratitude to the Ideology and Discourse Program at the University of Essex as well as the People, Organisations and Work group formerly at Swansea University for their inspiration. I also want to thank David Varley and everyone at Routledge who made this work possible.

Personally, I would also like to thank my family and friends: in particular, my parents, Martin Bloom and Phyllis Bloom, as well as Sara Cowan Bane and Fred Bane; my sister Jane Bloom and her husband, Jay Baur; my wonderful "Vanbrugh Hill commune"; and most importantly my amazing partner Sara Gorgoni and her family. Finally, thank you to my soon-to-arrive son Tomas—to whom this book is dedicated.

1 The Paradox of Neoliberal Ethics

In the 21st century, capitalism has supposedly invaded and colonized every aspect of human existence. Indeed, the current age is seemingly defined by the inevitable rise of "marketization." Once sacred public institutions, from education to transportation to healthcare, are being increasingly privatized. Even those that have escaped a direct market takeover are still being progressively subjected to the iron logic of profit and efficiency. In the new millennium, the public good is primarily a private interest.

Even more troubling is the perceived total intrusion of capitalism into our very consciousness and sense of self. It is not just that the market concretely dominates social, political and economic relations. It also extends and shapes the way we see the world, the way we reason and the way we make moral judgments. In the famous words of Jameson, "Someone once said that it is easier to imagine the end of the world than to imagine the end of capitalism" (2003: 76). Far beyond just being subjected to market excesses and exploitation, modern humans have now become fully thinking and acting market subjects.

We are thus living in the time of neo-liberalism. Amid global differences in culture, wealth, and privilege, there is a shared social evolution toward a capitalist reality that is increasingly total in its external and internal reach. Marketization is transcending its former limits as an economic system and cementing itself as the sole basis for organizing contemporary existence. According to Larner (2000: 6–7):

> New forms of globalized production relations and financial systems are forcing governments to abandon their commitment to the welfare state. Rather than formulating policies to ensure full neoliberalism employment and an inclusive social welfare system, governments are now focused on enhancing economic efficiency and international competitiveness. One consequence is the "rolling back" of welfare state activities, and a new emphasis on market provisioning of formerly "public" goods and services.

Under neo-liberalism, all things are judged in terms of their market worth. For anything to be possible it must first be fiscally viable. Achievement is a matter of advancing constantly upward professionally. Everything is a potential market opportunity. Entrepreneurship now trumps all other values. The epitome of leadership—whether political or economic—is that of a hard-charging, decisive and visionary corporate executive. Ethical value is firmly and almost completely determined by the dominant financial values of the age. According to Wendy Brown (2015), it signifies no less than the profound shift from "homo politicus" to "homo economicus" based on "the image of man as an entrepreneur of himself." She warns, hence, that "neoliberal reason, ubiquitous today in statecraft and the workplace, in jurisprudence, education and culture, and a vast range of quotidian activity, is converting the distinctly *political* character, meaning and operation of democracy's constituent elements into *economic* ones" (17).

Yet there is a profound but all too often ignored question that has arguably never been more crucial and urgent than in the present. Quite simply, how much has neoliberalism really transformed us ethically? Are people now more "capitalist" or "market oriented" in their core moral beliefs than ever before? Has neoliberalism in fact succeeded in fundamentally altering our individual and collective notion of the good to exclusively reflect market ideals?

The idea of social enterprises, cooperatives, the sharing economy and the demands for "work-life balance" would seem to suggest otherwise. Anecdotally, stories abound of co-workers banding together—even if only informally—to help each other cope with the increasingly unreasonable demands of the current neoliberal workplace. There is a renewed emphasis on the need for an "ethics of care" to counteract the negative impacts of an individualistic and market-oriented neoliberalism (see McDowell, 2004; Lawson, 2007). The rise of the so-called "sharing economy" reflects this rather contradictory character of present-day neoliberal ethics. According to the *Economist*: "This emerging model is now big and disruptive enough for regulators and companies to have woken up to it. That is a sign of its immense potential. It is time to start caring about sharing" (2013).

The financial crisis and decades-long problems of economic inequality and political oligarchy are, moreover, dramatically challenging this monopolistic capitalist paradigm. Movements across the world have arisen that call into question the previously assured inevitability of the global market. The Arab Spring, anti-globalization struggles, Black Lives Matter and the broader rebirth of progressive politics from Spain to the United Kingdom and Greece to the United States speak to this emerging shift against a world created by capital for the principal benefit of capitalists. Potentially arising is a new socio-economic order that places the needs of people over the demands of profits.

These developments raise an equally significant set of concerns. If we indeed have maintained a vibrant sense of nonmarket ethics, then how has neoliberalism turned this social consciousness paradoxically to its own advantage? How does capitalism ironically rely on us to be morally and practically non-capitalist in our everyday lives and relationships for its very survival and prosperity?

Aim

This book seeks to illuminate the role of ethics within contemporary neoliberalism. It is commonly assumed that increased marketization is linked to the growth of market-based values. However, empirical evidence at the social, organizational and individual levels profoundly put into question this belief. Societal pressure for "corporate social responsibility," professional demands for a more "humane" workplace and personal desires for greater "work-life balance" reveal the continued influence and prevalence of "nonmarket" moralities and ethics. It is important, therefore, to reconsider the relation of neoliberalism to ethics. Are these nonmarket ethics challenging the hyper-capitalism of neoliberalism? If not, how are they paradoxically strengthening this present capitalist order?

This book aims, therefore, to be a critical and comprehensive analysis of the "ethics of neoliberalism." Its main theme is that neoliberalism is creating "ethical capitalists." In doing so, it challenges assumptions associating the spread of the free market with the internalization of market values. By contrast, it hopes to conceptually clarify the paradoxical relationship of modern capitalist hegemony and non-capitalist ethics. Moreover, this work will attempt to empirically explore how non-capitalist ethics ideologically and structurally reinforce capitalism at the political, institutional and personal levels as well as theoretically illuminate the ways this "ethics of neoliberalism" shapes present-day capitalist subjectivity.

What Is Neoliberalism?

The term "neoliberalism" is now part of the popular lexicon. It symbolizes a general trend toward greater marketization and the upward transfer of wealth as well as power to the financial elite. Indeed, contemporary critiques of the current status quo are replete with broadsides against neoliberalism and all those who unfairly benefit from it. It is thought to proliferate and worsen a whole range of chronic social ills, from widening inequality (Chomsky, 1999; Lazzarato, 2009) to global underdevelopment (Ferguson, 2006; Navarro, 2007) to deepening political authoritarianism (Bloom, 2016a). The actual definition of neoliberalism—what exactly it conceptually represents and how it concretely orders socio-economic relations—remains ambiguous (See Brenner et al., 2010; Clarke, 2008; Mudge, 2008).

Traditionally, it is presented as principally an economic phenomenon. It has its historical roots in the free-trade theories of post–World War II European economists such as Friedrich Hayek and later Milton Friedman (see Mirowski and Plehwe, 2009). They laid out a vision that was distinctly opposed to the interventionist Keynesian thinking at the time—a view that would hold sway within elite policy circles for at least the next three decades. In this respect,

> [n]eoliberalism is in the first instance a theory of political economic practices that proposes that human well-being can best be advanced by liberating individual entrepreneurial freedoms and skills within an institutional framework characterized by strong private property rights, free markets, and free trade.
>
> (Harvey, 2007: 2)

Beginning in the 1980s, these theories would have the chance to be tested as a number of "radical" Conservative governments—notably Ronald Reagan in the United States and Margaret Thatcher in the United Kingdom—came to power promising a "market revolution."

The concrete realization of these ideals in policy has dramatically transformed the national and global economy. It has prioritized values of private ownership and competitiveness at the expense of public ownership and direct welfare provision. Specifically, this has meant a push for economic privatization, the weakening of collective bargaining and unions along with the reduction of taxes particularly for top earners. The objective, at least rhetorically, is to "free up" capital to spur investment and enhance efficiency (Hill and Kumar, 2008; Mensah, 2008; Passas, 2000; Steeger, 2005).

Its legitimacy is founded on its supposed discovery of "objective" economic laws—an economic dogma that is often portrayed as a science (see Clarke, 2005). The idea, for instance, that lower wages lead to higher mass welfare may seem counter-intuitive. Yet according to a neo-classical perspective, higher wages lead inevitably to increased prices, resulting thus in a worse outcome for society generally. Neoliberalism draws upon and takes this law-like approach even further. It contends that markets are inherently self-regulating and consequently any government intrusion is at best a necessary evil. It combines a simplified cause-and-effect paradigm of economic relations (e.g., more regulation equals a less competitive business environment) with progressively sophisticated statistical models for forecasting market behavior. Neoliberals portray their brand of economics as a social science—the technical application of incontrovertible truths to human affairs. The economist, here, is akin to a financial meteorologist—though it must be said with considerably less success at making predictions than the average weather person.

It is also a distinct ideology for structuring society from the top down. It presents government as a hindrance to economic growth and dynamism. In the famous—or, given recent history, infamous—words of Ronald Reagan, "government is not the solution to our problem; government is the problem." Despite its claims to empiricism and pretensions of objectivity, neoliberalism is fundamentally utopian in its theoretical foundations and social aims. It offers up a romanticized vision of a "free market" society—where an enterprising spirit is the key to unlocking individual upward mobility and competition is the sole path to collective progress and evolution (see Brenner and Theodore, 2002; Cohn, 2006; Giroux, 2003).

Neoliberalism's profoundly utopian spirit is reflected in its real-world manifestations. Neoliberalism prescribes "cure all" capitalist solutions to all socio-economic problems. Across contexts and amid vast differences in geography, culture and history, marketization and privatization offer a supposedly universal means of realizing prosperity. The common goal of all economics, no matter the size or stage of a country or region's development, is to privatize services, reduce government, enhance markets and empower the financial sector.

Yet this pronounced ideological character also produces a type of delusional quality shared by other fundamentalist religious or political "faiths." As will be discussed in detail later in this work, the refusal to abandon "austerity" even in the face of mounting empirical evidence of its disastrous impact exemplifies this tendency toward dogmatism (Atonakakis and Collins, 2014; Blyth, 2013). To this end, neoliberalism is at least partly, if not primarily, a market orthodoxy that seeks to shape an often unwilling and complex reality into its simplified hyper-capitalist image, regardless of the cost.

Approaching the "Moral Order" of Neoliberalism

It is perhaps natural to assume that the ethics of neoliberalism reflects its underlying market ideology and values. The ostensible aim of neoliberalism is to construct a thoroughly hyper-capitalist society. This expands beyond the global objective to turn every corner of the world into a fully marketable and marketized entity. It also extends inward—as part of a concerted effort to subjectively engineer "market subjects." Neoliberalism, hence, is as much an ethico-political project as it is an economic one.

For this reason, there is an increasing emphasis on the cultural and political dimensions of neoliberalism both conceptually and in practice. Significantly, neoliberalism offers a distinctive and exportable theory of government and governance, respectively. In terms of sovereignty, it has mentioned advocates for minimal public authority—limiting it only to "necessary" tasks such as defense. This has been referred to as the "watchman" approach to government, and it

> ranges over a wide expanse in regard to ethical foundations as well as to normative conclusions. At the one end of the line is "anarcho-liberalism," arguing for a complete laissez-faire, and the abolishment of all government. At the other end is "classical liberalism," demanding a government with functions exceeding those of the so-called night-watchman state.
>
> (Blomgren, 1997: 224)

Neoliberal governance, in turn, is primarily associated with an ethos of self-regulation and personal responsibility. Just as markets are seen to be the best source for their own management, so too are individuals. External interference by governments is not only ineffective but also socially harmful. To this end,

> neoliberalism is grounded in the 'free, possessive individual', with the state cast as tyrannical and oppressive. The welfare state, in particular, is the arch enemy of freedom. The state must never govern society, dictate to free individuals how to dispose of their private property, regulate a free-market economy or interfere with the God-given right to make profits and amass personal wealth. State-led "social engineering" must never prevail over corporate and private interests. It must not intervene in the "natural" mechanisms of the free market, or take as its objective the amelioration of freemarket capitalism's propensity to create inequality.
>
> (Hall, 2011: 10–11)

In theory, therefore, the task of governance is restricted to the realm of private actors—whether that be individuals or businesses.

This ethos of limited government and private governance contributes to a broader morality of what can be termed "market responsibility." While neoliberalism presents an image of a return to a Hobbesian society of "all against all," its notion of private governance is underpinned by a deeply moral notion of collective relations. It draws inspiration from Smith's concept of the "invisible hand" whereby individual self-interest produces generalized welfare. The attempt by governments to intervene is then an immoral intrusion and subversion of this delicate and implicit social compact—one that can have quite detrimental consequences for the very people it is said to be helping.

The political foundations of neoliberalism are therefore distinctly moral in character. They are built on a fear of authoritarianism and publicly enforced servitude. Hayek's influential book *The Road to Serfdom* exemplifies this moralistic justification for rejecting governments in favor of markets. Written in response to the expansion of the welfare state in the wake of the "great depression," Hayek writes of an ominous future in which public dependency is equated with political subjection. In his words:

> But when economic power is centralized as an instrument of political power it creates a degree of dependence scarcely distinguishable from slavery. It has been well said that, in a country where the sole employer is the state, opposition means death by slow starvation.

The modern liberal or socialist government is a replacement for the lords and kings who once dominated the population under feudal rule. It was only capitalism that could liberate individuals from this sovereign stranglehold over their life and actions.

These sentiments have been drawn on consistently to politically and ethically legitimate neoliberalism in the contemporary period. The introduction of regulation is at best a "necessary evil" and at worst the first step on a short road to social oppression. This is not to say that the government should have no role in the economy; rather, it is to say that the government should play the part of ensuring an orderly market society. As Biebricher (2015: 1) notes,

> neoliberalism is often understood as a synonym of the doctrine of self-regulating markets, but I don't think that's really appropriate. I think neoliberals are quite clear that states have certain functions to fulfil in order to make markets function. But they should only engage in certain kinds of actions, and these are particularly market-enabling actions.

Indeed, neo-liberalism has in many ways seen the rise of a new regulatory regime aimed, at least in part, at the unleashing of market forces (Levi-Faur and Jordana, 2005). The threat of government transcends its presumed simple inefficiency or a blind rejection of public intervention. It is directed instead at what it perceives as a profound moral threat to personal freedom and shared progress when it is aimed at repressing the expansion of markets.

Such moral underpinnings were and are reinforced through emotive depictions of non-capitalist alternatives. Politicians routinely invoke Orwellian images of a "Big Brother" to warn citizens of the mortal danger posed by "encroaching socialism." For neoliberals, this is no simple worry concerning the appropriate level of public intervention within a market economy. It is an existential struggle for the very fate of human freedom. The genuine historical terrors of really existing socialism were transformed into the key set pieces for a broader capitalist morality play. Any type of government action or assistance—with the exception of those required for defense, domestic safety or market expansion—is framed as a harbinger of Communist bondage. It is put forward as the surest and quickest route to modern slavery.

Neoliberalism represents a firm belief in the fundamental morality of the market. Above all else, it associates the marketplace with a fair and free capitalist society that must be spread and defended. There is a moral imperative to be ever vigilant against any and all threats to this moral order. The growth of government is a danger to the very future of human liberty. At its heart, it is a moral crusade for civilization itself.

An "Ethical Responsibility" to a Moral Market

The moral order driving neoliberalism also not surprisingly promotes, and to a certain extent demands, a sense of obligation to maintain these ideals. Indeed, a critical irony of neoliberalism is that the more self-regulating it desires society to be, the more self-regulated it expects and requires subjects to become. Amid different and often competing perspectives of neoliberalism is an implicit need for institutions and individuals to conform to market values and assumptions of "human nature." While neoliberals preach of the freedom it offers from government tyranny, it nonetheless comes with a range of informal demands for shaping external behavior and internal beliefs.

In this respect, neoliberalism attempts to concretely and ethically wed the full-scale embrace of the marketplace with the expansion of capitalism generally. The spread of wage labor and the power of capital as well as capitalists are legitimized as part of an ethical ahistorical vision of an efficient, productive and dynamic market society. Indeed, as Chaudhry (1993: 246) notes, "The reification of 'the market' as a neutral and natural institution, apolitical and ahistorical—as an end in itself rather than a means to promote social and individual welfare—has become common in academic and policy circles." Neoliberalism and its "explicit preference for private over public control" came "to dominate the global political economy," signifying "a dramatic break from post war policies" as an "economic policy, expression of political power and ideational hegemony" (Centeno and Cohen, 2012: 317). The extension of the market into all spheres of human relations is seen as crucial for ensuring a fair distribution of resources and encouraging constant innovation.

The subjective and ethical foundations of neoliberalism are reflected even in its most ostensibly objective economic models. The renowned Chicago school—perhaps the most influential source of "free market" thinking and policy in the world—is framed as a scientific explanation of how to most efficiently materially organize the economy and society. However, underpinning these "laws" is a clear set of assumptions of how individuals do—and, significantly, should—behave. In order for markets to operate efficiently, so must their participants. There is a belief that individuals and organizations will naturally seek to maximize their utility. While what this good ultimately is can be quite undefined—and even amoral—there is a clear ethos of instrumental rationality informing the pursuit of these ends.

These ethical foundations are also evident in the Austrian school of economic thought that was similarly influential for the formation of neoliberalism. In contrast to the Chicago perspective, Hayek and others focus on the construction of value within the marketplace. Subjective evaluation is crucial to this social process of valuation. While there is not an inherent basis for determining value, through the auspices of the market and the interactions of its members the worth of an object or service can be properly set and determined. Thus, according to von Mises (2012: 29), 'it is not the state but the common practice of all those who have dealings in the market, that creates money'. For this reason, the intrusion of the market would be an artificial and ultimately undesirable example of price manipulation. Yet for such a self-regulating market to work effectively, the subject must be willing and able to continually acquire information. While the assumption is that this is simply a natural human tendency and supposedly based on inductive observation, it points to the social need and potential ethical imperative for people to be effective information gatherers.

In an even more explicit vein, Freidman puts forward what can be considered a foundational ethics for neoliberalism. He directly contravenes traditional notions of social responsibility and morality prioritizing cooperation and charity. He proclaims that the only "social responsibility of business is to increase profits" (Friedman, 1970). In his view, if one is involved in a business relation with another, then there is an ethical obligation to maximize profits. This ethical prerogative transcends a mere rejection of the state or typical broadside against the dangers of Communism. Conversely, it is proffered as an ethical duty to maintain a moral market order. The failure to do so can lead to a lack of competitiveness, resulting in the loss of revenues and therefore jobs. Fundamentally, it threatens the very sanctity of human liberty. In this respect, Friedman declares, "Whether blameworthy or not, the use of the cloak of social responsibility, and the nonsense spoken in its name by influential and prestigious businessmen, does clearly harm the foundations of a free society" (Friedman, 1970: 1).

Neoliberalism, thus, has deep ethical roots, which are nourished by the bedrock belief in the morality of the market and the ethical responsibility of its members to live up to its cherished ideals. In order to maximize the benefits of this fully capitalist order—to optimize a "free society"—subjects must internalize its underlying assumptions of human nature and embrace its overarching ethos of private competitiveness. From these ethical roots would spring, in turn, a marketized public ethics of neoliberalism.

Toward a Market Ethics of Neoliberalism

The last decades of the 20th century witnessed the rise of neoliberalism. What was once a mere theory—and often a fringe one at that—became the dominant framework for organizing economic relations and society. Suddenly, all things that hindered the market were considered at best economically misguided and at worst a moral danger to a prosperous market order. In practice, this meant that the welfare state and labor unions were now seen as outdated parts of an outmoded system. Emerging as well was a novel ethical framework compatible with this new marketized social reality.

Crucial to this neoliberal ethics was the seemingly perverse reification of conventionally morally castigated values of greed and even gluttony. The heroes were Wall Street stockbrokers who could fully exploit a volatile and lucrative marketplace (See Guerrera, 2010; Winter, 2007). The pursuit of personal gain and profit above all other ethical considerations was justified as crucial to creating and maintaining a dynamic and ever-expanding capitalist society. This ethics was equally individually and collectively oriented (See Beeson and Firth, 1998). Personally, this full-throttle adoption of a competitive and atavistic mentality was imperative for individuals to "get ahead" and provide for themselves and their families. Collectively, mass consumption was critical to keeping businesses afloat, keeping employment up and keeping national economies growing (see, e.g., Midgley, 1992).

Rather ironically, this "revolutionary" market ethics was matched by a pronounced social conservatism. This combining of economic radicalism with cultural traditionalism will be analyzed in further detail later in this book. However, it is worth highlighting here that this "market" ethics was never pure and always partially legitimized by a range of conventionally nonmarket values. For this reason, the growth of neoliberalism was complemented by the rise of what has been referred to as "conservative capitalism." According to Hoover (1987: 245):

> Reagan and Thatcher have assembled a rationale and a series of policies for what I will identify as conservative capitalism. Rather than dealing incrementally within a general consensus on reformist policies, they have reversed the growth of taxation, shifted resources away from human service programs, resuscitated traditionalist prescriptions for personal behavior, and advanced the apparent substitution of the market for government as the key institution of the society.

Consequently, capitalism during this period morphed from a preferable economic system with clear moral implications (individualism, private enterprise, hard work, etc.) into an extreme market ethos for ethically overdetermining society and the individual. The famous quote by the fictional financier Gordon Gecko in the movie *Wall Street* that "Greed is good" came to define a generation and epitomize this ethical embrace of selfishness (see Tett, 2009) even as it politically espoused traditionalist values of patriotism and the "family" (Steeger and Roy, 2010).

Perhaps not surprisingly, this unadulterated reification of the market as the highest good was soon challenged by its less than ideal real-world consequences. Only half a decade after the triumphant elections of Thatcher and Reagan, in the United Kingdom and United States respectively, corporate scandals were rocking both countries' economic and moral sensibilities. The savings and loan scandal in the United States, for instance, cost taxpayers over $100 billion as well put temporarily to rest any notion that market principles were in any way ethically sufficient (see Day, 1993). They presented neoliberals with a crisis of market morality—one that would need to be addressed if its larger project of marketization was going to continue to be politically successful.

Arising out of this crisis was a reconfigured sense of ethical responsibility that simultaneously preserved the prominence of the market as well as recognized the need for some sort of ethical management of its excesses. Namely, it produced an increased call for "business ethics." The problem was seen as—rather than any immorality of the market or ethical lack in capitalism—a deficiency in knowledge. In other words, corporations and those in finance simply did not have an adequate set of ethical guidelines to direct their actions. Freshly chastened, they accepted the relevance of standard codes and practices for pointing them in the right ethical direction. Yet business ethics also continued to prioritize the overall moral desirability of a market system.

Essential for the implementation of this ethics was a notion of "self-regulation." Individuals and companies were expected to regulate themselves in order to act legally in their pursuit of profit (see Shamir, 2004). This ethos of personal accountability was translated into a broader theory of "corporate social responsibility." In this regard, it was corporations themselves who were charged with monitoring their own behavior and conforming to an increasingly standardized code of "business ethics." However, this did not alleviate the need for government or regulation. As an Oxfam paper on Globalization so presciently declared:

> At their best, voluntary codes of conduct can act as a guide to corporate practice and set standards for others to follow. . . . At their worst, they are little more than a public relations exercise. But the deeper point is that corporate behavior is too important for poverty reduction to be left in the field of voluntary codes and standards defined by the corporate sector itself. . . . What is needed is a set of verifiable and enforceable guidelines covering all aspects of corporate activity.
>
> (Shamir, 2005)

Rather, neoliberalism reframed governance as a process of encouraging governments to find new ways to incentivize such corporate social responsibility (Gond et al., 2011).

Present was an evolving and progressively sophisticated ethical vision of a "free market." The unbridled capitalism of the past was posited as naïve and dangerous. However, this ethical check on the market did not entail its full-scale rejection or even a return to social democratic principles. It promoted a model of ethics that combined the dynamism and liberty of capitalism with the regulation required to mitigate its worst characteristics. Freedom, hence, was slightly reworked to mean the granting of subjects the freedom to be ethical capitalists.

The Ethical Market Subject of Neoliberalism

The beginning of the new millennium witnessed the ascendancy of the market as an almost unassailable ideology. The fall of the Soviet Union and the conclusion of the Cold War ushered in the supposed "end of history," meaning the free market and liberal democracy could finally reign supreme. Capitalism had not just decisively defeated Communism politically, it had also claimed a moral victory over it as a social system. The concerns over the ethical deficiencies of a more market-oriented society were displaced and overshadowed by its global triumph. Consequently, the effort to morally reform capitalism's excesses was transformed into a diverse project to successfully instill a sense of market ethics into a new generation of individuals.

The overarching ethos of this period—particularly in the West—was one of being able to best take advantage of the opportunities offered by a market economy. The keywords of this era were "competiveness" and "efficiency" for both nations and individuals alike. Significantly, this was as much an ethical as it was an economic imperative. It was necessary for everyone to cultivate a proper mentality and set of skills in order to keep up with a rapidly globalizing and always dynamic free market. The good life was explicitly equated with private enterprise and an entrepreneurial spirit. The best people, companies and countries would all succeed by continually improving themselves and effectively adapting to a competitive international marketplace.

This was reflected perhaps most notably in the predominance of a widely embraced "market rationality." Individuals were charged with cultivating a strictly instrumental rationality for getting ahead in a cutthroat business environment (Kunda and Ailon-Souday, 2005). What was particularly interesting was that this calculating mind-set was framed as being not just economically necessary but also ethically desirable (see Sennett, 1998). To fail to think and act in such a way was to fail to maximize one's own talents and abilities. It was to deny oneself all the advantages and possibilities provided to one by the market—thus limiting one's own life choices (Bloom and Cederstrom, 2009). Moreover, it was viewed as a form of waste, a squandering of resources akin to throwing away perfectly good food when others were starving.

In a similar vein, entrepreneurship became a crucial and increasingly all-encompassing modern ethics. The business owner was lauded as the driver of innovation, the visionary who was singularly capable of radically transforming society (Allen, 1997; Anderson and Warren, 2011; Banfe, 1991). The tech "start-ups" and Silicon Valley were the epitome of the entrepreneur as revolutionary and even deserving of a religious-like devotion (Robinson, 2013). Underpinning such romanticized—and woefully simplistic—depictions of the power of private enterprise was an ethical prerogative to be constantly entrepreneurial regardless of whether one owned a business. It meant that one had to be eternally on the lookout for new business opportunities, fresh ways to make profit, previously undiscovered avenues for social advancement and economic exploitation.

Central to this emerging marketization of ethics was the emphasis placed on employability. Social worth was almost completely defined by its market value. Likewise, personal success and aspirations were judged according to their market attractiveness (Arthur and Rousseau, 1996; Hall, 2004). All activities were expected to be part of a broader "employment biography" that could enhance one's employability. The previous ethical prerogative of hard work and self-reliance was reconfigured into an analogous but updated demand that individuals perpetually acquire new skills to optimize their professional and personal prospects. To this end, employability aims to make "each worker a more aware and a more independent organizer of the succession of activities and commitments that, combined, constitute his/her working life" (Gazier, 2001: 23). The unwillingness to make the most of one's resume was seen as a sign of moral weakness rather than as a reflection of difficult and stressful economic realities.

This contemporary capitalist morality went beyond individual ethics. It was part of a collective obligation that people owed to their community. It was an essential component of an organization's broader strategy for wealth creation (see Hitt et al., 2001). To return to a previous point, to be entrepreneurial was necessary for ensuring national innovation and international progress (Williams and McQuire, 2010). The willingness to become more employable was critical for populations to "not be left behind" by the inevitable march of globalization (Lee and Peterson, 2001). It also placed fresh expectations on governments to help individuals adopt and maximize such a market ethics (See Feldman, 2014; Matthews et al., 1996).

Revealed, therefore, was how neoliberalism had constructed a new market-based conception of individual and collective ethics. It was not enough to accept the permanence of the market. Nor was it sufficient to defend it against "radical" threats such as socialism. The new expectation was that people fully invest in becoming the best market subjects they could be. This notion of the good conflated capitalist success with moral value—presenting market rationality, entrepreneurship and employability as inherently ethically worthwhile. Yet, as will be shown, neoliberalism also reinforced, or at a minimum was forced to accept, a wide range of decidedly—on the surface—nonmarket principles and normative ideals.

The Nonmarket Ethics of Neoliberalism

Neoliberalism ostensibly promotes the total marketization of socio-economic relations. This spread of the market supposedly extends into the inner recesses of the modern psyche—influencing ethics to reflect its values of competition, productivity and profitability. However, they are by no means exhausted. Arising alongside these explicitly capitalist ideals were conventionally nonmarket aspirations for personal freedom and well-being. Moreover, these desires were often publicly directed against the professional ethics seemingly so central to the establishment of neoliberalism. If in the 1980s "greed was good," by the beginning of the new millennium the longing for personal fulfillment in all areas of one's life suddenly reigned supreme.

The increasing calls for work-life balance exemplify this trend. One must not work to live or live to work but rather find equilibrium between the two. Having time to pursue your own interests was no longer considered a luxury. Rather it was an ethical obligation one owed to oneself. As such, the goal of achieving this balance was to 'enhance organizational structural and cultural/relational support for work, family and personal life' (Kossek et al., 2010: 4). Employers, in turn, were progressively tasked with finding ways to help individuals realize this mutually beneficial desire for balance (Byrne, 2005). Employees, for their part, were charged with changing their mentality and behavior in order to simultaneously meet their professional obligations and fulfill their personal desires (See Fleetwood, 2007).

Concretely, these values were translated into fresh demands for flexibility at work. Specifically, individuals sought to use digital advances to better control when and where they worked—as employees increasingly were subjected to an "electric panopticon" (see Lyon, 1993; Bain and Taylor, 2000). Reflected was a deep ethic for people to have the right and ability to manage their time in order to successfully realize their potential inside and outside of work. The idea of fitting everyone's work life into a one-size-fits-all schedule was progressively viewed as outdated and even immoral. Hence, initial employer-friendly conceptions of work-life balance were progressively criticized inasmuch that

> [i]n the work-life balance debate, over-work is perceived as the problem. Nevertheless, beyond working time and the provision of flexible working practices to enable child care, there is little in the debate but the need to change work per se. The debate also narrowly perceives "life," equating it with women's care work, hence the emphasis again of family-friendly polices.
>
> (Eikhof et al., 2007: 325)

The "good life" involved uniquely exploring one's own ideal "work-life balance."

Even more radical were popular calls to abandon the traditional "9–5 working" life. It was an ideal that now appeared "anti-ethical" to leading a rewarding and exciting existence separate from work. According to a 2008 Chartered Management Report:

> New strategies are needed to make greater flexibility work for us, rather than being at the mercy of forces that bombard us with information and choices, blur boundaries beyond comfort and manageability, or leave us at the mercy of machines that manage our lives for us.

In its place stood a new ethical vision of finding ways to earn more by working less. Individuals are encouraged to "escape 9–5, Live Anywhere, and Join the New Rich," while employees should learn "how to fill the void and [create] meaning after removing work and the office" (Ferris, 2007).This appeal to get rich quick was understood as a necessary part of being able to freely "follow your dreams."

Significantly, there was also a pronounced growth in discourses of personal wellness. Ideas of success began to exceed the boundaries of mere professional upward mobility. "Success" now included and increasingly prioritized emotional and physical health. Moreover, there was an enhanced emphasis on the pursuit of spiritual fulfillment (Burack, 1999; Byrne, 2005; Der Klerk, 2005). Employment was framed as both a barrier to and a potential positive force for the achievement of these wellness goals. The new capitalist ethic transcended traditional demands to work hard but now encompassed the ability to thrive spiritually and psychologically. The aim of employment, in turn, was to guarantee the "increased physical and mental health of employees," including their "advanced spiritual growth and enhanced sense of self-worth" (Krahnke et al., 2003: 397).

In a decidedly more political spirit, this ethos has focused on achieving social justice. Entrepreneurship is being redirected beyond objectives of mere profit maximization and toward aims of sustainability and social inclusion (see Dees, 2007). To this end, the negative social and economic costs of neoliberalism are thus transformed into an opportunity to use the market for the purpose of social progress (Nicholls, 2008). This ethical challenge is also witnessed in the rise of progressive political movements such as "Occupy Wall Street" and the "Arab Spring."

Indeed, from the ashes of Communism's defeat and capitalism's triumph arose a reinvigorated social consciousness. The total expansion of the market was met with a resurgent progressive ethics that reconsidered and resisted capitalist values of profit and competition at the expense of equality and welfare. Less asked but no less important, though, were questions of how such nonmarket ethics positively interacted with and even perhaps reinforced neoliberalism.

Uncovering the Paradox of Neoliberal Ethics

Neoliberalism was meant to fundamentally alter society and in the process completely marketize both the economy and ethics. The free market would shape morality and normative notions of the good, serving as a modern ethical framework for how one ought to work and live. Yet, as always, reality was considerably more complicated. Far from transforming traditional ethics, principles of personal well-being and social justice were if anything strengthened. Ironically, what the "free market" produced was in certain key ethical aspects a dramatically nonmarket ethical subject.

The romantic visions of an efficient, innovative and aspirational economy gave way to widespread concerns of financial crises, slow recoveries and precarious economies. In this light, the prospect of greater public investment and the rebuilding of a secure social safety net understandably took on a renewed attractiveness. Yet it also highlighted the resilience of these social democratic principles even after the three-decade onslaught of neoliberalism.

It revealed the fact that ideas of social fairness and individual welfare had not been completely extinguished in the ash heap of history.

There is thus a certain contradictory aspect to neoliberalism when it comes to ethics. On one hand, it purports to be the pre-eminent champion of the "free market." On the other hand, it often results in the perpetuation of perceptively nonmarket values and aspirations. The dual analyses by one of the foremost scholars on neoliberalism, William Davies (2014), highlight this bipolarity. In his seminal work *The Limits of Neoliberalism* he convincingly argues that a crucial consequence of this present system is to universalize the ethos of competition—to render it a central and constitutive feature of every social relation and institution. Yet this increasingly exportable competitive spirit is matched in intensity by a mass desire not for efficiency or productivity but rather for personal happiness. In his equally compelling book *The Happiness Industry*, Davies (2015) explores this dimension of neoliberalism—examining how this fully marketized global order is fueled as much by aspirations for well-being as it is profit.

The first and perhaps most obvious explanation for these, if you will, competing ethical impulses is one of simple exploitation and profit—or, to be more precise, the ability of neoliberalism to economically take advantage of core human emotions and longings. Rather than seeing the ineffable desire to be "happy," to find "satisfaction" and to achieve "fulfillment" as challenges, neoliberalism sees them as merely new revenue sources to be mined. Our emotional and spiritual well-being become commodities to be bought and sold. If capitalism is leaving you dissatisfied and unwell, it also produces the very goods and services that can allow you to purchase happiness. Like radicalism and creativity before it, neoliberalism has in part tamed anti-market desires through transforming them into an exciting market opportunity for producers and consumers alike.

There is also a less cynical and more psychological perspective for explaining this seeming ethical contradiction. The promotion of happiness and the possibility of a cooperative and socially just world is understandably and particularly comforting in a winner-take-all society increasingly ruled by the cutthroat logic of total competition. These distinctly nonmarket principles have arisen as a necessary coping mechanism—one that grows in urgency as neoliberalism appears ever more permanent and intractable (see Wren-Lewis, 2014). The ability to dream of a different reality and the celebration of small moral victories are an essential part of emotionally and ethically dealing with neoliberalism.

However, there is an alternative and potentially more profound understanding of this phenomenon. It is to reframe the question to ask, "What affective and structural role does nonmarket ethics play in reinforcing and reproducing neoliberalism?" The resilience of these values goes beyond simple economic opportunism or emotional endurance. It is a potentially key and fundamental element to its survival and perpetuation. The production of "ethical" neoliberal subjects, as this book will explore, helps cover over and temporarily resolve the deep structural contradictions of this hyper-capitalist system. It, furthermore, provides subjects with an appealing sense of self to embrace and cultivate that unconsciously binds them to this neoliberal order, even as they consciously morally question and even ethically oppose it.

This work therefore aims to uncover a central paradox of neoliberalism. The more marketized a society becomes, the more it ultimately concretely comes to depend upon nonmarket ethics. This does not mean that conventional market values are completely eschewed or that ethically inspired progressive political movements are insignificant or ineffective. By contrast, it highlights the very real and important function played by traditionally dismissed principles such as cooperation, interpersonal care and desires for public welfare to the functioning of neoliberalism specifically and capitalism generally. For a "free market" to socially exist and expand, it must continually produce ethically "nonmarket" subjects.

The Ethics of Neoliberalism

This book boldly proposes that neoliberalism strategically co-opts traditional ethics to ideologically and structurally strengthen capitalism. It produces "ethical capitalist subjects" who are morally accountable for making their society, workplace and even their lives "more ethical" in the face of an immoral but seemingly permanent free market.

Rather than altering our morality, neoliberalism "individualizes" ethics, making us personally responsible for dealing with and resolving its moral and structural failings. In doing so, individuals end up perpetuating the very market system that they morally oppose and feel powerless to ultimately change.

This analysis reveals the complex and paradoxical way capitalism is currently shaping us as "ethical subjects." People are increasingly asked to ethically "save" capitalism both collectively and personally. This can range from the "moral responsibility" to politically accept austerity following the financial crisis, to the willingness of employees to sacrifice their time and energy to make their neoliberal organizations more "humane," to the efforts by individuals to contribute to their family and communities despite the pressures of a 24/7 global business environment. Neoliberalism, thus, uses our ethics against us, relying on our "good nature" and sense of personal moral obligation to reduce its human cost. Ironically, in the new millennium, it seems the more ethical we are, the stronger capitalism becomes.

2 Producing the "Ethical" Capitalist Subject

The question of capitalist ethics has become increasingly urgent in the new millennium. The fallout from the 2008 financial crisis highlighted the excesses of marketization and the potential need for a new moral order. The images of "greedy" banker and "psychopathic" financiers threatened the economy and society—demanding a dramatic reconsideration of what was legally allowable and ethically permissible. Adding to this widespread ethical unease were feelings of increased anxiety and anger caused by heightened job insecurity, growing inequality and the undue influence of the "1%." The idea that the market was either sustainable or ethical looked progressively doubtful.

Underpinning these concerns are shared fears that people in general are becoming less ethical. There is a veritable "moral panic" about the present state of individual virtue (see Goode and Ben-Yahuda, 2009; Hier, 2011). Bemoaned is a lack of real community due to the digital revolution and, as such, the disintegration of genuine human connection (Hampton et al., 2011; Sum et al., 2008; Warschauer, 2004) and basic standards of decency (Gofin and Avitzour, 2012; Law et al., 2012; Topping, 2014). The predominance of the Internet has perpetuated a culture of online bullying and anonymous vitriol. Moreover, there are worries that this new generation has let go of traditional moral values in favor of snark and cynicism (Miller, 2014). More radically oriented is a common notion that markets, while efficient, have served to erode basic standards of human decency such as cooperation and care for the other (Falk and Szech, 2013).

These worries point to the need to reconsider the relation of ethics to capitalism. More precisely, they reflect the often contradictory, even schizophrenic, relationship between what it means to have a good society and a market economy. Traditionally, capitalists have explicitly championed compatible ethos with being a "good" employee—including being punctual, frugal and sober (see, e.g., Giorgi and Marsh, 1990). The ability to be professional was necessary both for the structural reproduction of a capitalist system and for individual upward mobility. The cultivation of a proper "capitalist" ethic was therefore crucial for ensuring national development, growth and competitiveness. In a less austere spirit, consumerism has produced its own ethos—one that revolves around consumption, desire for material goods and excitement at being able to follow the latest trends (see Campbell, 2005; Nava, 1987).

There is also a political dimension to this dynamic. Whether or not capitalism requires a certain ethics is of course debatable. What is clear, though, is that its political success relies on incorporating and even co-opting a diverse range of often competing cultural values. This includes rendering compatible resistance demands around the environment, social justice and economic equality with the perpetuation of a "free market" (Henderson and Sethi, 2006; Mackenzie and Lewis, 1999; Sunstein, 1999). This can be witnessed in ideas of the "green market," social entrepreneurship or even trickle-down economics. There is a certain ethical malleability to capitalism that ensures its legitimacy and continued relevancy within evolving and commonly unpredictable socio-political environments.

Yet these perspectives, valuable as they are, do little to fully address the broader role of ethics for structurally and subjectively reproducing capitalism generally and neoliberalism specifically. To this effect, they do not get to the heart of how nonmarket values are aiding and strengthening marketization across contexts. One means of approaching this issue is by highlighting the ethical agency created by existing capitalist relations. Put differently, how does the shift toward trends such as financialization and privatization allow for and rely upon the perpetuation of seemingly contradictory ideals of cooperation and pursuing a public good rather than merely private gains? How does it practically permit and even encourage individuals to act ethically in ways that are theoretically opposed to what would traditionally be considered a "self-interested" market subject?

This chapter reconsiders the ethics of neoliberalism. It reframes the traditional question of "How is neoliberalism shaping ethics to reflect market values?" to "How is it paradoxically enabling individuals to act ethically for its survival?" To do so, it will highlight the ways marketization offers individuals the opportunity to be good and do good. In this respect, marketization provides them the cultural resources as well as social expectation to be "ethical." From this foundation, it will present how this imperative to "do good" translates into an appealing identity as an "ethical subject" that ironically structurally reproduces this form of hyper-capitalism organizationally, politically and interpersonally. In sum, it reveals how the success of neoliberalism depends on its production of the "ethical" capitalist subject.

The Critical Morality of Capitalism

Capitalism and morality have a complicated and not altogether straightforward history. Indeed, for many the system represents the height of amorality. It is merely a set of economic relations whose inexhaustible demand for profit and productivity turns a blind eye to other considerations, such as personal well-being and social justice. Its supporters contend just as passionately that a "free market" society, in fact, reflects the height of human morality—the fullest realization of individual freedom and collective progress. Of course, both of these perspectives are necessarily incomplete and misleading. The former ignores the crucial role of social morality for capitalism—even if only superficially. The latter remains intentionally ignorant of the massive destruction caused by capitalism and capitalists both past and present.

Indeed, the introduction of capitalism as a dominant mode of production was intertwined with its promotion as part of a broader moral order. It was associated with emerging social discourses of "modernization" and "progress" (Appleby, 2011). Industrialization pointed to the possibilities of a novel economy and society, built on the ideals of efficiency and productivity rather than tradition and feudal rule. Consequently, "freedom meant prosperity; freedom meant progress; freedom meant having willing workers as opposed to unwilling ones" (Temperly, 1977: 109). Praised were principles of meritocracy, competition, innovation and increasingly personal freedom (see Hume, 1955; McNulty, 1967; Stigler, 1957).

The early association of wage labor, urbanization and industrialization to triumphant discourses of modernization and progress reveals capitalism's historically long-standing links to social morality. These connections were made even clearer in the association of such ideas of economic modernity to broader cultural notions of "being civilized." Countries that were undergoing this "industrial revolution" attributed their mass development to their own cultural superiority. Capitalism came to signify "civilization" in toto—signifying the capacity of certain societies to advance more rapidly than others (Dussell, 1997; Hamer, 1998; Stark, 2007; Tawney, 1926).

Hence, to be capitalist was as much a judgment or indictment of a society's essential moral character as it was a description of its economic system.

Not surprisingly, morality, in this respect, was deployed to justify prevailing economic inequalities and social oppressions. The industrialized use of slavery throughout the "New World" was legitimized through rendering non-Europeans as "uncivilized" (GoGwilt, 1995). The declaration of the 19th-century evolutionist Alfred Wallace Russell personified such views:

> The Red Indian in North America, and in Brazil; the Tasmanian, Australian and New Zealander in the southern hemisphere, die out not from any one special cause but from the inevitable effects of an unequal mental and physical struggle. The intellectual and moral, as well as the physical qualities of the European are superior. . . . If my conclusions are just it must inevitably follow that the higher—the more intellectual and moral—must displace the lower and more degraded races.
>
> (quoted in Stone, 1998: 164)

Racism became an important discursive tool for framing European colonialism and mass capitalist exploitation as not only economically necessary but also completely moral. In the words of one 19th-century socialist:

> Consider a population like ours, placed in the most favourable circumstances; possessed of a powerful civilisation; amongst the highest

> ranking nations in science, the arts and industry. Our task now, I maintain, is to find out how it can happen that within a population such as ours, races may form—not merely one but several races—so miserable, inferior and bastardised that they may be classes below the most inferior savage races, for their inferiority is sometimes beyond cure.

Just as importantly, they moralized the imperialism that was central to the spread of capitalism globally. Indeed, "For some, Britain's expansion overseas lifted anxieties about contemporary social trends, about luxury and immorality. It convinced them that, despite the encroachment of luxury and irreligion, the nation retained God's favour" (Harris, 1996; also see Bass and Cherwitz, 1978).

Such moral justifications also extended to individuals. The inability to be successful was due to personal irresponsibility rather than structural conditions. Whereas previously individuals were trapped in the condition given to them by their birth, the market—at least theoretically—permitted anyone to advance (see Aune, 2002; Honig and Black, 2007; Horowitz, 2006; McNamee and Miller, 2009). The failure to do so could be traced back to individual failings and in no way reflected on the overall morality of capitalism itself. Indeed, early capitalist society legitimized the use of quite strong forms of legal regulation and punishment to "correct" the bad behavior of its growing workforce (see Bloom, 2016b). Its common refrain of equality was, thus, ironically employed to ethically justify existing and often worsening inequality.

Just as significantly, the human cost of this new economic order was morally defended and even embraced as necessary to social progress. The 19th century witnessed the introduction and perpetuation of ideas of "Social Darwinism" that offered a moral framework for rationalizing the social and personal destruction caused by capitalism (See Hawkins, 1997; Hofstadtar, 1944; Moscovici et al., 1976). More than simply economically organizing society around the pursuit of profit, what markets supposedly did was the moral work of separating able and less abled individuals. Differentiating the so-called "wheat" from the "chaff" had an ethical importance that went beyond ensuring that a country was supposedly economically successful: it was put forward as vital to the evolution of the human species as a whole. In the words of the prominent late-19th-century philosopher and zoologist Conway Lloyd Morgan:

> In social evolution on this view, the increment is by storage in the social environment to which each new generation adapts itself, with no increased native power of adaptation. In the written record, in social traditions, in the manifold inventions which make scientific and industrial progress possible, in the products of art, and the recorded examples of noble lives, we have an environment which is at the same time the product of mental evolution, and affords the condition of the

> development of each individual mind to-day. . . . [T]his transference of evolution from the individual to the environment may leave the faculty of the race at a standstill, while the achievements of the race are progressing by leaps and bounds.
>
> (1896: 340, also quoted in Hodgson, 2005: 906–906)

The theoretical and concrete instantiation of capitalism therefore was intimately bound up with the spread of an official sense of morality. As will be discussed throughout this book, the "free market" was and is as much a moral and ethical project as it is an economic one. Such insights, however, are by no means original. What is crucial, though—and dramatically less understood—is to better understand how such morality and the ethics it prescribes contributes to the economic and political reproduction of a market-based capitalist society.

The "Necessary" Ethics of Capitalism

It is perhaps tempting to treat capitalism and markets as a purely economic phenomenon. There is a simplicity and elegance to the idea that markets operate according to objective economic laws. Further, there is a certain appealing security in the notion that individual and collective prosperity is attainable simply through implementing a system that best suits our human nature. Yet this scientific and essentialist perspective is belied by the past and present importance of ethics for the survival and reproduction of capitalism. The development, advancement and perpetuation of a "free" labor and consumer market requires the promotion of specific social values.

Arguably, the first and most famous discussion of the importance of ethics for capitalism was Weber's *The Protestant Ethics*. In it he argues that countries that have a "protestant work ethic" consisting of hard work, discipline and frugality will naturally have more success in introducing and profiting from a market-based economy (Weber, 2002; see also Furnham et al., 1993; Giorgi and March, 1990). The veracity of this insight notwithstanding, it was an early example of how fundamental ethics could be to the overall success of capitalism generally. It made an assumption that markets worked better when conjoined with specific principles and cultural values. The survival of markets was directly linked to the perpetuation of a complementary ethical system.

During the latter half of the 19th century, the creation of a capitalist society was married to an increasingly strict set of social morals. In England, for instance, the Victorian age was renowned not simply for its imperialism and rapid industrialization but also for its prudish cultural sensibilities (Grugel, 1979; Seaman, 2002). It was a time in which sex was treated as taboo and public morals were often strictly enforced (see Weeks, 2014). These conventions were responsible (ironically, as Foucault presciently observes) for reproducing this moral order—including those activities and identities it considered "deviant." To this effect, Victorians aimed not to suppress it, "but rather to give it an analytical, visible, and permanent reality" (Foucault, 1990 [1976]: 44).

Equally important is how it reflected a new capitalist ethics—promoting the image of the upright bourgeoisie citizen. Specifically, it framed the political, social and economic challenge posed by capitalists to the landed aristocracy as an ethical struggle to replace privilege, promote "sovereign" self-determination and protect the rights of property. According to Isin (1997: 127),

> The doctrine of the sovereign citizen capable of entering into contracts, pursuing his private interests, selling his labour power, and purchasing his sustenance, and capable of being governed, was the characteristic aspect of modern citizenship and reflected the demands made by the bourgeoisie on the landed aristocracy.

Such a strong ethical character was considered fundamental to the proper social shaping of this radically new economic system and those who populated it. It involved a process of public emulation (Harrison, 1999) and one that had a strong roots in the Enlightenment revolutions (see Kaplan, 2003). More precisely, it set the stage for the construction of a "public morality" that could be continually displayed and reinforced within an emerging civic sphere. Quoting Habermas (1991: 108) at length on this development:

> Political actions, that is, those referring to the rights of others, were themselves declared to be in agreement with law and morality only as far as their maxims were capable of, or indeed in need of, publicity. Before the public it had to be possible to trace all political actions back to the foundation of the laws, which in turn had been validated before public opinion as being universal and rational laws. In the framework of a comprehensively norm-governed state of affairs (uniting civil constitution and eternal peace to form a "perfectly just order") domination as a law of nature was replaced by the rule of legal norms—politics could in principle be transformed into morality.

In this respect, the public performance of morality—the continual presentation of the "lawful" politician and citizen—moralized an exploitive capitalist mode of production.

Likewise, in the United States, alongside the growth of chattel slavery and Native American genocide, there was a profound articulation of a "new" type of man for the supposedly "new" world. It was one committed to individual freedom and liberty from social hierarchies and government tyranny (Foner, 1999). It promoted a fresh ethos of entrepreneurism and the "self-made man" (see Wyllie, 1956). The rugged individualism of the frontier was transferred to the creation and reproduction of a competitive private economy. As with Weber, these claims did not perfectly mirror or even come close to reflecting the actual realities of the time. The growth of the capitalist economy in the United States was very much attributable to large-scale, though often invisible, government investments and interventions as well as the profit gained through the mass use of brutal human bondage. To this effect, in 19th-century America, to a

> government that was often most powerful in shaping public policy when it was hidden in plain sight. Such was the case when the government created and nourished a corporate-driven market, stimulated expansion by subsidizing exploration and removing the Indians and influenced trade patterns through communication and transport policies.
>
> (Balogh, 2009: 4)

Within the considerably different context of France, capitalism and the spread of markets was analogously part of a nation-building exercise that was very much associated with the spread of "modern" ethics. The introduction of markets as a central organizing feature of society was premised on a larger project of creating "Frenchmen" through the auspices of industrialization for the purposes of reflecting patriotic market values (Weber, 1976). This process has been referred to as no less than a process of "self-colonization" whereby

> roads and railroads brought hitherto remote and inaccessible regions into easy contact with the markets and lifeways of the modern world. Schooling taught hitherto indifferent millions the language of the dominant culture, and its values as well, among them patriotism. And military service drove these lessons home.
>
> (Weber, 1976: 493–494)

Foreshadowing neoliberalism and its teaching of "market ethics" to the general population, it signified an attempt to ethically transform French peasants into hardworking and civic-minded laborers of the emerging French nation.

Outside the nascent metropoles of capitalism, this ethical imperative would soon become a key feature of capitalism's global expansion through colonialism. The distinction between the supposedly "civilized" West and the rest of the world (considered to be at best exotic and at worst "barbaric") gave, as hinted at above, moral cover to imperial oppression and genocide (see Young, 2005). It reflected an emerging "Enlightenment narrative" portraying

> a descent from classical antiquity into the 'barbarism and religion', and the emergence from the latter set of conditions of a 'Europe' in which civil society could defend itself against disruption by either. This history

> had two themes: the emergence of a system of sovereign states . . . and the emergence of a shared civilization of manners and commerce.

However, it also reinforced the myth that these political and economic inequalities were the product of inherent cultural differences rather than conquest (Metha, 1999; Monasterios, 2008). This provided an incentive for "less civilized" nations to try to imitate their Western rulers. Additionally, it legitimized commonly severe moral codes of behavior between colonizers and the colonized (see McClintock, 2013; Stoller, 2002).

Moving forward into the early 20th century, capitalism was regularly associated with the promotion of new ethics for its reproduction. These included, for instance, the construction of the "corporate man" as well as the cultivation of a consumer subject. Tracing out the full scope of these ethical attachments would, of course, exceed the scope of this or any single analysis. What is crucial, however, was the understood central role played by such social values in spreading markets. Indeed, capitalism's ability to endure and prosper was, therefore, early and often linked to the championing of a "necessary" ethics.

The Ethical Politics of Capitalism

Yet capitalism was always much more than merely an economic system or moral order. It was also a political project. Specifically, it required serious and consistent political legitimization. For markets to be introduced and reproduced, there must be a concerted political effort on their behalf. While certain variants of capitalism, such as neoliberalism, eschew the state as the chief antagonist to freedom and progress—supportive pro-business governments are not only desirable but also absolutely necessary. Central to these public efforts was the sanctioning and cultivation of a complementary political ethics.

Traditionally, capitalism was meant to challenge the authority of the state and governments. This pronounced anti-statism signified as much a political ethics as an economic structural requirement. Proliferating alongside the rise of capitalism was a laissez-faire ideology that positioned governments as the primary barrier to the achievement of market progress (see Berend, 2006; Henry, 2008) Any and all legislation to deal with market excesses was derided as a precursor to tyranny (Friedman, 2009). It produced, in turn, a political ethos where market freedom must be defended at all costs against the intrusions of misguided and even malicious meddling governments.

However, the development of capitalism was aided by a strong—though, as mentioned, at times largely "hidden"—public intervention. This contradiction reflected a certain irony in terms of capitalism's politics. What it promoted ethically was often at odds with what it needed concretely. Indeed, the success of capitalists and capitalism was very much linked to strong public investments (Wolfe, 1977). Further, it demanded a concerted government effort to protect property rights as well as the new legal frameworks implemented to defend the other "rights" of capitalists (Murray, 1971). In this respect, the state political "ethics" of capitalism diverged from its concrete politics.

The state, moreover, was positioned as a crucial source for intervening to ensure the continued hegemony of capitalist class rule. Specifically, it created the institutions and policies that guaranteed businesses would continue to be privileged over workers or the emergence of a non-capitalist society. Indeed, a seemingly fundamental feature of liberal democracies is "the expansion of bureaucratic states as power structures maintaining police and military control over potentially rebellious populations and reproducing the conditions of capitalist accumulation" (Alford and Friedland, 1985: xiii). For instance, early factories in the UK required the presence of a strong public police force to regulate and keep in check employee demands and protests over their conditions (see Foster, 2003).

More broadly, the state can play a critical political function in organizing social relations for the benefit of capital. Indeed, it is worth highlighting that for early proponents of neoliberalism, the goal was not the elimination of the state but rather its redirected use to promote an orderly and prosperous market society. Hayek (2012: 110) thus declares that government was needed to "make competition as effective and beneficial as possible—and to supplement it where, and only where, it cannot be made effective." He tasks the state, therefore, with creating what he referred to as "competitive order," noting:

> What I mean by "competitive order" is almost the opposite of what is often called "ordered competition." The purpose of a competitive order is to make competition work; that of so called "ordered competition," almost always to restrict the effectiveness of competition.
>
> (Hayek, 2012: 111)

In actual fact, then, the relation of political ethics to capitalism and markets is decidedly more complex and ambiguous than is perhaps first assumed. Their interaction is far from clear-cut or always obvious. Indeed, as much as it can be said that capitalism has shaped politics, it also must constantly respond to an unpredictable and contingent cultural and political environment (Gramsci, 1971; Gill and Law, 1989). Indeed, perhaps the very reason that capitalists are so vociferous in promoting a market-friendly ethics is due to just how difficult it is to shape society and populations according to their desired specifications. A good example of this difficulty historically was in the attempt to legitimize wage labor in the 19th-century United States. Even for a country that very early embraced the dominance of markets and free enterprise, there were strong concerns that accepting a salary was anti-ethical to the nation's commitment to democracy and the personal, economic and political sovereignty of its citizens (Foner, 1999).

Namely, it was feared that to rely on an employer for one's own material reproduction was to create an unfair dependency that could unduly influence the citizenry's opinions and actions. Tellingly, the justification for these anti-capitalist arguments was premised on the very pro-market ethics of "rugged individualism" and "fair competition."

To account for this volatile reality, new critical perspectives have sought to move away from overly functionalist accounts of the dynamic between politics and capitalism; instead, they prioritize their mutually constitutive and context-specific character. Open Marxists, to this effect, stress how capitalists effectively aim to use politics to their advantage when confronted with inevitable economic crisis linked to the very real structural contradictions of a capitalist system (Bonefeld, 1992; Bonefeld et al., 1992; Burnham, 1994). In particular, they act to neutralize more revolutionary proposals and ideologies by emphasizing the possibilities of reform and market-compatible changes (Cleaver, 1992). Such views echo Gramsci's earlier distinction between an organic and political capitalist "crisis":

> [A] crisis occurs. . . . This exceptional duration means that uncurable structural contradictions have revealed themselves . . . and that, despite this, the political forces which are struggling to conserve and defend the existing structure itself are making efforts to cure them within certain limits, and to overcome them.
>
> (Gramsci, 1971: 178)

To this end, capitalism prospers through the political incorporation of often ostensibly opposing ideals. It must constantly legitimize itself in the face of both economic and political challenges to its principles. This process of ethical co-optation is not limited to times of crisis. It is a regular and significant feature of capitalism generally. In previous centuries, the brutal material costs of the Industrial Revolution and colonialism were covered over through normative appeals to the spread of democracy and civilization. More recently, the threat of globalization to people and the environment has been met with novel ideas of a "responsible market" and "sustainable capitalism."

The ethics of capitalism is thus produced in the heated and dynamic environment of its politics. Put differently, it must promote certain values not only for its economic reproduction but for its very political survival. Its ethics, therefore, is a by-product of its specific socio-political conditions. As such, they cannot be easily predicted or universalized. Nevertheless, this explicitly political reading runs the risk of ignoring a key dimension of this relation between ethics and the reproduction of capitalism and markets: ethics are not just a set of abstract values. They are practices that provide individuals and communities with particular types of social capacities and opportunities. This rather straightforward insight considerably complicates the critical question of capitalist ethics. At stake, in turn, is how the freedom and agency provided by ethics helps shape and reinforce capitalism in a given time and place.

The Ethical Agency of Capitalism

The conventional account of capitalist politics has been upended in the face of a much richer and more complex reality. Functional assumptions that the economy produces the politics it requires are challenged by the sheer diversity of political values and institutions existing alongside market systems. The question is, thus, transformed from "What politics is produced by what economies?" to "How does and can politics shape present and future economic arrangements?" Also crucial is the role that certain forms of political agency have for reproducing specific economic ideologies and practices—such as how the right to vote can serve as an electoral means of enacting pro-market policies. A similar transition must occur for understandings of ethics and capitalism. It is imperative to study how ethics influence capitalist relations and, just as importantly, how the freedom to act ethically critically reinforces this broader economic system.

One of the original touchstones for approaching capitalist ethics is through the concept of hegemony. Gramsci's non-economist treatment of capitalism opened up the possibilities to analyze the role of cultural forces for determining economic relations and not vice versa. For Gramsci (1971) and those inspired by him, ethics was itself a social construction that was instrumental in legitimizing given historical power relations. While it was true that the "ruling class" was able to solidify its power through promoting its ethics, it could also be challenged through the construction and cultivation of alternative ethical ideals and practices. To this end,

> ideas in the form of intersubjective meanings are accepted as part of the global political economy itself. This is significant because ideas, developed for example by key organic intellectuals, can play a crucial role in forging a hegemonic project in times of structural crisis.
>
> (Bieler and Morton, 2003: 480)

Ethics, in this respect, was fundamental to the constitution and development of the economy. Material relations were formed within the matrix of a broader hegemonic struggle between competing visions of social and personal ethics. This cultural battle was a key part of the spread of markets and reproduction of a capitalist economy across time and geography (see Commons, 1924; Lindblom, 1982; Wernhame, 1968). Indeed, the periodization of capitalism can be traced in no small part according to which ethics were at that point dominant and which remained marginalized (or even invisible). Consequently, the Keynesianism of the postwar era (see Anderson, 2012;

Wilensky, 1974) reflected an ethos of ensured public welfare and social cohesion that is at odds with the anti-statist and individualistic values that supposedly characterize contemporary neo-liberalism.

Despite the move away from economism, Gramsci and his acolytes continued to prioritize class. Post-Marxist things go even further in this regard, treating social reproduction as a continual but never determined struggle between discourses for dominance. Notable among these are Laclau and Mouffe's theories of discursive hegemony, in which prevailing sets of ideas and practices regulate individual and collective understandings and actions. In the words of Howarth (2000: 102), it is an attempt to "weave together different strands of discourse in an effort to dominate or structure a field of meaning, thus fixing the identities of objects and practices in a particular way."

Arising from these critical insights is a more complex perspective for approaching capitalist ethics. It highlights how alternative values can serve to strengthen a status quo, even when they are ostensibly seeking to resist it. There is a paradox of order at play (Bloom and Dallyn, 2011) whereby stability is security through keeping certain cleavages open over others. Specific to themes of capitalist reproduction, for instance, are the ways passionate partisan battles between social liberalism and conservatism can detract from their larger shared pro-market policies and their marginalization of less capitalist-friendly views in the public debate.

Yet there is also another level at which capitalism and ethics operate. It involves the degree and ways that subjects affectively invest in their discourses. Drawing on psychoanalytic perspectives, critical theorists emphasize how capitalism depends on a range of context-specific market fantasies (see Bloom and Cederstrom, 2009; Contu, 2008; Roberts, 2005; Vidaillet, 2007). These cultural fantasies shape individual desires and help constitute their social realities—promising them psychic harmony and more fundamentally ontological security. To this effect, fantasy is

> the element which holds together a given community that cannot be reduced to the point of symbolic identification: the bonds linking together its members always implies a shared relationship to the Thing, toward enjoyment incarnated. . . . If we are asked how we can recognise the presence of this Thing, the only consistent answer is that the Thing is present in that elusive entity called our "way of life."
>
> (Žižek, 1993: 201)

This affective dimension represents the ethical commitment individuals and communities invest in their capitalist subjugation. The fantasy of entrepreneurship exemplifies the power of this affective capitalist ethics, as the romanticized vision of controlling one's own destiny, "being your own boss," means much more to subjects than any straightforward belief in the moral righteousness of "being entrepreneurial" (Jones and Spicer, 2005).

Rather, it is a broadly defined but emotionally resonant discourse for ordering and securing identity.

These perspectives offer a crucial advancement for critically understanding capitalist ethics. However, they overlook a key aspect of this relation—the importance of social agency. Typically, within much of the post-structuralist literature, agency is consigned to being either a complete illusion on the part of the subject or occurring in the breakdown of their social regulation. Yet these views elide how empowered subjects are within these culturally prescribed systems as well as the disciplining function of such empowerment. As Ortner (1995: 86) critically observes, "every culture, every subculture, every historic moment, constructs its own forms of agency, its own modes of enacting the process of reflecting on the self and the world and of acting simultaneously within and upon what one finds there."

Foucault is especially instructive on this account. He portrays discursive power as primarily a productive force—one that creates but can never fully contain a range of social capabilities. In this respect,

> [w]hat makes power hold good, what makes power accepted, is simply the fact that it doesn't only weigh on us as a force that says no, but that it traverses and produces things, it induces pleasures, forms of knowledge, produces discourse. It needs to be considered as a productive network that runs through the whole social body, much more than as a negative instance whose function is repression.
>
> (1982: 225)

The early capitalist emphasis on producing a strong body for meeting its industrial and military needs is a case in point. Although the impetus for this corporeal disciplining was to create a healthy workforce, its results far exceeded these original intentions. It gave birth to an entire social ethics of literally and figuratively increasing one's strength through building up one's own body. Again turning to Foucault:

> discipline increases the force of the body (in economic terms of utility) and diminishes these same forces (in political terms of obedience). In short, it dissociates power from the body; on the one hand, it turns it into an 'aptitude', a "capacity," which it seeks to increase; on the other hand, it reverses the course of the energy, the power that might result from it, and turns it into a relation of strict subjection. If economic exploitation separates the force and the product of labour, let us say that disciplinary coercion establishes in the body the constricting link between an increased aptitude and an increased domination.
>
> (Foucault, 1977: 138)

This focus on agency reframes the question of capitalism and ethics. It places at the forefront how agency is strategically produced and directed in support of reigning ideologies and status quos. Foucault's above differentiation of the economic from the political is particularly instructive. The crux of domination thus is not in repressing individuals but in expanding their capabilities in ways that reinforce as opposed to challenge a prevailing hegemony. Specifically, relevant to the theme of capitalist ethics is how a diverse array of seemingly contradictory and even competing values and practices can be cultivated and marshaled for culturally reproducing a market system. To give but one example that is only indirectly related to capitalism, soldiers are often encouraged to be socially undisciplined, to "let off some steam," alongside their otherwise quite regulated social existence as a way to cope with the traumatic experiences of battle.

Consequently, scholars increasingly highlight the significance of this socialized agency. They stress how certain social identities grant subjects specific forms of freedom to act and enact social change (see especially Scott, 1990). They chart out how daily forms of transgression and resistance reflect and can reinforce prevailing hegemonic discourses (see Bloom and White, 2016; Hollander and Einwohner, 2004). In tracing out such critical forms of agency, it becomes possible to illuminate how they challenge or entrench dominant discourses and power relations. For this reason, Abu-Lughod (1990: 420) calls, in this regard, for a "diagnostic of power" in order "to show how in the rich and sometimes contradictory details of resistance the complex workings of social power can be traced." Particularly relevant to the study of capitalist ethics is the need for a diagnostic account of the agency subjects are given to reproduce the very system that is exploiting them.

Ethically "Fixing" Capitalism

Understanding capitalist ethics is thus by no means a straightforward or even immediately obvious proposition. It is clear that certain skills and values are more compatible with wage labor and market relations than others. Not surprisingly, the ability to be productive and the ability to be efficient are desirable traits across capitalist contexts both historically and geographically. Similarly, an entrepreneurial spirit is more often than not always welcome. Yet these like-for-like examples only scratch the surface of the broader existence and function of capitalist ethics. It is in the continual presence of seemingly anti-market principles and practices that the complex interplay of capitalism and ethics emerges.

A fruitful entry point for undertaking such an analysis is through the concept of "problematization." From a purely critical viewpoint, it demands inquiring into how and why capitalism relies on such a wide range of what appear to be contradictory sets of ethics. The notion of problematization, accordingly, refers to "[not] behaviours or ideas, not societies and their 'ideologies' but the problematization through being offers itself to be, necessarily thought—and the practices on the basis of which these problematizations are formed' (Foucault, 1985: 11–12). It is to ask familiar questions anew to shed light on heretofore-unseen social dimensions, providing, in turn, fresh insights on old and supposedly commonly understood concerns. Returning to the present research, it is to question the sheer ethical diversity of capitalist values in practice in order to further a general theoretical understanding of this relation. According to Althusser (1969: 164):

> [T]o say that this is a problem implies that we are not dealing merely with some imaginary difficulty, but with a really existing difficulty posed us in the form of a problem, that is in a form governed by imperative conditions.

Nevertheless, there comes to pass an interesting and revealing opportunity to "problematize" problematization—specifically, to interrogate how the social itself is discursively framed in precisely such problematic terms. Or to put it differently, to shed light on how economic, social and political systems are portrayed as a "problem" to be "fixed." This reading goes against a conventional assumption that social reproduction and domination depends precisely on rendering a status quo as "unproblematic." The invoking of a political logic, the politicization of the socialized, if you will, is conceived to quite a large extent as a matter of making problematic that which is considered natural and permanent (Glynos and Howarth, 2007).

This prevailing view has much to recommend it. It offers an elegant explanation of how the social is instituted and made available to change over time. Unfortunately, though, society is rarely ever so simple or uncomplicated. For this reason, it is very much worth exploring a perhaps initially counter-intuitive line of inquiry—how can the act and process of problematizing the social be used in fact to bolster a prevailing hegemony?

Drawing again on Foucault, he points exactly in such a direction, in his relating of the production of the subject to an evolving set of underlying historical problematics. These can range from the sudden "problem" of the "mentally insane" (which required a vast array of institutions and treatment) to the need to deal effectively with "criminals" and "criminality" (leading to a culture of legitimized punishment and incarceration). In terms of capitalism, the market is portrayed as a necessary but commonly flawed mechanism for achieving individual and collective progress. Its dominant framing, then, was as a perfectable system that, although overall beneficial, required ongoing improvement. This overarching reforming discourse was what drove, for instance, the early efforts to address industrial poverty caused by the first wave of mass industrialization in the 19th century. Such an ethos reflects, in turn, what Marx refers to as the "reactive" quality of capitalists, whereby they reacted to working-class demands for change with compromised reforms (see especially Choat, 2010).

Yet it was also spearheaded from the top down, so to speak, and focused on supposedly pressing non-economic concerns. The strong Victorian code against "indecency" was, in this respect, the articulation of a serious moral "problem" arising from modern social conditions involving sex, perversity and promiscuity that must, it was argued, be urgently and systemically addressed. Foucault, rather notably, depicts this period as a time not just of repression but also one of social expansion. Importantly, these new prohibitions against sexual deviancy and perversity "did not aim to suppress it, but rather to give it an analytical, visible, and permanent reality" (Foucault, 1990: 44).

This process of naming, categorizing, cataloging, educating, regulating and repressing was, hence, born out of a desire to "fix" this moral "crisis."

Significantly, this call to action provided social actors a range of new forms of agency to solve these imminent "problems." The squalid living and working conditions of the proletariat catalyzed a revitalized interest in the idea of "public health" (see Szreter, 1988). It also spurred novel urban building projects from public gardens to modern pedestrian squares, creating the "infrastructure state" (Guldi, 2012). It also led to the formation of incipient employee unions and, consequently, the promotion of industrial democracy—along with the "reactive" local and national police force to repress them. On the cultural front, there was a pronounced cultivation of an appropriate moral and civic spirit. Individuals were literally taught how to be proper "citizens"—necessitating in part the creation of a public educational system (see Weber, 1976).

The durability of capitalism, hence, is directly and indirectly attributable to its discursively propagated need to be "fixed." This reading puts in a somewhat new light Harvey's (2001) astute concept of the "spatial fix" of capitalism. His use of the concept speaks primarily to the geographic establishment of capitalism—its physical "fixing" in a certain time and place (see Herod, 1997; Jessop, 2006). The instantiation of concrete "market places" allows for the material affixing of a capitalist mode of production within specific contexts. The emphasis, here, by contrast, is on the discursive "fixing" of capitalism. This is meant in two ways, reflective of both meanings of the term "fixing." On one hand, it refers to the dominant "fixing" of capitalism to a certain set of social, political and economic problematics. On the other, it involves the emerging forms of capabilities and agencies granted to subjects as a way to "fix" these "problems"—thus turning them into dynamic agents of capitalist reform and reproduction.

Producing the "Ethical" Capitalist Subject

The survival and prosperity of capitalism rests in no small part on its social production of an "ethical" social subject. While its raison d'être is primarily economic—focused on finding ever new ways to extract profit—its resilience and continuation is largely politically and culturally derived. More precisely, its sustainability is found in the creation and recreation of a capitalist self that is not only amenable to market demands but also able to find innovative ways to cope with and resolve its social and economic shortcomings. Its regenerative capacity is grounded in its ability to manufacture a subject that can adapt and help overcome its social contradictions without fundamentally threatening its hegemony.

This discussion draws upon the distinction—originally made by Foucault—between subjection and subjectification. Simply put, the former is the practical inscription of subjects to dominating and disciplining discourses. The latter is the knowledge of self that emerges from this subjection. Butler (1997: 2) describes it as a process that "consists precisely in this fundamental dependence on a discourse we never chose but that, paradoxically, initiates and sustains our agency" and, as such, "signifies the process of becoming subordinated by power as well as the process of becoming a subject." Crucially, this is not to imply that subjectification is completely synonymous with consciousness per se; rather, subjectification forms the social basis for such conscious development as a thinking and acting intentional subject (see Bhabha, 2012; Heller, 1996).

Such theories point toward the ethical subjectification at the heart of capitalist subjection. The material regulation to a capitalist economy required a compatible "ethical" self. It demands subjects who is committed to "fixing" themselves and society. The introduction and perpetuation of a market system lays the cultural groundwork for dynamic subjects that can temporarily find ways to resolve and cope—even if only temporarily and in an ad hoc way—with their individual and shared exploitation. The "self," in this sense, is a profound "problem solver" who has the agency and capabilities to deal with capitalism's imperfections.

The actual manifestations of this "ethical" capitalist subject are necessarily historically contingent and are as such quite diverse. They are a response and a reaction to existing capitalist conditions—though never in an easily or always predictable over-determined fashion. Instead, they are produced through and by prevailing discourses. At stake is to constantly and rigorously identify what are the perceived "problems" of capitalism and how are they being "addressed." To do so, demand, in this spirit, the critical application of a "diagnostic" approach to ethics, power and capitalism.

These insights both echo and critically expand upon existing understandings of the relation of capitalism to ethics and morality. Boltanski and Chiapello (2005), perhaps most prominently, stress the need for market relations to be adequately "moralized"—to be seen as contributing to a broader common good. For capitalism to work, it must be perceived to be fair and to benefit rather than merely exploit the public—or at least be perceived as having the potential to do so. Consequently, Boltanski and Chiapello refer to the "spirit of capitalism" as "the ideology that justifies engagement in capitalism" (2005: 8). This analysis introduces what can be referred to as the "ethical spirit of capitalism"—highlighting that it is precisely within this ethical gap between capitalist ideal and reality that capitalist self appears and flourishes.

The success of capitalism as a social system is intimately connected to its production of an evolving ethical subject. Thus, the crux of capitalist reproduction is discovered in this seeming ethical reversal, whereby the desire for markets to be moral is transformed into a demand and duty for subjects to make it more ethical. It is exactly in this respect that the affective dimension of capitalism becomes so powerful and attractive. The capitalist subject "makes sense of" and comes to embrace his or her economic subjection through psychically investing in this "ethical" reforming self. These fantasies of social improvement and self-improvement reproduce not only the capitalist self but also the system more generally. In the next chapter, we will turn our attention to the specific production and power of the "ethical" neoliberal subject.

3 The Ethical Power of Neoliberalism

Neoliberalism is commonly portrayed as completely devoid of ethics. It is caricatured as a total embrace of the market without any consideration of its human costs. The threat of global marketization is bound up with serious fears that profit will be prioritized over people. It conjures up images of present-day oligarchy, job insecurity and a future rule by corporate overlords. These worries have only been exacerbated in the aftermath of a financial crisis that left the world on the brink of economic ruin followed by a recovery that saved the banks—"Wall Street"—while doing little to help "Main Street." Yet this moral critique, while certainly justified, overlooks the profound importance of traditional ethical values and practices for reproducing and in fact reinforcing neoliberalism at all levels of its operation—from the political to the institutional to the personal.

Thus far, this book has sought to put forward the broader significance of ethics for the survival of capitalism. The perpetuation of wage labor and markets as a central organizing feature of social relations is made possible through a discourse of needing to continually "fix" capitalism. This "problematization" of capital, in turn, produces "ethical" subjects able to innovatively, though only within rather narrow ideological confines, improve on capitalism. Although this reformist ethical imperative is arguably close to universal, its manifestation is historically and contextually specific. Each era and place articulates this "problem" of capitalism differently and as such its socially approved solutions.

The question then of this work is "What particular ethics does neoliberalism crucially depend upon structurally and subjectively for its continuation and expansion?" It is not a coincidence that the "inevitable" advancement of the market into every corner of the globe and all spheres of social existence has been met with a flowering of desire for social justice and greater individual morality. To reiterate a point made earlier in this analysis, the most obvious explanation is that it is simply a reaction to this almost complete capitalization of society. Nevertheless, this understanding ignores the critical function of these ethics for concretely preserving the continued dominance of neoliberalism. If neoliberalism is totally devoted to the market ideologically, it nonetheless requires a decidedly nonmarket "ethical" subject for its perpetuation. More precisely, it benefits from the construction of individuals and groups who, due to their ethical beliefs, are willing to regularly and creatively help resolve the deep moral problems caused by neoliberalism.

In order to properly understand this seemingly rather contradictory phenomenon, it is essential to place it within its proper historical context—in particular, to trace out how the "problem" of capitalism has discursively evolved from its beginnings in the early 19th century as a dominant economic system to its contemporary promotion of neoliberalism. Such a time-sensitive account would reveal, in turn, specifically how, in the current era, capitalism is being framed, as well as the prevailing sets of practices and aptitudes that are developing to effectively deal with these issues. Specifically, it highlights the shift from a collective "political ethics" and "organizational ethics" (representing liberalism and unions, respectively) of the recent past to the individual "moral ethics" of today.

This historical perspective, in turn, highlights a fundamental, though heretofore almost wholly ignored, contradiction of neoliberalism: notably, that the more marketized an economy and society becomes, the less its population necessarily embraces the market ethically. Indeed, it is exactly because the market is so prevalent that individuals and communities must adopt an alternative ethical framework, one in which values of reciprocity, care and social responsibility are prioritized as a means of adequately coping with and to some degree "fixing" the subjective and material problems caused by neoliberalism. As the traditional political and organizational options gradually erode, people progressively turn to moral "solutions" for improving capitalism both in their own life and more widely. The power of neoliberalism flows from its capacity to channel this nonmarket ethics for the purpose of ultimately achieving market ends.

The Historical "Problem and Solution" Capitalism

A rather dominant framing of capitalism since its inception has been its ultimate rightness. In other words, the introduction and advancement of a capitalist economic system has been legitimized as both objectively correct and historically necessary. It is socially justified as simply following "economic laws" that conform to human nature. In much the same way, it is put forward as an imperative force for realizing social, political and material progress. More recently, corporate globalization is considered, often in quite lamentable terms, to be "inevitable" and accordingly irreversible (see Bloom, 2016a; Deudney and Ikenberry, 2009; Spicer and Fleming, 2007; Steger, 2005a).

Yet there is a simultaneous discourse that emphasizes the precise opposite of this vision of capitalism as theoretically and historically unassailable. A persistent feature of a market society is the injunction to "fix" capitalism so that it can work better. This consistent desire for improvement provides a general ethical framework that is both universal in its scope and contextually specific in its application. It is worth highlighting that industrialization emerged just in the aftermath of the first blush of the Enlightenment. It adapted, in this respect, the prevailing language of "progress" characteristic of the times. According to Nisbet (1979: 7):

> The essence of the Western idea of progress can be simply stated: mankind has advanced in the past, is now advancing, and may be expected to continue advancing in the future. . . . We find the perspective of progress used, especially in the modern world, to give substance to the hope for a future characterized by individual freedom, equality, or justice. But we also find the idea of progress made to serve belief in the desirability and necessity of political absolutism, racial superiority, and the totalitarian state. In sum, there is almost no end to goals and purposes which have been declared the fulfillment or outcome of mankind's progress.

This notion of progress would develop into a potent and sustaining capitalist idea. However, it would undergo a key evolution: whereas in its early form it emphasized the undeniable evolutionary progress brought about by the forces of capitalism, as this system matured and its less than ideal consequences appeared, this emphasis on progress was transformed into an overriding imperative to "improve" capitalism. There was, in this respect, a subtle and gradual transition from "capitalist progress" to "progressing capitalism." Of course, this movement does not follow a strict historical chronology. Rather, it represents a general narrative dynamic for discursively understanding the ethical legitimization of capitalism within and across its distinguishable eras. Neoliberalism, for instance, first trumpeted the objective necessity of marketization for social progress only to ultimately come to promote how marketization can progressively be used for achieving socially just ends in the present.

Significantly, the emphasis on "progressing capitalism" can be traced back to the 18th century. The earliest debates on free trade revolved around how it could be properly regulated for the public good (see Semmel, 1970; Williamson, 1996). Already, then, there was a perceived ethical need to "fix" capitalism. This ethics would remain into the 19th century but would undergo a dramatic reversal. Suddenly, it was free trade that would help morally "correct" an international order plagued by destructive military rivalries between nations and misguided protectionism (Bell, 2006; Sen, 2004). Hence, there was an optimistic sentiment that "prosperity, peace and progress were about to supplant the un-holy trinity of protectionism, militarism and imperialism" (Schuyler, 1921: 539). This is a first glimpse at how nonmarket values of cooperation and reciprocity would be strategically marshaled in support of markets.

There was also a quite vibrant political ethics that emerged alongside economic capitalism. Mass politics, at least partly, arose as a means of harnessing the advantages while effectively addressing the threats of capitalism. While industrialization was still on the horizon, the late-18th-century revolutions were inspired to a certain extent by the desire to ensure the freedom of individuals against the tyranny of not only the ancien régime and monarchy but also the greed of "money men" (Hunt, 2004; Morris, 1962). Just as significantly, the explicitly political character of these revolutions reflected an embryonic attempt to completely transform and "fix" the economic and social ills of society. "The social distinctions and economic deprivations that today we think of as the consequences of class divisions, business exploitations, or various isms—capitalism, racism, etc.," observes Gordon (2011: 5), "were in the 18th century usually thought to be caused by the abuses of government." Likewise, the less "class-based" American revolution, at least compared to its French counterpart, nonetheless had an often decidedly egalitarian spirit aimed at "solving the problem" of material inequalities and their associated power imbalances. As one prominent contemporary declared at the time, "liberty of the press, the Habeas Corpus writ, even Magna Charta itself" were

> inferior considerations when compared with a general distribution of real property among every class of people . . . let the people have property and they will have power . . . the liberties of America therefore, and her forms of government, stand on the broadest basis.
>
> (quoted in Bailyn, 1962: 374)

This early framework provided the foundation for a more universal and hegemonic capitalist political ethos. Initially, the Industrial Revolution was posited as an ipso facto driver of social and economic progress (see Cannadine, 1984; Griffin, 2010; Teich and Porter, 1996). In the famous words of Engels in the late 19th century, humans were experiencing "an industrial revolution, a revolution which at the same time changed the whole of civil society" (1987). In this respect, economic development was prioritized, framed now as a reforming force for culture and politics as a whole. To this end, "freedom meant prosperity; freedom meant progress; freedom meant having willing workers as opposed to unwilling ones" (Temperly, 1977: 109). However, as the actual consequences of the "revolution" became clearer, it was increasingly seen as a type of "new barbarianism" (Coleman, 1992: 35).

Government in general was thus presented, hence, as a prime mechanism for reforming the economy and society. It framed social progress as something that must and could be popularly decided upon for the benefit of both personal and collective interests. The revolutions of 1848, in particular, fundamentally questioned ideas that industrial development could be considered just or progressive without associated political, social and economic rights (Hobsbawn, 2010; Jones, 1991). Sovereignty, in this regard, became a continual means of shaping the economy and addressing the problems of a rapidly emerging market society. For this reason,

> liberal market logics and technologies have historically presented biopolitical problems for state authorities and populations, including social infrastructure, labor availability and vitality, poverty and consumer demand. Correspondingly, biopolitical concerns have shaped market logics and operations, particularly by prompting state intervention, regulation or military actions, thereby altering the trajectories of liberal governmentalities.
>
> (Nadesan, 2010: 46)

Democracy, specifically, was presented as a political force for "fixing" the economy and society. It was a regularized referendum on the "state of capitalism." The rise of "social democracy" in the 1890s epitomized this shift in governmentality, as it represented a new political "responsibility" to "extend the democratic principle of equality from the civil and political spheres to the entire society and the economy" as well as to "champion gradual constitutional reform instead of revolution" (Kloppenberg, 1988: 6).

The market, therefore, was increasingly framed as if not necessarily permanent then at least permanently reformable. The politics of capitalism, of course, is neither singular nor straightforward. It is characterized by a large amount of cultural and economic diversity (Dore et al., 1999; Hall and Soskice, 2001; Rueda and Pontusson, 2000). Yet a key dimension that deserves to be highlighted is the continual use of social discourses of "progress" to propagate the market economically. There is a widespread and diverse effort to diagnose the exact "problem" of capitalism and how it can be resolved. This insight draws upon a rich vein of critical scholarship that reveals the central role of crisis for capitalist reproduction (Bonefeld, 1992; Bonefeld et al., 1992; Burnham, 1994). However, it stresses not the "boom and bust" dynamic of these "catastrophic" events but the regularized and rather permanent "problematization" mentality that recognizes this system's deficiencies while creating an ethical demand and desire to find ways of improving it. Capitalist politics, then, is largely founded upon a constant articulation of this "problem" and a debate as to how it can be most adequately addressed.

This general dynamic provides for a plethora of culturally and historically influenced manifestations. It can, and often has, been reversed so that the focus rests on the social and political barriers to maximizing market prosperity. It can also turn its gaze on issues that are not explicitly capitalist in character, such as "racism" and "moral decay." These systems are commonly marked by wide-ranging "moral panics" (Critcher, 2003; Thompson, 2005). There is, furthermore, commonly a combining of issues and their cause and effect—so that racism, for instance, is attributable to a past and present legacy of capitalism (Marable, 2015; Pitcher, 2012). Historically, this has also entailed more darkly the attempt to literally "solve" the supposed "race" problem for the stated goals of advancing a prosperous and "civilized" market society. In this respect, in the early 20th century, both in the United States and (to a slightly lesser extent) in Europe, a politics of eugenics was promoted based on the widespread social "alarm about racial degeneration," including "that higher birthrates among the genetically inferior would lead to a 'menace of the feeble-minded' and that inferior nations, men, and races would, through immigration, undermine American supremacy" (Soloway, 1990: 245).

Central to this "problem and solution" politics has been the concurrent ethical imperative for individuals and groups to adopt the proper attitude and actions necessary for meeting these challenges. Thus, the urgent need to address the systemic problem of "racism" has become a personal demand for people to become "tolerant." This "trickling down" of ethics from the social to the subject holds true across the ideological spectrum. Fascism and Communism both famously fixated on the creation of a "new man" who could best realize their respective utopian visions (Griffin, 2007; Mosse, 1966; Săndulescu, 2004). For neoliberalism this has meant, as will be explored in greater detail later, the production of a more entrepreneurial and "rational" market subject.

Just as importantly, it is a catalyst for creating new resources for subjects to undertake these improvements. Progress is not simply on the horizon but also made possible in the present. It means that personally and collectively new forms of agency must be encouraged and perpetuated for advancing these diverse reforms. The "progressive" reforms of the early 20th century revolved around the granting of governments fresh abilities and technologies to directly regulate a capitalism "run amok" (Davis, 1984; Wiebe, 1989). It was an age characterized by a "fierce discontent" driving progressives to champion

> a stunningly broad agenda that went well beyond the control of big business, the amelioration of poverty and the purification of politics to embrace the transformation of gender relations, the regeneration of the home, the disciplining of leisure and pleasure, and the establishment of segregation. Progressives wanted to not only use the state to regulate the economy: strikingly they intended nothing less than to transform other Americans, to remake the nations feuding, polyglot population in their own middle class image.
>
> (McGerr, 2005: xiv)

The ethics of capitalism is, hence, at its heart one of "problem solving"—finding novel methods and capabilities to cope with and if possible overcome its shortcomings. The image of the conformist capitalist subject, the Chaplinesque figure of modern times, is belied by the dynamic reality of market exploitation and capitalist reproduction. Rather, they are characterized by one of continual innovation, directed toward the daily and long-term maintenance of a "problematic" system. With particular regard to neo-liberalism, contemporary subjects must utilize their moral agency to effectively deal with and to help better what is understood to be a profoundly immoral modern-day capitalism.

The Ethical "Problem" of Neoliberalism

In the 1980s, neoliberalism delivered on its promise to completely revolutionize economic relations. The election of first Thatcher in 1979 and then Reagan in 1980 signaled a new era of privatization and competition (see Bishop et al., 1994; Henig, 1989; Tingle, 1988). Not surprisingly, Reagan and Thatcher framed the "problem" of current capitalist society as one of public bureaucracy and the inefficiencies of the welfare state (Kaufmann, 2015; Pandey and Scott, 2002). It was a distinct, though by no means unprecedented, reversal of the demand to "liberalize" capitalism in the initial postwar eras. Now it was markets coming to the rescue rather than needing to be rescued. This discursive strategy drew upon a rich vein of Conservative political rhetoric prevalent during the Cold War—in which markets must be protected against the threat of Communism so that the "free world" can be saved. Here though, it was government itself that was being put on trial, portrayed as a dangerous and corrupting influence on the implementation of an idealized free market.

This neoliberal "problem" took on a decidedly moral character as a key part of its political success. The welfare state represented something much more sinister than simply a lack of economic productivity and growth. It pointed to a severe perceived moral decay to traditional social values (Adonis, 1994; Edwards, 1999). Economic restructuring was, in this respect, joined to cultural conservatism so that the "free market" was officially framed as a return to "the way things were and ought to be." Ideals of patriotism and "family values" were, thus, championed as a way to depict all those who did or may oppose marketization as "radicals" and "social deviants" (Samuel, 1992). Unions, liberal social groups, environmentalists and leftists of all stripes were dangerous "idealists" who were either naïve in their understanding of actual human nature or nefarious in their efforts to play on the population's "better angels" for their own manipulative ends. Figures like Thatcher were reified as saviors, idolized for leading the country out of socialist-created economic deprivation. As one British newspaper recently declared:

> Thatcher rescued us from all that. She broke the power of the unions and gave birth to a brand new era of British entrepreneurism, where individuals were left to profit from the fruits of their labour and ingenuity.
>
> (quoted in Beynon, 2014: 214)

The appeal to tradition was to some extent simple power politics. It was a reframing of the "enemy" to garner popular support for radical economic "reforms." The enemy, in this case, was that "ill-assorted mix of elitists and special-interest groups who see government as the principal vehicle of social change, who believe that the only thing we have to fear is the people, who must be watched and regulated and superintended from Washington" (quoted in Weiler and Pearce, 1992: 237). Theoretically, this highlights the integrative, contingent and complex ways an ethics of capitalist "problematization" plays out politically within the broader course of history. Additionally, it shows how important the discursive articulation of this "problem" is for strategically ushering in such a radical capitalist change such as neoliberalism. If a crucial potential critique of full-scale marketization was its inhumanity, the depiction of those who oppose it as "immoral" pre-emptively turned these criticisms on their head. Suddenly, it was people calling for a social safety net or greater workers' rights that were to be feared and held in suspicion (Handler and Hasenfeld, 2006, Jones, 2012; Sidel, 1996; Withorn, 1996). Further, it recast the country's "moral failings" as associated not with capitalist greed or imperial ambitions but rather with the celebration of "immoral" modern lifestyles and the loosening of cultural mores overall.

This potent political mixing of the old and the new echoed the "gilded" free-market eras of the last 19th century. The railing against social and economic immorality in both periods legitimized the hoarding of wealth and power by a narrow financial elite. The bourgeois sexual prohibitions of this time were intimately connected to fears of female liberation, so that "a woman who moved into the realm of paid employment implicitly compromised her sexual reputation" (Wood, 2005: 16). Similarly, "the scientific discourse of criminology virtually excluded discussions of theory and methodology . . . and concentrated on the description and tabulation of the physical, physiological, and moral characteristics of the 'criminal man'" (Leps, 1992: 3). In the same spirit, the growing masses displaced by industrialization and alienated from this new economy were demonized as dangerous "vagrants" (see Bratlinger and Ulin, 1993). The "vagrants" of the Victorian age were updated to the "homeless" and "lazy unemployed" and "chavs" (in the UK) of the late 20th century (Owen, 2012). The sexual "perverts" now comprised "gays" and all others who rejected the nuclear family as a lifestyle. The revolutionaries, anarchists and organized labor were contemporized as "Communists," "terrorists" and unions.

However, this critical interpretation elides a central, though all too commonly overlooked, dimension: this moralization also provided individuals and communities with the agency and capabilities to address these problems. It was up to people themselves to work hard and be responsible for their economic destinies. In this regard, the external forms of demonization turned decidedly inward, becoming a self-disciplining discourse. Thus it was proclaimed that "thoughts are vagrants which must be diligently watched for, caught, examined, whipped and sent on their way" (quoted in Hill,

1964: 131). Tellingly, the emphasis shifted from security and welfare to competition and employability (see Bloom, 2015). Hence, in this new age, "to be employed is to be at risk [and] to be employable is to be secure" (Hawkins, 1999: 8). It, moreover, granted the government fresh powers to "discipline and punish" social threats such as criminals, unions and cultural transgressors (see Bloom, 2016a; Giroux, 2004).

These initial ethical problematizations of neoliberalism would create a general discursive framework for its legitimization and reproduction up to the present. This framework positioned capitalist subjects themselves as responsible for sustaining and overcoming the ethical and economic problems of capitalism. Unemployment and racism, hence, were not systemic problems of capitalism but rather attributable to individual moral failings. The best way to address social and cultural issues was through self-regulating one's own moral behavior. Emerging, in turn, was the creation of the "ethical" neoliberal subject.

Making Capitalism Morally Responsible

The rise of neoliberalism was linked to the growth of a new cultural, political and economic ethics. Alongside its promises of market prosperity was a distinct moral vision of a "traditional" and individually "self-reliant" society. This vision focused on an ethics of responsibility—shifting, in turn, the duty of pursuing socio-economic improvement from governments to organizations and individuals. Reflected was a pronounced privatization of ethics. Consequently, people would be made to feel increasingly ethically responsible for their own economic fate and progressively the morality of capitalism generally.

From its inception, neoliberalism sought to portray the market as an idealized solution rather than a potential problem. This belief was buoyed by the market's dramatic political successes in the early 1980s. By the middle of the decade, though, the market was once again under attack. A range of scandals had highlighted the ugly consequences of leaving it unregulated (see Day, 1993; Calavita et al., 1997). Jimmy Carter's declaration in the late 1970s that there was a "crisis of confidence . . . that strikes at the very heart and soul and spirit of our national will" increasingly extended to corporations and the financial sector (see Lipset and Scheider, 1983). It became politically and economically necessary to forge a new ethical discourse of reform that could simultaneously recognize the "problems of capitalism" without returning to past Social Democratic answers.

A focus on ethics was a key element of this strategy. The issue was not with markets per se or with capitalism as an economic system. Instead, the blame was fully on the immorality of the individuals who were responsible for such misdoings. Optimistically, the problem was dismissed as a case of a "few bad apples" (see Hansen and Movahedi, 2010). Yet, as the rot spread wider, the demand for genuine reform became louder and more urgent. The call for regulation and prosecution loomed larger and threatened to unravel the entire neoliberal project. In this spirit, the problem of capitalism was reframed and repackaged for mass consumption. It was now attributed not to structural contradictions of the market or personal moral failures of its leaders but rather to a lack of knowledge by executives concerning ethical behavior. Business ethics were then introduced as a meaningful way to "fix" this issue. The attempt to apply ethical values to the business world was, of course, not novel in and of itself. It followed a long historical tradition (see De George, 1987; Velasquez and Velazquez, 2002).

What was distinctive was the assumption that its injustices and wrongdoings could be almost wholly traced back to a lack of ethical knowledge by its leaders. The moral intent was there, yet in the fast-paced competitive world of the modern marketplace, ethical guidance was sorely needed. New ethical frameworks for "guiding" business leaders' decisions were proposed and popularized throughout this period, including most famously by Ferrell and Gresham (1985), Trevino (1986) and Jones (1991). These discourses undeniably played into the market-friendly ideologies of the time. The introduction of business ethics would allow companies to become ethically responsible free from government interference. Further, it would reinforce a "new ideology of corporate governance" that was focused almost exclusively on maximizing shareholder value (Lazonick and O'Sullivan, 2000).

Yet it also shifted the debate regarding the exact fundamental problem of capitalism. The present emphasis was on "responsibility"—more precisely, on who was "responsible" for ensuring that the market was ethical and how far this "responsibility" extended. In this respect, the ethics of neoliberalism was not only privatized but also personalized. Specifically, this was manifested in an ethical requirement for individuals and organizations to balance market demands with broader moral obligations. There was a certain, though not entirely unpredictable, irony in the fact that as the state seemed to retreat in its responsibilities, ethics became increasingly sovereign in its focus. The initial onus was establishing and ensuring elite "responsibility." Appearing informally was the creation of an incipient social contract between financial leaders and the population as a whole. It was suddenly imperative that these new economic "masters of the universe" should rule ethically within a dynamic and often anarchic marketplace.

This heightened emphasis on business ethics gave rise, therefore, to a pronounced sense of "corporate responsibility." Previously, the obligations of executives were limited to legally maximizing their profit. Yet in the increasingly charged atmosphere of the 1980s, this would expand significantly. It was now expected that capitalists morally account for themselves publicly. In particular, how were they personally and organizationally acting ethically? This represented a transference of the "great power" ethics so prominent in the postwar era, especially in the United States. Just as a superpower must simultaneously act ethically while not cowering from the "realism" of geopolitics, so too must corporations be morally responsible while not abandoning their economic duty to be profitable.

Confronting the Ethical Paradox of Neoliberalism

This enhanced call for responsibility reveals a commonly overlooked and rather ironic aspect of neoliberalism related to ethics. As discussed on p. 000, the destruction of the welfare state led to increased public demand for private responsibility. Initially this demand was placed onto organizations and corporate executives. They were expected to mix their competitive profit-making ethos with at least a modicum of concern for the morality and social consequences of their actions. Fueling this ethical demand was a pronounced feeling of social lack—a sense that someone was going to have to take moral responsibility for the market if the state was now unable to. Put differently, neoliberalism at once freed the market from public intervention or democratic restraints while producing a resurgent ethical desire for it to be moral and its leaders to be ethical.

Thus it is not precisely accurate to say that neoliberalism produced an unregulated or non-governed market system. Rather, it created an alternative paradigm and rationale for shaping private economic relations. For this reason, scholars have increasingly linked neoliberalism to a new era of regulation (see Broomhill, 2001; Jessop, 2002; Levi-Faur and Jordana, 2005). Normatively, it was crucial that corporations become properly governed. Of course, the exploitive reality of these companies belied their often ethical public relations. Charges of corporate "greenwashing"—in which companies would use token environmental gestures to distract from their overall poor environmental record—were symptomatic of the all-too-common divide between word and deed (Bowen and Aragon-Correa, 2014; Cherry and Sneirson, 2011; Laufer, 2003; Vos, 2009).

Nevertheless, the very fact that companies were expected to "self-regulate" reveals the widespread ethical pressures being placed on them. Even more significantly, it belied a fresh rationale for capitalist governance. Corporations were suddenly legally people—fully capable of moral agency and ethical responsibility. Firms were, thus, expected to take it upon themselves to ensure that markets were "good" (see Graham and Woods, 2006; Jenkins and Unies, 2001; Maxwell et al., 2000; Parker, 2002). The limits of this moral imperative were quite narrow, of course, and conformed to conventional capitalist norms. Fairness was defined almost exclusively in terms of meritocracy and not equality, for instance. Underneath this hyper-capitalist veneer, however, existed a fresh logic for market governance whereby problems of the market could be reduced to issues of personal immorality and in which it was, therefore, up to individuals and organizations, themselves, to find ways to ensure capitalism's overall morality.

Neoliberalism, hence, posed a new set of moral problems and ethical possibilities for society. Crucially, it started from the founding proposition that free markets were both economically desirable and undeniable. In other words, the initial blush of neoliberalism witnessed the justification of marketization in largely utilitarian terms—it was an antidote to the inefficiency of the welfare state and a bulwark against the tyrannical threat of Communism. Yet as it progressed, this idealism evolved into a pronounced pragmatism: globalization and the spread of markets was "inevitable" and could not be realistically challenged or altered. Consequently, subjects were responsible and empowered to discover innovative means for rendering this "unchangeable" marketization as ethical and just as they could.

In this respect, the potential for moral and political agency was exchanged for ethical capability. It was impossible to change the system—socialism had been relegated to the ash heap of history, and capitalism uncontrovertibly reigned supreme. The desire for a different type of society was either "idealistic" (of the variety that "Communism works in theory but not practice") or completely blind to contemporary realities. At stake instead was how individuals and organizations could "self-regulate" to resolve the problems of the market. Fundamental to such a proposition was the creation of a diverse array of regimes and aptitudes for permitting subjects to proactively meet this moral imperative.

Revealed were two profound ethical paradoxes resulting from and ultimately confronting neoliberalism. First and foremost, the stronger the ethos of marketization became, the greater the desire and demand to make it moral. If the free market was here to stay, then it must be directed toward ethical and socially just ends, not just profit-making. Economic hegemony became a social imperative. The second, and in many ways just as significant, paradox was that the more disempowered individuals became as historical and political agents, the more they became empowered as ethical agents. More precisely, as individuals came to terms with the fact that they were at "the end of history" and that they could not replace capitalism, they were granted new opportunities and, in principle at least, greater latitude to find innovative ways to make it more ethical. Indeed, the advent of neoliberalism would see the creation of a vibrant new ethical subject.

Producing the Ethically Responsible Neoliberal Subject

Neoliberalism brought with it an entirely new ethical framework for organizing social relations. The supposed amorality of the market was supplanted with a resurgent desire for moral responsibility and an ethical subject who could realize this demand. These calls were exacerbated by historical challenges facing neoliberalism as it morphed from a viable economic project into an all-encompassing ethico-political system. The cultural consensus that emerged around capitalism and markets produced, in this sense, an array of discursive contradictions and practical problems. Notably, if the market was permanent and not to be fundamentally altered or even necessarily reformed, than how could governments, organizations, communities and citizens pursue a common good that transcended profit maximization?

These deep, and very real, concerns were transformed into an articulate and urgent "problem" associated with neoliberalism. The material benefits of free markets were perceived to be unquestionable. What was less assured was how traditional and still-longed-for social and ethical forms of justice could also be achieved within this broader system. From this foundation, the attempt to make capitalism "moral" metamorphosed into an overwhelming injunction to make subjects morally responsible. It was, in turn, the obligation of individuals and communities to find expansive ethical solutions within these narrow ideological limits of capitalism.

At first glance, this appears to be a rather traditional story of subjugation and complicity—a pragmatic acceptance of a dominant economic and social order that subjects can at best change around the edges but do little to fundamentally resist or transform it. There is undoubtedly an element of this despair in the pronounced ethical response to neoliberalism. If corporate globalization is akin to gravity, we can do only so much to lessen its heavily weighted effect upon us. Yet such judgments miss a critical part of this neoliberal dynamic: namely, its ability to empower subjects—to grant them the opportunity to, if not systemically change capitalism, discover unique and interesting ways to render it more ethical. It is part of a broader development agenda associated with hyper-capitalism in which "Poor people are not liabilities, but assets. We have to invest in them, and empower them. Development must not be done to them, but by them" (Wolfensohn, 2001: 4). In this spirit, there is a definite need for capitalists to "invest" in individuals for their ethical empowerment.

Neoliberalism, accordingly, operates by providing individuals and communities the motivation and techniques to "fix" capitalism's seemingly infinite moral shortcomings. At stake is a broader modern "politics of empowerment" (Sharma, 2008) which former US president Bill Clinton famously described as "shifting the emphasis from dependence to empowerment." In this respect, "tempered by the current emphasis on dismantling welfare, exerting fiscal discipline, and privatizing state services, the neoliberally imagined empowerment project seeks to enable grassroots actors . . . to fulfill their own needs through market mechanisms instead of relying on state largesse" (Sharma, 2008: xvi). Consequently, empowerment discourses combined with ideas of "ownership" as individuals and communities were increasingly expected to take personal and collective responsibility for their personal and social outcomes (see Taylor, 2004).

This potential for enhancing one's ethical capabilities—and, to a certain extent, moral agency—was manifested in a range of context-specific ways. However, it played into a generalized ethos of personal responsibility that was integral to the success and reproduction of neoliberalism overall. There was, to this end, a common framework for organizing and managing the subject inexorably linked to a governmentality of "self-regulation" and personal responsibility. Namely, what techniques could subjects—whether individually or organizationally—adopt to successfully cope with and prosper in a competitive and often uncaring free-market environment?

Such an ethos of personal responsibility had significant historical antecedents. It echoed ideas emanating from the 1970s on "responsible autonomy" (Friedman, 1977). The aim was to strike a balance between reduced regulation and enhanced employee productivity. The championed "silver bullet" was a sense of individuals taking accountability for their actions—an acceptance of their own ownership of their efficiency and work regime that would eliminate the need for strict or regular managerial oversight. It was precisely this social logic that permeated and informed the previously discussed corporate responsibility. As such, freedom was progressively understood as a "choice, autonomy, self-responsibility and the obligation to maximize one's life as a kind of enterprise" (Rose et al., 2006: 91). The weakening power of the state created an ethico-political void filled by self-regulating firms. In the wake of scandals, there was increased public pressure for companies to practice such "responsible autonomy." Yet this demand would expand and reach downward, soon becoming an overarching ethical imperative for employees and citizens alike.

The seemingly all-encompassing and inescapable call for "fiscal responsibility" exemplified this trend. Finance transitioned from an exciting opportunity to become wealthy into a moralized injunction to conform to its economic demands. To be "responsible," hence, was disassociated (at least temporarily and partially) from conventional ethical and social desires of fairness, justice and equality. They were instead repurposed toward economic ends of being profitable and fiscally solvent (see Gilbert, 2013; Palley, 2007).

It was, furthermore, a duty that was not just reserved for corporate executives and firms. It was also an obligation that individuals were expected to embrace and mold their behavior to. This meant choosing a professional career wisely, as well as finding ways to help your employer "do more with less." However, it even extended into the personal realm, with the social insistence that individuals take responsibility for their own finances—to be "fiscally responsible" especially as the safety net of the welfare state was being gradually eroded and in some instances outright destroyed. It represented

> a new market-centric "politics"—struggles over political authority that share a particular ideological centre or, in other words, are underpinned by an unquestioned 'common sense'. On the elite level, neo-liberal politics is bounded by certain notions about the state's responsibilities (to unleash market forces wherever possible) and the locus of state authority (to limit the reach of political decision-making).
>
> (Mudge, 2008a: 703)

The progressively disciplining and coercive ethos of "personal responsibility" increasingly shaped the everyday economic and personal decision-making of neoliberal subjects. The discourse of "employability" was indicative of this pervasive form of neoliberal governance. Problems of unemployment and lack of opportunity were channeled into an appealing but also anxiety-inducing project of becoming "employable." Through ensuring that one was ever more attractive to employers, one could realize one's professional and personal ambitions. With full employment an ideal relegated to the past, "employability" took its place as a romantic vision of personal achievement and even freedom. Accordingly, employability served to "indicate how people should behave and what their responsibilities are" (Field, 1997: 62) and therefore signifies a new form of capitalist self-disciplining (Dean, 1994, 1998; Cruikshank, 1999). Yet it also served as a moral barometer of an individual's ethics and "goodness"—the inability or unwillingness to be "employable" was widely condemned as a failure of character rather than as a reflection of difficult economic realities.

Consequently, the economic and the ethical became unified and in a real sense socially interchangeable. "Personal responsibility" was attached to being fiscally astute and economically successful. Morality was associated with one's financial practices and outcomes—did individuals have a job? If so, did they maximize their skills for getting the highest-paid job available? Were they spendthrifts, or were they literally and figuratively accountable for their lifestyle? Did they help the organization save money, or were they a drain on its resources? These questions, as will be explored throughout this analysis, would have a broad influence on ethics, ranging from the political to the organizational. Just as important, though, was how the economic also became moralized—or, to put it more precisely, how this demand for personal responsibility would transform into a desire, and then an expectation, for neoliberal subjects to make capitalism "ethical."

The Power of Neoliberal Ethics

Thus far, this analysis has introduced the crucial ethical dimension to the legitimization and reproduction of capitalism generally and neoliberalism specifically. Notably, that is sustained as vital through its discursive framing as a problem to be fixed. The articulation of this capitalist "problem" is contingent and particular to its specific time and place. The solutions provided and made allowable are similarly diverse and evolving. There is thus a fundamental reforming ethic at the heart of capitalism. This ethos takes form on many different and interlocking fronts—ranging from political to organizational to personal. At stake, accordingly, in this work is to ask and attempt to answer a critical question of our time—"What precisely is the ethics of neoliberalism, and, just as importantly, what is its power for shaping the contemporary subject and ordering social relations?"

This key ethical aspect of neoliberalism constitutes and is constituted by its power relations. The expectation and demand for ethical responsibility, while by no means neutral, is also not necessarily aimed at securing elite rule and profit. Indeed, as discussed, its initial impetus was born out of a desire to hold corporations and their leaders publicly accountable for the adverse effects of marketization. However, these to a certain extent progressive beginnings gradually gave way to an encompassing ethical system for strengthening neoliberalism and its elitist power regimes. In this regard, it created a forceful justification for this system through redirecting personal responsibility for its morality to the majority of subjects themselves.

The most obvious—and arguably most fundamental—power relation at the core of neoliberal ethics is the social determination of who is "personally responsible" for capitalism. A crucial, and much lamented, aspect of neoliberalism is its "trickle-down" dynamic (see Aghion and Bolton, 1997; Arndt, 1983; Talbott, 2011) in which, ideally, elite wealth will gradually create mass prosperity. This top-down paradigm is not reserved solely for wealth distribution. It is also central to its conception of ethics and governance. Yet it is one strewn with irony—in so much as this critical "trickle-down" logic produces a bottom-up effect. More precisely, responsibility ultimately shifts from the government to corporations to, in the final instance, the individual. However, it also highlights a key (though too often ignored) element of neoliberalism. It economically promotes elite capitalists as the primary drivers of the economy, whose bounty must be passively accepted by the vast majority under them. Ethically, by contrast, it champions those on the bottom as the principal catalyst for creating a more decent capitalism. Consequently, it is assumed that if the majority of people behave justly, this will rise up to make the system as a whole more just. Significantly, it assumes that the people most ethically responsible should be those who would seem to have the least political, social or economic agency—everyday workers, the unemployed and vulnerable minorities, for instance. They are now tasked with acting ethically and, in so doing, ensuring capitalism's morality overall.

This complex and largely paradoxical ethical power relation has dramatic disciplining effects. It produces a pervasive and almost inescapable ethical demand that equally shapes expectations and behavior across politics, economics and society. It is not simply that subjects are now meant to act ethically; they must be innovative and persistent in their pursuit of playing their part in creating a morally upright market. The failure to do so, the inability to live up to this mandate, is a catalyst for social castigation and even legal punishment. Hence, it is assumed that "The state must never govern society, dictate to free individuals how to dispose of their private property, regulate a free-market economy or interfere with the God-given right to make profits and amass personal wealth" (Hall, 2011). However, the price of this freedom is a "popular authoritarianism" aimed at policing (either personally or by the state) one's behavior to reflect these market values (see Gamble,

1979; Gordon, 2005; Harvey, 2005; Wacquant, 2010). The common adage "you get what you deserve" has in the contemporary period been imbued with profound collective implications. If the market is materially and morally destructive, its victims really only have themselves to blame.

This widespread disciplining ethos of personal accountability reflects the deep structural function that this ethics of neoliberalism serves for the reproduction of capitalism generally. Power is directed at pressurizing and requiring subjects to manufacture solutions to its ethical problems. In doing so, it creates a rich human resource for displacing and temporarily resolving its concrete contradictions politically, organizationally and socially. It is worth noting even how issues of the market are phrased, in this regard—as excesses to be reined in and restrained rather than as inherent flaws that must be systemically and fundamentally addressed. Accordingly, it directs the energy of subjects toward "saving" capitalism rather than transforming it. It limits the cultural imagination to ethical innovations instead of a more profound radical rethinking. Of equal significance, it produces the very ethical "problem solvers" who can keep this hyper-capitalist ideal "working" in reality.

Crucial to this ethical power is the role of empowerment. This process must be understood as one of subjection and subjectification, not simple subjugation. This ethical injunction is framed as an opportunity for agency. It stands as a profound possibility for subjects to establish their unique selves and shape their environment at the supposed capitalist "end of history." It, furthermore, offers hope of changing, reforming and redirecting a seemingly "inevitable" and entrenched free market. This agency produces a range of capabilities for subjects to draw upon in order to make this system more ethical. It can manifest itself as a set of daily coping techniques for lessening the human and social costs of these policies or as a more collective effort to maximize the moral potential of neoliberalism. The ethical power of neoliberalism is, thus, grounded in the empowerment it offers individuals to politically, institutionally and even personally ethically reform this system even as they are denied the opportunity to fundamentally alter it or replace it.

4 The Political Power of Neoliberal Ethics

In 2008, the world of neoliberalism came crashing down, seemingly destroyed by the financial crisis. The near-global economic collapse challenged the entire economic system. Politicians and citizens from around the world sought to assign blame: who was responsible for this crisis? What could be done to fix it? What type of new order can be created from its ashes? Present was an opportunity for a drastic cultural rethinking—one that could lead to a profound economic, social and political transformation. What emerged instead was a concerted effort at neoliberalism's rehabilitation and recovery.

This was not just an economic crisis but also importantly a political crisis. A key tenet of neoliberalism was the idea that government should be limited as much as possible. Public interference in the economy strangled entrepreneurial freedom and innovation. There was a constant need to be vigilant against the threat of an overreaching state. The entire basis of neoliberalism depended on public recognition that governments were to be feared and markets to be celebrated. However, this popular narrative was difficult to maintain in the face of the near economic collapse, particularly because governments were needed to ensure the very survival of capitalism through massive public bailouts.

Significantly, this turn of events represented much more than a mere short-term political crisis for neoliberalism. It went to the core of what neoliberalism stood for and why it should be sustained. Conventionally, the political problem of neoliberalism was governments themselves. It was the state that imperiled progress. It was inefficient bureaucracies that threatened global progress. It was a nefarious and "rent seeking" public sector that prevented countries from developing prosperously. However, now the reverse seemed to be true. Financial markets and elites threatened the very material existence of people across the world. It was the free market and corporate globalization that was suddenly the real worry. The problem of neoliberalism became not governments or the state but the danger posed by markets and those who profited the most from them.

Tellingly, this revitalized political critique of neoliberalism took on decidedly ethical overtones. A crucial component of neoliberal political discourse was to focus on ethical as opposed to structural shortcomings. Unemployment was blamed on the lazy poor. Poor public management was blamed on the unmotivated government worker. Labor struggles were blamed on the greedy desires of unions. Massive national debts were blamed on the parasitic benefit cheats fueled by an immoral welfare state. The 2008 crisis turned such ethical scapegoating on its head. The near-total economic collapse was attributable to the avarice and immorality of financial elites. It was their lack of ethics and personal self-interest that now must be reined in.

Although on the surface this signified a profound challenge to neoliberalism and neoliberals, it simply continued in a new radicalized form an old neoliberal governing logic. It retained an underlying notion that politics was principally concerned with questions of personal ethics. Desires for more systemic change were overshadowed by the call for the elites who caused the crisis to be ethically rehabilitated or replaced. Underpinning such popular ethical demands was a continuing romantic vision of a self-regulating free market that simply required some occasional moral guidance. If there were problems, they could be traced back to individuals' immorality, not deep structural flaws. The role of government was to ensure that those who were acting unethically were suitably punished and chastened. Thus, while the political emphasis may have changed since the crisis, its primarily ethical rationale remains largely the same.

This primarily ethical reaction had profound political consequences. The immediate desire was to focus on determining the best ways in which to morally reform capitalism. There was an attempt to save the market from itself. Specifically, solutions were concentrated on its supposed worst excesses. Economic exploitation, inequality and mass insecurity were merely the outcomes of the financial system that needed some commonsense reforms as opposed to far-reaching transformation. Nevertheless, these initially minor institutional changes aimed at recovery were transplanted by a public outcry to hold the bankers and finance directors who caused the crisis accountable. At stake was a new and politicized neoliberal ethics of personal responsibility.

This ethos of responsibility would discursively reframe the very political problem of neoliberalism. What was crucial was to ensure that subjects acted ethically. The crisis was first and foremost a moral failing of all those associated with the market. Critically, this initially radical sentiment would come to legitimize a decidedly conservative market politics. It would be strategically marshaled to once again scale back the regulatory power of the state. Instead, it would justify the need for austerity based on the belief that it was an overspending public greedy for benefits they could not afford that created this economic catastrophe.

Yet it also revealed a more fundamental neoliberal ethics. In particular, it once again framed capitalism as an ethical problem to solve. Political empowerment was found in having the moral agency to fix the market and oneself. The political rationale of austerity was not just a matter of repression, nor was it a simple expression of individual and collective sacrifice. Rather, at its core, it played into a broader ethico-political discourse in which subjects were made responsible for capitalism's material survival and moral goodness. This ethical imperative would also be witness to the rise of innovative but ultimately reformist ideals like social entrepreneurship and financial social responsibility.

The Political Crisis of Neoliberalism

The 2008 financial crisis conjures up images of mass economic fear bordering on hysteria. With it, the romanticized vision of a free-market society was almost instantaneously punctured. It also, almost overnight, transformed the government from market villains into heroes. More precisely, it pointed to the continuing need for public intervention in order to sustain neoliberalism economically. Yet it raised difficult political questions for neoliberalism. Why were popularly elected governments bailing out the financial elite? Whose interest did they represent? And was the social and economic cost of neoliberalism too expensive to maintain?

There was little time to consider the content of these concerns in the immediate aftermath of the crisis. Major banks and financial institutions publicly accepted and lobbied for public funds merely as a matter of survival (Acharya and Richardson, 2009; Chari and Cristiano, 2008; Congleton, 2009; Roubinin et al., 2010). Nevertheless, it did point to the intimate relationship between government and financial elites. They were part of a "new aristocracy of finance" (Hager, 2015). It revealed their strong ties and perhaps more importantly the ways "financial measurements, ideas, processes, techniques, metaphors, narratives, values and tropes migrate beyond the financial sector and transform other areas of society" (Haiven, 2014: 1).

The role of government, it appeared, was primarily to protect the market and those who most profited from it. As the initial shock of these events began to wear off, it was increasingly unclear what the purpose of the financial sector was and why it should be maintained. Seemingly emerging was a form of governance that prioritized the "way that the state calls upon and organizes populations so that some will flourish and others will be left behind" (Martin et al., 2008: 123). In this respect, "What we have witnessed over the past few decades is therefore not the retreat of the state at the behest of the market," observes van der Zwan (2014: 117), "but rather the emergence of an altogether different type of state intervention." The crisis spotlighted how new regulatory structures were actually being created and deployed to advance the interests of these financial and corporate elites at the expense of the rest of the population (Daguerre, 2014). It thus left a growing number of people to wonder whose interests the government served.

Capitalism, of course, always operates according to a boom-and-bust dynamic. Economic boom times are invariably followed by down times.

As Rothbard (2009: 7–8) observed, "we used to suffer nearly periodic economic crises, the sudden onset of which was called a 'panic' and the lingering trough period after the panic was called a 'depression'" (also see Calomiris and Mason, 2003; Gorton, 1988; Kelly and Grada, 2000). Neoliberals originally blamed this volatility on undue government interference; more recently, it has been ascribed to "rational bubbles" inherent to a dynamic free-market economy (Blanchard and Watson, 1982; Diba and Grossman, 1988). Neoliberalism is perhaps particularly susceptible to these inherent market fluctuations (see Kotz, 2009; Perez, 2009). It creates exciting bubbles based on sophisticated financial models that promise ever-rising profits and infinite growth before coming suddenly and disastrously to an end. According to Tett (2009: 1):

> During the past two decades, a wave of innovation has reshaped the way markets work, in a manner that once seemed able to deliver huge benefits for all concerned. But this innovation became so intense that it outran the comprehension of most ordinary bankers—not to mention regulators. As a result, not only is the financial system plagued with losses of a scale that nobody foresaw, but the pillars of faith on which this new financial capitalism were built have all but collapsed. That has left everyone from finance minister or central banker to small investor or pension holder bereft of an intellectual compass, dazed and confused.

Capitalist politics and, therefore, neoliberal politics plays a significant role in mediating these crises. It must find ways to save capitalism without sacrificing the primacy of the market for organizing social and economic relations. To this effect, it highlighted the existence of a pro-finance "political bubble" in which free-market orthodoxy created a false sense of optimism (McCarthy et al., 2013). After this political bubble popped, new stories were required to revive the legitimacy and appeal of neoliberalism. In the Netherlands, for instance, elites reported to policymakers that they were no longer investing in "bad" US-type financial securities but rather "good ones" (Engelen, 2015). Developed economies, in general, drew on novel elite discourses of a reformed financialization to convince a skeptical public (Froud et al., 2007).

To this end, there was an urgent call for recovery. Across the ideological spectrum, there was a recognition that dramatic changes were needed. Even the famously pro-market and politically conservative *Wall Street Journal* declared that "the crisis compels economists to reach for a new paradigm" (Whitehouse, 4 November 2009). The need to pump trillions of dollars into the financial sector just to prevent it from crumbling and bringing the world down with it seemingly made it clear that the status quo was no longer working. Then President-elect Barack Obama (2009) proclaimed that the free market was based on a "failed philosophy" while former cheerleader of financialization Tony Blair (2009), the British prime minister, admitted that "What is plain is that the financial system has altered in its fundamentals and can never be the same again." However, it was precisely the sense of uncertainty and insecurity that emboldened political elites to once more publicly champion neoliberalism.

Most obviously, Obama and Blair warned of the danger of rejecting the long-term benefits of neoliberalism in the face of short-term losses. There was a popular tendency to throw the proverbial baby out with the bathwater in this respect. One of the key initial roles of the government thus was to prevent such popular overreaction. Outgoing US president George W. Bush thus stated that "this problem did not develop overnight and it will not be solved overnight"; nevertheless, "with continued cooperation and determination it will be solved" (Office of the White House Press Secretary, 2008). In one of his first major addresses to a joint session of Congress in February 2009, Obama optimistically proclaimed:

> But while our economy may be weakened and our confidence shaken, though we are living through difficult and uncertain times, tonight I want every American to know this, We will rebuild, we will recover, and the United States of America will emerge stronger than before.

Here the specter of insurrection loomed large. The fear of mass violence and potential global instability was used to reframe neoliberalism as a source of social security. Hence, preserving the strength of the financial sector was required to ward off economic social and political chaos. Within a year, neoliberalism had been discursively transformed from a system that caused the breakdown of social economic order to the very thing that would preserve it in the midst of troubling and uncertain times.

Such sentiments play directly into conventional political discourses of needing to maintain law and order. As previously discussed, the neoliberal retreat of the government from the economic sphere was often matched by its expansion in the social sphere. Security discourses were a central feature of this proliferation. Economic vulnerability both individually and collectively was assuaged by public promises to fight criminality and make people feel safe in their communities and homes again. These deep insecurities were transferred onto a rising tide of social deviants that only a strong state response could stem. Internationally, a similar rhetoric was turned outward as foreign policy became an exercise in fighting extremists who threatened international stability and domestic safety. The free market, in turn, was associated with the comforting vision of social stability and economic prosperity (see Bloom, 2016c).

At a structural level, it reflected the role of government in protecting capitalist interests. This was never more critical than in historical moments of recession and even depression, in which structural flaws were brought into the public light. The often hidden public securitization of the market, through such measures as mass government contracts, was progressively revealed. The rise of a neoliberal "power elite" (Rothkopf, 2008) and the "cognitive closure" (Albers et al., 2011) by policymakers around its ideologies were a global threat. Significantly, worries of instability became indelibly linked to the market and its principal actors. They were seen as the producers of an unpredictable economic system that threatened individual well-being and social as well as political security. The function of politics in this regard is twofold—it is to publicly give vent to such frustrations and popular concerns in a way that does not fundamentally threaten capitalism and markets.

The early post-crisis championing of neoliberalism as a potential safeguard against mass disorder fulfilled both these political functions. It recognized the need for vibrant public action while reaffirming the ideological and concrete supremacy of the free market for organizing socio-economic relations. The critical ethos in this regard was one not of transformation but of recovery. With only a few commonsense regulations—it was proclaimed—the free market could be saved, and international and national progress could be once again guaranteed. There was no need for an overreaction or an ill-advised fundamental rejection of financialization. Returning to the optimistic ideals of Obama, by early 2009 he was already stating confidently that

> a crisis like this reminds us that we just have to put in some commonsense rules of the road, without throwing out the enormous benefits that globalization has brought in terms of improving living standards, reducing the cost of goods, and bringing the world closer together.
>
> (Obama, 2009)

As appealing as this discourse may have sounded, this was at best a temporary solution. Significant political crisis remained. Specifically, what was the actual role of government in the economy? If the status quo was to be largely preserved, how could the government prevent another crisis from occurring? These worries were exacerbated by the seeming inability of the state to address chronic economic problems associated with neoliberalism, such as growing inequality, failing infrastructure and mass unemployment as well as underemployment. Just as important, it raised economic questions about the desirability of financialization and marketization. It was no longer so clear that the financial sector should have such economic power and influence. Nor that market practices and values led to the best social and economic outcomes. Indeed, neoliberalism faced its most profound and threatening political test to date. To meet this challenge, it would turn to ethics.

The Ethical "Problem" of Neoliberal Politics

The relation of ethics to politics is one of the most studied and arguably least understood in all of social science and theory. It can refer to the proper conduct of leaders and citizens or to the principles and practices that constitute an ideal political culture. Within neoliberalism, these very questions risk becoming redundant. If the market is the highest good, then political ethics is merely a matter of using state power and molding popular culture to reflect these values and protect them from ideological and material threats. However, neoliberalism has gone considerably further, in this respect. It has transformed these conventional philosophical and normative concerns into a potent political discourse for ethically reproducing a hyper-capitalist system. Specifically, it has made ethically "fixing" the free market into both an individual and a collective political moral responsibility as well as an exciting and empowering opportunity to make a "real" difference.

Historically, this pointed to a supposedly more fundamental capitalist politics of crisis. Issues with the market are channeled into a broad panic concerning the threats posed to society as a whole. If economically capitalism is in constant fluctuation from boom to bust, politically it finds its ironic stability in the perpetuation of a consistent threat that can fix the population's attention and mobilize them into action explicitly or implicitly in support of the market (Bloom, 2016c). It should not be surprising, then, that in the wake of the current crisis there was a renewed embrace of neoliberalism (De Cock et al., 2013; Halsall, 2013). This insight challenges conventional readings of the crisis character of capitalist politics. It highlights instead how a market society depends on manufacturing a permanent sense of crisis for its continual dynamic revitalization. Politics, in turn, morphs into a repeated experience of permanent capitalist protection and reform rather than systemic transformation.

For neoliberalism, politics was almost always first and foremost a problem of individual morality. It was assumed that the fundamentals of the economy were already set in stone. In this regard, the crisis was framed as a need to, above all, return to "business as usual." This "framing" was primarily predicated on a

> "fear and insecurity" scenario—economic insecurity in particular—which is forging a new category of the "nervous public at risk." This is being played out to suggest that the main task is to re-secure the stability of the financial system, stimulate the consumer boom once again ('maintain aggregate demand'), re-capitalize the banks so that they can reestablish credit circulation and, hopefully, re-stimulate the housing market.
>
> (Thompson, 2009: 521)

Globalization and the onward march of the free market were still as unalterable as the laws of gravity (Fleming and Spicer, 2007). Any negative outcome or deviation from its rosy expectations must be personal rather than structural in nature. It was considered a moral failing not to adapt and be productive in this emerging world of capitalist possibility. Such an inability to evolve with the times imperiled not only one's own fate but the prosperity of society as a whole.

Yet this ethos of fighting moral decay could also be directed toward those who were successful when necessary—though not in a way that fundamentally questioned the moral goodness or economic necessity of the free market. In the aftermath of the breakup of the Soviet Union, for instance, the promise of liberal democracy and capitalist freedom was tainted by the predatory actions of its new oligarchic elites (Hanley, 1999). Such behavior was explained away as a matter of their own immorality rather than characteristic of the free market itself. Put differently, although capitalism could never be blamed, capitalists certainly could and often were sacrificed on the altar of public opinion in the name of preserving this higher ideological ideal (see Enriques, 2003). The issue was one of "infectious greed" on the part of financial actors rather than anything systemic about contemporary capitalism generally or essentially (see Partnoy, 2003).

This potent mixture of moralization and realpolitik also was used against the political establishment. The free market, at least theoretically, was meant to reduce the presence of political corruption (Gerring and Thacker, 2005). Nevertheless, the undue influence that money bought the financial sector and corporations created widespread public backlash and consternation. Of course, the corrupting effect of money in politics is a long-standing concern for liberal democracies (Fiss, 1997). Yet it took on an even heightened character in the wake of the financial crisis. Specifically, it challenged the very cherished ideal that the United States and other democracies were in fact still actual democracies. To this end, it raised profound questions:

> Why does it cost nearly a billion dollars to mount a successful presidential election campaign? Why are corporations considered citizens and entitled to unlimited political spending? Will unaccountable organizations such as superPACs and dark money groups become more influential than political parties? How have citizens been forced into the role of consumers in a political marketplace? Indeed, how did democracy become a market?
>
> (Kuhner, 2014: ix)

These concerns were only exacerbated by the revelation that the United States was now officially an "oligarchy" (Anon., 2014).

However, again, the fault lay with corrupt politicians who placed personal interests over the public good. The issue of the ethical politicians was once more obviously not new. It took on a particular significance after the crisis. Quoting Birch and Allen, "Maintaining the trust of the citizenry is crucial for politicians if they are to retain legitimacy, yet there is ample evidence to suggest that political leaders are very often believed by those they represent to behave unethically" (2015: 43). Contemporary political leaders were demonized into an "us vs. them" public discourse at "folk devils" (Flinders, 2012; Hatier, 2012). Required, in turn, were leaders with stronger moral fiber who would resist such gilded temptations.

This elite problem extended to society as a whole. It was filled with people who were trying to take advantage of the system and take advantage of their fellow citizens. It was the fault of "benefit thieves" and "welfare frauds"—a discourse that increasingly influenced not just everyday citizens but also elite decision-makers (Conor, 2010; Garthwaite, 2011; Slater, 2012; Wiggan, 2012). This was epitomized in Britain, for instance, in the distinction between positive "strivers" and negative "skivers." In the words of prominent Conservative politician Iain Duncan Smith (2012):

> Those we should have been helping . . . the *strivers*, the tryers . . . too many of them lost most of every pound they earned in work . . . I ask you, what kind of message does that send out? I will tell you—that it's not worth working—that it's not worth trying—that you're better off *playing the system and taking the money*. Shameful!

This further played into a type of "poverty porn" in which popular media shamed those made vulnerable as selfish and scheming rather than as victims of an unjust system (Jenson, 2014).

There was, therefore, an interesting political dynamic at the core of neoliberalism. Neoliberalism championed an economy of competition and self-interest yet promoted a politics that at the very least sought to minimize such a mentality. The obvious conclusion is that the oft-revered market must be morally reined in through the auspices of the state. Yet this option is relatively impossible in the context of contemporary neoliberalism. It is therefore modified to reflect its unwavering commitment to the free market. Rather it is the fault of immoral politicians who must be reformed. Similarly, it focuses its attention on the morality of the population at large, seeking to channel their market impulses into socially appropriate avenues.

What is revealed is a transformed and reenergized capitalist political discourse. The problem of capitalism was not material or even ideological; it was, at its essence, moral. People gave in too often to their worst moral instincts, leading them to embrace an unethical lifestyle in toto. This politically manufactured ethical crisis concentrated the public's attention and efforts on rooting out and punishing such unacceptable behavior. It isn't surprising, for instance, that in the 1990s, a decade filled with financial deregulation and conflict around the world, the most consequential political scandal was one that revolved around an American president's extramarital affair while in office. The problem was not the free market; it was the immorality of the few who ruined it for everybody. The government had a sacred duty to stamp out this moral decay. And it was the deep political responsibility of all citizens to solve this ethical crisis of neoliberalism.

Taking Ethical Responsibility for Neoliberal Politics

The theme of responsibility has a long history of importance for capitalism. In order for factories to run, in order for offices to work and in order for industry to profit, subjects must be at the very least made professionally responsible. The ideal worker is one who can be held personally accountable for doing his or her job in a competent, efficient and productive manner. Yet this discourse transcends merely ensuring the daily economic functioning of capitalism. It is also imperative to its social and political reproduction. Structural faults become inexplicably linked to personal irresponsibility—from the lazy worker to even the greedy businessperson.

Significantly, it is imperative to historically trace out how political responsibility is assigned within the broader evolution of capitalism. Democratic elections have to a large degree revolved around the perceived and often varying sense voters have regarding the responsibility of politicians for controlling and improving economic outcomes. Indeed, a crucial distinction between liberalism and neoliberalism is its assumption of whose duty it is to fix the market. For the former, it is a combination of public intervention and private initiative (see Skocpol and Finegold, 1982). In the immediate postwar era, there was a range of ways in which parties and politicians could undertake this political responsibility relating to the economy (Lowi, 1972). Neoliberalism, conversely, devolves such political obligations to the individual. Hence,

> politics everywhere are now more market driven. It is not "just" that governments cannot manage their national economies; to survive in office they must increasingly "manage" national politics in such a way as to adapt them to the pressures of transnational forces.
>
> (Leys, 2001: 2)

Neoliberalism thus privatized not only the economy but also politics.

Interestingly, this neoliberal politics of responsibility resonates with perhaps its most implacable political enemy—scientific Marxism. Much, though perhaps not enough, has been made of how both conceive of economics and social change as law-like and inevitable. For Marxists, it is the inherent contradictions of capitalism that will lead to its ultimate and unavoidable downfall (Bullock and Yaffe, 1975; Chesnais, 1984; Clarke, 1994; Mattick, 1981; Yaffe, 1973). For those who have a fervent faith in the free market, they are simply embracing the natural order of things as borne out in human nature itself (Bernstein, 1971; Tipps, 1973). This teleological ideal of market and democracies became even more heightened and aggressively pursued under neoliberalism:

> With modernization discredited and no single overriding narrative of progress to replace it, neoliberals took the field with their own promises

> of accelerated, benevolent change. . . . Neoliberalism, in other words, prevailed precisely because it revived a vision of the global mission of the United States and made the same sort of transformative claims that modernization had.
>
> (Latham, 2011: 158)

What is not commonly—if truly at all—highlighted is that each also displays a similar deep understanding of political responsibility. Marxism proclaims that the historical and inevitable role of the proletariat is to rise against capitalism and dialectically transform the system into Communism. Less romantically or explicitly, neoliberalism expects all its subjects to adapt to and help destroy all vestiges of the liberal social democracy it is seeking to replace. The supposedly unstoppable character of these changes also demands that individuals assume personal responsibility for keeping up with, protecting and even prospering in this new free-market reality. This extended to politics as well. Just as in really existing socialism, in which revolutions were followed by expectations of strict moral conformity to Communist principles, the same occurred in neoliberalism. It was the civic duty of every citizen to be fiscally responsible and maintain the moral order.

This neoliberal promotion of responsibility is, as alluded to above, reflected in and linked to the economic notion of fiscal responsibility. Here economics, politics and morality merge, as the imperative to be profitable is granted an explicitly normative character. The need to be fiscally solvent is transformed into an ethical pre-condition for determining social and economic policy. Thus, the very idea of such "financial responsibility" has its historical roots in the 1970s, when elites were confronted with the perceived inability of traditional liberalism to resolve chronic social and economic issues (Krippner, 2012). It was enshrined first not in the neoliberal hero of Reagan but in his predecessor, Carter, who made a formal commitment to achieving a "balanced budget" (Stein, 2011). This discourse of "responsibility" was instrumental in creating an elite form of modern governance in which "corporations, financial institutions and governments worked hand-in-hand with conservative political leaders to construct a "free market" regime of accumulation designed to bolster business profits at the expense of all other social groups" (Block, 2010: 31). To ignore such capitalist concerns is a total dereliction of one's moral duty; the ethics of capitalism now applied to politicians. Political leaders, democratic or otherwise, must first and foremost, like corporate executives, prioritize the fiscal survival and prosperity of the country above any and all other concerns (McGregor, 1993; Doornbos, 2003).

This increasingly marketized discourse of political duty would become an entire global governing paradigm. Indeed, in the 1990s, the emphasis for international development took a decidedly political turn (see Bloom, 2016a). As United Nations secretary-general Kofi Annan declared, "good governance is perhaps the single most important factor in eradicating poverty and promoting development" (Annan, 1998). It went from "getting prices right" to "getting institutions rights" (Choudhary, 2007). Consequently,

> [f]rom the early 1990s onwards, the call for less state has gradually been substituted by a call for a better state. This new approach should not be confused with a plea for a return to the strong (Keynesian or socialist) state. Rather it implies better and transparent governance of what is left of the state after neoliberal restructuring has been implemented.
>
> (Demmers et al., 2004: 2)

The stress was on good governance and having the moral commitment to adhere to such "right" principles against often difficult obstacles and temptations to do otherwise. The primary obligation of politicians was to the higher good of protecting the market, which it was blissfully assumed coincided with the interests of the population in the long term despite temporary setbacks and short-term suffering. Such moral political responsibility went beyond one's own national borders, as the failure to be fiscally responsible could negatively impact one's neighbors and the international economy as a whole.

There is a common critique that neoliberalism is governed by technocratic rule, that the strongest demand placed on governments is to adhere to a free-market dogma, that such orthodoxy restricts democracy and leads at times to downright repression (Allier Montaño and Bloom, 2014). Undoubtedly neoliberalism is spread through the framing of politics, at its ideal, as a technical exercise in implementing "correct" economic policies. However, it is also forced to be highly creative and innovative, as leaders must ever find new ways to politically convince subjects to be fiscally responsible citizens as well as show how this inviolable commitment to the market can be deployed to more just ethical ends. The defining feature of neoliberal politics is hence not the cold, dry application of technocratic mandates. Instead, it is the dynamic and constantly renewable effort to morally and ethically "fix" capitalism by producing politically "responsible" subjects.

Producing the Politically Moral Neoliberal Subject

A critical political concern for neoliberalism, both genuinely and strategically, is "How and in what way can this new hyper-capitalist order be made moral?" The initial justification of neoliberalism largely ignored such concerns of social justice. It was good simply by virtue of its promotion of the free market. It was a tautological political proposition: as the market was by its very nature morally just, then ipso facto any political program that championed and protected it was also just. However, the economic and social devastation caused by neoliberalism over its first three decades dramatically challenged the proffered essential righteousness of the market. These concerns were only exacerbated by the financial crisis. The political task of neoliberalism and neoliberals was to publicly and politically remoralize neoliberalism.

The initial and understandable response to the near economic collapse of 2008 was the desire for a total economic and political reboot away from neoliberalism. Bankers and financial actors were castigated for their dodgy deeds and practices, credit givers were condemned for their failure to uphold standards and politicians were accused of profiting and abetting this widespread economic malpractice (Gray et al., 2009; O'Malley and Fitzgibbon, 2014; Sim and Sumikawa, 2012; Vertinsky, 2009). Although these critiques were originally leveled at the capitalist class, they transformed into a broader challenge to the system's institutional immorality.

This mixing of the systemic and moral was witnessed in the depiction of executives and financial actors as psychopaths (see, e.g., Babiak, 1995; Babiak and Hare, 2006; Clarke, 2005). Boddy (2011: 256) describes them as "sociopathic" monsters, as "they ruthlessly manipulate others, without conscience to further their own aims and objectives. Although they may look smooth, charming, sophisticated and successful, Corporate Psychopaths should theoretically be almost wholly destructive to the organizations they work for." The important assumption is that a more moral version of capitalism could emerge if only it wasn't filled with so many pathologically dangerous leaders. Perhaps the better question, though too often left unasked, is how to create a less psychopathic economic politics and society in the first place.

The policy of austerity hoped to resolve this moral problem of neoliberalism. Firstly, it returned to a common political discourse that the issue was fundamentally moral in nature (see Brash, 2003). Bad behavior and unethical desires were to blame. It also completely neutralized the radical content of this moral critique. Stopping the spread of an ill-conceived and totally fiscally irresponsible politics of public overspending was now an urgent priority. Conservative elites continuously intoned that the real cause of the crisis was a political culture that was addicted to spending of public funds beyond what the country could afford. In this way, the crisis has been ideologically reworked, at least in the UK and much of Europe, from an economic problem (how to "rescue" the banks and restore market stability) to a political problem (how to allocate blame and responsibility for the crisis): a reworking that has focused on the unwieldy and expensive welfare state and public sector, rather than high-risk banking strategies, as the root cause of the crisis (Clarke and Newman, 2012: 2).

Tellingly, this solution of austerity to structural problems of capitalism turned into a profound moral imperative. From its earliest manifestations this rather technocratic and largely counterproductive and highly unproven economic programmed was transformed into an ethical prerogative. It brought together national populations through appeals to shared sacrifice (Levitas, 2013). According to Mitrea (2015: 1):

> austerity is more than a material condition and policy orientation, it is a moral-affective economy in which the politics of consumption are negotiated through material, temporal, and moral logics. "Moral austerity" inculcates and shapes a common sense around individual responsibility for reduced consumption as a virtuous necessity.

Yet, as the crisis receded, the promised recovery of austerity continued at only at trickle, and for many was completely invisible. The previously temporary sacrifices of austerity transformed, in turn, into a permanent moral obligation to be "fiscally responsible."

Significantly, it was not a matter of temporary sacrifice and unifying for the common good; it was a personal reflection of one's moral fiber. Were you willing to sacrifice the future of the country and upcoming generations for short-term prolificacy? Moreover, it was rationalized in exactly these moralistic individual terms. It was akin to a person who, as a result of spending too much money, left himself and his family drowning in bills. Quoting Brown (2016: 4) at length, in this regard:

> Here a logic unfolds that blends the hardheaded approach to human capital of any successful firm with a national-theological discourse of moralized sacrifice, a sacrifice required for the health and survival of the whole. Moralized sacrifice finesses the paradox of unrewarded conduct normatively prescribed by neoliberalism. Examples of this paradox include the promulgation of hard work in vocations or industries suddenly outsourced or eliminated; the promulgation of "ownership societies" that cast hundreds of millions into underwater mortgages or foreclosure; the promulgation of privately invested retirement funds that melt in finance capital crises or downturns; the promulgation of college education and other kinds of training that fail to yield jobs or a "college premium."

Critically, this completely incongruent comparison between the personal and national economy resonated even more forcefully due to the rising and increasingly inescapable reality of individual debt. In this respect, the collective appeal of cutting back on unnecessary expenditures made progressively more sense. The majority of people were having to tighten their belts to ease their debt burden; why should the country do any differently? This reflects what Lazzarato (2012) famously refers to as a modern-day "debt morality," in which once-enshrined social rights such as to healthcare and employment are made into social debts that we must all "repay" through being fiscally and socially responsible in our everyday behavior.

Crucially, neoliberalism used one of its most destructive and problematic aspects—the use of easily available credit to replace higher wages—to its political advantage. As personal debt increased exponentially, so too did the idea that it was a unquestioned principle of public reality to actively address this issue through greater austerity. Consequently, fiscal responsibility transformed into a resurgent political responsibility. According to Forkert (2014: 41), "The present moment is marked by anxieties about society falling apart, and nostalgia for a lost era of social cohesion," and

> [t]hese anxieties shape the dominant narrative about the causes of the recession—which are seen as resulting not from the excesses of the financial sector but from a profligate welfare system and an overly permissive immigration system, which has given the wrong people access to public services—the unemployed, the disabled, single parents and immigrants.

It was thus the moral duty of governments not to overspend—to in fact cut back on public investment so the country could meet its debt obligations and remain fiscally solvent. Of course, this austerity was selective in its targets for "necessary reduction"—defense and security remained vital and largely untouched parts of the national budget and indeed, if anything, increased. However, welfare benefits, it was argued, must be scaled back to the extreme—they were simply unaffordable and immoral.

The framing of welfare as a form of personal responsibility, not necessity, reflected the personal character of this political morality. It was not a question of whether one wanted to invest in public infrastructure or helping the needy or seeking to create jobs. These things were simply fiscally unfeasible, and to argue otherwise was to be not only economically misinformed but individually unethical. Accordingly, citizens were made into "nationally abject subjects"—figures to be constantly morally judged as failing to live up to these socially prescribed expectations of responsibility (Tyler, 2013). It was akin to desiring a bigger entertainment system or expensive car when one was on a tight budget. It exemplified a type of wanton disregard for the economic well-being of the country and must be directly rejected and morally condemned. To this end, spending cuts were legitimized by non-elites as ethically necessary based on a shared mentality of "you reap what you sow" (Stanley, 2014). Voting for austerity was more than economically sensible, or so conservative experts and leaders proclaimed; it was the only truly moral political choice in the long run.

Such normative appeals also had profound political implications on the collective level. The moral severity of the consequences of this selfish overspending could not be exaggerated. It would lead to a crisis in the short term and complete fiscal ruin for future generations. Indeed, there was a public obligation to the population at large to not surrender to such temptations. The eurozone crisis of Greece exemplified the simultaneous fiscal and moral nature of this responsibility (Bloom, 2016a; Papadopoulou and Sakellaridis, 2012). As the world economy became ever more international, so too did a country and its citizens' political moral responsibility to each other. More than just the need to have good governance, there was now a shared obligation to embrace austerity.

In this spirit, the political morality of neoliberalism, especially after the financial crisis, reflected its deep ethos. Notably, it expected and demanded subjects themselves to solve the problems of capitalism politically. More precisely, there was a personal and collective duty to fix the market. This meant holding true to the deeper morality of the free market, even if it looked utterly heartless in practice. One must ignore food banks, rising inequality and poverty in order to stay strong in one's moral commitments to frugality and being fiscally responsible as both an individual and a nation.

Solving the Ethical Problem of Neoliberal Politics

The efforts to link fiscal solvency to a broad ethical political morality was a strategy that infused traditional liberalism with a new sense of personal and collective ethics. It provided a vision of a nation that was committed to the common good of freeing a state and its citizens from the heavy weight and entanglements of debt. It is akin to the personal sense of relief and renewed agency that is felt in accounting for one's own debts in creating a clear path for repayment. Yet the moral legitimacy of these strategies was severely tested by their often destructive social and economic consequences. Although most may agree that it was morally good to meet one's financial obligations—whether they be personal or national in nature—this should not be done at the cost of going hungry or letting fellow citizens unduly suffer. The uneven effects of these policies in which the poor were the hardest hit put these moral concerns into even sharper relief. This catalyzed a profound political challenge that threatened to expose the material and class-based interest driving these moralistic appeals.

Hoping to publicly cope with these challenges was a resurgent political discourse emphasizing the ability of governments, corporations and private individuals to direct the free market through their own entrepreneurship to more ethical social ends. Tellingly, these discourses sought to negate and overcome the negative results of "actually existing neoliberalism" (Cahill, 2010). They did not shy away from discussions of rising inequality, employment, unemployment and poverty. By contrast, they recognized that the status quo could be dramatically improved and that to do so formed a key component of contemporary social and personal ethics. However, these critiques are not systematic and instead focus their criticality almost solely on the lack of ethics and ethical initiative of the people involved themselves. It was seen as a deficit of imagination and the general unwillingness for political and organizational leaders as well as everyday citizens to discover how the dynamism of the market could also be best used for socially just purposes.

Imperative was the need to creatively fix present-day capitalism without fundamentally altering its market principles. As such, economic justice was an ideal that must be achieved through a neoliberal logic of efficiency—a rationality Grindle (2007) calls "good enough governance." At the global level, international organizations like the OECD championed "social inclusion" strategies that focused on empowering marginalized populations by promoting "commitments to global economic integration and openness, market led growth, 'good' and 'joined up' governance, local and social partnerships, and wide-ranging activation of 'capacities'" (Porter and Craig, 2004: 387). Similarly, and even more explicitly, there were increasing efforts among both the scholarly and policymaking community to link the realization of millennium development goals to policies of "financial inclusion" (Chibba, 2009).

Emerging was a problem-solving ethos in which the very issues caused by marketization and financialization turned into an opportunity for them to be reformed. Significantly, this in no way mitigated the deep moral political responsibility one had to be fiscally responsible. Rather, it was a chance and an increasing demand to figure out how to balance this economic accountability with a broad commitment to achieving the public good. The failure to find such "silver bullet" solutions represented not a substantial flaw of neoliberalism but a deficiency in the subjects themselves.

At the more concrete political level, this problem-solving was witnessed in the rise of what can be termed "fiscal social responsibility." There was suddenly an explosion in new initiatives utilizing financial mechanisms for achieving social and economic justice. This echoed a common assumption linking corporate governance to broad goals of corporate social responsibility (Jamali et al., 2008). Yet it was also more expansive than this rather limited conventional notion. Instead it advocated for a "creative capitalism," trumpeting the use of leveraged investments to mobilize the talents of the world's poor and solve its pressing problems (Haiven, 2014).

This ethos thus devolves such ethical demands to the personal level. Specifically, it was down to individual agents to help lessen the social and environmental costs of the free market. However, this problem-solving mentality extended firmly into the sphere of political ethics. The early 21st century saw the huge growth of social entrepreneurship (see Cho, 2006; Peredo and McLean, 2006; Seelos and Mair, 2005). The market was now a fertile site for creating a better world, one often disconnected from the traditional business demands and desires of profit. There was additionally a dramatic cultural cachet in being such a social entrepreneur—it was suddenly hip to be able to combine market savvy with a sense of social consciousness.

This new entrepreneurial ethos directly and indirectly challenged the perceived inherent immorality of capitalism generally and neoliberalism specifically. Emerging critiques of the free market were replaced in the popular imagination with the promise of "ethical capitalism" (Barry, 2004) and "green capitalism" (Smith, 2011; Tienhaara, 2014; Watts, 2002). The struggle was not to make capitalism viable or normatively desirable but for good ethical social entrepreneurs to triumph over bad greedy capitalists.

Appearing, therefore, was "a state that performs the role of a market actor and regulator at the same time" (Livne and Yonay, 2015).

The role of government, in turn, was to encourage such an ethical market "sea change." It was to publicly endorse and fiscally support entrepreneurship as a means of addressing and providing business friendly solutions to chronic and worsening social problems. The state was expected to take an active role in formally and informally promoting such ethical market behavior. Specifically, states were expected to creatively "manage" globalization and the free market to combine economic growth and social justice (Kaul et al., 2003). Significantly, governments were tasked with working within and through their own context-specific cultural constraints to realize this simultaneous social and fiscal responsibility (Engelen et al., 2008; Kalinowski, 2013). Their shared aim was to undertake, within certain ideological limits, "national and regional experiments with global ideas" (Roy et al., 2006) for this purpose.

In this spirit, capitalism became a system that required ethical solutions. To do so, it needed to foster its citizens' critical problem-solving abilities. Accordingly, it shifted the ethical focus away from the system itself. There was thus a fresh expectation for individual communities to become not just economically moral but actively engaged in social reform. They were charged with taking political responsibility for resolving the ethical problems of neoliberalism.

The Political Power of Neoliberal Ethics

The political ethics of neoliberalism of course reflects broader discourses and regimes of contemporary political power. The governing logic of personal responsibility has been strategically marshaled in support of elites as a means of reinforcing existing and even relations of power. That is arguably its most obvious and most significant aspect. Yet it also critically produces a new discourse of moral power and ethical empowerment that strengthen the hyper-capitalist order both ideologically and in practice. In doing so, it has limited the horizon of the population's political imagination and therefore capacity for envisioning and implementing radical change.

In principle, such demands for ethical responsibility were egalitarian in the extreme. Everyone had to be morally and ethically accountable. Perhaps not surprisingly, though, the reality of this ideal was dramatically different; such responsibility traveled from the top down. The elites were routinely forgiven for their transgressions, while those who were most powerless and vulnerable were more and more called upon (often coercively so) to personally and collectively politically save capitalism.

Revealed, in turn, is a tragic irony of neoliberalism that extends across its political organizational and interpersonal relations. Those with the least transformative agency are often expected and demanded to do the most to impossibly change the system. In this respect, the responsibility for creating a moral and ethical free market routinely was replaced to those members who suffered the most under it and have the least capacity to reform it, let alone transform it.

In perhaps an even further irony, this trickle-down ethical politics requires an active and expansive state: specifically, an institutional commitment from government to help encourage and produce such moral and ethical market subjects. As Gilbert (2013: 9) notes, neoliberalism

> advocates a programme of deliberate intervention by government in order to encourage particular types of entrepreneurial, competitive and commercial behaviour in its citizens, ultimately arguing for the management of populations with the aim of cultivating the type of individualistic, competitive, acquisitive and entrepreneurial behaviour which the liberal tradition has historically assumed to be the natural condition of civilised humanity, undistorted by government intervention.

To this end, the government proactively helps to socially manufacture individuals to become personally responsible for the economy as a whole. Job centers' strict "work to benefit" requirements highlight the role of the state and surveilling and guiding individuals to personally regulate their immoral tendencies and become economically productive members of society.

Critically, this once more shows the potential ethos of neoliberalism both literally and figuratively. It was the responsibility of the state to guarantee that subjects become personally responsible for themselves and, even more significantly, for capitalism itself. The chronic lack of good employment and the low wages that are the hallmark of neoliberalism were displaced by a state-mandated ethical demand for individuals and communities to resolve these problems on their own or else suffer the economic and social consequences.

The state was, therefore, expected to educate the population through its institutions in this shared civic morality and sense of private-market ethics. This was manifested in the suggested connecting of risk assessment to conventional goals of social protection in a manner that "emphasizes the double role of risk management instruments—protecting basic livelihood as well as promoting risk taking" (Holzmann and Jørgensen, 2001: 529). Likewise, academics have proposed a "pro-poor (public) expenditure index" that would serve as a conditionality for countries receiving further aid while granting donors "greater flexibility" in auditing and punishing those who fail to keep these commitments (Mosley et al., 2004).

Theoretically, this harkens back to Althusser's notion of ideological state apparatuses (1974). Institutions were formed and maintained in order to inform and shape subjects to reflect this ideological reality. Schools, governments and churches produce an "imaginary" world through their discourses and practices that help make capitalism feel more concrete and real. In the case of neoliberalism, its institutions incentivize individuals to learn and perform their moral and ethical civic and economic duties. All that was needed for society to become just was a population invested in their own moral reform and ethically committed to personally and creatively fixing the system.

However, it is important not to overlook the productive aspects of this political power. These ideological accounts risk focusing on the repressive aspects of this ethical hegemony at the expense of its more enabling and empowering dimensions. In the face of a seemingly permanent and unalterable corporate globalization, the ability to do something—anything—can feel incredibly powerful. What neoliberalism politically offers subjects is a profound, though restrictive, sense of moral agency. To wit, it may be impossible to change capitalism, but one can engage in individual collective moral transformation.

There is thus an appealing, though deceiving, political dynamic at the heart of neoliberalism: if you morally and ethically improve yourself and your communities, so too will the market magically improve. Moreover, government elites must use financialization to realize the public good. And when it all invariably comes crashing down, it will be up to the rest of society to account for their failures and pick up the pieces. The power of neoliberalism is such that the majority of citizens have a political responsibility to morally and ethically save capitalism for the ultimate benefit of those who profit the most and are ultimately the least accountable for the economic and social damage they cause.

5 The Institutional Power of Neoliberal Ethics

Neoliberalism, at least at first glance, can appear to be the antithesis to sustainable and successful organizations. Its focus rests almost solely on demonizing collective achievement in favor of extolling individualism and enterprise. Institutional life is depicted as the epitome of regulation and conformity. The free market, by contrast, represents the liberation of subjects from these oppressive organizational bonds. Just as with politics, the very idea of institutions and collective organization seem to be the enemy of neoliberal freedom and progress.

Neoliberalism in fact relies on even more monopolistic and monolithic economic organizations. It is not a coincidence that its spread coincides with the growing power of corporations and the corporatization of society. According to Harvey (2007: 36):

> The corporatization, commodification, and privatization of hitherto public assets have been signal features of the neoliberal project. Its primary aim has been to open up new fields for capital accumulation in domains formerly regarded [as] off-limits to the calculus of profitability.

Almost half a century after the beginning of its ideological reign, neoliberalism has given birth to a world where large-scale and apparently faceless organizations dominate. It is no exaggeration to claim that

> [b]y many measures, corporations are more central players in global affairs than nations. We call them multinational but they are more accurately understood as postnational, transnational or even anti-national. For they abjure the very idea of nations or any other parochialism that limits them in time or space.
>
> (Barber, 1995: 23)

Nevertheless there is much truth in the claims that neoliberalism weakens if not completely severs conventional organization ties. Putnam's (2001) famous characterization of contemporary life—at least within much of the so-called developed world—as individuals consigned to an existence of "bowling alone" may be exaggerated for many but only slightly so. Economically, neoliberalism has slowly but surely reduced the sense of shared belonging. It has supposedly produced a modern employee who embraces market rationality rather than institutional loyalty or community investment. Its policies have created a profound feeling of economic insecurity fostered by its cutthroat culture coupled with an increasingly uncaring state.

However, there is an alternative organizational reality that decidedly complicates this bleak picture. Even as corporations conquer all corners of the world in the name of the financial bottom line, internally they perhaps have never been so human-focused, and, if its rhetoric is to be believed, almost downright caring (Velasquez, 2002). This new mentality is one of using human resource thinking and practitioners for maximizing employee potential. The aims of modern companies transcend profit motivations and are focused instead on the "increased physical and mental health of employees," including their "advanced spiritual growth and enhanced sense of self-worth" (Krahnke et al., 2003: 397). It is a culture that increasingly asks its workforce to simply "be yourself" (Fleming and Sturdy, 2009) and is committed to principles of allowing them to achieve a healthy work-life balance.

It may come as no surprise that there is a serious dark side to this corporate humanism. The appeal to balance often leads to an ironic recommitment to work as a necessary part of a healthy life (Alvesson, 2000; Caproni, 2004; Healy, 2004; Rego and Cunha, 2008; Scholarios and Marks, 2004). The possibilities for freedom and flexibility commonly result in greater work intensification as well as a more managerial-driven "Results Only" firm that is even less forgiving of employees who fail to deliver on time. This fresh corporate humanity is as much a strategy of regulation and control as the more explicitly coercive and conforming ideologies that preceded it.

Nevertheless, these valuable critiques risk hiding the deeper function of this new organizational ethics. It is one in which nonmarket values, desires and impulses are used to paradoxically institutionally reproduce this hyper-capitalist order. Analogous to its political demands, organizational subjects are expected to take personal responsibility for normatively saving economic organizations enthralled to the seemingly inescapable forces of global finance. Specifically, the problem that afflicts corporations is a reflection of the broader systemic and structural flaws of neoliberalism. It is found again in the moral and ethical deficiencies of employers and employees. They therefore have a personal responsibility to fix this problem either through their own moral improvement or through the discovery of innovative ethical institutional solutions for overcoming these market-created shortcomings.

The Institutional Crisis of Neoliberalism

Neoliberalism, as mentioned above, would appear almost as a matter of principle to be anti-organizational. Indeed, whereas it is easy to conjure up an image of the neoliberal individual, it is much harder to do the same for the neoliberal institution. Nevertheless, it does continue to perpetuate a rich organizational discourse. In fact, neoliberalism has

> been a political project concerned with institutional change on a scale not seen since the immediate aftermath of the Second World War and it is a project that has attempted to transform some of the most basic economic and political settlements of the post war era including labor market accords, industrial relations systems, redistributive tax structures and social welfare programs.
>
> (Campbell and Pederson, 2001: 1)

For this purpose, it reifies ideals of flexibility, dynamism and efficiency. Similar values seem to inform its organizational ethos—championing at its most forward thinking the ability to work flexibly rather than oppressed by institutional regulation. However, this vision would be challenged by chronic issues increasingly asking employees "to do more with less." This would also put into question its broader institutional morality. Neoliberalism would progressively face a veritable crisis regarding its organizational ideals and practice and even its sustainability.

A crucial but commonly overlooked part of neoliberalism is its promise to transform contemporary organizations. Traditional hierarchy was meant to be replaced by more productive and less rigid post-bureaucratic institutions (Barry et al., 1996; Rose, 1999; Cruikshank, 1999; Miller and Rose, 2008). They represented a profound "anti-bureaucratic" evolution from an outdated organizational past, in this regard (Baker, 1992; Bartlett and Ghoshal, 1997; Peters, 1992; Quinn, 1992; Savage, 1996). There is thus a mainstream view in which "post-bureaucracy" is put forth

> as a form of organization that has made a distinct break with the bureaucratic legacy. For the sake of flexibility post-bureaucracy is alleged to emancipate individuals from the formalistic constraints of bureaucracy, arranging them instead in organic and fluid networks.
>
> (Maravelias, 2003: 547)

The new bywords of these institutions are "innovation" and "competitiveness" (Adler, 2001; Handy, 1997). The modern organization was going to break through these hierarchal procedures and regulations to produce a more horizontal, dynamic and therefore more "moral" institution (Kanter, 1990; Savage, 1996).

The aspirations were underpinned by pioneering perspectives of neoliberal management, whose goal was to put market values at the heart of modern institutions. "Neoliberalism, in this sense, has become the name for a set of highly interested public policies," notes Ferguson (2009: 170), "that have vastly enriched the holders of capital, while leading to increasing inequality, insecurity, loss of public services, and a general deterioration of quality of life for the poor and working classes." As expected, this revolves around maximizing efficiency and productivity. Yet it also was founded on doing away with unnecessary institutional regulation. It was a "lean organization" philosophy, preaching the virtue of an institution that could streamline its processes and enhance its structures in order to unleash individual innovation (Basin and Burcher, 2006).

However, for many people, this promise soon turned into an organizational nightmare. Being lean meant trimming unnecessary fat that devastated workers and regulated them even more in the name of being productive and efficient. Under the guise of granting employees greater freedom, it implemented a more sophisticated regime of managerial control (Barley and Kunda, 1992; Carr, 2000; Delbridge, 1995; Jermier et al., 1994; Ogbor, 2001; Willmott, 1993). Also, rather ironically, given its legitimization as a paradigm of efficiency par excellence, due to mounting inefficiencies companies were left to complete basic functions with dramatically lessened resources. Accordingly, "the ability to engage more creative and innovative ways of delivering and managing services rest on exploiting economies of scope. This exploitation rests on establishing a minimum scale of operations and not on post-bureaucracy" (Budd, 2007: 537). Increasingly, basic security and well-being were out of reach for most of the workforce and managers, because, more and more, they were asked to do the impossible—increase performance with significantly decreased resources.

Such pressures were only exacerbated after the 2008 financial crisis. Suddenly, already cash-strapped firms were on the brink of complete and total financial ruin. What once appeared to be smart fiscal sense, now seemed to be the latest cost-cutting measures imposed by senior management that organizations and their employees can ill afford. Public services often faced a similar fate, as there simply weren't enough resources to run them properly. This played on a deeper discourse of neoliberal intuitionalism—one that was opportunistic and inexorably linked to the threat and presence of "crisis":

> Far from providing a functional(ist) resolution to these crisis conditions, early-stage or "roll-back" neoliberalism was very much a creature of the crisis. With significant and enduring consequences, its favored strategies heaped the burden of economic adjustment on the working class, the unwaged, the social state, and (even) on domestic manufacturing capital, while failing to initiate generalized and sustainable economic development. . . . On the face of it, hard choices called for decisive actions from conviction-driven politicians. But the barely concealed *reality* of "actually existing" neoliberal restructuring programs is that they were always pragmatic and opportunistic, that they involved muddling through and experimentation.
>
> (Peck and Tickell, 2007: 30)

Nevertheless, the 2008 crisis increasingly challenged rather than emboldened the quintessential institutional logic of neoliberalism. Efficiency progressively appeared to be a code word for having the bare minimum the absolute maximum. The focus was always on stretching institutional development and finding innovative ways to pick up the slack. Profit was the first and last priority, and after the crisis its prioritization was framed as simply a matter of survival. The idealized image of the post-bureaucratic neoliberal organization faded, leaving in its wake crumbling institutions whose members were left to continually grasp for their own fiscal lives.

The Ethical "Problem" of Neoliberal Institutions

Neoliberal organizations are confronted with unprecedented challenges to their values and practices. In the rosy imaginings of their supporters, new companies were meant to be lean institutions optimizing activity and streamlining the practice. In reality, it is making do with an ever smaller set of resources coping with change-management strategies that place profit over people and even plausibility. The demand that employees achieve unfeasible if not impossible management prerogatives marks modern-day organizational existence. Modern organizational life feels less streamlined and more increasingly squeezed.

The backdrop of neoliberal discourses associated with organizations was the gray and coercive visage of large 20th-century public and private institutions. It is an understandable response to the repressive realities of modern times. The emphasis on rules, roles and regulations created dehumanizing institutional cultures in which individuals were little more than disposable cogs in the machine. It also reflected an organizational ethos that was centered on and obsessed with the exploitation of the natural world and its inhabitants. In the impassioned words of Arundhati Roy (2001),

> the transfer of productive public assets from the state to private companies. Productive assets include natural resources: earth, forest, water, air. These are the assets that the state holds in trust for the people it represents. . . . To snatch these away and sell them as stock to private companies is a process of barbaric dispossession on a scale that has no parallel in history.
>
> (also quoted in Harvey, 2007: 36)

These negative connotations were enhanced by the increasing reliance on scientific management notably associated with Taylor and Ford. They were put forward as a "rational" and evidence-based paradigm for organizing employment relations (Boje and Winsor, 1993; Littler, 1978; Maier, 1970; Waring, 1994). It, moreover, pointed toward the complete "bureaucratization of society" (Höpfl, 2006).

Human relations perspectives arose precisely as a supposed counter to this dehumanizing situation. The treatment of humans as mere interchangeable parts spurred demand to create a more directly and indirectly human-centered approach. The original impetus of the state was to propagate laudable and not immediately obvious market values of democracy, inclusion and emotional well-being, in which

> the person, not the organization, is managing. It consists of all the person's varied experiences in education, training, work in several organizations, changes in occupational field, etc. The protean person's own personal career choices and search for self-fulfillment are the unifying or integrative elements in his or her life. The criterion of success is internal (psychological success) not external.
>
> (Hall, 1976: 201)

This reflects a broad "crisis of bureaucracy in the age of enterprise" (Courpasson and Reed, 2004: 7). In the "post-bureaucratic" era, firms would be "structured to increase flexibility, with less formalization and more decentralization than in the traditional bureaucratic organization" (Contu and Grey, 2003: 935).

Importantly, these high-minded principles reflected and underlined capitalist ideology. This human perspective was adopted in the service of maximizing profits and productivity. Hence, as Courpasson and Clegg (2006: 319) note, "bureaucracy far from being superseded, is rejuvenating, through complex processes of hybridism in which supposedly opposite political structures and principles, the democratic and oligarchic, intermingle and propagate." Notably, it represented an aspiration for a win-win situation for employers and employees. What was good for employees was put forward as also being beneficial to companies. Left unsaid but absolutely critical is that these ideals were legitimate only if they contributed to the bottom line.

At first, the spread of neoliberalism appeared to—if not wholly, then at least largely—reject human resource management. It replaced traditional bureaucracies with even more scientific forms of management. People were consigned to variables in highly complex efficiency models. The human aspect was relegated to minor concerns in the all-consuming desire for greater productivity and profit. Further, it was rooted in a "language of commercialization, privatization, and deregulation and . . . within the language and images of corporate culture, citizenship is portrayed as an utterly privatized affair that produces self-interested individuals" (Giroux, 2002: 425). Subjects' personal development was ignored in the name of sophisticated "objective" strategies to maximize their functionality. In this respect,

> Corporate Culturists commend and legitimise the development of a technology of cultural control that is intended to yoke the power of

> self determination to the realization of corporate values from which employees are encouraged to derive a sense of autonomy and identity.
> (Willmott, 1993: 563)

This efficiency-at-all-costs ethos was not sustainable. Official avenues such as individual democracy and collective bargaining may have been in decline, but this did little to mitigate the organizational problem that resulted from maximizing not just profit but also capitalist exploitation. Materially, issues of overworked employees as well as widespread economic anxiety (Da Costa et al., 2015; Pilkington, 2016). It thus also put these modern organizations in a rather new moral light. Instead of being an improvement on the bureaucracies of old, allowing them to become better service providers, they were seen as creating institutions that prioritized profit over the needs of those they employed and the people they were meant to serve. Privatization transformed from a force for revitalization to a seeming guarantee of destruction.

The mounting popular organizational challenges to neoliberalism revealed a growing systematic critique of its market values and capitalist objectives (Carroll et al., 2010). Anti-globalization movements focused on the new and soulless corporations rapidly taking over the world. Within popular culture, "corporation" began to become a byword for greed and quite literally inhumanity (Eubanks, 2005). It was widely acknowledged that

> [c]orporations rule. No other institution comes close to matching the power that the 500 biggest corporations have amassed over us. The clout of all 535 members of Congress is nothing compared to the individual and collective power of these predatory behemoths that now roam the globe, working their will over all competing interests. The aloof and pampered executives who run today's autocratic and secretive corporate states have effectively become our sovereigns.
> (Hightower, 2002: 1; also quoted in Eubanks, 2005: 180)

Moreover, increasingly "the economic, political and cultural signs of corporate colonization are increasingly abundant and readily visible in the contours of daily life" (Boggs, 2000: 9). The ceaseless drive for profit overshadowed basic human wants, leaving communities devastated and individuals increasingly desperate and hopeless.

In the wake of the financial crisis, there were even greater calls to move away from organizational models built on marketization and financialization. According to Rodrik (2011: xvi–xvii):

> The intellectual consensus that sustains our current model of globalization had already begun to evaporate before the world economy became engulfed in the great financial crisis of 2008. Today the self-assured

> attitude of globalization's cheerleaders has all but disappeared, replaced by doubts, questions and skepticism.

They were renewed aspirations for more sustainable and environmentally friendly egalitarians and even democratic firms. It was increasingly questioned whether the hyper-capitalism of neoliberalism was compatible with these broader social needs and desires. According to an October 2008 UN Global Compact statement, "the financial crisis and the subsequent economic downturn represent a significant upheaval in the evolution of markets and the private sector."

Critical to undermining these emerging systematic critiques was to reframe these concerns as principally moral and ethical issues. The problem was that those leading corporations and the financial sector simply did not act morally. It was imperative not to replace these capitalist institutions but to force them to undergo a dramatic and substantive normative transformation. The problem in this regard was no longer capitalism per se but rather corporate bad behavior. In this spirit, world leaders convened in France with the purpose of creating a "New World, New Capitalism" immediately after the near-global economic crash. There were also a series of G20 summits with the aim of implementing reforms while retaining a firm "commitment" to financial capitalism. This was epitomized in the global leaders' opening statement at the 2008 Washington Summit:

> We recognize that these reforms will only be successful if grounded in a commitment to free market principles, including the rule of law, respect for private property, open trade and investment, competitive markets, and efficient, effectively regulated financial systems. These principles are essential to economic growth and prosperity and have lifted millions out of poverty, and have significantly raised the global standard of living.
>
> (Office of the White House Press Secretary, 2008)

This challenge would have profound political implications for its proposed solutions. The overriding emphasis was on how to ethically fix firms. Suddenly, corporate ethics took on decidedly greater importance. The so-called triple bottom line (Elkington, 1997; Hall, 2011; Normal and McDonald, 2004) that accounted for social, environmental and financial parts of the organization took on increased importance (Savitz, 2012; Willard, 2012). After the crisis, corporations attempted to take the lead on effectively spreading Environmental, Social and Governance Values in their own operations and practices as well as within wider society. To this effect, the UN Global Compact declared, "The global crisis in financial markets raises a range of important and urgent questions regarding corporate sustainability generally . . . For this reason it is more important than ever to build market legitimacy and political support based on sound ethical frameworks."

Tellingly, this concept of ethically fixing organizations operated on three interrelated levels. The first was in fixing companies to better take into account demands for social and environmental justice. The second was the perhaps less obvious, but no less vital, discursive fixing of corporations as permanent parts of the economy and society, entities that could be reformed, certainly, but never fully replaced. Lastly, it located the problem of labor and work squarely in the realm of morality and ethics—marginalizing and co-opting the desire for more systemic solutions to this problem.

Consequently, profit incentive remained prominent and prioritized. However, it was now forced to share the stage with traditional nonmarket ethical goals. The key question was thus "Who should take moral and ethical responsibility for these capitalist organizations?"

Taking Ethical Responsibility for Neoliberal Institutions

Neoliberalism confronts individuals with a rather strange institutional contradiction. On one hand they appear to be the epitome of personal accountability. Each person is supposedly responsible for his or her own economic destiny. On the other hand, neoliberalism appears to ascribe little to no sense of institutional accountability. Instead the organizations and their senior leaders are merely responding to capitalist conditions. Globalization is inevitable, fiscal viability is a necessity, and all else is simply details or idealism. However, as even the most capitalist firms have been forced to respond to the financial crisis and embrace (at least publicly) broader nonmarket values for their survival, the question of who was and should be most responsible for realizing these ideas loomed ever larger. While initially threatening to hold corporations formally and informally responsible for their actions, this quickly transformed into new moral and ethical demands placed on individuals within these institutions. According to Salsman (2012):

> The original and still most fundamental responsibility of corporations is to maximize profit. This all-consuming focus on the bottom line went beyond simple economic rationality. It morphed into the moral duty of organizations. Hence, If, instead, "austerity" means something rational and commonsensical, something that benefits our life instead of ruining it, something that entails not our sacrifice but our *self-interest*, then we've got a truly *moral* meaning. . . . To produce income for oneself and one's family, to plan ahead and budget, to borrow only when it makes sense to do so (but not in order to consume beyond one's current or near-future means), and to save some income, are all acts that are rational, selfish, and thus moral. None of them is an ascetic act of self-sacrifice performed from duty or a hatred for one's life, self, health or wealth. . . . To live this way used to be called "economizing." It was the life of the virtuous—and prosperous.

It may appear to be coldhearted, but shareholders and CEOs were adamant that it was absolutely more than just. In the corporate world of the free market, only the financially strongest could survive. This played into a rhetoric of "economic correctness" that was used to "sell the free market" (Aune, 2002).

This moralization of the overriding profit motive extended far beyond the private sector. It evolved into a universal moral obligation for all institutions. Public-sector institutions were forced to adopt strict fiscal discipline as a matter of principle (see Burchell, 1993; Garland, 1996; O'Malley, 1996). The need to show that they were not wasting taxpayer money superseded providing for the public good. This sense of financial morality even infected NGOs, as charities were forced to prioritize efficiency over the good illustrated to do (Gideon, 1998). Tellingly, the lack of money for the services and charities was rarely a catalyst for questioning the system overall. Rather it was seen as a sign of the organization's overall ethical character.

What is particularly revealing about this discourse did not reflect an explicitly capitalist ethos. For corporations, the need to embrace profit-making was directly linked to the pressures of market competition. Instead it was a more prerogative to be fiscally solvent for its own sake. Consequently, it produced modern organizational subjects

> whose moral quality is based on the fact that they rationally assess the costs and benefits of a certain act as opposed to other alternative acts. As the choice of options for action is, or so the neo-liberal notion of rationality would have it, the expression of free will on the basis of a self-determined decision, consequences of the action are borne by the subject alone, who is also solely responsible for them.
>
> (Lemke, 2001: 201)

It was one made all the more pressing by a seemingly permanent condition of austerity that marked the economy and politics (see Giger and Nelson, 2011; Pierson, 1998). The lack of available funds was thus not a temporary condition but a universal and eternal economic reality all organizations must adjust to.

In this respect, the very notion of organizational accountability revolves around maintaining a sense of moral fiscal responsibility. CEOs and managers were portrayed as having to make difficult choices to save money and, as such, the company (Unger, 1987; Sutton, 2004). Shutdowns, layoffs and cutbacks were framed as necessary evils. Even these were framed as having a moral dimension—as they were hard choices necessary to long-term progress. They might have appeared to be discompassionate but would actually lead ultimately to a better world. The rejection of such measures was seen as failing to face this difficult market reality and being idealistic to an unethical degree. It was an altogether immoral selling of people a false sense of hope.

This dismissal missed the crucial structural condition that corporations cannot so easily ignore. It was ironically more socially acceptable for corporations to care only about profit when there was a viable social safety. However, with the destruction of this public support system, there was a more urgent need to address or at least explain the destructive economic effects of these policies. The need to do so was even greater as the promise of the free market and corporate globalization was increasingly challenged both socially and materially. Thus, if governments were no longer responsible for the public good, who precisely was?

Filling this structural and political gap were discourses of corporate citizenship. Firms were put forward not just as moneymaking enterprises but as having a positive stake in the local, national and global community. Corporations were now citizens with basic obligations toward other fellow citizens that transcended profit motive. This pointed to the theorized and socially expected enhanced moral agency given to the unregulated firms in a "free market" (Carroll, 1999; Castro, 1996; Albrow, 1997; Roberts, 2003; Paul, 2004; Besmer, 2006; Conley and Williams, 2005). It also spoke to the creation of "value-oriented" cultures within firms to better address these issues (Vogel, 2005).

Of course, the reality of the citizenship left much to be desired. Issues of taxation, environment degradation and financial improprieties continue to proliferate and, if anything, seem to have worsened. Nevertheless, there was an expectation of social ethics placed on firms. The very invoking of the term "citizen" implied a previously unseen sense of accountability for their actions and their broader consequences. Specifically, it proffered a notion of greater organizational responsibility for these problems—playing on idea that Shamir (2008: 7) refers to as neo-liberal "responsibilization," a "technique of governance" that is fundamentally premised on the construction of moral agency as the necessary ontological condition for ensuring an entrepreneurial disposition. Neo-liberal responsibilization is unique in that it assumes a moral agency that is congruent with the attributed tendencies of economic rational actors: autonomous, self-determined and self-sustaining subjects.

Yet it also helps discursively fix this problem as one principally of morality—revolving around corporations' struggle to fulfill their fiscal and social duties. More precisely, the retreat of the state has increased pressure on companies to fill this ethical "gap" (Howitt, 2002; Rowe, 2005; Shamir, 2004, 2005, 2008; Sklair, 2002; Winston, 2002; Wright and Rwabizambuga, 2006). In other words, the question of citizenship was increasingly framed not in terms of whether marketization was moral but in terms of how firms and individuals alike could achieve just outcomes within an entrenched free-market system.

Human resource management, as mentioned, was crucial to conceiving and implementing such a romantic vision of a win-win solution. At first glance, this constituted the potential for a serious critique of the neoliberal organization. However, at its core it revolved around the institutional cultivation of a moral self used to cover over its systemic problems. Moral responsibility was promoted as neither a force to reorder institutions nor find innovative ways to literally reconfigure society and economy but as a concerted attempt to find ways to make this permanent demand of being profitable coincide with the corporation's other nonmarket responsibilities to its employees and the wider community (see Preuss et al., 2009; Sharma et al., 2016;

The upshot of this human-oriented neoliberalism, at least within organizations, was a complete transformation of employment relations. For employers, it meant creating strategies that maximized performance of their workforce while respecting their personal and professional needs as well as aspirations. More importantly, it entailed the creation of an organizational regime dedicated in large part to the task of ensuring that employees took responsibility for their own actions. In practice, this translated into renewed and progressively coercive calls for individuals to take personal moral responsibility for their jobs and organizations, starting first and foremost with themselves.

Producing the Institutionally Moral Neoliberal Subject

A crucial concern for the 21st century is how to make organizations more humane. Key in this respect is creating a workplace that cares about more than simply profit and efficiency and is attuned to the personal, professional and even spiritual needs of its workforce. It also entails a common institutional commitment to the public good. Yet within this paradigm, to achieve such goals requires neither a fundamental transformation of capitalist firms nor the prioritization given to financial values within the public sector and NGOs. It is a concerted strategy to create more moral institutional neoliberal subjects.

Crucially, neoliberal institutions are fundamentally framed as moral communities rather than simply as economic organizations. They eschew hierarchy and bureaucracy, at least ostensibly, and promote instead a post-bureaucracy based on an "organic" communitarian system (Heckscher, 1994; Kanter, 1990). It is composed of deeply intertwined emotional relationships between members that are reinforced by a common normative commitment and shared sense of organizational identity (Bartlett and Ghoshal, 1997; Volberda, 1998). Notably, it relies on a strong sense of trust and reciprocity, in which flexibility and autonomy are possible only through a firm faith that oneself and others will do what is necessary for the institution's success. Tellingly, this more "free" culture requires management to take a leading role in helping individuals realize and embody this proper moral attitude (MacGregor and Burns, 1978; Bass, 1985; Bartlett and Ghoshal, 1997).

The lean performance–oriented corporation of the early millennium thus increasingly gave way to a more human-centric organization. The bywords of "efficiency maximization" and "profit" suddenly sat alongside such ideals as freedom, flexibility and caring. The goal for employees was to establish a proper work-life balance—an ambition that demanded the support of their employers (Burack, 1999; Byrne, 2005; Der Klerk, 2005; Mattson et al., 2000). Workers desired more from their jobs than just a salary. To this effect, they longed to "Stop Living Your Job and Start Living Your Life"—their aspirations went "much deeper than simply choosing a satisfying career" and included adding meaning to their lives (Malloy, 2005: 24). They hoped for a sense of purpose and even spiritual fulfillment. For the growing number of individuals subjected to precarious employment, there was a substantial desire for an organization that cared as much about their well-being as the bottom line, an organization in which they could "develop their complete self at work" (Mitroff and Denton, 1999).

This desire was recognized in a range of new organizational policies. The aforementioned work-life balance produced such innovations as on-site childcare and flexible working (Kossek and Friede, 2006; Lewis et al., 2003; Lewis et al., 2007). Specifically, it subverts "basic assumptions about the ideal worker who is "unencumbered' by family or other non-work commitments" (Kossek et al., 2010: 9; also see Kanter, 1977; Rapoport et al., 2002) which aims to "enhance organizational structural and cultural/relational support for work, family and personal life" (Kossek et al., 2010: 4). These trends were enhanced by the introduction of new technology that allowed individuals to work outside the traditional office setting (Leclercq-Vandelannoitte, 2011). Consequently, employees had greater ability to work anytime and anywhere (Sennett, 1998).

This new human-centered culture led to often perverse results. Its actual consequences were less time and freedom, and then they were greater anxiety and work intensification. Individuals often faced increased time pressures in "flexible" firms (Van Echtelt et al., 2006; van der Lippe, 2007). In this vein, the promise of greater flexibility is currently being challenged as a modern "empowerment myth" (Boreham, 1992; Harley, 1999) masking the continued prevalence of quite regulative workplace regimes. Self or mutual "concertive" types of control (Hales, 2000) are now replacing coercive control. This has led to new questions regarding autonomy and flexibility for contemporary employees. Hutton (2003), for instance, maintains that the contemporary work subject is confronted with a profound "time squeeze," leading "more and more people" to ask "fundamental questions about whether they have struck the right balance between the demands of their job with other no less valid, important or simply pleasurable claims on their time."

Individuals, moreover, were subjected to a seemingly all-pervasive form of electronic surveillance. Within post-bureaucratic companies, Frenkel et al. (1995) talk of an "info-normative" control strategy, in which the promotion of autonomy is associated with managerial desires for increased performance standards. Digital advances in communication and surveillance technologies mean that many employees no longer have to be in the office to do their job effectively (Green, 1999; Bain and Taylor, 2000). In the face of such physical distance, managers rely on employees' "inner panopticon" for disciplining "workers who are removed from the immediate sphere of influence of management and co-workers" (Jackson et al. 2006).

It was not only that employees *could* work anytime and anywhere, but rather that they were *expected* to. Organizations became borderless, because the separation between work and home was rapidly dissipating. The firm now extends outside established boundaries of the workplace (Kunda, 1992; Barker, 1993; Casey, 1995; Perlow, 1998). Indeed, in post-Fordist organizations, individuals are often expected to spend more time at the office or doing their job away from the workplace, so that "work becomes home and home becomes work" (Hochschild, 1997). In this shifting capitalist environment, individuals are "simultaneously always a producer, always a family member and always a consumer" (Fleming and Spicer, 2004: 90).

The quintessential ethos of the so-called compassionate firm was one of personal responsibility. Specifically, it played on the notion of the responsible corporation. The aim, at least in theory, was to create an enhanced form of accountability for companies and their members. It was how to ensure that faceless organizations, preoccupied with profits and abstract models of efficiency, could be forced to take responsibility for the consequences of their actions. There was a radical impulse behind such desires, one aimed at making CEOs answerable for corporate scandals. Perhaps a key part of this discourse was the demand that companies and governments not outsource the responsibility for the devastating consequences of free-market globalization by claiming they were merely inevitable and unstoppable. Within the firms themselves, it was an attempt to feel as if those in charge cared about the fate and well-being of those who worked for them.

Underpinning this challenge to corporate power was a potentially dangerous governing logic. It created a precedent for demanding moral responsibility of those within organizations. This originally upward-looking regime of accountability was soon applied not just to those at the top but to all employees. Promised is a new organization where work-life balance can be realized through time arrangements that help improve employees' quality of life in accordance with their changing needs and preferences, while at the same time being reconcilable with employers' increasing needs for flexibility and with collective interests of economic sustainability (European Foundation for the Improvement of Living and Working Conditions, 2003).

At the most basic level, the attention shifted to the need for individuals to act ethically in their everyday actions, so that the increased pressure placed on organizations to fulfill social goals (Davies, 2003; Logsdon and Wood, 2002) is now effectively placed onto employees themselves. In this respect, "effective corporate social responsibility policies are a requirement for today's companies. Policies have not only to be formulated, they also have to be delivered by corporate employees" (Collier and Esteban, 2007: 19).

Corporate mission statements highlighting ethical codes and stated moral values put the onus on workers themselves to live up to these ideals. This moral conformity can range from the small-scale of complying with the official recycling plans to the large-scale of doing one's part to create a non-discriminatory workplace. In this spirit, individuals must constantly take on and perform the role of moral organizational subjects.

Human resource managers play a critical role for this moral discipline. Their function is to a large extent to monitor employees to ensure compliance with these normative standards (Sharma et al., 2009). This reflects the broader role of human resource practitioners and practices within contemporary strategies of managerial control (e.g., Barratt, 2002; Casey, 1999; Fleming and Spicer, 2003, 2004). The goal is to use continual processes of evaluation and appraisal in order to discipline the workforce to meet organizational prerogatives and norms while further crafting them as "appropriate individuals" conforming to institutional values (Alvesso and Willmott, 2002; Covaleski et al., 1998; Maravelias, 2009; Ogbor, 2001; Willmott, 1993; Ogbor, 2001). This has extended into shaping individuals' sense of "corporate social responsibility" (Preus et al., 2009) as a matter of "co-creation" between CSR and human resource management (Jamali et al., 2008). Thus, the compliance officer nominally empowered to investigate wrongdoings at all levels focuses unsurprisingly on the behavior of staff members, not senior management. The importance of human resource management signaled a pervasive regulatory ethos in which individuals were the primary targets of this moralizing campaign as well as the creation of an empowered regulatory regime to ensure their conformity.

Tellingly, such values evolved into essential criteria upon which to judge, categorize and reward the broader workforce. People were hired and fired on the basis of whether they simply displayed a collaborative and cooperative attitude to interview panels. Once they were employed, their willingness to help others and "go the extra mile" was a vital part of their job prospects. This was further used and linked to a broader attempt to increase employee organizational commitment to their workplace (Backhaus et al., 2002 Brammer et al., 2007; Turban and Greening, 1997). More explicitly, there was a desire and increasing demand for employees to bring their ethics into the workplace. This mentality was also witnessed in the drafting of individuals into voluntary community-service activities and programs linked to their jobs.

Solving the Ethical Problems of Neoliberal Institutions

A quintessential, though perhaps rather ironic, component of neoliberalism is its role in the social production of moral organizational subjects. This institutional phenomenon echoed and in many ways mirrored that of the similarly constructed fiscally moral political one. However, as shown, this individual discourse of moralization went even further in the case of contemporary institutions. Notably, within the political sphere, financial responsibility was made into a morally righteous position. Likewise, even those working within public organizations and NGOs faced a moral imperative to be "fiscally responsible." Significantly, there was also an increasing demand that those within organizations of all types use their creativity to resolve the ethical institutional issues directly and indirectly associated with neoliberal restructuring.

A critical part of most modern organizations is the need to be "visible" in order to prove one's commitment. Consequently,

> human resource management (HRM) techniques established among employees a desire to be recognised as a trustworthy member, on the one hand, and a constant fear of being unseen, on the other. This drove employees to continuously take initiatives that placed them in a self-regulating limelight.
>
> (Maravelias, 2009: 349)

This reflected a broader "problem-solving" culture endemic to neoliberal institutional life. Unlike bureaucracies, with their stifling hierarchies, post-bureaucracies encourage continual innovation that allow their members to creatively solve any issues that they confront (see Maravelias, 2009). As such, it grants them the resources and freedom necessary to cope with the unpredictable and dynamic world of global capitalism (Andler, 2001; Handy, 1997). In this spirit, there was an increasing call—implicitly and explicitly—for employees to be "visible" ethical problem-solvers within their organizations and beyond.

Particularly significant in this regard, thus, was the transformation of such moral organizational responsibility into a broad and active organizational ethos. As mentioned, the greater freedom granted to employees linked to flexible working and work-life-balance policies was coupled with a high ethical trust culture. Such notions, however, have evolved from ensuring that individuals use their time efficiently to encouraging and even mandating that they deploy it ethically (Cederstrom, 2011). The conventional "visibility-vulnerability spiral" now centered on successfully establishing a strong work-life balance (Bloom, 2016c). Cutting-edge "human" workplaces like "Googleplex" expected individuals to exploit their gym facilities or simply socialize with other employees on the corporate campus. Modern employees were, furthermore, encouraged to contribute to the company's ethical brand through using their time off to do good and publicize it (Land, 2010).

The ethos at play was one that mixed organizational commitment to employee well-being with a commitment by employees to not let these non-market desires interfere with their capitalist duty to optimize their performance. Individuals had become ethical innovators—personally figuring out ways to balance professional obligations with personal convictions. They were tasked with being "super navigators"—crafting a unique path for the realization of their own work-life balance. This entails disciplining individuals to take control of their own schedules to simultaneously meet their personal and professional needs. Flexible workers are charged with reflexively organizing their time so that each person can successfully achieve what is "best for me" (Bruckner and Mayer, 2005). Working life, in this respect, is expected to be "flexibly" brought into line with one's own non-employment goals (Giddens, 1991). Yet this was primarily not a matter of self-interest but one of pursuing a more ethical life. This was witnessed in growing ideas of "life navigation," understood as "an innovative work culture concept, which builds on a company's need for increased productivity and the human being's need for life quality" (Super Navigators).

Cultivated accordingly was a self-directed neoliberal organizational subject. This played into the broader promotion of self-directing workers who could take responsibility to manage themselves. It emphasized an ethos of "self-management" (Luthan and Davis, 1979; Manz and Simms, 1980; Mills, 1983) and "self-leadership" in the name of achieving ethical goals. Reflecting monogamous logic, employees were increasingly asked to use and flexibility to achieve a healthy professional life. This highlights the evolution of institutional regulation to encompass the "whole person." The social construction of capitalist identity has expanded in scope, in this respect, to shape the "complete person" (Mirvis, 1994), thus extending to an individual's working and non-working life. This includes the moral disciplining of the work subject both inside and outside the boundaries of the firm. Termed "bio-morality," these 24/7 companies help employees be personally and professionally "healthy" so that they can lead an overall "ethical" life (Cederstrom, 2011).

This reveals a specific but rather underexplored form of the contemporary organizational governance revolving around ethical compliance. In particular, there was a fresh demand on individuals to creatively solve the normative issues affecting contemporary institutions. They were increasingly solely responsible for making ethics work in the workplace. The recognition that each employee has unique needs and personal pressures was transformed into an individualized imperative to find ways to make one's organization more moral.

This problem-solving ethos served ultimately to reinforce neoliberalism. To this effect, Platman (2004: 594) refers to the current "paradox of freedom" whereby the release from organizational restraint was, itself, restraining since individuals were now responsible for their own welfare. "On the one hand there was a 'positive' pole of self- expression, independence and control; on the other, a 'negative' pole of fragmentation and uncertainty." This also points to an ethical paradox of neoliberal institutions—the freedom offered by flexibility was a freedom to innovatively "fix" capitalist organizations ethically. Accordingly, the focus was placed firmly on how to ensure as much as possible a win-win situation between the nonnegotiable demands for capital on one hand and the nonmarket human desire for ethics on the other. This resonated with the results-only strategy increasingly adopted by many organizations. There was a general sense of combining total freedom with total responsibility. Individuals were liberated to discover ethical solutions that did not challenge and even enhanced the organization's fiscal bottom line (Hawkins, 2006; Van Marrewij, 2003; Visser, 2014; Vogel, 2005).

Such ethical governance also shaped more collective organizational relations. A key feature of neoliberalism is an urgent and supposedly necessary institutional restructuring to meet difficult fiscal realities. These change strategies presented employees with a hard ethical quandary. In the face of such mandated transformations, individuals and teams were forced to take on the contradictory task of providing the same quality and quantity of service with considerably less resources for doing so (Gilley and Rasheed, 2000; Janssen and Estevez, 2013; Winslow, 2016). This also applies to ethical concerns, as workers were charged not just with doing what was necessary for economic survival but also with figuring out how to stretch these more limited funds to compassionately care for their colleagues and the stakeholders they were charged with serving.

It is precisely this quality that influences and catalyzes the rising ethics of care that is ironically at the heart of many neoliberal organizations (Groenhout, 2004; Simola, 2010). This is built on managerial discourses that attempt to incentivize employees to "help" one another when necessary to complete their job (Baker, 1992; Drago and Turnbull, 1988; Holmstrom and Milgrom, 1991; Itoh, 1992; Lazear, 1989; Ramakrishnan and Thakor, 1991). Here, the caring firm was redirected downward to the workforce itself. Employees were explicitly and implicitly expected and encouraged to take care of each other and customers regardless of the obstacles created by neoliberalism. In particular—and this is absolutely crucial—individuals and teams were tasked with solving ethical problems experienced by neoliberal institutions. They were now required to "go the extra mile" and be the ultimate ethical "corporate citizens" (see Bateman and Organ, 1983; Bolino and Turnley, 2003). It was their responsibility to creatively and innovatively address these organizations' systemic moral shortcomings.

Employees were thus progressively tasked with taking more responsibility for the organization. The public call to create corporate citizenship turned into a disciplining discourse producing moral employee citizens. In a quite real sense, this represented the effort to recast systemic problems of capitalist exploitation and discrimination into an issue of individual immorality. As such, there was an underlying expectation that it was the duty of employees themselves to moralize neoliberalism and its institutions.

The Institutional Power of Neoliberal Ethics

The production of moral and ethical organizational neoliberal subjects had definite power implications. They represent discourses for normatively disciplining institutional subjects in such a way to strengthen existing power relations and hyper-capitalist ideologies of financial position and marketization. Institutions morally regulate and ethically empower subjects to cope with the human issues generated by neoliberalism. They were charged, as such, with caring for those within them and those that they serve, despite unavoidable fiscal pressures.

As was the case with politics, this involved the trickling down of the moral and ethical responsibility for neoliberalism from those at the top to the workforce more generally. To this end, it is not about ensuring that corporations are better citizens but rather about ensuring that their members are more morally and ethically committed. This shifted the focus from leaders, thus neutralizing the radical element of this discourse of individual accountability by concentrating most strongly on those at the bottom. These insights raise immediate questions regarding the scope and extent of organizational control. Maravelias (2003: 560), for his part, challenges ideas of "total control," arguing that such complete regulation is "exaggerated and implausible." However, it is precisely the incompleteness of neoliberalism that creates its totalistic and often seemingly paradoxical managerial desires. The fact that current firms are expected to be ethical in a hyper-competitive financial environment that makes doing so almost impossible creates an even stronger incentive to place this demand and responsibility on less powerful employees.

This ethos continues, furthermore, an imperative component of neoliberal organizational culture. It not only encouraged and mandated this behavior but also helped teach it. This was witnessed in the proliferation of teamwork seminars and ethical training and the use of institutional sensitivity training to educate members on the values of respect and care (Paluck, 2006; Holladay and Quinones, 2005). Although such principles are undeniably important, they can be turned into an ideological apparatus that institutionalizes this mounting ethical duty of individuals toward the workplace. Significantly, nonmarket ideals of compassion and care were discursively framed and concretely deployed in such a way as to reinforce rather than challenge the hyper-capitalist status quo.

Just as critical was how this neoliberal ethics expanded the scope of managerial power. Far from putting checks and balances on the actions of senior management, the strive to moralize institutional regimes became another set of criteria to monitor and, if necessary, punish employees. Failures on the part of the workforce would be used against them—for example, to prevent their promotion. This placed the onus for realizing such squarely on the shoulders of employees, not employers. If a company fails to meet its ethical commitments, it is the fault not of the company but of its workforce, who should be reprimanded accordingly. Employees are normatively disciplined to use their power of autonomy properly in compliance with established managerial prerogatives of productivity, efficiency and, increasingly, ethics. Freedom, thus, is not a right but an "opportunity" and a privilege that should be treated respectfully (Maravelias, 2007: 568). The failure to do so leaves employees open to informal condemnation and formal disciplining.

To this effect, there is clearly an oppressive component to such moral managing. It is an attempt to coerce employees into being ethical—more precisely, an effort to guide them into becoming responsible corporate citizens. This means doing whatever is necessary to make organizations more ethical. If this entails working longer hours, covering for an overworked colleague or finding ways to watch your children while also completing a job task, then so be it. One must be prepared to do anything and everything for this higher moral purpose.

Such ethical managerial prerogatives could also be empowering, though in a quite limited and restricted ideological way. The demand to be ethical is an opportunity for cultivating new capabilities. It entails learning how to multitask more effectively. It is to be again a "super navigator" in order to find creative ways to balance profit with care. The previously referred to "empowerment era" organization also extends to ethics—as firms empower individuals both formally and informally by supporting them in being good and productive. This dual imperative to be at once ethical and profitable was likewise expansive rather than restricted in character. There was an increased demand for existing and potential employees to contribute to the company's ethical brand.

Key in this regard was a strategy of using employee innovation to solve the neoliberal problem of institutional ethics. It socially produced the ethical problem-solving organizational subject. The goal was the always-unfinished task of combining the higher ideal of citizenship with the always-pressing bottom line. To do so required the creation of an ethically empowered and ever more morally managed capitalist workforce.

6 The Personal Power of Neoliberal Ethics

Neoliberalism is, above all, the reification of the individual. Beneath its technocratic models lies a free-market dogma that symbolizes an earnest and deep belief in freedom from collective tyranny. Neoliberalism vehemently scoffs at ideologies that proclaim the state as a force for common good. It historically points out that the road to hell is paved by good intentions. In contrast, it revels in the capacity of individuals to do great things once liberated from unnecessary collective bondage. To paraphrase Marx for a hyper-capitalist age, "individuals have nothing to lose but their collectivist chains."

The ethics and the celebration of individual rights within neoliberalism may be straightforward. It's a total embrace of the market: a veritable love affair with buying, selling, inventing, and capitalizing on opportunities. Once emancipated, individuals are free to pursue profit to their heart's content and to work to defeat any and all competitors. In the famous words of Margaret Thatcher:

> there's no such thing as society. There are individual men and women and there are families. And no government can do anything except through people, and people must look after themselves first. It is our duty to look after ourselves and then, also, to look after our neighbours.

However, the reality of individual existence within neoliberalism is dramatically different. To be sure, there's an almost overwhelming imperative to make money and optimize one's economic prospects. Yet there is an equally growing and seemingly unquenchable longing for personal ethical and spiritual fulfillment. This visage of the ultra-successful capitalist, for whom life is dedicated solely to the pursuit of money, is held up as a warning, not a model. In its place are renewed aspirations of personal well-being and a just society. The golden calf of neoliberal freedom, it would seem, leaves most people much to desire. Hence, "Not only is success in the world empty when it is purchased at the cost of spiritual fulfillment but spirituality makes good business sense" (Dean, 2009: 60).

At first glance, it is easy to mistake these conflicting aspirations as an understandable response to the close to pathological focus of neoliberal society and economics on profit and self-interest. Undoubtedly, this feeling of lack plays a role. Where is there room for family adventure, spiritual enlightenment or even just relaxation in the non-stop world of 24/7 capitalism? Yet it also stands as an increasingly pressing and damning ethical problem for neoliberalism. It highlighted its incompleteness and insufficiency for meeting the full needs of individuals. Its promise of individual liberty par excellence was revealed as morally shallow and ethically empty. It spoke little to concrete efforts and possibilities of living a good life.

However, as was the case with politics and institutions, neoliberalism has effectively turned this individual ethical problem on its head and to its advantage. The personal crisis of the neoliberal subject evolved into a means of distracting from the deeper systemic issues afflicting the contemporary hyper-capitalist order. The moral and ethical critiques levied at neoliberalism transformed into moral and ethical demands placed on individuals themselves. It was a force for personally disciplining the modern individual to not only save themselves but also capitalism as a whole.

The Existential Crisis of Neoliberalism

Since its inception, neoliberalism has championed and embraced a market ethics for individuals, an ethics that has elevated capitalist values such as entrepreneurship, competition and profit maximization to a higher moral good. At first blush, this would appear to eschew conventional ethics of cooperation, caring and community. It reified individual self-reliance and a free-market economy—presenting it not as a path to a "nasty, brutish and short life" as posed by Hobbes but as one of personal liberty and, through indirect interventions of the "invisible hand," common prosperity. Although this vision of the moral life may suffice in theory, in practice it was increasingly found to be profoundly incomplete.

Tellingly, capitalism as a dominant economic system has always fostered a culture that carefully balances individualism and community. It relied heavily on the family and an emerging civic community. It concretely combined myths of personal heroism with a quite regulated form of collective existence—culturally exalting figures from the "soldier-hero" (Dawson, 2013) to the all-conquering business executives (Bloom and Rhodes, 2016). This reflects a deeper sociology of celebrity that is embedded within the very DNA of capitalism itself. This can be traced back, at least theoretically, to Weber's initial ideas of class, status and party (1966), along with his focus on personal charisma as a basis for modern power (see Ferris, 2007). This culture of capitalist idolization ranges from the modern celebration of entertainers and athletes as potential "role models" (Lines, 2001; Kellner, 2001; Fraser and Brown, 2002) to the desire for strong leaders to be figureheads for social and political movements (Meyer and Gamson, 1995). It furthermore, represented a contemporary capitalist "pathology" (Lowenthal, 1961; Monaco, 1978) and commodification (Cashmore, 2006; Dyer, 1979; Marshall, 1997 of the individual.

Critically, it additionally provided capitalist subjects with a range of nonmarket values to invest in and care for. A well-established trope of market society is the willingness and desire of subjects to surrender to the pleasurable escapism of mass entertainment—allowing them, in turn, to be more easily manipulated and regulated by elites (King, 1992). The love of god, country and family acted as potent soporifics to ease the often harsh ethical and material realities of the market society. Hence, patriotism and the emergence of nationalism was not simply a rational justification for the spread of capitalism and empire; it was also an emotionally resonant "imagined community" (Anderson, 1991):

> What the eye is to the lover—the particular, ordinary eye he or she is born with—language—whatever language history has made his or her mother-tongue—is to the patriot. Through that language, encountered at the mother's knee and parted with only at the grave, pasts are restored, fellowships are imagined, and futures dreamed.
>
> (Anderson, 1991: 154)

Similarly, even the most intimated human relations—those revolving around romantic love—became a cultural support for capitalism (see especially Illouz, 1997).

Under neoliberals, these commitments to such nonmarket values ironically seemed to intensify. Despite their rhetoric of revolution, the rise of Reagan and Thatcher was propelled by a fresh sense of patriotic revival. It was a new morning for America, where it was all right to believe in traditional ideals of God and country that had been so supposedly debased by the cultural liberalization permeating the postwar era (see Himmelstein and McRae, 1984; Lugg, 1996). This played out into a larger morality tale between a "free" West struggling against the tyranny of Communism. In this spirit, "Free enterprise, free elections and the personal freedoms of movement, speech, religion, and political participation came to epitomize the West's cherished values . . . Anticommunism united the nations of Western Europe and the free world" (Appleby, 2010: 304). Not surprisingly, this also appealed to the imagined community of the nation that was also present in the right-wing US-backed neoliberal dictatorships spreading across the world, particularly in Latin America (see Ramsdell, 2004; Stern, 2009).

It was, furthermore, witnessed in a less reactionary and more liberal manner with the emergence of the "third way" in the late 20th century (Giddens, 2013). Here, social justice, civil rights and political care for others was tied into an idea of a more ethical form of marketization. The neoliberalism of Clinton, Blair and others gave way to a pronounced neoconservatism (Brown, 2006; Dorrien, 2013; Mearsheimer, 2005). This shift represented an even deeper ideological commitment to protecting the free market by any and all means necessary, both at home and abroad. President George W. Bush thus proclaimed, speaking of the "War on Terror," that "this struggle has been called a clash of civilizations. In truth, it is a struggle for civilization . . . free markets and free trade and free societies are proving their power to lift lives" (Bush, 2002).

It reflected a shared desire for something to believe in in the midst of the late-20th-century and early-21st-century corporatization. As Aune (2002: 168) astutely notes:

> No human being can live for long live solely as *Homo economicus*, so all free-marketers end up with elements of irrationality in their systems: disciples of Mises become radical neo-confederates, Randians become love junkies, and libertarians become racists or gun fanatics.

The result was the resurgence of fundamentalism of global proportions between the market faithful and religious zealots, with the expected deadly consequences rather famously prophesied by Barber (1992: 53) in his idea of "Jihad vs. McWorld":

> The tendencies of what I am here calling the forces of Jihad and the forces of McWorld operate with equal strength in opposite directions, the one driven by parochial hatreds, the other by universalizing markets, the one re-creating ancient subnational and ethnic borders from within, the other making national borders porous from without.

In the wake of the disastrous War on Terror and near-global economic collapse, neoliberalism suffered a dramatic personal crisis for individuals. Its once-vaunted discourse of progress and patriotism ran progressively hollow. Again, on a more individual level, the triumph of freedom of the market was displaced by feelings of economic insecurity and political alienation. One could not buy personal fulfillment nor substitute real community bonds with empty and romantic platitudes of the free market.

The Ethical Problem of Neoliberal Individuals

The aftermath of the financial crisis brought to the fore deep concerns over neoliberalism's impact on individuals. The trumpeted freedom of the market gave way to general feelings of economic insecurity and anxiety. Work was the cause for no longer happiness but burnout. Even those privileged fortunate and hardworking enough to be successful often found such victories to be hollow—coming at a high personal cost. Neoliberalism is confronted, then, with a crisis that is as much economic as it is existential.

The immediate and most obvious critique emerging from these fundamental challenges to neoliberalism was that it simply was unable to provide individuals with a sense of the good life they longed for. This was reflected, for instance, in the growing promotion of well-being by human resource management professionals and thinking that could literally be described as a new "technology of happiness" (Hamburg Coplan, 2009). This can be considered a new stage of the past capitalist effort to create what C. Wright Mills (1959: 176) referred to as bureaucratically regulated "cheerful robots." Yet this was undermined by a "joyless economy" that seemingly prioritized economic rationality and desires over all other human wants and needs (Scitovsky, 1976).

The pursuit of happiness would, nonetheless, increasingly take center stage within neoliberalism. Traditionally, market values of enterprise and productivity were explicitly linked to broader attempts to achieve personal well-being. According to Binkley (2011: 382):

> No longer an instrument for the strengthening of the state, happiness is today more than ever tied to economic freedom and the inclination to act in one's own self-interest. The logics composed through contemporary technologies of happiness and practiced by individuals in their own self-government are specifically centred on the production—or more accurately through the inducement to self-production—of a distinct form of enterprise.

Reflected was a contemporary "logic of happiness" that reinforced not only market principles but also prevailing power relations. All unhappiness and dissatisfaction was reduced to a lack of a positive attitude. Accordingly, "The real conservatism of positive psychology lies in its attachment to the status quo, with all its inequalities and abuses of power" (Ehrenreich, 2009: 170). Happiness could easily be found in this hyper-capitalist order, if only people worked hard enough to personally achieve it. Deeper questions as to the ethics or justness of this system were marginalized as either unimportant distractions or, more ominously, evidence of a negative mentality that could be psychologically detrimental to one and those around one. Hence, "when happiness is assumed to be a self-evident good then it becomes evidence of the good" (Ahmed, 2010: 13).

Despite these growing demands to "be happy," such contentment remained elusive for many. It revealed a mainstream discourse that was clearly out of touch with the dispiriting effects of people's everyday economic and social struggles. It meant being subject to a capitalism but "never sleeps" and was therefore "always open" and inescapable. "24/7 markets and a global infrastructure for continuous work and consumption have been in place for some time," notes Crary (2013: 3–4), "but now a human subject is in the making to coincide with these more intensely." This speaks to an emerging 24/7 capitalist reality in which corporations never stop, businesses are forever open and workers are eternally "on call" to the needs of management—one in which "it's all work and all play, all the time" (Shelton, 2010), and, even more troubling, employees are "always still at work" (Brocklehurst, 2001). Central to this shift is the ways increased "flexibility" makes it harder for individuals to separate their work from their life (Cohen, 2010). At a deeper level, there is a rising expectation that individuals are required as part of their jobs to work "Anytime, Anywhere"

It was deemed close to impossible in such a context to truly find genuine well-being and fulfillment. Market freedom seemed to dissolve into a mocking mirage of overwork, economic insecurity and personal dissatisfaction. Underneath the supposed freedom of always "being your own boss" was the reality of fiscal precarity mixed with never having enough time for yourself. To this effect, companies reserved the right to invade and organize employees' identity, time and actions inside and outside the job site (van der Lippe, 2007). As a result, many "flexible" workers complain that their newfound flexibility has led them to be "always on call" to their employer (Fraser and Gold, 2001: 692).

Present were authentic and rising concerns about whether it was even feasible to live a truly moral life in such a free-market world. In an age of zero hour contracts, slowly decaying postindustrial communities and massive personal debt, how could anyone really do any more than to sensibly survive as best they could? Quoting Agger at length, in this respect:

> There is a telling dialectic at play here: as people enjoy shorter working hours they must be encouraged to use those hours in a way that benefits capital. Hence we watch television and surf the internet, bombarded by advertising images of the "goods" life. Yet the intensification of time robbery and time administration robs people of the time and space in which to recreate themselves, reproducing their selfhood and identity in ways that allow them to function during public time.
>
> (Agger, 2007: 222)

Revealed, moreover, was a prevailing feeling that neoliberalism confined individuals to a fate of perpetual anxiety and dissatisfaction.

This bleak outlook was added to by growing cultural critiques. The new age was characterized by disconnected virtual interpersonal interactions and consumerism. The explosion of social media was perceived as the death knell of genuine human contact and with it a culture of compassion. This spreading sense of human disconnect was captured in the lament that

> [t]echnology has created tremendous opportunities for the world to shrink through communication, collaboration, and cloud-based productivity tools. But it has created immense opportunities for wasting time. On Facebook. On Twitter. On stupid games. Human beings are losing their ability to communicate in person. To smile at each other. To converse. To enjoy a meal together without looking at their smartphones. To look into each other's eyes. To touch. To honor food that someone

> else has cooked with love and care. To be present in the moment without interruption. This is a tremendous loss that cannot be quantified.
>
> (Mitra, 2015)

Further, the daily news reported on cities afflicted by drug use and violence, mass shootings and riots. The list of modern ills appears almost endless, with little truly positive news to offset it. Traditional social bonds were disintegrating rapidly, with nothing sustainable to replace them.

Tellingly, these cultural fears translated into a renewed search for authenticity. It was a common longing to return to something more real to get back in touch with our humanity. This could take the form of those who literally sought to reconnect and rediscover a time that in the romantic haze of hindsight appeared less artificial (Gilson, 2016; Tudor, 2012). Such critical desires were borne out of a feeling that community was being lost to the tide of modern progress. This is also witnessed in the resurgent—and increasingly commodified—search for spiritual fulfillment (Lau, 2000; Redden, 2016). Consequently,

> [f]or many people, spirituality has replaced religion as old allegiances and social identities are transformed by modernity. However, in a context of individualism and erosion of traditional community allegiances, "spirituality" has become a new cultural addiction and a claimed panacea for the angst of modern living. Spirituality is celebrated by those who are disillusioned by traditional institutional religions and seen as a force for wholeness, healing and inner transformation.
>
> (Carrette and King, 2005: 1)

It also led individuals in many communities that felt left behind by the mainstream to embrace their traditional culture and seek at all costs to preserve it. In the United States, the rise of the Tea Party, therefore, spoke to desires to restore "national values" (Skocpol and Williamson, 2012). According to one of its leading spokesmen, "Conservatives are trying to conserve America's liberating values . . . the Tea Party unites around three values: limited government and the rule of law, free market capitalism and personal responsibility" (quoted in Parker and Barretto, 2014: 1). More generally, witnessed by many was the rapid demise of an economy in which hard work and education were a sure ticket to economic security and domestic stability. Instead they were rewarded with oppressive debt and chronic underemployment. The once envied and ridiculed middle-class suburban lifestyle was receding from history for all but a rich few.

These feelings were increasingly laid at the feet of a society that worshipped financial success above all else. It rallied against a culture that appeared to lack any fundamental sense of morality and decency. Everything truly good—community well-being—was sacrificed for the lucrative but hollow pursuit of personal profit. Graham Music (2014) referred, accordingly, to present-day consumerism's creation of "our divided brain," in which individuals are torn between care for others and self-interested capitalist wants. This notion builds on work by scholars like Kasser (2003) who note how the present market society creates extrinsic desires for consumption, individual economic success and social status that supersede intrinsic desires for community, family and the well-being of others. Neoliberalism, therefore, was creating a world where the non-compassionate and even psychopathic person came out on top while those who refuse to embrace such immorality seemed destined to lose. Needed instead was an emphasis on "sustainability for wellbeing" (O'Riordan, 2013).

This shifting discourse following the financial crisis revealed this potentially ethical turn. Additionally, the blame was placed on financial barons and the need for the very few to pay for what they had done and their fair share. As time progressed, the lack of widespread recovery exposed an increasingly ethically bankrupt modern system. It was attributed to a lack of social cohesion and polarization caused by such factors as social media and the failure of a common moral purpose. It was a critical updating, in a certain sense, of Reinhold Niebuhr's idea of the "moral man and immoral society":

> Individual men may be moral in the sense that they are able to consider interests other than their own in determining problems of conduct, and are capable, on occasion, of preferring the advantages of others to their own. They are endowed by nature with a measure of sympathy and consideration for their kind, the breadth of which may be extended by an astute social pedagogy. . . . But all these achievements are more difficult, if not impossible, for human societies and social groups. In every human group there is less reason to guide and to check impulse, less capacity for self-transcendence, less ability to comprehend the needs of others and therefore more unrestrained egoism than the individuals, who compose the group, reveal in their personal relationships.
>
> (Niebuhr et al., 1963: 1)

Similarly, it was increasingly believed that while people on their own may be moral or compassionate, in neoliberal society this impulse was too often stifled or, even worse, eliminated.

There is thus a clear and present ethical problem with neoliberalism. Despite its ostensible pursuit of liberty, it produced a seemingly selfish and demoralized population who could find neither happiness nor justice in an increasingly oppressive and expanding "free market" reality.

Taking Personal Responsibility for Neoliberal Ethics

The new millennium was ushered in with a seemingly clear sense of moral purpose. There was a widespread embrace of liberal democracy and free-market economics. On a personal level, it meant the move to create a simultaneously more tolerant and more entrepreneurial society. Less than three decades on, it is clear that we are in an age marked by enhanced militarism, economic inequality and rising extremism. Personally, it is not at all clear how to ethically respond to a morally corrupt status quo and a society committed to profit above all other human considerations. The main question facing the free market was the evolution of how to make it more responsible to who should take moral and ethical responsibility for it. Specifically, there was a widespread and quite universally spread blaming of individuals for this system's moral failures.

As hinted at in the above section, there was a clear generational divide, in this respect. There was a pervasive sentiment that young people, the so-called millennials—were the embodiment of all that was ethically wrong with the modern world. They were viewed as entitled, lazy and completely disconnected from the struggles of the "real world." In this regard, they were referred to popularly as the global "me, me, me generation":

> Each country's millennials are different, but because of globalization, social media, the exporting of Western culture and the speed of change, millennials worldwide are more similar to one another than older generations in their own nations. Even in China, where family history is more important than any one individual, the internet, urbanization and the one-child policy have created a generation as overconfident and self-involved as the Western one. And these aren't just rich kid problems: poor millennials have even higher rates of narcissism, materialism and technology addiction in their ghetto fabulous lives.
>
> (Stein, 2013: 2)

Obviously, such rhetoric was over the top in its fears and in many cases was outright misinformed. What was true, though, was the fact that, due to the financial crisis in particular and neoliberalism generally, the youth of today were "faced with grim job prospects and saddled with record student loan debt, millennials are experiencing levels of stress way beyond those experienced by baby boomers and the mature generation" (Ferri-Reed, 2013: 22). Tellingly, they were also exhibiting increasingly and unpredictably as a generation nonmarket characteristics. Evidence showed that

> [a]s a group, Millennials are unlike any other youth generation in living memory. They are more numerous, more affluent, better educated and more ethnically diverse. More important they are beginning to manifest a wide array of positive social habits that older Americans no longer associate with youth, including a new focus on teamwork, achievement, modesty and good conduct.
>
> (Howe and Strauss, 2000: 4)

Conversely, the older generation was held personally accountable for holding back progress due to their outdated views on pressing economic and social issues as well as their inability to adapt to the benefits of contemporary technology. The so-called baby boomers were commonly derided as "digital immigrants," as opposed to their "digital native" offspring (see Bayne and Ross, 2007). Their former radicalism had been displaced by a profound economic, political and even personal culture of "selfishness" (Karisto, 2007). As such, "Baby boomers (in addition to being older) are constructed variously as lucky, selfish, conservative, risk averse, blocking access to jobs for young people and in a privileged financial position" (Pritchard and Whiting, 2014). Even worse, they had "sold out" the future with their journey from idealists to cynical and naïve promoters of the market. To this end,

> [the radicalism of the 1960s] decayed fast. It decayed not because it was groundless, but because it was not grounded. What began as the most radical sounding generation for half a century turned into a random collection of youthful style gurus who thought the revolution was about fashion; sharp toothed entrepreneurs and management consultants who believed revolution meant new ways of selling things; and Thatcherites who believed that freedom meant free markets, not free people. At last it decayed into New Labour, who had no idea of what either revolution or freedom meant, but rather liked the sounds of the words.
>
> (Beckett, 2010: ix)

There was a similarly existing and to some extent mythical "urban and rural divide" (Scott et al., 2007). Those in the country were stereotyped as being "socially backward" and misinformed. "The media play a key role in stereotyping as 'ignorant and uncouth hillbillies' people who live in rural US communities," observe DeKeseredy et al. (2014: 179). "As well, since the early 1970s, popular films frequently portray rural areas as dangerous locations, places where urban people are at high risk of being savagely killed and tortured by demented, in-bred locals without conscience or constraint." In the 21st century, it was assumed that society would be magically cured of all its hated intolerance if only those in the "uninformed" rural and Rust Belt areas would get in step with these more multicultural and cosmopolitan times. Conversely, those living in the city were supposedly out of touch with "real" values. They were smug and superficial in their Liberal beliefs. Their existence represented the height of hypocrisy—living a life of privilege while decrying from their comfortable place in the world all its injustices.

This prevailing culture of blame also permeated class lines. The class struggle was reframed as who was most to blame for the current economic malaise. The much maligned 1% were justifiably lambasted for their role in turning democracies into oligarchies (Calhoun, 2013; Gitline, 2012). Renowned former World Bank chief economist Joseph Stiglitz wrote explicitly of the "1 percent's problems":

> In a society in which inequality is widening, fairness is not just about wages and income, or wealth. It's a far more generalized perception.

> Do I seem to have a stake in the direction society is going, or not? Do I share in the benefits of collective action, or not? If the answer is a loud "no," then brace for a decline in motivation whose repercussions will be felt economically and in all aspects of civic life.
>
> (Stiglitz and Blimes, 2012)

The poor, for their part, were portrayed as ignorant, the personification of a population "dependent" on welfare that would rather get something for nothing rather than work for a living (see Desante, 2013; Seabrooke and Thomsen, 2016).

Alongside these class tensions were a resurgence of xenophobia and racism. The killing of black Americans by law enforcement was met by Conservative demands that the focus be shifted from systematic police brutality to ending the "endemic of black on black" crime. Similarly, the legacy of past and present racism created by the forces of yesterday and today's colonialism was reduced to needing to ensure that individual white people became "woke." Moreover, it was asserted that immigrants and refugees should enter the country responsibly and legally, no matter the economic and political hardships they were facing.

Fundamentally, these diverse discourses represented a shared ethos centered on individuals taking personal ethical responsibility for their own fate and that of their wider society. More precisely, subjects were held personally accountable for their own condition caused by the forces of neoliberalism. It was up to them to morally reform themselves and ethically improve their communities in order to create a more fulfilling and righteous existence and world.

Producing the Personally Moral Neoliberal Subject

A chief complaint against neoliberalism, both in theory and in practice, was that it condemned people to an unrewarding life full of financial strain and little else. Fundamentally, the core question was whether it was possible for even the most successful neoliberal subject to be moral. These normative critiques coincided with a material reality in which job insecurity, economic anxiety, social alienation and political disempowerment combined to make morality too often a mere afterthought. In response, neoliberalism has directly addressed this righteous despair. Arising, thus, are fresh demands on subjects to become personally moral, in the process shifting responsibility from the current capitalist system and its institutions to individuals themselves.

A prime contemporary concern is how individuals can overcome the challenges of neoliberalism to actually forge for themselves a moral existence. These desires were countered by what appeared to be little opportunity to have any control over their economic, collective or ethical fates. This reflected a deep sense of what Costas and Fleming (2009) refer to as "self-alienation," in which individuals realize that it is society and its organizations—such as the workplace—that shape one's identity and beliefs rather than oneself. People, in this regard, often feel as if they are "strangers" to themselves (Leidner, 1993; Sennett, 1998). A continual neoliberal struggle waged by people is therefore to protect their "inner selves" from being completely overdetermined (Mumby, 2005; Tretheway, 1997).

As mentioned, discourses such as employability ostensibly were meant to fill this gap. They offered an appealing vision of contemporary empowerment that affectively combined the professional and personal. In the words of Cremin (2010: 131), "the subordination to capital (the material fact of labour) is defused by the sense we have of our independence from the employer (an identification that is not associated with the act of labour)." In this respect, to be employable was to have the power to lead a life of one's choosing and in a way that was in line with one's own moral beliefs. This championing of "freedom" reveals the shifting character of "neo-normative" control (Fleming and Sturdy, 2009). Workers find themselves "existentially empowered" to make their own personal goals and create their own professional boundaries so that they may better embody their "authentic selves." The call to be "autonomous" offers individuals the "freedom," and therefore the expectation, to lead "healthier" and more "fulfilling" lives, both professionally and personally.

These romanticized visions, though, were assailed by the concrete hardships individuals suffered at the hands of a hyper-competitive free market. The promise of reversing the damage of corporate globalization simply through enhancing personal employability rang increasingly hollow. "Citizens are now abandoned to their own cunning and guts while held solely responsible for the results of their struggles against adversities not of their own making," according to the pioneering sociologist and theorist Zygmunt Bauman (Bauman: 65). This was especially so in the wake of the financial crisis. The emancipatory hope brought by Obama—one built tellingly on a "new era of responsibility" in which individuals were urged and inspired to help personally bring about positive social change—was, after he assumed office, "sabotaged by a movement of centrists, lobbyists, market fundamentalists, militarists, and right wing ideologues who believed there was no longer any need for either hope or struggle once their expertise was called upon" (Giroux, 2015: 12). On the everyday level, there was a "crisis of faith" with finance and its morality (Riles, 2013). As Nobel Laureate economist Robert Shiller acknowledged in 2015,

> [s]ince the global financial crisis and recession of 2007–2009, criticism of the economics profession has intensified. The failure of all but a few professional economists to forecast the episode—the aftereffects of which still linger—has led many to question whether the economics profession contributes anything significant to society. If they were

> unable to foresee something so important to people's wellbeing, what good are they?

Tellingly, this discourse of economic responsibility was soon reaffirmed as a personal moral imperative. The failure to find a job, or at least a job that matched one's skills and talents, was laid at the feet of individuals rather than the system. There was an ever-growing call for individuals to "help themselves" (see Kenny and Bell, 2014). Indeed, "over recent years the concept of 'self-care' has been mobilized by policy makers and governments in the deepening of neoliberal objectives to dismantle public welfare resources and shift responsibility for care onto individual citizens" (Ward, 2015: 45).

It was framed above all else as a matter of individual initiative or, as more usually was the case, a lack thereof. The "neoliberal self" was wrapped up in a disciplining ethos of financial "self-help" representing

> [a] set of discourses, practices, techniques, interactions and objects through which people make sense of and attempt to transform their social planning and behavior, their social positions, their goals and their selves. . . . Most financial self-help books do not provide simple formulas to "get rich quick" but rather suggest to readers that there might be something fundamentally wrong with the core of who they are as a person, and that they have to endure a long and challenging self-transformation to correct it.
>
> (Fridman, 2016: 5)

If neoliberalism once promised to create an aspirational society, it now morally castigated people for not doing enough to realize such aspirations.

This moral calling also extended similarly to an "individual" personal morality. Neoliberalism was accused of threatening the very comprehensive well-being of modern subjects through a combination of overwork and anxiety that could be traced back to a corporate reality obsessed with profit and productivity. Countering this systematic problematization was a pronounced discursive emphasis on individual "wellness." Being physically and mentally healthy has become a mantra bordering on the pathological. Cederstrom and Spicer (2015) refer to this condition as the "wellness syndrome":

> Today wellness is not just something we choose. It is a moral obligation. We must consider it at every turn of our lives. While we often see it spelled out in advertisements and life-style magazines, this command is also transmitted more insidiously, so that we don't know whether it is imparted from the outside or spontaneously arises within ourselves. This is what we call the wellness command. In addition to identifying the emergence of this wellness command, we want to show how this injunction now works against us.

Individuals are thus asked to take personal moral responsibility for coping with neoliberalism.

This plays into a profound expectation on individuals to secure for themselves their own "professional" and "personal" borders, involving almost a heroic effort to find innovative ways to meet one's professional and personal obligations (see Ford and Collinson, 2011; Meriläinen et al., 2004). In particular, there was a moral expectation that individuals take it upon themselves to find time for their family, friends and own needs. To treat their career with the appropriate respect, as simply one part of their larger "healthy" existence (Bloom, 2016c).

These interconnected discourses point toward a more pernicious ideal of neoliberalism revolving around personal morality: any and all systemic constraints can be overcome with the proper moral willpower. Present, for instance, was a disciplining form of "neo-liberal feminism" demanding that each woman accept "full responsibility for her own well-being and self-care, which is increasingly predicated on crafting a felicitous work-family balance based on a cost-benefit calculus" that paints the West as a bastion of gender progress against less developed cultures around the world (Rottenberg, 2013). More broadly, the mass media perpetuated the dominant idea that morality, such as paying one's debts on time, was linked to personal responsibility (Turken et al., 2015). In particular,

> popular culture such as reality TV shows that proclaim the need for individuals to "take responsibility for yourself" act to ". . . construct templates for citizenship that complement the privatization of public life, the collapse of the welfare state, and most important the discourse of individual choice and personal responsibility."
>
> (Ouellette, 2004: 232)

Therefore, it was not the economic order that was held accountable but rather the people who were subjected to it and suffered under it. In this spirit, it was individuals, not neoliberalism itself, that were expected to be "good" and "just." Any hint of being unwell was, therefore, chalked up to personal moral failings or mental health problems. Further, it was individuals' responsibility to address these medical issues through consuming the proper drugs and seeking out the appropriate treatment (Esposito and Perez, 2014). According to Esposito and Perez (2014: 418):

> Presupposed by all this is that the market reality within which individuals exist (and uphold through their daily actions/interactions) is beyond question. Therefore, mental health must necessarily be conceptualized and achieved within the ontological confines of the marketplace.

It was thus simply to be accepted that life was difficult. Yet people must find ways to cope in a healthy way with these obstacles. To give in to addiction, obesity or anti-social behavior generally was to harm not just oneself but society as a whole. Indeed, such moral duty went beyond the mere personal—it was part of an obligation to society's health generally.

Personally Solving the Ethical Problems of Neoliberalism

The perhaps most common expectation for the future of individuals in neoliberalism was prominently defined by a combination of self-interest and competition. The hope and fear was that society would be suddenly full of aspiring "masters of the universe." However, a quite interesting thing happened on the way to this proffered market utopia. Individual ethics of cooperation and social care seemed to proliferate exponentially across society. Ideals of giving and generosity reflected this growth in nonmarket values. The most charitable reading of this ethical shift is that individuals were "freed" by the market to "do good." Yet a more accurate assessment is that they are being shaped as "problem solvers" to "fix" the ethical problems of this contemporary hyper-capitalist system.

One of the most revealing examples of this rather ironic modern ethics is the widespread call for "mindfulness" and enhanced spirituality. The rise of 24/7 capitalism and financialization came with a high human cost, characterized by a marked increase of "toxic stress" (Taylor, 2014), anxiety and even loneliness. "Though our wellbeing is inextricably linked to the lives of others," notes George Monbiot (2016: 1), "everywhere we are told that we will prosper through competitive self-interest and extreme individualism." Nonetheless, there has been little opportunity to challenge and transform these underlying values. In the impassioned words of Naomi Kline (2016: 1):

> Here is what we need to understand: a hell of a lot of people are in pain. Under neoliberal policies of deregulation, privatisation, austerity and corporate trade, their living standards have declined precipitously. They have lost jobs. They have lost pensions. They have lost much of the safety net that used to make these losses less frightening. They see a future for their kids even worse than their precarious present. At the same time, they have witnessed the rise of the Davos class, a hyper-connected network of banking and tech billionaires, elected leaders who are awfully cosy with those interests, and Hollywood celebrities who make the whole thing seem unbearably glamorous. Success is a party to which they were not invited, and they know in their hearts that this rising wealth and power is somehow directly connected to their growing debts and powerlessness.

Thus, the focus has been on self-improvement, specifically cultivating a "healthy" approach to one's work and life. Doing yoga and buying products associated with this practice became its own moral imperative—a way to improve oneself and the world (Lavrence and Lozanski, 2014). Similarly, new technologies and initiatives promoting "digitized health" focused healthcare on individual self-management rather than broader social factors—

especially linked to an unhealthy neoliberal environment (Lupton, 2013). Tellingly, these calls to be "mindful" transcend such high-minded language. Mindfulness is a set of concrete techniques individuals can draw upon to better deal with the stresses of present-day reality. The use of meditation, yoga and assorted breathing methods are concrete means of decreasing the mental and physical toll of by neo-liberal existence. Even the aging body, wracked by years of capitalist labor and stress, is socially constructed as a site for preparing for retirement by being a good and healthy "neoliberal citizen" (see Rudman, 2015).

However, mindfulness goes beyond personal solace and betterment; it was also a pathway for actively improving one's performance and moral duty as a social and economic citizen. Mindfulness was used to help US marines become better soldiers, to give one prominent and especially telling example. In fact,

> [a]fter eight weeks of meditating for just 15 minutes a day, the soldiers are far better at dealing with anxiety, stress, depression and insomnia. It helps them stay calm and focused in the thick of battle, while improving overall mental and physical fitness.
>
> (Penman, 2012)

More generally, this reflected the creation of social technologies of "responsibilization" (Rose, 1996; Shamir, 2008) for individuals to effectively "take responsibility" for the simultaneous greater volatility and reduced public welfare within neoliberal societies (Bennett, 2008). Inner peace was put forward as a catalyst for external social justice and economic prosperity. The desire and opportunity to be mindful progressively became a diverse and enforced civic obligation to others in one's organization and community (see Cook, 2016). In this spirit, the deep structural issues of neoliberalism were reduced to the championing of a corporate friendly "McMindfulness" (Walsh, 2016) that focused individuals' attention squarely on the need to adopt a more spiritually healthy perspective.

This reflected a broader ethos of ethical self-improvement that was presented as the key force for achieving social progress generally. Specifically, the use of mindfulness to champion the existence of a supposedly "responsible, autonomous, choosing individual" engaged in a "life-long project of self-improvement and self-discipline" (Arthington, 2016: 87) was transferred into the socio-political realm. In this sense, individuals are expected to use such conventionally economic methods and mentalities—such as continually auditing their performance and processes of constant self-accounting—for maximizing positive outcomes for themselves and others (see Gidden, 1999). In the words of Clarke (2005: 451), they are

> the bearers of responsibilities as well as rights . . . [entrusted with] the responsibility to produce the conditions of one's own independence—ideally by becoming a "hard working" individual or family. . . . Citizens

> must manage their lifestyles so as to promote their own health and well-being. Members of communities must eschew anti-social behaviour so as to promote harmony. . . . Parents must take responsibility for controlling and civilizing their children.

Significantly, this neoliberal notion of personal responsibility existed alongside other less market-based notions of being responsible, such as care and established notions of the liberal social contract. Yet while these were often seen as competing with or tempering these neoliberal discourses, they can also complement and serve to ironically reinforce them. Hence, individual charity and community engagement were promoted as the primary methods of positively addressing mounting issues of material and cultural deprivation (Ilcan, 2009; Kapoor, 2012; Raddon, 2008; Weiss, 2001). This extended also to interpersonal relations—as helping others was championed as the best salve for curing one's own feelings of social alienation. This belief was witnessed, for instance, in the popular idea of "paying it forward." It was also firmly on display in the use of social media to highlight the random good deeds of strangers—such as giving a waiter or waitress an exorbitant tip to help him or her in his or her life. There was, moreover, a growing emphasis on "crowd funding" for realizing personal and collective moral aims—from raising enough money to get an operation one could otherwise not afford to starting a community-based initiative (Bellefamme et al., 2014; Mollick, 2014).

More directly political in character, individuals were encouraged to use their economic power to push for broad socio-economic justice goals. On first glance, this may appear to be simply an innocent use of market forces;

> [b]ut there is also a more disturbing possibility: that the critique of individualism and monetary calculation is now being incorporated into the armoury of utilitarian policy and management. One of the key insights of behavioural economics is that, if one wants to control other human beings, it is often far more effective to appeal to their sense of morality and social identity than to their self-interest.
>
> (Davies, 2015)

Thus, a novel type of personal civic "engagement" was encouraged that actively consulted and included citizens in democratic discussions and decision-making for combining policies that enhanced the social welfare while also cutting the financial budget (Lee et al., 2012). This could be seen in citizens' ability to support firms with similar ethics to their own while boycotting those that differ from their own (Clarke et al., 2007; Micheletti and Stolle, 2004; Neilson, 2010). This strategy was used, for example, to help encourage corporations to adopt "fair trade" practices. Similarly, individuals are now progressively engaging in ethical investment (Michelson et al., 2004; O'Rourke, 2003; Sparkes, 2001). There is even a special stock listing for firms that are categorized as "ethical"—a designation earned by not being involved in the weapons trade as well as by being "green."

Such ethical commitments additionally touch and shape daily life. Indeed, lifestyle choice is put forward as a significant, if not the most important, means of achieving progressive social change. It is not simply good enough for an individual's morals to be "well"; there is also an ethical responsibility to use this healthy perspective to improve one's community and the world at large. Doing yoga became a competitive activity to see which organization could involve the most people during a global event like "Yoga Aid World Challenge" (Koch, 2015). This lifestyle politics could range, moreover, from eating organically, to growing a community garden, to taking public transportation or riding a bicycle to work. It also meant creating ethical consumer lifestyles. The social enterprise "me to we" exemplified this trend:

> As a "new kind of social enterprise," "Me to We" "transforms consumers into world changers, one transaction at a time"; it promotes a way of being good in the world as a consumer identity: "Every trip, t-shirt, song, book, speech, thought and choice adds up to a fun, dynamic lifestyle that's part of the worldwide movement of *we*." The business model for the enterprise, the consumer is assured, "measures the bottom line, not by dollars earned, but by number [sic] of lives we change and the positive social and environmental impact we make." Such a model has significant implications for both humanitarianism and global education.
>
> (Jefferess, 2012: 18)

The individual, therefore, becomes the focal point for pushing forward the cause of social justice and environmental sustainability. The problem of neoliberalism may run deep, but its solution was surprisingly simple. All that was needed was a personal commitment to cultivating a spiritual and psychologically healthy attitude and leading a "good life" for oneself and in service of others. The onus of ethically reforming neoliberalism primarily started and ended with the individual.

The Personal Power of Neoliberal Ethics

Neoliberalism, arguably above all else, is meant to empower individuals. At least theoretically, its overwhelming priority is to liberate all people from their collective bondage so that they can fully embrace their personal freedom. This involved directly taking aim at the threat of "Big Government" and socialism. In reality, individuals en masse lost much of their economic and political power, yet they were simultaneously empowered as moral and ethical subjects. Through their own moral improvement and ethical initiative, they could personally "fix" neoliberalism.

Significantly, the expectation for individuals to "be good" became a social duty. It was only in modifying their behavior that this hyper-capitalist system could be reformed. Accountability was focused firmly on people to themselves—shifting attention away from governments and corporations as well as capitalism itself. Hence, responsibility was devoted to individuals first and foremost.

The power effects of this enhanced culture of personal moral and ethical responsibility were manifest. In principle, they applied to all subjects equally. In fact, their most obvious targets were political and economic elites, deemed for obvious reasons the most responsible for current injustices and therefore those who should have the most responsibility for resolving them. However, over time this moral duty trickled down to the middle and lower classes. For those who could afford it, it was a non-binding opportunity to make the world better. For those closer to the bottom, it was another coercive social force to hold them personally accountable for their unfortunate condition. Consequently, it was a process of shifting blame onto those who were least responsible and who could least afford it.

This, of course, did not absolve the duty of the state and other institutions for this normative purpose. Instead, they were transformed into a vehicle for encouraging and if necessary enforcing this personal moral and ethical duty. Media continued to highlight stories of citizens performing good deeds while warning daily of how bad moral choices by much of the population were bringing the country to the brink of ruin. Governments were committed to clamping down on drug users and benefit thieves while turning a relative blind eye to corporate malpractice and tax evasion. Organizations took it upon themselves to police their workforce, blaming employees for not realizing the impossible goal of being an ethically "always on" worker.

Nevertheless, this moral and economic demand provided subjects with a serious sense of, once again, if not agency then at least personal empowerment. In the face of incredible odds and a system not in their favor, the idea that if they just changed their behavior then professional success, personal well-being and social justice could be realized was understandably downright liberating for many. Behind their everyday experiences of frustration, there was a profound "secret"—all they needed to do to be happy, achieve their dreams and make the world a better place was simply adopt the "right" positive attitude. And confronted with a global order that put profits over people, having the power to "live right" for oneself and others felt not only personally emancipating but collectively required.

The true consequence of this empowerment, however, was to limit individuals' social imagination for change. Little could be done to substantively transform society away from the powerful grip of neoliberalism and its hyper-capitalist values. Corporate globalization was a fact of life—permanent and irreversible. Yet by turning inward, individuals had a chance of making it just a little bit better for themselves and even others: ensuring that it was not quite so devastating, that its human costs were not so destructive. Real change, it was continuously proclaimed, starts with oneself. In neoliberalism, there is no changing the world, but there is a chance to improve it—a process that always begins and ends with "me."

7 The Subjective Power of Neoliberal Ethics

Neoliberalism is often described as amoral and unconcerned with any values except for advancing the free market and maximizing profit. However, this pronounced economic focus is underpinned by a comprehensive and dogmatic moral vision for society. It promotes the conception of a society in which the entrepreneurial spirit is permitted to rule unbound by government regulation or the threat of collectivist ideologies. It envisions liberated people who are simultaneously self-reliant and competitive. However, as this book has sought to reveal, in fact this romantic capitalist ideal is sustained through the strategic promotion of a decidedly nonmarket set of ethics. Indeed, as will now be explored, neoliberalism affectively grips present-day subjects via an appealing fantasy of a moral capitalism.

Discussions of morality and ethics typically revolve around questions of right and wrong or, more critically, themes of social conformity. Moral norms are framed as a constraining force repressing material and social impulses for the purpose of realizing justice and a larger personal and public good. Nevertheless, this seemingly rather commonsense paradigm poses a distinct and not easily overcome issue for neoliberalism. Neoliberalism must assert that its market principles are simply adhering to a deeper human nature while also constructing a quite comprehensive system for encouraging and enforcing these supposedly natural desires and practices (see Ubel, 2013). It is further complicated by the need to somehow find ways to mitigate its human and material cost without fundamentally challenging its underlying beliefs or power relations.

Critical, in this respect, is to understand neoliberalism as not merely rational. Neoliberalism is also inspiring, even if often ultimately disappointing. Its trumpeting of the market rationality is at least as much a fantasy as it is a rational theory. It relies upon an idealistic discourse of moral capitalism that frames the exploitive and insatiable demands of a capitalist system as part of a project of building a just and sustainable society.

It is precisely such fantasies that help individuals create a stable sense of self within this often capricious and uncaring free-market order. The ideology and long-term effort to be good and contribute to a better world shapes and helps psychically ground present-day identity. It is a fundamental subjectivity that ironically reproduces a modern hyper-capitalist subject. It offers individuals ontological security in the never-finished and thus always-present attempts to become a more moral and ethical capitalist.

The Subjective Crisis of Neoliberalism

The rise of neoliberalism coincided with the supposed end of history. Just as liberal democracy was considered the unquestioned stopping point in humanity's political progress economically, the free market represented the apogee of its economic achievement. Consequently, it painted an appealing picture of a material prosperity to come. The secret laws of personal and collective prosperity had supposedly been discovered and needed only to be applied judiciously to become a reality.

These ideas contrasted heavily with previous expectations of the unregulated market as the height of economic stability due to its actual cycle of boom and bust (Van Barbera, 2009). The attractiveness of Communism and Keynes in the past was in no small part linked to a mass longing for security in the face of this up-and-down and seemingly arbitrary nature of the market when left to its own devices. Indeed, the initial appeal of neoliberalism was a promised escape from this boom-and-bust fate. British chancellor to the exchequer Gordon Brown thus proclaimed in 1997 at the beginning of New Labour's reign:

> Stability is necessary for our future economic success. The British economy of the future must be built not on the shifting sands of boom and bust, but on the bedrock of prudent and wise economic management for the long term. It is only on these firm foundations that we can raise Britain's underlying economic performance.

A decade onward, and only one year before the financial crisis, he triumphantly repeated such desires:

> Now tonight I am here speaking to you ten years on to say that we need to forge that same shared national purpose for the years ahead, this time to meet and master the new and even more profound challenges ahead—mastering the pace of change and the relentless competition we face from every part of the world in every area of our economy, now affecting every area of your business. And building on our hard won stability, our mission now and for the years ahead is to make Britain the dynamic and competitive enterprise economy we all want it to be.

A critical problem facing capitalism, whether of its neoliberal variant or otherwise, is thus how to provide individuals a sense of security in the midst of this large and small-scale market uncertainty.

Neoliberalism in this regard has seemingly been especially marked by a culture of crisis. The most obvious manifestation of this crisis culture was associated with the near-global economic crash of 2008. This catastrophic event produced a strong collective desire for recovery—not just a longing for a return to economic security, but also a desire for greater psychological and social security. Optimistically, it gave rise to fresh longings for radical and progressive socio-economic transformation. "For me, the most fundamental question in economics still concerns the 2008 crisis. Was this event the last in a series of shocks needed to allow a third technological revolution to take off?" asked Paul Mason (2015),

> Or was it evidence that capitalism's tendency to adapt and reshape in response to technology has stalled, or is even finished? That is the shadow we have to jump over in economics. Amid a mania for "new economic thinking," it is what we need to think hardest about.

In this spirit, the threat of such volatility and the prospect of ending it permanently catalyzed quite innovative and previously unthinkable solutions—even from rather mainstream sources. For instance, there was increasing talk of moving toward a "cashless economy":

> In this futuristic world, all payments are made by contactless card, mobile phone apps or other electronic means, while notes and coins are abolished. Your current account will no longer be held with a bank, but with the government or the central bank. Banks still exist, and still lend money, but they get their funds from the central bank, not from depositors. Having everyone's account at a single, central institution allows the authorities to either encourage or discourage people to spend. To boost spending, the bank imposes a negative interest rate on the money in everyone's account—in effect, a tax on saving.
>
> (Leaviss, 2015)

Reflected, in turn, was a more profound crisis narrative in which instability and panic caused by the economic meltdown produced an intensified psychic longing for stability manifested in discourses of "recovery" (Bloom, 2014a). This culture of instability extended to all levels of neoliberal existence, from the political and institutional all the way to the personal. Here, traditionally

> castaway categories—unemployed youth left adrift, the beggars and the homeless, aimless nomads and drug addicts, postcolonial immigrants without documents or support—have become salient in the public space, their presence undesirable and their doings intolerable because they are the living and threatening incarnation of the

> generalized social insecurity produced by the erosion of stable and homogenous wage work . . . and by the decomposition of the solidarities of class and culture it underpinned within a clearly circumscribed national framework.
>
> (Wacquant, 2009: 4)

Yet almost regardless of position or social sphere, neoliberal actors are constantly under a threat of financial and increasingly ethical pressure. Their very material survival is always potentially dramatically imperiled. Indeed, having to make "tough decisions" in "insecure times" is the very "nomos" of neoliberalism (Cooper, 2004).

Such economic anxiety fosters a crisis mentality that is a structurally endemic and subjectively internalized feature of the contemporary free market. Every institution, every individual and every event, it seems, is only one poor quarterly report or lost paycheck away from falling off the proverbial social cliff. The unquenchable thirst for capitalist profit within neoliberalism is given as much by fear as it is greed.

More generally, the 21st century has ushered in what literally appears to be the soon-to-come end of history. Total economic collapse and ecological disasters are looming on the horizon. There is a decidedly apocalyptic undertone to actually existing neoliberalism (Fisher, 2010; Hardt, 2010; Williams, 2010). It can be seen, for instance, in the popularity of apocalyptic literature such as those prophesying the Christian rapture (Strombeck, 2006). More critically, the figure of the "zombie" has become predominant within popular culture to allegorically explain humanity's current condition—one composed of "a determinate political subjectivation—and the representation of the end of the world so common in these narratives" (Lanci, 2014: 26). Its consequences go beyond poor growth rates and unfair debt burdens and rather point toward the potential total planetary extinction of humanity (Blouin, 2016; Felli, 2014; Gorman, 2004; Peeple et al., 2014).

Specifically with the fall of the Soviet Union came renewed promises for a prosperous global future. Political and economic freedom would sweep the world. Liberty would finally reign unopposed (Fukuyama, 1989). These romanticized visions—whatever their concrete impossibility—and ideological narratives were psychologically powerful. Economically, such optimism spurred the "dominance of a market oriented approach to the question of national development and the willingness of governments to follow the policy dictates of international finance organizations based on this perspective" (Portes, 1997: 229). At the more personal level, neoliberalism offered the emotionally resonant prospect of guaranteed success if one were only willing to work hard.

The seemingly never-ending and utterly destructive actuality of neoliberalism, coupled with the financial crisis and now the threat of authoritarian populism, has transformed these dreams into what appears to be hollow nightmares. Its shared collective and personal fantasy of a free-market utopia was rapidly crumbling—robbing individuals of not only the promise of material wealth but also their very ontological security and sense of self.

The Ethical Problem of Neoliberal Selfhood

Neoliberalism is trumpeted as being based first and foremost on an economic rationality; it is actually deeply tied and reliant upon a diverse range of cultural fantasies. Originally this revolved around the freedom that would be unleashed with the destruction of constraining bureaucracy and the bondage of public welfare. Every individual could dream of personal economic prosperity and social mobility. As these realities began to disintegrate, a new fantasy was required to maintain this appealing neoliberal delusion.

In this respect, neoliberalism faced a serious ethical problem—though not the one that is usually immediately imagined. Instead it was how to ensure that individuals continue to ethically invest in this hyper-capitalist system. Such an investment does not necessarily reflect its established market ethics. Rather, it signified the effective ways they related to the order that preceded and shaped the rational understandings and actions within it. Imperative, in this respect, was the presence of an appealing capitalist discourse. Drawing again on the insights of the psychoanalyst Jacques Lacan, such a neoliberal fantasy can be viewed not as rational per se but rather as representing that

> which holds together a given community [that] cannot be reduced to the point of symbolic identification . . . the bonds linking together its members always implies a shared relationship to the Thing, toward enjoyment incarnated. . . . If we are asked how we can recognise the presence of this Thing, the only consistent answer is that the Thing is present in that elusive entity called our "way of life."
>
> (Žižek, 1993: 201)

This was in fact exemplified in the "fantasy of market rationality" that largely drove neoliberalism forward. It represented that exciting prospect of a "boundaryless career of inter-organizational job-hopping" (Kunda and Ailon-Souday, 2005: 210). It was, furthermore, put forward as a way for historically disenfranchised groups to "take control" of their career and broadly their economic fate (Ackah and Norma, 2004). To this end, it provided a strong neoliberal narrative of success, signifying in this function "a plot comprising causally related episodes that culminate in a solution to a problem" (Boje, 1991). In doing so, it served to "impose a formal coherence on what is otherwise a flowing soup" (Weick, 1995: 128) of a fragmented neoliberal existence. For this reason, according to Bloom and Cederstrom (2009: 167),

> market rationality can be understood as the new fundamental fantasy of capitalism. Its construction of a stabilizing, beatific, fantasy associated

> with notions of a "boundaryless career" increasingly structures what individuals principally desire from their economic and social relations replacing previous fantasies centred on the security found within an organization community or "family." Indeed, the organization is now dominantly seen as an obstacle, as opposed to the almost exclusive vehicle, for attaining a sense of existential completeness in contemporary capitalism.

However, this fantasy was getting ever harder to maintain alongside the actual economic, political and environment disasters caused by 21st-century capitalism. Put differently, as neoliberalism itself was increasingly revealed to be unable to meet the present-day needs of individuals, what was left to socially and psychologically invest in? The construction of a new romantic discourse of moral capitalism was crucial for filling this ethical gap. It provided a fresh ideal and motivation for being. Capitalists effectively believed in and ethically invested in the possibility of using the market to create a better world in a better self. It held the promise the free market could be perfected. All was not lost, social and economic progress was still within our personal and collective reach.

Consequently, this new affective "moralization" of capitalism, at least in the short term, benefited both the system itself and those still subjected to it. To this effect,

> [w]hat psychoanalysis can do to help the critique of ideology is precisely to clarify the status of this paradoxical jouissance as the payment the exploited, the served received for serving the master. This jouissance of course, always emerges within a certain phantasmic field; the crucial precondition for breaking the chains of servitude is thus to "transverse the fantasy" which structures our jouissance in a way which keeps us attached to the Master—makes us accept the framework of the social relationship of domination.
>
> (Žižek, 1997: 48)

In the present context of neoliberalism, this echoes the win-win claims of early human resource management–inspired post-bureaucratic organizations whereby employee profits and employee well-being were not at odds absolutely fundamental to each other's achievement. There is an analogous movement happening on the deeper psychic economy of present neoliberalism. More precisely, the rather fresh ethical paint applied to this hypercapitalist system provided it with a temporary social legitimacy while granting subjects a renewed sense of ontological security.

This arising cultural fantasy of moral capitalism was crucial for covering overcoming the subjective and structural contradictions of neoliberalism. Fantasy acts to cover over the "real" fragmentary and contingency character of both one's self and, by association, society (Glynos and Howarth,

2007; Stavrakakis, 2007). Returning again to Žižek (2004: 94), it signifies that which is "objectively-subjective." Specifically,

> [t]he ontological scandal of the notion of fantasy resides in the fact that it subverts the standard opposition of "subjective" and "objective." Of course, fantasy is, by definition, not "objective" (in the naive sense of "existing independently of the subject's perception"). However, it is also not "subjective" (in the sense of being reducible to the subject's consciously experienced intuitions). Rather, fantasy belongs to the "bizarre category of the objectively subjective—the way things actually objectively seem to you even if they don't seem that way to you.

In this respect, at stake was no longer whether the free market can result in a just and sustainable society. Rather it was how this could be realized. Such ethical critiques once were meant to be sympathetic of the undesirability of marketization and financialization. Yet it was now a symptom gesturing toward its eventual rehabilitation and cure. Accordingly, neoliberalism was increasingly presented not as an economic powerhouse but, ironically, as an ethically perfectible system.

It is precisely these ethical imperfections that temporarily ensure its survival. In this spirit, neoliberalism exhibited a profoundly paradoxical psychic character. Its continual ethical failures were strategically used to reinforce its future success and reproduction. At stake was an affectively "gripping" possibility and demand to morally and ethically improve neoliberalism and the subjects who populate it. The question hence was who could and would be responsible for undertaking this eternal project of making the free market more moral and ethical. As will be seen, it was exactly this question of ethical perfectibility that sustained and imprisoned individuals as modern subjects of neoliberalism.

Taking Subjective Responsibility for Neoliberalism

As this analysis has sought to highlight, neoliberalism is fundamentally an ethical system. At its most ideologically pure, it promotes a distilled market ethos of competition and profit maximization. In its concrete actualities, it strategically demands a strict and creative conformity to traditionally non-market values of cooperation, care for others and sustainability. Even more deeply, it requires a strong affective investment in these ideals—whose grip extends far beyond the economy or a simple market rationality. It is not simply that neoliberalism produces an ethical subject not in its direct image for its ultimate advantage; this fantasy of a moral and ethical capitalism increasingly frames the very basis of contemporary subjectivity and therefore "reality" itself.

Just as with any dominant social system, there's a need to critically distinguish the ideal neoliberalism from its concrete and complicated existence.

In stark contrast to its promise of a free-market paradise, neoliberalism has led to massive material deprivation and psychological stress. This begs the obvious question, then, of "What is the real neoliberalism?" The answer, to some extent, is clear: it is an order marked by marketization and financialization that results in economic and political oligarchy (Bloom, 2016a; Demmers and Jilberto, 2000; Karadag, 2010; Snyder, 1999). Thus, despite its emancipatory claims, "in actual fact what has occurred is a Weberian combination of neo-liberal management joined to bureaucracy to regulate and de-democratise and create increasing un-freedom for the many and freedom for the elite" (Hanlon, 2016a: 186; also see Biebricher, 2015; Müller, 2015; Streeck, 2014).

However, the ideology underlying this progressively dystopian order is dramatically less solid than it may first appear. Neoliberalism may have a clear set of associated meanings, like all social discourses, yet it consists of a "real" nature that is inherently empty of permanent meaning or content. Hence, from a Lacanian psychoanalytic perspective,

> [i]n our attempts to keep the ground from shifting, we idealize our objects, clothes, people and thoughts with the ideological garments of the good, the true and the beautiful. But the unconscious moves as desire, tracking evacuated jouissance, as desire and sexuality dancing in the endless vacillation around a void.
>
> (Ragland, 2013: 100)

The construction of an attractive reality, then, is a primal effort to secure identity in order to bring coherency to this inherently fragmented psychic and material existence. It is always and continually a form of "failed identification," a constant and eternal effort to achieve this long for psychic harmony associated with a cultural fantasy. In the words of Lacan (2002: 262):

> This limit is present at every instant in what is finished in this history. It represents the past in its real form; it is not the physical past whose existence is abolished, nor the epic past as it has become perfected in the work of memory, nor the historical past in which man finds the guarantor of his future but rather the past which manifests itself in an inverted form in repetition.

Bringing this analysis back to neoliberalism in particular—neoliberalism offered subjects a range of fantasies for securing their identity. One of the most notable was that linked to discourses of entrepreneurship. The strive to be entrepreneurial provided individuals with a "script" for their life, even if this term was often rather ambiguous in meaning and impossible to fully achieve. According to Jones and Spicer (2005: 237),

> [o]ne secures identity not in "being" an enterprising subject but in the gap between the subject and the object of desire. Not only does it not

> matter that the object is unattainable. This lack is central to maintaining desiring. And, as Lacan indicates, if we ever achieve the object of desire, it collapses—it falls apart and is changed inexplicably into a gift of shit.

More recently, selfhood has revolved around the phantasmastic desire to achieve an ever-elusive "work-life balance" (Bloom, 2016c).

These fantasies offered a distinct sense of ontological security for modern subjects. The shaking of these subjects' psychic stability was a traumatic occurrence—not just that it put into doubt once-solid "truths" about the world and a guaranteed pathway to prosperity, but also in its disintegrating of the very existence of contemporary reality itself. This of course sounds quite dramatic, and it is important not to overstate the case. Rather, it is necessary to highlight that 2008 represented an existential crisis as much as it was an economic crisis. The ethical turn of neoliberalism was therefore an attempt to cover this secure sense of identity in the face of a seemingly failing free-market system.

Renewed focus—for instance, on environmental, social and governance values—may be interpreted as merely a cynical expression of corporate co-optation. Yet it also stands as the reconstruction of a comforting affective neoliberal reality. The notion of responsibility stands crucially as a discursive means of continually shoring up this precarious but relatively stable capitalist existence. It provides subjects with a constant fantasy with which to form themselves around. In taking more responsibility for themselves and ethical responsibility for neoliberal society generally, subjects are in fact reaffirming their identity as capitalist subjects and recapturing their fragile psychic security.

This promotion of an ethically responsible neoliberal subject expands upon the previous romanticization of the entire neoliberal project. The potential to be ethical is a gift—signifying the possibility to be a self. Specifically, it offers an emotionally resonant discourse to build a personal and collective subjectivity around. It provides an ideal to strive toward, for failure is understood to be an opportunity for improvement rather than a sign of total social and psychic disintegration. Critically, morality and ethics, in this regard, exist as a "symptom" of capitalism's problems—the very thing that invites and sustains subjects into a coherent neoliberal reality. "What we must bear in mind here," according to Žižek (1989: 75),

> is the radical ontological status of the symptom: symptom . . . is literally our only substance, the only positive support of our being, the only point that gives consistency to the subject. In other words, symptom is the way we—the subjects—"avoid madness, the way we choose something (the symptom-formation) instead of nothing (radical psychotic autism, the destruction of the symbolic universe), through the binding of our enjoyment to a certain signifying symbolic formation which assures a minimum of consistency to our being-in-the-world.

This rather perverse demand placed on individuals to be moral subjects in an immoral society reflects an underlying sense of psychic enjoyment in this perceived unethical system. Ironically, it is the exact ethical deficiencies that make it so psychologically attractive. It signals once more what Stavrakakis (1999) refers to as an appealing form of "failed identification." Hence, the moral and ethical lack at the heart of neoliberalism serves to ironically sustain it as a secure system, specifically relating to its production of a stable, ethically lacking neoliberal subject.

The increasing expectation for individuals to morally and ethically "fix" capitalism represents the social construction of a new desiring subject of neoliberalism. It fixes identity around the ongoing responsibility to be more moral and ethical. Subjectivity is found in this ongoing and necessarily always unfinished righteous struggle. The promotion of a fantasy of moral capitalism creates the conditions for individuals to ethically reinvest in this market system. While at present it is riddled with injustice, it is morally and ethically salvageable—a promise that serves to structure modern identity and life practices.

Producing the Moral Neoliberal Subject

The morality of neoliberalism is an often contentious topic. The concerted critiques of the system as amoral miss the complexity of this relation. The attempt to consign its morality to simply the moralization of market values also ignores the richness and often counterintuitive character of actually existing neoliberalism. Instead, as this work has sought to critically show, contemporary efforts to spread this hyper-capitalist order rely heavily upon conventionally nonmarket norms and ideals. This is arguably especially so at the subjective level. Neoliberalism is sustained not only through its social framing as a moral system but also in the affective production of moralized capitalist subjects.

A crucial component of the social legitimization of capitalism are, as discussed, processes of cultural moralization. Specifically, the introduction of wage labor and the prioritization of the market has been both directly and indirectly associated with moral improvement. These evolving discourses reveal the affective moral character at the heart of capitalist reproduction historically (see Maitland, 1997; Picchio, 1992). In this respect, markets historically have been framed as "civilizing" forces (Dekker, 2016; Hirschman, 1982; McCloskey, 2006). Much of its social appeal and justification lies in its ability to promise individuals the opportunity to become better people. Indeed,

> there is a long tradition within economic discourse of explicit praise for the moral benefits of market society. The precise benefits vary. The *doux commerce* tradition is carried forward by arguments that the market nourishes personal virtues of honest behavior, civility, and cooperation.

> Others have seen markets as a necessary condition for freedom in other aspects of life, most prominently in politics and in the cultural realm. A final tradition, represented today by the bulk of prescriptive macroeconomics, emphasizes economic growth as a condition for human progress, and it is best encapsulated by Keynes's comment that economists are the "trustees, not of civilization, but of the possibility of civilization."
>
> (Fourcade and Healy, 2007: 287)

Within the contemporary context, the negative effects of the free market are creating the conditions, supposedly, for a shift from a strictly materialist economic rationality to a more norm-driven "substantive rationality" first proposed by Weber. As such, it is not simply that markets moralize subjects but rather that markets themselves must be moralized (Best, 2003; Fuller, 2006). To this effect,

> [a] trend toward a moralisation of the markets implies that the regulation of market relations in response to major societal changes makes reference to, and is based on, constraining notions such as fairness, authenticity, compassion, goodwill, fear, guilt, solidarity, sustainability, revenge, age, exclusivity, reciprocity, duty, benevolence, and many other moral concerns, including political considerations, that not only negate any persistent short-term pursuit of the best buy, but also take a long-term perspective on board.
>
> (Stehr, 2008: 151)

In the case of neoliberalism, its survival and prosperity are increasingly linked to the cultural fantasy of moral capitalism and with it the ongoing potential for creating a "better" free market. The question of "is moral capitalism even possible?" is replaced with a new affective discourse trumpeting not only its plausibility but also its urgent necessity both for individuals and society. Hence it is proclaimed that

> [s]eeking market profit through business and the professions is honorable and worthy . . . each of us can indeed go to work everyday for any business, great or small, feeling genuinely happy and proud of our career commitment. From those more removed from careers in business . . . (affirmed is) a vision of social justice: that moral capitalism is the most appropriate means by which our modern, global human civilization can empower people and enrich their lives materially and spiritually.
>
> (Young, 2003: 1)

However, as perhaps can be expected, such affective promises of moral capitalism are not so easily rationally or meaningfully pinned down. Its definition is discursively ambiguous and fundamentally ideologically contradictory. Nonetheless, its success rests not on putting forward a clear set of values or propositions but on offering the possibility for subjects to strive to be moral subjects. According to Stehr and Adolf (2010: 215):

> defining questions of a single, universal morality cannot be answered. The consequence of this indeterminacy for markets is that in modern societies, there is a plurality of markets, within which the trend to moralization is put into practice by consumers and producers in various methods, and in accordance with a variety of values. Although set moral imperatives that determine action—such as sustainability, for example—are not being completely realized, these values are changing the market and social life. The fact that they can hardly ever be completely realized, however, does not mean that fairness or solidarity become mirages or illusory values.

Accordingly, it is not primarily a matter of subjects merely embracing the righteous possibilities of the present-day free market. Instead, it is the ability of subjects to attain ontological security by the continuing enjoyment they gain in their status as perversely morally lacking subjects within this market system. Using slightly different language, it is the pursuit, rather than the achievement, of being moral that nourishes and perpetuates a modern capitalist subjectivity.

One of the most obvious examples of this arising neoliberal moral subjectivity is that of the moral leader. There is a growing body of evidence showing the need for ethical leadership in the public and private sector. In this respect, there's an emphasis on being able to personally exhibit moral values (Gini, 1997; O'Toole, 1996; Whetstone, 2002). What is especially significant is that the moral lack of neoliberalism is transferred into the possibility of a "transformative" individual to make up for these deficiencies (see Caldwell, 2016; Daft, 2014; Johnson, 2013; Schuh et al., 2013). If capitalism appears overcompetitive and ultimately amoral, then its leaders must simply work even harder to reflect opposite values of cooperation and care for others. They must put themselves forward as a public example that this is not only necessary but utterly possible. Increasingly, this leadership mentality is not reserved for elites but universally expected—as everyone must in some capacity take responsibility for becoming a "moral leader" for themselves.

Of course, exhibiting such morality can be quite hard and even dispiriting in the face of the seemingly unmovable forces of the free market. The perverse enjoyment to be had in being a moral leader can also quickly fall apart in its repeated daily futility. Fortunately for neoliberalism, this frustration has produced a perhaps even stronger subjective attachment to this hyper-capitalist order, notably in the identity found in the pain of having to be amoral despite one's best intentions. Cederstrom and Grassman (2008) speak in this spirit about the enjoyable masochism of the modern workplace, where

> [f]ar from trying to lull their employees into comfortable and imaginary worlds, certain organizations take the opposite tack: they make a proud case of not providing security, meaning or happiness to their employees. Instead, they offer high salaries and a reflexive organizational culture, perceived to resonate better with the "harsh reality" in which they operate. The employees of these organizations are not only allowed to openly despise their occupation, but they are even compelled to exhibit this disdain in order to fit in . . . at the same time as the employees are well aware of their misfortunate situation, they derive some form of enjoyment from it.
>
> (42)

Likewise, a definite strain of "moral masochism" permeates neoliberal existence. More precisely, there is a continuous refrain of highlighting how much one has sacrificed in order to be moral against formidable odds. Additionally, there's a certain enjoyment gained from emphasizing one's own moral failings—epitomized in such laments as "I would love to be more caring, if only I could in this forsaken world." Selfhood is, therefore, often made possible and strengthened in the pain of having to be moral and the failure to ever fully be so in the midst of an immoral neoliberal order.

This points to a novel type of subjective moral relation of subjects to modern capitalism. It speaks to the much-discussed and, for many, maligned present-day cynic (see Sloterdijk, 1984). Here, subjects, despite not being able to change the system, believe they can mentally and ethically see through its absurdities. This is often referred to as a contemporary form of "enlightened false consciousness," as it accepts the unchangeable character of the system even as it expresses the dramatic sense of this identification toward it. Represented, in fact, is a deep psychic investment in this hegemonic identity.

In the context of the workplace, this can even lead to greater actual conformity with managerial desires (Bloom, 2008; Fleming and Spicer, 2003). In a similar way, neoliberalism helps produce moral subjects who gain a sense of continual perverse enjoyment in accepting the immoral status quo with a knowing, world-weary laugh. They go to work, follow orders and accept the free market with the secret belief that they are morally superior to his hyper-capitalist order that they must invariably and unstintingly submit to.

The lifeblood of neoliberalism, thus, is the manufacturing of a moralized subject. It involves the crafting of subjectivity around the belief in one's own existing potential moral value. Far from just preaching the righteousness of the market, it encourages a selfhood that invests in the nonmarket ideals of cooperation, care and social justice—even if they are only an impossible dream to strive toward. Ontological security is paradoxically found and maintained in this ironic enjoyment of the system's moral failings and the eternal capacity of subjects to correct it.

Ethically Solving the Problem of the Neoliberal Subject

A constant lament posed against neoliberalism is the way it makes being ethical a seeming impossibility. How can one personally fight global environmental destruction and sweatshop labor? How can any single individual improve collective job security, make governments less corrupt, make communities safer or make our world more peaceful? A distinct sense of ethical hopelessness plagues neoliberalism. Yet, as with morality, it is precisely this seeming intractable ethical lack that psychically sustains it.

The current reality of neoliberalism is decidedly different than it was originally envisioned. In its initial idealized imagining, the free market was put forward as the struggle of valiant entrepreneurs against parasitic governments and populations (Williams and Nadin, 2013). This was exemplified in the mythical heroic American entrepreneur:

> The older and still dominant American myth involves two kinds of actors: entrepreneurial heroes and industrial drones—the inspired and the perspired. In this myth, entrepreneurial heroes personify freedom and creativity. They come up with the Big Ideas and build the organizations—the Big Machines—that turn them into reality. They take the initiative, come up with technological and organizational innovations, devise new solutions to old problems. They are the men and women who start vibrant new companies, turn around failing companies, and shake up staid ones. To all endeavours they apply daring and imagination.
>
> (Reich, 1987)

In truth, neoliberalism is taking a rather strange turn. It is a different battle, composed of well-meaning people who struggle ceaselessly in the face of an apparently unjust and uncaring hyper-capitalist system. Boutillier and Uzinidis (2014: 15), for instance, discuss the transition from the "heroic entrepreneur" of the 19th century who "built a new economic and productive logic on the ruins of the feudal system" and the "socialized entrepreneur" who

> devoted himself to the development of managerial capital, that is an economy dominated by large firms and in which the role of the State is preponderant both in terms of its own functions as an entrepreneur (notably during the period 1950–1970) and in the reduction of uncertainties (since the 1980s).

Added to this in the 21st century can be the "social entrepreneur" who is responsible for using his or her market skills and innovative talents to make capitalism more socially just (Alvord et al., 2004; Austin et al., 2006; Mair and Noboa, 2006). As such, the fundamental tension of our time has become how to be and do good while remaining financially solvent.

Underpinning the fantasy of a potentially more righteous free market is a renewed sense of ethical responsibility. This can often mean directly contravening existing social and organizational norms and prerogatives. There is an increasing emphasis on the idea of "moral subversion (Folger et al., 2013), representing situations in which employees or citizens must, as a matter of principle, subvert codes of conduct in the name of a higher moral good. This is further witnessed in the notion of " 'constructive deviance," in which employees violate norms but "in doing so contribute to the wellbeing of an organization, its members or both" (Galperin and Burke, 2006: 333). Likewise, recent critical perspectives of leadership argue the need for a "progressive pragmatism," allowing both leaders and implicitly followers to "bend the rules" to achieve broader emancipatory aims (Alvesson and Spicer, 2012).

Tellingly, this creates a new disciplining regime based on a sense of personal ethical responsibility. It is not enough to merely follow the rules. Rather, one must be willing to "fight the power" for the sake of social justice. This places the onus of the system's overall goodness squarely on the shoulders of individuals. This plays on a more critical "ethics of rule breaking" (Weiskopf and Willmott, 2013), in which subversion is encouraged as a means of revealing the unethical aspects of a prevailing status quo. Yet it also points to how individual subjects are made ethically complicit and therefore accountable within these orders. If neoliberalism is inherently immoral, then the blame lies at least partially if not mostly with the failure of people to normatively challenge and reform it.

Here, what is crucial is to personally use neoliberal techniques of accounting and optimization in order to continually enhance one's individual positive impact on the community and world. To maximize one's ethical effect, it is necessary to move beyond the conventional limits of a market ideology. In this respect,

> [w]e are in an age of heropreneurship: everyone wants to *be* a social entrepreneur. The myth of the entrepreneur creates a false hierarchy with *start-up founder* at the top. We foster this obsession in our education, our funding, our awards, and our media. But we don't just need more founders. We need more positive social impact.
>
> (Tackling Heropreneurship website)

The ultimate goal, then, is to align one's professional goals and accomplishments for optimal social impact.

However, the failure to fulfill this duty brings with it a paradoxical subjective stability. It is, again, a form of "failed identification"—an ideal to consciously keep striving for and unconsciously to continually form oneself around. As such, "ethically branded" companies such as The Body Shop involve individuals in a continual normative "struggle" to use commodities and their own consumerism as a force for social justice. Thus, "with its progressive objectification, promotional culture has become reflexive, often directly thematizing the role of advertising and the notion of the commodity, addressing and indeed soliciting consumers' capacity to appropriate commodities in personalized ways" (Sassatelli, 2009: 230). Thus, there is also a profound and perverse psychic enjoyment found in often being haplessly in this ethical situation.

Accordingly, the more neoliberalism suffered from an ethical crisis, the greater the psychic stability and ontological security it provided subjects. The presence of permanent crisis became an invitation to reconnect with one's identity as an ethical capitalist subject. Arising simultaneously are the "ethical citizen" and "moral neoliberal," exemplifying how structural problems of social welfare and justice are placed firmly upon the ethical innovation and commitment of the individual. Helping others is a form of "gifting" oneself to others—replacing the previous sense of collective reciprocity between the state and its citizens. Here,

> the ethos of the gift rather than disappearing under neoliberal conditions, is reappearing in its most ideologically heightened form—as magnanimous, selfless, unrequited acts of voluntary generosity performed by what appear as disembedded individuals—that is to say as charity.
> (Muehlebach, 2012: viii)

> Cremin (2011, 2012), in this respect, refers to the existence of a psychically enjoyable "crisis capitalism." Thus, the looming threat of climate change and economic disintegration ironically is exactly what offers individuals the greatest safety as neoliberal subjects. Cremin highlights how modern identity is linked to a "crisis industry" that allows individuals the opportunity to constantly secure their identity by ethically bettering neoliberalism in both big and small ways. Such insights are complemented by Costas and Grey's (2014) association of modern selfhood with a fantasy of the "future oriented self." Here, the imagining of a better tomorrow helps individuals cope with the difficulties and oppressions they are experiencing in the present. Specifically relevant to themes of neoliberalism and ethics, subjects affectively deal with the problems of the contemporary free market by continually fantasizing about the possibilities of its ethical improvement in the long term.

Often a strong feeling of pride accompanies this identity. Indeed, the capacity to do one's part is translated into a proud ethical self even at the most unethical of times. Hence, the world may be universally imperiled by fossil fuel–induced environmental collapse, but one can still buy a hybrid car. Coming from a perhaps different ideological direction, even in the face of moral decay and the loss of traditional social bonds, one can still be a strong and ethical family man and member of the local community. Such an attitude is often lampooned as mere smugness. Nevertheless, it is indicative of a psychic economy in which there is enjoyment to be had and security to be found in the effort to be ethically responsible in any way one realistically can.

This modern striving for the achievement of an always-elusive fully ethical capitalism points to the critical subjective relation between neoliberalism and ethics. In particular, the use of mindfulness and other wellness techniques, as previously discussed, represents a type of surrendering oneself to this hyper-capitalist reality. It is a substitution of personal spirituality for true collective transformation. The "good" organization, for instance, is not the one that does not exploit its workers but rather helps its employees peacefully survive the suppression. Quoting a recent *New York Times* article:

> the advice to be more mindful often contains a hefty scoop of moralizing smugness, a kind of "moment-shaming" for the distractible, like a stern teacher scolding us for failing to concentrate in class. The implication is that by neglecting to live in the moment we are ungrateful and unspontaneous, we are wasting our lives, and therefore if we are unhappy, we really have only ourselves to blame. . . . This is a kind of neo-liberalism of the emotions, in which happiness is seen not as a response to our circumstances but as a result of our own individual mental effort, a reward for the deserving. The problem is not your sky-high rent or meager paycheck, your cheating spouse or unfair boss or teetering pile of dirty dishes. The problem is you.

Yet this disciplining appeal for personal betterment goes even further to be the primary force for bringing about social justice. It is not simply that one must be mindful; one should use this more enlightened way of being to "think positively" about how to balance market demands with ethical desires.

Identity, hence, is secured in this attractive means of coping with the neoliberal system through this ongoing attempt to personally make it more ethical. It is an inscriptive fantasy of the entrepreneurial problem-solving subject who heroically is able to moralize the market—even if only in rather limited ways. Consequently, it is the duty of all people to make the world a better place. And it is in the attempt to personally overcome this ethical lack that neoliberalism is reproduced and the modern market subject maintained.

The Subjective Power of Neoliberal Ethics

The power of neoliberal ethics extends beyond the mere material or rational. As this analysis has sought to illuminate, the higher moral ideals that organize politics, institutions and even interpersonal relationships within this ostensibly free market system also subjectively shapes the very formation of the modern self. The contemporary subject is one forever seeking to be and do good in a bad neoliberal world. In this respect, while individuals may be economically exploited by the free market, they are also continuously and constantly affectively gripped by its nonmarket ethics.

This ethical grip may be universal, but it is far from equal in its actual social effects. Theoretically, this responsibility should fall most heavily on those with the most power within contemporary capitalism to change things. Corporations and politicians can appear to be the most able to realize these widely shared ethical aspirations and should therefore be held most accountable when they fail to meet these goals even if only partially. As already discussed, the consequences of such an ethical lack are most severe for those with the least concrete ethical agents. Socially oppressed groups, often with the least material means, are commonly the most ethically held responsible.

At the affective level, this means that those who are most materially and socially vulnerable are often the most psychologically anxious and fearful. Whereas this chapter has spoken consistently about the idea of enjoyment, this should not be confused with pleasure. Rather, it is symptomatic of the ways subjects cling to an identity and agency that is not of their own making but is all that is socially available to them. Predictably, this enjoyment takes its greatest psychological toll on those at the bottom. Communities separated by culture and geography commonly experience shared psychological distress associated with the forces of marketization and financialization. This includes greater mental health issues, drug addiction and—in some parts of the developed world, for the first time in generations—declining life expectancy (Moncrieff, 2006; Teghtsoonian, 2009).

This reflects more fundamentally the hysterical nature of modern capitalism. According to Lacan, the discourse of the hysteric represented the almost pathological and obsessive seeking of confirmation by individuals from an idealized authority or Big Other. In the context of present-day neoliberalism, it is a constant questioning of oneself as to whether one is morally and ethically good enough, reflected in the form of what Roberts (2005: 637) refers to as a "self-defining discipline" revolving around

> the processes of comparison, hierarchization, differentiation, homogenization and exclusion that Foucault observes as the objective mechanisms of discipline have as their necessary correlate similar process[es] "within" the ego as I seek to fix identity in the essentially competitive space of the mirror of my own and other's objectifications.

In the current context, this is translated in the continual attempt by individuals to live up to such demanding normative standards. More precisely, subjects compare themselves to the tantalizing but impossible ideal of ethical capitalist upon which they judge their own worth and that of others.

The figure of the perfectly moral and ethical market subject is thus both grounding of identity and fundamentally disciplining. Neoliberalism confronts subjects with the difficult choice of either giving up their own morality and embracing the immorality of the free market or doing all they can to maintain their sense of moral integrity in what appears to be a losing battle. The only solace is the personal and collective ways they can work to make this hyper-capitalist system even slightly better—an effort that can take quite a large psychological and physical toll.

To this end, neoliberal identity is reduced to a diverse undertaking of ethical coping, doing whatever one can to persevere and maintain one's own moral integrity within an uncaring and uncompassionate market order. Neoliberalism was initially premised on the principle of survival of the fittest. This meant doing whatever it takes to materially survive and thrive. Yet the latest stop in its evolution is one in which individuals are simply trying to use ethics to personally and collectively persist, to create in the midst of an increasingly competitive unregulated world a small semblance of righteousness.

Yet far from liberating individuals, this call to be ethical leads them to invest even more deeply in neoliberalism for their very psychic survival. It traps subjects in its fantasy of a moral market. Accordingly, it renders them disciplined subjects of desire to neoliberalism, aspiring and taking it upon themselves to maintain their own morality and find ways to achieve social justice despite the negative effects of the market. The inability to do so, fully and often partially, simply makes individuals feel guiltier about their own ethical lack and intensifies their desires to overcome it. Such "blaming the victim" is commonly and tragically the only thing preventing modern subjects' psychic disintegration even as the world disintegrates ever faster around them. At stake, then, is how and who can fight this material and subjective power of neoliberal ethics.

8 Fighting the Power of Neoliberal Ethics

Morality and ethics are often counterposed to neoliberalism. They are framed as its polar opposite—the desire for good against an increasingly exploitive system. They are, therefore, often turned to as social forces for challenging this hyper-capitalist system. As shown throughout this analysis, however, it is actually nonmarket values and desires that ironically reproduce neoliberalism and keep its subjects under its concrete and affective rule. It is imperative to understand how to fight this power of neoliberal ethics.

There is a certain and alluring psychic safety found in continually investing in modern capitalist ethics. It holds out the tantalizing possibilities that the free market is fixable—it can be cured and made better. Indeed, the history of neoliberalism is the continual promise of individual and collective progress. Initially, this meant simply replacing "big government" with the free market. When this failed to fully deliver on its promises of prosperity, it evolved into a positive discourse of the saving power of "good governance." This was captured when the president of the World Bank, Barber Conable, optimistically proclaimed, "If we are to achieve development, we must aim for growth that cannot be easily reversed through the political process of imperfect governance" (quoted in Doornbos, 2003). Yet now a new ideal has taken hold—that of a moral capitalism created by ethically committed neoliberal subjects. Consequently, it is assumed that there is no need to stretch our political imagination or revolutionize economic relations. Instead, the current capitalist status quo requires only some minor moral tweaks for the creation of a just global market.

It is a safe identification linked to a shared cultural fantasy that at least temporarily offers a secure reality that can offset the threat of material and psychic insecurity. Nevertheless, this ethical neoliberal dream is becoming harder and harder to sustain. The lack of agency and empowerment individuals encounter translates into something much greater than the enjoyable identity founded on being an ethical subject in an unethical world. It signals the complete disintegration of this romantic vision. It is no longer a productive subjective experience of ethical life. It is now close to becoming a complete systematic breakdown in the face of the impossibility of being moral and ethical in this system.

These internalized demands to be good and do good increased as recent economic pressures and political problems mounted ever higher. Neoliberal politics has always contained within it a certain liberal cynicism in which

> the possibility of a dramatic revolution seems naive and idealistic in the extreme. Such thinking, they hold, totally ignores the cold, hard realities of the 21st century, where the free market is here to stay, globalisation cannot be turned back and the best that could be hoped for are minor improvements. At the heart of this cynicism lies the pragmatic belief in the possibility of bettering society. Proponents want to achieve reforms despite an entrenched economic and political system. They had long ago given up on winning the war but there were still real and important battles to be fought.
>
> (Bloom, 2016c)

However, for many this cynicism is turning into a full-blown sense of nihilism, in which the slim veneer of ethics covering neoliberalism's exploitive core is being torn off for all to see, leaving people with little if anything to continue to believe in.

Neoliberalism is, hence, facing a profound ethical and therefore existential crisis. It is less and less possible and desirable to fix one's individual and collective identity to the potential of a moral free market. People are left with little normatively or psychologically to invest in. Even partial victories are increasingly denied to them, engendering despair rather than a longing for more. The status quo is progressively being rendered irredeemable and not worth concretely or ethically saving.

There is, of course, much to be excited about in this growing rejection. The ethical fantasy of neoliberalism was always one that would lead to disappointment and the ultimate limitation of social possibility. It kept the market firmly in place as an effort, thus reinforcing its ideological dominance and social domination. However, there is a need to be cautiously optimistic on this front, as it creates new forms of old repressive politics that deepen values and practices of authoritarianism and oppression rather than transcending them. This chapter explores, accordingly, the growing desire to move beyond the fantasy of neoliberal ethics.

The Ethical Crisis of Neoliberalism

For much of its short history, neoliberalism has been confronted with a dramatic ethical crisis: how to legitimately present the free market as moral. Critical in this respect was the creation of a multifaceted cultural fantasy of moral and ethical capitalism, one that combined ideologies of marketization and financialization with broad-based desires for social justice and environmental sustainability. However, such an impossibly romantic vision was getting increasingly harder to justify or sustain. Despite hopeful claims of elites, the ability to actively make the current market system more righteous increasingly seemed like an officially propagated delusion. As the recovery following the financial crisis failed to appeal for the majority of people, the possibility of ever creating a more ethical market, whether on a large or small scale, appeared ever more remote and was hardly secure enough to concretely or psychologically invest in.

Ethics, here, served as perhaps one of the primary components of contemporary subjectivity. There was a strong identification with being a good person, both in terms of values and actions—of being able to lead a "good life" within or even despite these neoliberal conditions (O'Flynn and Peterson, 2007). It was a response to the increasing conscription of individuals into a hyper-capitalist system. This reflected the previously discussed distinction between subjection and subjectification. Respectively, they represent, in this regard, the ability to transcend a "subjectless" capitalist existence in which one is reduced to being a cog in the machine or, in more modern terms, demand number and a performance spreadsheet. Consequently, the "self" emerges as a "spectral shadow" symbolizing the desire to be a unique person in this often dehumanizing system (see Tie, 2004; Bloom, 2016b). For this reason, Žižek (2001) contends that the self is always experienced as "exceptional"—the exception to a regulative and overdetermining ideological order.

In the present day, this striving for selfhood is commonly seen in the popularity of self-help discourse or, conversely, the figure of the antihero (Tie, 2004). Both offered symbolically a sense of freedom and agency in an otherwise subjectively stifling system. It, further, bonded individuals together in a shared sense of identity. As Davies (2014: 13) presciently notes,

> for social relations to be organised into reasonably persistent, reliable and peaceful institutions, at some point there must be a shared sense of normativity, a shared basis on which to distinguish between people and between things and make evaluations of their relative worth.

The opportunity to be ethical was a potent opening to personally and collectively experience such neoliberal selfhood. At stake was the feeling of being able to shape, rather than being shaped by, the free market. It is represented, for instance, in the appeal of employability, which offers individuals a supposed means of controlling their economic destiny. To be moral and ethical was a daily affirmation that one still had some agency in which to socially maneuver. The idea of self-help spoke to such longings in that it signifies an identity that was not completely overdetermined but still had a potential to develop and grow independently.

Such agency became to a large degree principally a means for, again, daily coping with neoliberalism. The much-critiqued and already discussed perceived inevitability of the system consigned the possibility of radical change to appear at best idealistic and at worst nefarious. This moral agency is a far cry from conventional ideas of self-determination. Rather, it revolves around an ethos of self-care (Lemke, 2001) and, in a quite restricted capacity, self-creation. The ability to "be yourself" at work or to craft a unique identity—within very narrow limits—was a means to "distract" from the oppressive character of modern life and organizations (Houlihan, 2002). It was an allowance of people to "be monstrous" in their contravention of superficial social and institutional norms in order to mask the deeper moral and ethical monstrosity of a free-market existence (Bloom, 2014b).

This restricted sense of freedom continues a tradition of capitalist agency—the capacity to shape one's own destiny without fundamentally altering the economic order. Principles such as democracy and even a welfare state pushed back against such limitations and, in a restricted sense, expanded individuals' social freedom. By contrast, in the context of neoliberalism there is much less ideological scope for change. What is available is a form of caretaking freedom that motivates and encourages individuals to deal with this permanent free-market reality in the most healthy and most ethical way that they can.

These insights reflect and, to an extent, dramatically extend emerging critical understanding of contemporary capitalism and freedom. Whereas neoliberalism ideologically champions individual liberty, actually existing neoliberalism is a much more restricted experience. Even more fundamental than the modern freedom of the market is the construction of an ethical subject that was able to be and do good without putting into jeopardy the underlying ideological or concrete dominance of his hyper-capitalist order. More precisely, it is the opportunity to ask, "What are the possibilities to create a fiscally viable ethical society?"

However, the possibility of moral capitalism is increasingly being profoundly challenged. The potential to even cope with this market order appears progressively nonexistent. The latent morality and justice of this system furthermore appeared increasingly as a mirage. It is more and more seen as simply impossible to live a life of care toward oneself, others and broader society. Consequently, a dramatic crisis of neoliberal ethics is emerging at all levels of social existence.

The Problem of Neoliberal Ethics

The chronic issues associated with neoliberalism profoundly put into question its ability to realize the common good or allow for individual morality. A critical part of its social legitimacy was its diverse and evolving capacity to be moralized. This historical evolution stands in stark contrast to the original justification trumpeting the free market's inherent moral goodness. Rather, it is increasingly put forward as an explicit and implicit opportunity for subjects to realize broader nonmarket goals of social justice and personal righteousness. This ethical fantasy offered, in turn, ample opportunities for subjects to innovate in the name of achieving these normative goals. However, as this hyper-capitalist order revealed itself to be fundamentally illusory, such ethical desires shifted to how to resist neoliberal ethics themselves.

To this effect, ethics is commonly posed as a form of resistance to neoliberalism. If the free market emphasizes the bottom line at the expense of all other priorities and values, then it is up to individual communities and even governments to fight for values of public welfare and environmental sustainability. This runs counter to the traditional notion of the "moral" neoliberal subject—construed as "an individual who is morally responsible for navigating the social realm using rational choice and cost benefit calculations grounded on market-based principles to the exclusion of all other ethical values and social interests" (Hamman, 2009: 37). In this respect, traditionally,

> [t]he model neoliberal citizen is one who strategizes for her- or himself among various social, political, and economic options, not one who strives with others to alter or organize these options. A fully realized neoliberal citizenry would be the opposite of public-minded; indeed, it would barely exist as a public. The body politic ceases to be a body but is rather a group of individual entrepreneurs and consumers.
>
> (Brown, 2005: 43)

However, in the current age, this individualist and explicitly market account of the subject is complicated. Rather, individuals stake their sense of self to using these techniques and finding the wherewithal within themselves to forge a more ethical society out of a rather disjointed and repressive neoliberal existence.

The longing for a more ethical capitalism is itself an ultimately safe form of resistance. Resistance is, significantly, always

> inherently to a degree "safe." Safety here is understood not principally as "making safe" the position of "power holders"; rather, it refers to the ontological security resistance grants individuals. This "safety," in turn, deepens our identification with prevailing power relations, our identity secured in our production as resistance subjects.
>
> (Bloom, 2013: 232)

Specific to neoliberalism, this ethical resistance was "safe" in two distinct but interrelated ways. It was a stable foundation upon which to construct and continually reconstruct identity. It had the benefit of simultaneously providing individuals ontological security as subjects of the market without having their sense of self completely defined by its prerogatives. Subject were able to accept the inevitability of neoliberalism while still embracing their limited ability to change it. Reflected was a form of what Contu (2008) refers to as "decaf resistance" on a massive level.

However, it was also safe to the maintenance of the hyper-capitalist system. This is not always the case per se. This ethical safe resistance served to reinforce the prevailing hegemony of the free market. Notably, it contributed to the market being not only a dominant discourse but a dominating one as well. Foucault (1988: 3), in this respect, defines social domination as "When an individual or social group manages to block a field of relations of power, to render them impassive and invariable and to prevent all reversibility of movement . . . we are facing what can be called a state of domination." Key, here, is the irreversibility of neoliberalism reinforced by this ethical orientation. Here, the capacity for neoliberalism's limited reform was exchanged for the possibility of its radical transformation. Thus, ironically, its survival came from the moral confession of its own citizens and its openness to being morally perfected as long as its core capitalist values remained sacred. It was hence a rather strange cultural orthodoxy—a dogmatic belief not in the ideological order itself but rather in its restricted ethical redemption.

The vitality of neoliberalism therefore flowed from its approved struggle for goodness. It was continually revitalized by subjects' attempts to morally reform themselves and ethically reform the system in small everyday and potentially even bigger ways. Hence, in the contemporary period, it is precisely such a "safe" resistance that makes capitalism progressively dominant. It formed an emotionally resonant narrative "script" (Domagalski, 1999; Fineman, 1993) for individuals to continually structure and restructure their existence around. It represented a tantalizing vision of being able to simultaneously embrace and contest an invasive and scripted neoliberal order.

Revealed was complex character of fantasy, composed dually of a positive and a negative component (see Bloom and Cederstrom, 2009):

> On the one hand, fantasy has a beatific side, a stabilizing dimension, which is governed by the dream of a state without disturbances, out of reach of human depravity. On the other hand, fantasy has a destabilizing dimension, whose elementary form is envy. It encompasses all that "irritates" me about the Other, images that haunt me about what he or she is doing when out of my sight, about how he or she deceives me and plots against me, about how he or she ignores me and indulges in an enjoyment that is intensive beyond my capacity of representations, etc.
> (Žižek, 1998: 192)

In the case of neoliberalism, this fantasy was even further complicated. The market took central stage in both its positive and negative affective dimensions. The "enemy" was an unethical free market, a capitalist order unrestrained by conventional norms or a sense of right or wrong. By contrast, what was desired, what was being continually undermined, was a more ethical and caring form of capitalism. Critically, this did not mean a return to the "welfare state"; it meant the freeing of individuals to personally and collectively find innovative ways to improve their communities.

In practice, this official resistance fantasy found itself in peril. This appealing opportunity to be ethical became, in time, a disciplining machine in its own right. It was more and more a moral duty to be a righteous market citizen. It was part of one's personal ethical responsibility to fix the free market. This once-precious ability to ethically cope with the free market was transformed gradually into a repressive force. Empowering discourses such as work-life balance, mindfulness or social entrepreneurship were commonly perceived to be quite regulative and constrictive. It is perhaps therefore not surprising that people sought to escape and push back against these oppressive ethical demands. The expectation to be healthy in mind and body and spirit was met with novel forms of committed anti-ethical resistance. The elusive search for balance lent itself to a renewed obsession with one's career (see Bloom, 2016c). Tellingly, the financial crisis gave birth to a type of "cyber-hedonism" in which the desire for pleasure took the place of longings for social change. Specifically,

> [i]n authoritarian countries with rising living standards—such as Russia and China, until recently—official tolerance of cyber-hedonism has been a sort of Faustian pact offered by the authorities: we will let you enjoy yourselves, in new and unconventional ways, if you keep off politics. But now that economies have turned sour, will the young go on keeping their side of that bargain?
>
> (*The Economist*, 2009)

Of course, as will be explored, this also catalyzed renewed forms of social radicalism. Nonetheless, this "Faustian bargain" symbolized a generalized way that neoliberalism was seeking to legitimize itself in a world where its own ethical potential was increasingly shown to be null and void.

Consequently, ethics was no longer a catalyst for resistance so much as the very thing that had to be resisted itself. The increasing discipline and character of this striving to be moral revealed its discursive and hegemonic ideological underpinnings in serving the interest of profit over people. As such, the fundamental tension plaguing neoliberalism was no longer how it could be made more ethical. Rather, it was whether morality, justice and even human survival was even sustainable within such a system. The neoliberal fantasy of a moral capitalism failed in this respect to fulfill its crucial function in covering over the ideological and concrete contradictions of the system through an affectively appealing shared cultural identity.

It was thus no longer secure to anchor one's identity to the possibility of a moral free market. The attractiveness of ethics was, in a perverse sense, always in its impossibility. It was a becoming that was never to be realized and always on the horizon. There was a desire for a moral capitalism fueled by an ethical market. These romantic dreams were quickly fading in the face of a normative promise of justice that was never going to arrive. Such moral notions of neoliberalism had always been utopian, but now they were revealed as an imaginary "no place" that was little more than an illusion. Neoliberal ethics no longer offered the promise of failed transcendence; it simply offered one of failure.

The reproduction and sustainability of the free market therefore required a loyal moral opposition. It depended, more precisely, on the construction of a safe ethical resistance. However, actually existing neoliberalism was far from stable—as its issues of deprivation, poverty and subjective insecurity could not be resolved with a mere minor ethical fix. Thus, to continue to believe in this fantasy was to consign oneself to a dim fate based on false promises. Accordingly, it was now neoliberal ethics that must be resisted.

Taking Responsibility for Resisting Neoliberal Ethics

In recent years, people are increasingly rejecting not just the formerly sacred tenets of the free market, but also the disciplining demand for them to morally and fix an unsalvageable hyper-capitalist system. Rather than being seen as something to be fixed, it is progressively viewed as a rigged march of folly leading in the present to oligarchy and authoritarianism and going forward to the potentially even worse fates of kleptocracy and fascism. The non-market ethics that ironically sustained this corrupt capitalist order are now increasingly synonymous with an inscriptive attempt to redeem that which is fundamentally socially unredeemable. It appears to be a cosmetic fix to a progressively understood to be systemic problem.

Accordingly, such ethics were considered to be a barrier rather than a gateway to freedom. In its most cherished presentation, the ideas of cooperation, care and justice were invested in as a means of altering the excesses of neoliberalism and finding some small spaces of agency against the domination and immorality of the free market generally. However, they were more and more culturally perceived as the exact obstacles to actually achieving a semblance of social freedom and being oneself—let alone as a genuine means of changing the system in any serious way.

The perceived unscrupulous principles of political correctness and wellness were perhaps high-minded ideals, but in practice they were felt by many to be more socially regulating than emancipatory (see Barnes, 1992; Sparrow, 2002; Williams, 1999). This was reflected, for instance, in a new "vice" lifestyle that eschewed any sense of political correctness or conventional moral decency. Here,

> the trangressive [*sic*] gesture in consuming shock media derives its *jouissance* from a temporary negation of the ego ideal of self-realisation that underlies contemporary modes of being at work and in consumption. The transgressive gesture in consuming the imagery of shock media amounts to the positing of an alternate ego ideal, which takes the shape of the cynical hipster for whom the notion of others' estimation of them is anathema. Work identities are here met with a nihilist indifference to

> organizational "committed" selfhood and careerism, while "authentic" consumer identities are interrupted by an obscene preoccupation with sleaze, hedonism and sarcastic denunciation.
>
> (Hoedemaekers, 2016: 61)

Thus a new fantasmatic narrative was arising—one in which the free market was a "malicious" barrier to progress. Originally, it was government, bureaucracy and socialism that threatened the very existence and sanctity of freedom. Now, in a somewhat ironic twist of history, it was nonmarket ethics that were preventing people from breaking free from the hegemony representing a barrier to their prosperity and liberty. Elliott (2013: 83) notes, thus, that neoliberal subjects encounter a form of overwhelming "suffering agency" representing "the experience of agency as an omnipresent and overriding burden for the neoliberal subject."

In its place emerged alternative utopian discourses. Although often ideologically opposed, they shared a growing rejection of the system and its associated disciplining moralities. To this effect, it was observed within the mainstream press that "Economic woes create anti-establishment movements around the world" (Allen, 2016). As even the *Economist* presciently noted, "Large parts of electorates across the West are fed up with traditional political parties that have in recent years presided over rising inequality and slow economic growth" (*The Economist*, 2016). This was witnessed, furthermore, in the rise of populist conservative movements like the Tea Party and more recently the election of Donald Trump as US president and the Brexit referendum. It was also apparent in the rebirth of grassroots progressive movements across Europe, the United States and the world, directly challenging the hegemony of capitalism and a global free market.

Reflected in this respect was the emergence of the anti-ethical neoliberal subject. In particular, it was a push against the disciplining moralizations that marked neoliberalism in favor of more radical solutions. The seeds of this revolt could be, as previously mentioned, observed in the popularity of the antihero in early 21st-century culture (Tie, 2004). The crafting of identity against prevailing currents of neoliberal ethics was an evolution of this antihero ethos. It was the appeal of a new subjectivity that associated identity with the ability to actively question and undermine the false moral fixes of the free market. It attracted a growing number of people who saw through the sanctimonious but ultimately exploitive discourses of ethical capitalism and were suddenly desperately seeking something to replace it.

Underlying this anti-ethical orientation was a fresh discourse of responsibility. Neoliberal ethics at first rested on empowering the subject to responsibly govern themselves politically, institutionally and personally. It further provided an emotionally resonant opportunity to moralize capitalism itself—a prospect especially attractive given its often-stated increasingly perceived inevitability regardless of its social or human cost. This transformed, however, into another way neoliberalism disciplined and coerced people into accepting this limited market reality.

It reflected, significantly, the renewed desire to find someone or some group to blame for these economic and social failures. This revealed the beginnings of a new fantasy—one premised on the evil "neoliberal" other. There was a growing paranoia about malicious elites and thieving "minorities" aiming to literally steal the prosperity of ordinary "people" and figuratively prevent them from enjoying the psychic harmony of being a secure market subject. This echoes the demonization of Jews in the 1930s, in which

> from being the positive cause of social antagonism, the "Jew" is just the embodiment of a certain blockage—of the impossibility which prevents the society from achieving its full identity as a closed, homogenous totality. . . . Society is not prevented from achieving its full identity because of Jews: it is prevented by its own antagonistic nature, by its own immanent blockage, and it "projects" this internal negativity into the figure of the "Jew." In other words, what is excluded from the symbolic (from the frame of the corporatist socio-symbolic order) returns in the Real as a paranoid construction of the "Jew."
>
> (Žižek, 1989: 127)

The question at present, then, is twofold: firstly, who should be made accountable for these market failures, and who should be responsible for making those people accountable? Obligation, hence, revolves around the duty to individually and collectively challenge neoliberalism and its moral inscriptions. It was upon these foundations that a new responsible morality and ethics would diversely manifest itself against the free market.

Morally Resisting Neoliberal Ethics

In the current era, there is a growing desire to resist and ultimately do away with the neoliberal status quo. These desires for change, of course, hold much optimism for the future. Yet they also can manifest in some quite unappealing and troubling ways. Indeed, the free market is being threatened with replacement by the revival of a populist authoritarianism dangerously characterized by racism and xenophobia. Moreover, there has been a dramatic resurgence of a progressive politics containing a genuine mass critique not just of neoliberalism but potentially of capitalism more generally. Despite their differences, both of these movements have quite similar historical roots. Each stands as a passionate political response against, at least in part, the disciplining morality of neoliberalism.

It is tempting, and not altogether misguided, to understand these competing but similar sentiments from the right and left in almost purely negative terms. They represent a total longing to break with the prevailing socioeconomic order that has dominated the past several decades. As mentioned, the results of Brexit and the shocking election of Donald Trump as US president revealed a widening gap between political elites and those they were meant to respectively speak for and serve. In particular, it revealed a broader "anti-establishment" ethos that crossed class, ideology and ethnicity. According to Leslie (2016):

> It is traditional for the left to decry the establishment, but these days the right also rages against the machine. Nigel Farage—stockbroker, alumnus of Dulwich College—has built a successful political career on his anti-establishment credentials. Across the Channel, Marie Le Pen is doing something similar. Hatred of the establishment is one of the few unifying forces in an age of polarisation. Anti-establishment leaders from either end of the ideological spectrum pay coy tribute to one another. In America, both Donald Trump and Bernie Sanders have achieved outrageous success by running against their respective party establishments.

These feelings of alienation were traced back to a deepening sense of political disempowerment and economic anxiety. The primary target of these attacks was not a particular idea or policy per se; it was the entire system itself. "In an age in which progressive social movements have largely crumbled," notes Malik (2016), "and in which people feel they have lost control of the forces that govern their lives, political anger often finds expression not through opposition to a particular policy or a government, but through a generalised hatred of everything and everyone in power." Thus,

> while Trump himself may be unique—and distinctly American—Trumpism and the politics of alienation he represents are not. In fact, most of the forces fueling Trump's rise are already wreaking political havoc around the world—including in a Europe buffeted by its own volatile mix of institutional malaise, economic stagnation, and demographic change.
>
> (Benaim and Cammack, 2016)

This spreading full-scale rejection of neoliberal morality left a distinct moral vacuum. Such desires transcended mere nihilism. It was not that people believed nothing—only that a growing number no longer accepted and embraced what was currently on offer, so to speak. Created, in turn, was a discursive space for alternative moral frameworks to entrench themselves and fight for hegemony. Nevertheless, the lack of ideological imagination produced by the decade-long supremacy of neoliberalism further complicated this already dislocating situation. There were few cultural resources to envision a radically transformed future. All that could be done was to hold off the looming free-market apocalypse through the rehabilitation of once-cherished past moralities.

However, before fully delving into these competing revitalized discourses of neoliberal resistance, it is crucial to pause to highlight their profoundly anti-neoliberal character. On the surface, neoliberalism could credibly theoretically claim the mantle as a dramatically anti-elitist ideal. It railed against the individual as well as social bondage caused by stifling collectivist institutions—promising to help humanity break free wholesale from a culture of conformity and thus liberating people to pursue their own enterprises and dreams. Yet such rebellious beginnings where progressively undercut by the fact that neoliberalism had become the establishment itself and, moreover, that its vaunted freedom seemed almost solely reserved for elites.

By contrast, these movements challenged this individualist mentality with a new populist emphasis on the "people." In the words of Zakaria (2016: 9):

> What is populism? It means different things to different groups, but all versions share a suspicion of and hostility toward elites, mainstream politics and established institutions. Populism sees itself as speaking for the forgotten "ordinary" person and often imagines itself as the voice of genuine patriotism.

This echoed a conventional populist trope of investing in a state or authority to fight for the "small man" or the "little guy" against corrupt and unjust socio-economic forces (see Varley and Curtin). Critically, it must be noted that

> populism today has changed and developed from its earlier iterations, embedded as it is within a rapidly shifting political and media communications landscape . . . we need to move from seeing populism as a particular "thing" or entity towards viewing it as a *political style* that is performed, embodied and enacted across a variety of political and cultural contexts.
>
> (Moffitt, 2016: 3)

Such historically infused populism has taken on a number of different forms. Perhaps most notably, it has coalesced around a shared cry for greater economic security. There was a common, quite understandable demand for governments to bring back lost jobs. It harkens back to a rosy vision of the recent postwar past, in which it was supposedly possible for everyone to work and in doing so achieve a middle-class life. To this end—speaking of the European context but potentially with wider international implications—

> although the threats to security and economic stability that have rattled Europe in the past few years may have spurred the current populist surge, they did not create it. Its origins lie further back, in the structural shifts in European society and politics that began in the 1960s. Because so much commentary on contemporary populism overlooks its deep

> historical sources, many observers fail to appreciate the durability of today's populist appeals and the likely staying power of the parties built around them.
>
> (Mudde, 2016: 25)

An abiding theme, thus, of this politics was that many had been left behind by globalization and free-market progress. It was an effort to bind together individuals to combat a status quo that served profits over people. In this respect, it represented the rise of a "protest populism" that had the potential to fundamentally restructure the main political divides of the 21st century (Kriesi, 2014)—notably around the neoliberal "establishment" and the people.

This popular challenge to neoliberalism could lead to quite right-wing and conservative places politically. It associated the entirety of the liberal political project—including values of multiculturalism, diversity and social equality—with a corporate agenda of free-market globalization. According to Seib (2016):

> In Europe and the U.S., the movement is fuelled by middle-class economic insecurities, exacerbated by fear of immigrants arriving to steal jobs or soak up welfare money and other taxpayer dollars. Those anxieties are overlaid with an absolute conviction among many citizens that existing political leaders either don't understand or, worse yet, don't care.

It reinvigorated nativism, turning on a widely held notion of the nation as a home whose borders and citizens must be protected from the threat of foreign invasion. In place of a more global order was a patriotism celebrating the sacredness and centrality of the nation and its people. Not surprisingly, such

> economic vulnerability is conducive to in-group solidarity, conformity to group norms, and rejection of outsiders. When threatened, groups are thought to seek strong, authoritarian leaders to protect them from what are perceived as dangerous outsiders seen as threatening jobs and benefits. Anxiety arising from contemporary events—boatloads of migrants and refugees flooding into Europe, images of the aftermath of random acts of domestic terrorism in Paris, Brussels, and Istanbul, and austerity measures—are blamed for exacerbating economic grievances linked with rising income inequality, the loss of manufacturing jobs, and stagnant wages.
>
> (Inglehart and Norris, 2016: 11)

It has also rather predictably taken on a more explicitly racist turn. The onslaught of a disconnected and exploded globalization has produced a pronounced modern tribalism (Bloom, 2017).

Emerging from the opposite end of the ideological political spectrum, however, was a revitalized progressive grassroots movement. It was likewise populist in composition and revolutionary in its rhetoric. Yet it promoted an inclusive class-based agenda. In the main, it railed against the power of the 1% and the decade-long destruction of the middle class by neoliberalism and free trade. Yet it also contains a sustained critique of liberalism and its seeming obsession with so-called identity politics. Quoting one prominent US commentator in the wake of Trump's election:

> We need a post-identity liberalism, and it should draw from the past successes of pre-identity liberalism. Such a liberalism would concentrate on widening its base by appealing to Americans as Americans and emphasizing the issues that affect a vast majority of them. It would speak to the nation as a nation of citizens who are in this together and must help one another. As for narrower issues that are highly charged symbolically and can drive potential allies away, especially those touching on sexuality and religion, such a liberalism would work quietly, sensitively and with a proper sense of scale.
>
> (Lilla, 2016)

Arising was a fresh political morality combining economic and social justice without sacrificing either one for the other.

Significantly, despite their vast difference, both of these emerging anti-neoliberal moralities strove to move forward by looking backward, so to speak. The nationalism and racism of the right idealized a past of more homogenous communities. Likewise, the progressives of the 21st century focused on an attempt to recapture past social economic gains snatched away by neoliberalism in the present age. Revealed was a mass desire for an ontological and psychic security that neoliberalism and its attendant moralities could no longer provide. The "hope" of liberalism, with its emphasis on diversity, and of neoliberalism, with its prioritization of the free market, were quickly wilting. It also pointed to the need for a more radical ethical break with a liberal past and a neoliberal present.

Solving the Problem of Neoliberal Ethics

At the heart of neoliberal ethics beats an ethos of problem-solving. Notably, it is the responsibility of individuals and communities themselves to fix the issues that afflict contemporary capitalism. Initially, this meant being self-reliant economically—breaking the personal and collective dependence on government welfare. However, this analysis has sought to show that this transformed into a moral obligation to resolve neoliberalism's ethical problems. Yet this has further evolved as a way to solve the problem of neoliberal ethics itself.

The principal problematic of the contemporary free market is how to balance the drive to maximize profit with social, economic and environmental justice goals. For this purpose, it generated an ongoing demand for subjects to devise innovative ways to deploy financial techniques for these righteous ends. Increasingly, this imperative is morphing into an ethos that aims at directly challenging and transcending neoliberalism. The ethical question has therefore shifted—in the wake of the financial crisis, people are increasingly asking "How can we break free from the power of markets and corporate rule?"

This points to a potentially profound, yet unexplored, dialectic at the core of neoliberal ethics. Marx's original dialect, philosophically drawn from Hegel, grants the proletariat a historically overdetermined role in revolting against capitalism and bringing about communism. Specifically, the unquenchable demand but ultimately shortsighted desire for ever-greater profit would lead to mass unemployment and the drive for an entirely new social economic system. The critiques of essentialism that have been levied against traditional Marxist theories of the dialectic are well served and seemingly borne out historically.

Nonetheless, neoliberalism displays an interesting dialectical movement of its own. The ethos of problem-solving that is so crucial to its survival and maintenance will provide subjects with the potential ability to fundamentally and finally solve the problem of neoliberalism itself. This is especially important in light of how limited the current social political imagination is, due, as discussed, to the enduring ideological hegemony of the free market. This means taking a more dynamic critical view of neoliberalism, one that does more than simply critique it as a hyper-capitalist system par excellence. Importantly,

> [t]he instrumental use of notions of governmentality to sustain theories of neoliberalism and neoliberalization supports a two-dimensional understanding of political power—which is understood in terms of relations of imposition and resistance—and of geographical space—which is understood in terms of the diffusion and contingent combination of hegemonic projects. Theories of neoliberalism provide a consoling image of how the world works, and in their simplistic reiteration of the idea that liberalism privileges the market and individual self-interest, they provide little assistance in thinking about how best to balance equally compelling imperatives to respect pluralistic difference and enable effective collective action.
>
> (Barnett, 2005: 7)

Thus, while politically neo-liberalism has arguably robbed individuals of the ability to fully envision the possibilities of an utterly different more emancipated world, it has unintentionally given these very same subjects the innovative skills required to potentially concretely transform it.

Indeed, in the wake of the financial crisis there was a new emphasis on credibly thinking beyond neoliberalism and its market paradigm. Calls to break up the banks responsible for the financial crisis and to implement former social democratic reforms bore early witness to such radical aspirations. Likewise, there were fresh opportunities to reverse capitalism's dramatic environmental degradation by moving "beyond green capitalism":

> The reign of capital has now been thrown into disarray not only by financial chaos, but also by the shrinkage and disruption of the natural infrastructure which serves not only the survival needs of the human species but also the particular requirements of the capitalist ruling class. . . . While there are obvious points of convergence between different green agendas, it eventually becomes clear that any full merger between an agenda that is insistently capitalist and one that accentuates the green dimension is impossible. On the other hand, immediate pro-ecology steps are urgently needed, irrespective of their sponsorship. The resulting dilemma is one that the Left must face without delay, as an integral step in developing whatever more radical strategy might be possible for the longer term.
>
> (Wallis, 2009: 33)

Crucial was the idea that nonmarket ethics was being put to use, at least rhetorically, not in the service of strengthening neoliberalism but as its social antidote. Arising was a new and radicalized version of the neoliberal citizen. Specifically, activists and community members began to appropriate these discourses for their own progressive purposes—deploying notions such as "growth" and "inclusion" as a means of furthering publicly funded and democratically organized cultural and anti-poverty initiatives. Investigating such actions within the context of the small Canadian city of Peterborough, Changefoot (2007: 130) notes that

> resistance reveals activists creatively and vibrantly responding to the effects of neoliberal policy in ways that reflect the attributes of a critical movement. Activists more than respond to neoliberal policy; they perform with it and with the local state. Cultural and antipoverty activism in Peterborough reveals intricate movement undertaken by activists. They "perform" good neoliberal citizenship by discursively appropriating neoliberal goals of economic growth in the case of cultural activism, and constructions of the self-sufficient, autonomous citizen in the case of antipoverty activism.

These broader movements represented only the tip of the proverbial iceberg of this potentially revolutionary problem-solving ethos. Significantly, concrete solutions originally meant to support neoliberalism by helping people cope with its worst effects are increasingly pointing the way toward its transcendence. The previously disciplining and exploitive use of mindfulness is a case in point. Such mental techniques can also be used for profoundly more transformatory purposes. Activist lawyers in California, for instance, have deployed "mindfulness" as a means of constructively guiding their progressive undertakings. According to Harris (2007: 2076):

> Founded in 1988 by law students from U.C. Berkeley's Boalt Hall School of Law, EBCLC is the largest provider of free legal services in the East Bay and a nationally recognized law and justice clinic. EBCLC's community economic justice work does not fit comfortably into any one scholarly model, however. Some of the Center's efforts could be described as "institution-building" and others as "movement-building." The lawyers at EBCLC find themselves playing multiple roles as they work toward aligning law with progressive social and economic change. At times, EBCLC lawyers have adopted a conventional technical assistance model of lawyering, and at other times a distinctly unconventional model of equal partnership with clients. The common thread throughout EBCLC's work is the approach it takes toward reconciling personal and professional roles. The best word we have found to describe that stance is "mindfulness."

The sharing economy is a further example of this potentially revolutionary neoliberal dialectic. Practices such as time-breaking and cooperative social organizations revealed the broader possibilities of nonmarket existence. "The sharing economy is the latest example of the internet's value to consumers (see Free exchange)" observed the *Economist* in 2013. "This emerging model is now big and disruptive enough for regulators and companies to have woken up to it. That is a sign of its immense potential. It is time to start caring about sharing." At the macro level, a sharing economy held out the possibility of positively resolving one of the modern economy's main threats—that of automation (which, with almost perfect historical irony, was also what Marx predicted in the 19th century would be the catalyst for capitalism's downfall). According to a recent World Economic Forum report:

> The truth is that the Internet-led process of exploiting under-utilized resources—be they physical and financial capital or human capital and talent—is both unstoppable and accelerating. The long-term benefits consist not just in efficiency and productivity gains (large enough to show up in macro data), but also in much-needed new jobs requiring a broad range of skills. Indeed, those who fear the job-destroying and job-shifting power of automation should look upon the sharing economy and breathe a bit of a sigh of relief.
>
> (Spence, 2015)

In the immediate present, the sharing economy offered the opportunity for "bottom up self-regulation rather than top down government control" along with the resources to "provide an environment for platforms to develop private solutions" (Allen and Berg, 2014). It also represented, in this regard,

> a radical shift toward new ways of organizing economic activity that will become increasingly dominant in this century. The assortment of behaviors (and organizations) that many of us optimistically call the "sharing economy" are early instances of a future in which peer-to-peer exchange becomes increasingly prevalent, and the "crowd" replaces the corporation at the center of capitalism.
>
> (Sundararajan, 2016: 3)

Similarly, the gig economy was initially a means for people, particularly those in the younger generation, to materially cope with a seeming permanent economic reality of job insecurity and chronic unemployment. It also gestured toward a tantalizing glimpse of a society free from capitalist ownership. "Imagine life without ownership," proclaimed Tarnaff (2016),

> You own nothing. You rent everything. You do this because it's cheaper: you pay pennies or fractions of pennies per day to have a bed to sleep in, or a winter coat to keep you warm. You scale up or down as needed: maybe your partner moves in, so you swap your single bed for a queen, or winter ends, so you return your coat. You are always optimizing. You pay for precisely what you use, when you use it—and nothing more.

Perhaps even more radically have been forward-thinking theories of a liberated non-capitalist future. Ideas of "post-capitalism," most notably proposed by Paul Mason (2015: xvi), exemplify such visions:

> Almost unnoticed, in the niches and hollows of the market system, whole swathes of economic life are beginning to move to a different rhythm. . . . New forms of ownership, new forms of lending, new legal contracts: a whole business subculture has emerged over the past ten years, which the media has dubbed the "sharing economy" . . . . Today the terrain of capitalism has changed: it is global, fragmentary, geared to small-scale choices, temporary work and multiple skill-sets. Consumption has become a form of self-expression—and millions of people have a stake in the finance system that they did not have before. With the new terrain, the old path is lost. But a different path has opened up. Collaborative production, using network technology to produce goods and services that work only when they are free, or shared, defines the route beyond the market system.

This is also witnessed in the potential use of distributed manufacturing to produce a society of material abundance. To this effect,

> digital fabrication offers the prospects of on-demand, mass personalisation, with more localised, flexible and sustainable production. These

> technologies have the potential to disrupt the organisation of manufacturing and the ways in which companies . . . create and capture value.
> (Technology Enterprise Group, Cambridge, 2016)

In a decidedly more utopian spirit, notions of "fully automated luxury communism" promise an exciting tomorrow "where machines do the heavy lifting and employment as we know it is a thing of the past" (Merchant, 2015). New policies are arising, moreover, that are bringing these dreams closer to a practical reality. Such efforts are seen in the introduction in some cities of the national basic income and free public transportation. The prospect of a post-work and post-capitalist society progressively seems more like an exciting promise than an ominous threat.

Fighting the Power of Neoliberal Ethics

This chapter has sought to highlight the growing resistance not only to neoliberalism but also to its disciplining regime of nonmarket ethics. It represents the effort to move beyond neoliberalism politically, institutionally and personally. It sheds light on how the ethical demand to reform the free market became a flashpoint for resisting neoliberalism more generally. This has the potential to completely reorder the social and, as such, move away from this hyper-capitalist ideology and ethical practices that sustain it.

Crucial in this respect is the bottom-up aspect of these desires. Theoretically, neoliberalism is egalitarian, because anyone with enough enterprise and dedication can succeed. Its ethics, likewise, are not inherently unequal. It is supposedly everyone's personal responsibility to morally and ethically fix the free market when necessary. However, such obligations almost always trickle downward to the most vulnerable. It was the largely powerless majority, not the elites, on whom the duty to normatively save capitalism fell. This imbalance in moral and ethical responsibility, in turn, has given birth to diverse movements seeking to profoundly reverse this dynamic. It is now the people who are demanding that those at the top be held accountable. This led to an often dangerous discourse of trying to assess who is at fault in a way that further victimizes historically disenfranchised groups and individuals. Reflected, though, is a concerted attempt by people to no longer blame themselves or feel ethically obligated for fixing the free market.

This has produced a new ethical imperative to challenge and replace this current capitalist status quo. Referendums and elections have turned into opportunities to directly take on the establishment. These were political wake-up calls putting elites on high alert. Represented were aspirations to root out the corruption not of an overbearing state but of the corporate oligarchy aided and abetted by the free market and its political supporters.

It is, further, imperative to create a more positive movement to cultivate noncapitalist forms of agency empowerment. Significantly, neoliberalism produces a range of capitalist capabilities—notably around the potential for moral improvement and ethical innovation. However, it also creates potent opportunities for the more expansive use of these abilities. To this end, the study of neoliberalism opens the space for critically exploring what agency exists within the otherwise constricting ideological and concrete domination of corporate globalization (Hay, 2002). Nevertheless, such accounts must go beyond simple critique. Indeed,

> in so far as theories of neoliberalism dismiss considerations of rational action, motivation, and decentralised coordination as so much "ideology," they remain chronically constricted in their capacity to reflect seriously on questions of institutional design, political organisation and economic coordination which, one might suppose, remain an important task for any critical theory.
>
> (Barnett, 2010: 3–4)

It is critical to transcend and even directly confront neoliberalism ideologically and in practice. To this end, the problem-solving ethos used so strategically and effectively by neoliberalism has become its potential undoing.

This opens the space for new and more emancipatory social discourses to emerge. They serve to affectively counter neoliberalism and provide fresh romantic visions of society for contemporary individuals to subjectively invest in. Significantly, at present these remain at the level often of populist revolt—an anti-elitist uprising attacking the current establishment from both the Right and the Left. Yet what should not be ignored is the distinct longing to go beyond neoliberalism, leaving it behind to the ash heap of history.

9 The Ethics of Neoliberalism

The Business of Making Capitalism Moral

This book attempts to answer what may appear to be a rather straightforward question—namely, "What is the ethics of neoliberalism?" The obvious response, perhaps, is that it is an ideological celebration of the free market that naturally exalts in the inherent normative goodness of conventionally market values such as enterprise and the pursuit of self-interest above all else. And yet the answer is, in fact, much more complicated and paradoxical by far. In point of fact, neoliberalism has propagated and heavily relies upon conventionally nonmarket ideas of personal care toward oneself and others as well as an abiding commitment to social and environmental justice. Critically, neoliberal subjects are made personally and collectively responsible for continually morally and ethically fixing the free market. Hence, the more capitalist a society becomes, the more ethical and, to a certain extent, less capitalist its population must become in order to ironically maintain this system.

This reading hopes to turn the traditional account of capitalism and ethics on its proverbial head. The established narrative is, of course, that the reproduction of a market economy requires the matching social construction of a suitably ethical market subject. The need for a reliable workforce thus gives birth to a moral emphasis on punctuality, frugality and sobriety. The complexity comes for sure in the competing demands of capitalism—as the priorities of profit-making can simultaneously desire both a puritanical workforce and hedonistic consumers. Nevertheless, its very existence depends on creating a population in its own ideological image.

However, this analysis asserts something profoundly different. It highlights the ways capitalism generally and neoliberalism specifically are made socially possible through the promotion of a decidedly nonmarket ethics. The free market is maintained through producing subjects who normatively embody often precisely what the free market claims not to be—notably other-minded and justice-oriented. This echoes the early Marxist reminder that "self-interest" is as much a social construction as any other type of morality. Indeed,

> private interest is itself already a socially determined interest, which can be achieved only within the conditions laid down by society and with

> the means provided by society; hence it is bound to the reproduction of these conditions and means. It is the interest of private persons; but its content as well as the form and means of its realisation, is given by social conditions independently of all.
>
> (Marx, 1973: 156)

Likewise, its ethical counterpart—that of care, compassion and a desire for collective welfare—is a cultural reflection of this socially created "privatized" world and one that has the capacity for strategically reinforcing it.

Crucial to these insights is the study of neoliberalism not as a coherent ideology or theory but rather as it concretely exists and evolved in practice. This reflects emerging ideas of an "actually existing neoliberalism" (Crouch, 2011; Ong, 2006; Overbeek and Van Apeldoorn, 2012). To this effect,

> neoliberalism has been a victim of its own success. A growing tide of conceptual critiques has begun to probe its usage and meaning . . . can neoliberalism as a broad, catch-all term adequately serve so many different phenomena and theoretical conceptualizations? Largely as a result of this growing conceptual ambiguity, neoliberalism is now widely acknowledged in the literature as a controversial, incoherent and crisis-ridden term, even by many of its most influential deployers.
>
> (Venugopal, 2015: 166)

This discursive ambiguity has led, predictably, to desires to give it some analytical clarity. The focus here, by contrast, is on the dynamic rather than stagnant character of this system. By highlighting and not seeking to do away with such contradiction, what becomes exposed is the "actually existing" diversity of neoliberalism across different contexts. Yet neoliberalism's seeming contradictions can similarly serve a reinforcing function. In the case of ethics, the success of neoliberalism lies in its apparent profound moral tensions—as it simultaneously embraces the free market as a bedrock guiding principle while fostering a nonmarket ethics for concretely supporting and reproducing these hyper-capitalist ideals. This study will further, it is hoped, shed greater light on the actually existing power of neoliberalism. To this effect, the emphasis is usually reserved for the ideological hegemony of the free market or the growing sovereignty of corporations. There is an understandable concentration on the spreading influence of the aptly named 1%. This book highlights, however, the multiple ways this paradoxical ethics strengths the systematic domination of neoliberalism and its elites. It does so by first disciplining subjects to accept moral and ethical responsibility for this order, as well as empowering them with new capabilities and social opportunities for undertaking this duty.

The ethical power of neoliberalism, moreover, is simultaneously structural and subjective in character. The free market on its own is unable to provide the mass welfare and social infrastructure necessary for providing for a sustainable community. Although Marx may have been wrong in his historical analysis of capitalism's overdetermined future downfall, he was eerily prescient about its deep structural contradictions. The pursuit of profit does create the conditions for its long-term ruin even as it can provide short-term gain. The neoliberal creation of ethical subjects thus turns its potential gravediggers into its greatest moral saviors. They are tasked with no less than innovatively solving the fundamental problems of capitalism—in the process preserving it despite their best and its worst nature.

Ethically Saving Capitalism

Neoliberalism finds itself at a historical and existential crossroads at the present. On one hand, it remains the dominant social economic ideology across the world. Moreover, it serves as the foundation for modern subjectivity and the current limit of the political imagination. Quoting Mark Fisher (2009: 8–9),

> In the 1960s and 1970s, capitalism had to face the problem of how to contain and absorb energies from outside. It now, in fact, has the opposite problem; having all-too successfully incorporated externality, how can it function without an outside it can colonize and appropriate? For most people under twenty in Europe and North America, the lack of alternatives to capitalism is no longer even an issue. Capitalism seamlessly occupies the horizons of the thinkable. Jameson used to report in horror about the ways that capitalism had seeped into the very unconscious; now, the fact that capitalism has colonized the dreaming life of the population is so taken for granted that it is no longer worthy of comment.

On the other hand, capitalism is being widely assailed for its economic volatility and propensity for mass social destruction. Popular movements have arisen internationally, from both the left and the right, protesting globalization and the spread of the free market. The very existence of neoliberalism—and, broadly, capitalism—appears to be at stake. Critical, in this respect, is the attempt to create disciplined moral subjects who can ethically fix the system.

Traditionally, it is assumed that the survival of capitalism depends on the concrete social manufacturing of individuals and communities that can morally and ethically reflect market values. To be sober and hardworking reinforces capitalist prerogatives of efficiency and productivity. In this respect, individuals are not merely "rational" economic subjects; they are socially produced as part of a wider and more encompassing cultural system of capitalism (see Meiksins-Wood, 1991; Wallerstein, 1995). This nevertheless implicitly places the responsibility for the continued existence of this cultural system on capitalist subjects themselves. To this effect, "the unbounded drive of modern capitalism undermines the moral foundations of the original Protestant ethic that ushered in capitalism itself" (Bell, 1976). Hence, if they were not disciplined and imbued strongly enough with the Protestant ethic, then the market—and, with it, all hopes for the perpetuation of individual freedom and collective prosperity—were doomed.

Neoliberalism acted to make this once tacitly assumed obligation into an explicit and even more forceful moral duty and ethical responsibility. Yet, in doing so, it has significantly reversed this initial normative dynamic of the market. Is no longer the case that subjects should be motivated to mirror the economic principles conducive to a smoothly running free market. Such a commitment is merely assumed as a matter of survival. Rather the social impetus is directed on individuals' own moral improvement and willingness to dedicate themselves to creating a more ethical market society. If, therefore, the market was less just than desired, then the fault was now assumed to lie with the individuals and groups who populated this capitalist system itself.

Reflected was an emerging ethos of capitalist care. Neoliberalism is often critically associated with ideas of self-care (see Ward, 2015). In this respect, subjects are expected to govern themselves—to manage their own affairs is a modern governmentality that Foucault refers to as shaping the "conduct of conduct" and one that places the responsibility for the self primarily on the subjects themselves (Hamman, 2009). Neoliberalism's governance is not exclusively or even necessarily primarily based on an ethos of self-interest. Conversely, it revolves around a deep and abiding sense of care. What is often completely overlooked, however, is how this culture of care is consistently and increasingly transferred onto relationship with others in the community. To this end, self-care is augmented and in some sense replaced by a social mission for fixing capitalism's inequities. Significantly, as Rose and Miller (1992: 175) observe, the "problematic of government" is one that involves both "political rationalities" and "governmental technologies" that

> represent the complex of mundane programmes, calculations, techniques, apparatuses, documents and procedures through which authorities seek to embody and give effect to governmental ambitions. Through the analysis of intricate interdependencies between political rationalities and governmental technologies, we can begin to understand the multiple and delicate networks that connect the lives of individuals, groups and organizations to the aspirations of authorities in the advanced liberal democracies of the present.

To this extent, the economic rationalities of neoliberalism have given birth to and concretely interact with "ethical technologies," a set of evolving and continually emergent methods and knowledges for balancing these competing demands of profit and morality. These ethical technologies entail finding ways to ensure that the free market is made more compassionate and directed toward the common good. There is a duty not simply to maintain the capitalist machine but to morally and ethically repair it.

These insights reveal a rather ironic component of actually existing neoliberalism. As this analysis and others have taken great pains to show, neoliberalism is an ideological rather than a purely "rational" paradigm, as its proponents commonly and loudly proclaim. Yet the attention understandably has focused predominantly on the economic aspects of this system—more precisely, on its spread and domination to all spheres of humanity's social and interpersonal existence. Poulantzas (1978: 210) speaks, in this regard, of ideological "regions which can be characterized, for example as moral, juridical and political, aesthetic, religious, economic, philosophical ideologies." He argues, moreover, that "Without going more deeply into this problem, it must also be said that in the dominant ideology of a social formation, it is generally possible to decipher the dominance of one region of ideology over the others." In the case of neoliberalism, though, it has progressively been the moral ideological "region" that has come to the fore despite its traditionally considered economic focus. The market is simply accepted as a permanent social fact of life—what requires attention is ensuring and promoting personal morality and social justice within this inevitable reality.

Hence, individuals are constantly tasked with saving capitalism. In this respect, "contemporary tendencies to economize public domains and methods of government also dialectically produce tendencies to moralize markets in general and business enterprises in particular" (Shamir, 2008). Structurally, it encourages and increasingly disciplines subjects to cover over and resolve the contradictions caused by neoliberalism. If streamlining is expected to make service providers more responsive but in fact renders them unable to meet the needs of those they are meant to serve, then it is up to individuals and groups themselves to use their ingenuity and sense of ethical commitment to fill this gap. If cost-cutting is leading to enhanced financial pressures, forcing people to fulfill the seemingly impossible task of being both fiscally responsible and morally good, then informal, often slightly subversive, behavior must be undertaken to help others so as to temporarily fix this otherwise untenable situation (see Bloom and White, 2016). Similarly, if the free market and corporate globalization is driving us to the edge of total extinction, it is the personal and collective responsibility of everyone to morally and ethically prevent this from happening without sacrificing our sacred faith in the market itself.

Subjectively, this struggle to create a better neoliberalism provides subjects with a much-desired sense of ontological, and therefore psychic, security. It does not fundamentally risk their fragile psychic stability; rather, it offers them a new capitalist fantasy to believe and normatively invest in. It is an appealing narrative—the aspiration and struggle to produce a more moral and caring market world. Moreover, it is a struggle that is full of emotionally resonant drama—representing an eternal and, at times, fevered dream to forge a path that is both fiscally and ethically responsible. It is an attractive script for structuring personal and shared existence. The longing for and effort to achieve this righteous, caring capitalism is the dominant cultural plot of the new century.

This disciplining modern governance of saving capitalism is thus both structural and subjective in character. It poses pressing questions, such as "What type of market world would one like to inhabit? A bleak and overly competitive and anxious one, or one that, even if not perfect, still reflects the highest values of humanity?" The choice, it is assumed, is completely up to each of us individually and therefore a matter of utmost personal and collective urgency. This timely ethical decision is exacerbated by the almost apocalyptic consequences of doing nothing. Without the presence of a more moral and ethical population to responsibly guide it, the free market would lead to social and environmental catastrophe. Just as it was a matter of survival to be fiscally solvent, it is now a matter of existential and social survival for subjects to take moral and ethical responsibility for capitalism.

The Global Spread of Neoliberal Ethics

Ethics form a core component of neoliberalism, and the production of good subjects is an essential feature of its structural production and social legitimization. Rather than just serve as an opportunity to make profit, the modern free market thrives as a constant site for pursuing moral improvement and ethical reform. To this effect, its attractiveness and desirability is rather paradoxically linked to this moral lack, as well as the continual possibility of at least partially overcoming it. As shown, this ethical fantasy has become an increasingly disciplining modern capitalist reality. Even more troubling is that this inscriptive and oppressive neoliberalism is rapidly spreading across the globe.

Globalization, in contrast to many initial predictions, has created a new need for governance internationally. As discussed, it has catalyzed a strong emphasis on the importance of good governance, symbolizing a dogmatic adherence to pro-market values of privatization and—increasingly, in the modern era—policies of austerity. Revealed is a new and multifaceted form of international rule that combines self-disciplining and, if necessary, coercion. Individuals and states are expected to be fiscally responsible or else face the consequences. In the words of Harvey (2007: 23):

> Neoliberalization has in effect swept across the world like a vast tidal wave of institutional reform and discursive adjustment. While plenty of evidence shows its uneven geographical development, no place can claim total immunity (with the exception of a few states such as North Korea). Furthermore, the rules of engagement now established through the WTO (governing international trade) and by the IMF (governing

> international finance) instantiate neoliberalism as a global set of rules. All states that sign on to the WTO and the IMF (and who can afford not to?) agree to abide (albeit with a "grace period" to permit smooth adjustment) by these rules or face severe penalties.

Such disciplining authority extends to neoliberal ethics. More precisely, individual states and regimes must responsibly address the social and political problems caused and exacerbated by neoliberalism. The emphasis was placed not on rebalancing power relations but on empowering countries to help themselves. More broadly,

> [t]he traditional assumption that markets will lead to democracy has been transformed into a twenty-first century story of authoritarian progress, where a fiscally self-disciplining state and disciplining international institutions will use their power to ensure that countries around the world develop and prosper. Required is not democracy, deliberation, debate, experimentation or a rethinking of core values. Instead all that is needed is for governments and IFIs to rule populations with a firm and "responsible" hand. . . . Significantly, the more capitalism grows globally, the more it relies on the power of governments for its survival. . . . A crucial component of neoliberalism is the channeling of this agency into an authoritarian mandate for governments to police themselves and their citizens in line with capitalist values.
>
> (Bloom, 2016a)

This extended to the international promotion of an empowering ethical capitalism. The use of microcredit entrepreneurship schemes to assist economically under-invested and exploited developing countries worldwide exemplified the discussion. These countries' problems were framed as being primarily moral and ethical rather than systematic or economic in character. Global elites enthusiastically embraced the need to spread "well being" to all corners of the world (Cederstrom and Spicer, 2015).

These ideals were both universal in their reach and colonizing in their effects. Ethics, of course, was far from a one-size-fits-all proposition. By contrast, it is crucial to remember

> that the market economy, like representative democracy or any other abstract institutional conception, is institutionally indeterminate: it lacks any single natural and necessary institutional form. The narrow repertory of variants of the market economy now established in the rich North Atlantic economies is made up of institutions and practices that have shown themselves more innovation- and growth-friendly, and more hospitable to free political institutions, than many others now or once on offer. Nevertheless, we read the lessons of experience wrongly if we suppose them to teach that these arrangements represent the

> inevitable outcome of a halting but relentless convergence toward the necessary, or even the best, form of the market. . . . Instead of seeing every disturbance of the present course of market oriented reform as the trumping of the market by a non-market based form of resource allocation, we must learn to recognize in some such disturbances early moves in a campaign to reorganize the market.
>
> (Unger, 2001: lxxxvi)

What was held to be shared, by contrast, was the demand placed on citizens and countries to find ways to balance these fiscal and ethical responsibilities. Accordingly, they were expected to constantly innovate to encourage economic growth in a market-friendly brand of social and environmental justice. Its colonizing aspect was not merely in its perpetuation of unequal power relations forged by imperialisms both old and new. It was also in the use of this ethical agenda to further internalize neoliberalism domestically and internationally.

Reflected was a new global fantasy of moral capitalism and the production of the ethical market subject. Presented was an alluring vision of the international free market that had the potential to more normatively empower individuals, communities and nations in order to achieve economic, social and environmental goals. It also linked the laudatory aims to broader agenda of neoliberal ethics. Specifically, it represented a tantalizing notion of cultivating global wellness. In particular, it promoted an individual reality focused on the overriding imperative to marshal the innovation and dynamic character of the free market for creating a more socially just world.

Significantly, this fantasy of moral capitalism is at least as much disciplining as it is empowering. This in the global free market is made and maintained through the shaping of subjects' knowledge and practice. Using the case of Cambodia as a prime example, Springer (2010: 931) reveals that

> [a]s disciplinary rationalities, strategies, technologies, and techniques coagulate under neoliberal subjectivation in contemporary Cambodian society through the proliferation of particular discursive formations like good governance, the structural inequalities of capital are increasingly misrecognized. This constitutes symbolic violence, which is wielded precisely inasmuch as one does not perceive it as such.

This also applied to discourses of a moral capitalism and the production of the ethical subject. Subjects were charged with being self-disciplined—at the personal, community and national level—to be both fiscally and ethically accountable.

Moreover, it created a universal ethos that stretched across national, cultural and geographical contexts to ethically save capitalism. It produced a new mandate for developing countries and regions to find a means of easing the historical legacy of colonialism and the present negative effects of corporate globalization. It is an evolution of colonialism for the modern age whereby the exploited nations and people must not only endure an unequal global economy but shoulder ethical and political responsibility for fixing this international order.

It reinforces the supremacy of hyper-capitalist principles of marketization, privatization and financialization interspersed with the social agenda of using capitalist techniques for achieving conventionally nonmarket aims. It championed a fantasy, again, of moral capitalism—promising to ethically empower individuals and communities to create a more just form of marketization internationally. It is a disciplining ideal that is becoming an ever more repressive global reality.

Ethically Perfecting Neoliberalism

The strength of neoliberalism is found not just in the supposed truth or prosperity of its economics but rather progressively in the possibility of ethics—more precisely, its promise to allow subjects to use the market to be good and do good. The actual social reality of this fantasy can take a multitude of forms. It can range from the romantic vision of a socially just and environmentally friendly capitalism to the perverse ability to find ways to achieve these laudatory normative goals individually or collectively despite the restricting and exploitive character of actually existing neoliberalism. The vitality of neoliberalism thus flows from its continual potential for moral and ethical improvement.

Ideological domination is commonly conceived, in this respect, as being characterized by a sense of discursive closeness. Particularly, it is conventionally theorized as hinging on the ability of a prevailing set of ideas and practices to become entrenched as a seemingly permanent and unalterable social reality (see Freeden, 1996, 2009). Nonetheless, recent critical scholarship has revealed the more dynamic nature of this relationship. This includes a larger awareness of how such success necessarily always involves a degree of openness and pluralism (Bloom and Dallyn, 2011). The previously discussed notion of safe resistance (Bloom, 2013) critically attests to this dynamism. Key is how a given status quo strategically allows and encourages setting limits to rather than merely trying to eliminate the possibility of social transformation.

The promotion of neoliberal ethics exemplifies this at first glance rather paradoxical quality of ideological domination. Ethics is often linked to a type of openness—a force for challenging hegemonic discourses and power relations within organizations and society. However, it also stands as the quintessential means of delimiting the scope of this change. Specifically, it fixes attention on the moral character of subjects and their ethical actions for doing good rather than the system that is responsible for producing these problems in the first place.

Here, neoliberalism relies on ethics to protect and perpetuate its ideological free-market imaginary. This overarching capitalist discursive horizon opens a space for a wide range of moral and ethical innovations. Crucial to this ethical hegemony are discourses of perfectibility—in particular, the capacity to continually improve, reform and transform a prevailing social order. The theoretical inspiration for such a politics is indeed quite radical. It is a critical intervention primarily attached to values of democracy that point to the eternal possibility of challenging a dominant status quo and its underlying ideological hegemony. According to Derrida (1997), it is a "promise" for a perfection that is always out of reach but nevertheless eternally "to come":

> The idea of a promise is inscribed in the idea of a democracy: equality, freedom, freedom of speech, freedom of the press—all these things are inscribed as promises within democracy. Democracy is a promise. That is why it is a more historical concept of the political—it's the only concept of a regime or a political organization in which history, that is the endless process of improvement and perfectibility, is inscribed in the concept. So it's a historical concept through and through, and that is why I call it "to-come."

Yet these longings for perfection can also critically be exploited for less radical purposes, specifically if subjects fixate on trying to fix the current system at the expense of more transformative alternatives. Democratization, for instance, is put forward as an opportunity to constantly reform democracy, in the process marginalizing socialist ideals. According to Allier Montaño and Bloom (2014),

> dominant ideologies and practices can be strengthened through a continual appeal to aspirations of democratic perfectibility. They support these structures by an always present promise of further democratization, an aspiration whose continual appeal lies exactly in the fact that it remains eternally unfinished and eternally subject to improvement. In this way, democracy stands as a socio-political fantasy for strengthening oppressive political rule, even those that have nominally achieved democratic transition.

Likewise, neoliberalism relies on discourses of ethical perfectibility to maintain its dominance.

There is thus a structural and subjective paradox at the heart of the seemingly progressive demand for a more ethical market. Concretely, it restricts the achievement of social justice to market means. Put differently, while the possibilities of building a better world may be relatively infinite, they can only be realized by a decidedly narrow set of techniques and methods. Hence, the more one strives for such an ethical outcome, the more one reinforces the exact obstacles to these goals. This points toward the subjective aspect of this paradoxical relationship. The greater the perceived ethical lack, the more pressing the need to ethically save capitalism. The failures of the free market are the foundation for stabilizing this ethical identification—offering subjects at one and the same time ontological and psychic security as ethical neoliberal subjects.

To this effect, individuals and communities remain subjectively imprisoned as modern-day capitalist subjects of desire. They are affectively enthralled to the eternal promise of being able to personally and collectively morally and ethically perfect a free-market society. Any discursive possibility that moves away from this vision is a challenge to one's very psychic existence. It is, hence, the enduring normative imperfections of neoliberalism—its social injustices, as well as its dramatic human and environmental costs—that pressure and discipline people to find the salvation presently available to them. To let go of this illusion, no matter how seemingly oppressive and anxiety inducing, would be to give up on the possibility of saving not only themselves but also the world.

The Ethical End of Neoliberal History

Neoliberalism is meant to represent the apogee of human civilization. It gives full expression to the creation of the most natural and rational society supposedly possible—that of the free market. All else is a deviation from the correct path to individual prosperity and collective progress. Yet this triumphant account is dramatically challenged by its actual reality. The global wave of marketization, financialization and privatization left many behind to suffer great material and mental hardships. In theory, this of course is completely expected and allowable. Within an ideal neoliberal paradigm, only the truly exceptional and hardworking deserve to succeed. Any gesture toward the equality of outcome is fundamentally dismissed as an obstacle to innovation and is therefore socially, however high-minded it may seem, socially regressive and repressive. Yet such justifications increasingly ring hollow in practice. The concrete sight of people plagued by the material and emotional pain caused by these "fair" policies challenged the rationality of such capitalist approved inequality. This growing ethical critique was only heightened by the fact that increasingly it was privilege, not merit or ability, that determined who got ahead and stayed on top. The intensified moralization of the free market expanded the possibilities of this otherwise permanent social economic order while closing off the potential for more radical progressive solutions to these problems.

As the 20th century came to a close, liberal democracy and the free market represented the end of history. Consequently, it was assumed that there was nothing that lay beyond neoliberalism. Neoliberalism was the ultimate height of what humanity could achieve economically. It held out a romanticized vision of a fully rational and prosperous free market that would profit both individuals and the world. This, nevertheless, posed two distinct problems for the social legitimacy of neoliberalism. Most immediately, it left neoliberalism with little discursive room to strategically maneuver.

More broadly, it linked the fate of capitalism to a hyper-capitalist version of itself—granting it little obvious opportunity to innovate or reform itself in the face of popular opposition. Its religious devotion to free-market orthodoxy therefore spelled its potential historical doom.

An initial means of overcoming these historical challenges was to stress, as discussed, neoliberalism's unquestionable inevitability. The market was put forward not only as an expression of human nature but in fact as a manifestation of the natural world and its physical order. It was akin to gravity and operated in the same law-like way of the planet's rotation. To resist such truths was to reveal one's ignorance of the unalterable trajectory of human evolution. Nevertheless, such assertions lost credibility as the full effects of neoliberalism came to light. After decades of growing inequality and a dramatic financial crisis, people en masse revolted against the demand that they morally and ideologically accept the weight of modern capitalism's gravitational pull.

The potential for moral improvement and ethical reform offered a way out of this apparent neoliberal historical dead end. Especially, it encouraged rather than repressed calls and longings for change, albeit within quite ideologically narrow parameters. It provided subjects with the opportunity to engage proactively with history and progress rather than merely be subjected to an economic and cultural fate that was historically predestined and not of their choosing. This insight echoes earlier discussion of the ways ethics granted individuals a sense of identity in an otherwise "subject-less" hyper-capitalist system. According to Dupuy (2009: 157–158):

> For Hayek, spontaneous order . . . signifies an emergence, an effect of composition, a system-effect. The "system" is obviously not a subject, endowed with consciousness and will. The knowledge that the system exploits is irreducibly distributed over the set of its constituent elements: it cannot be synthesized in one place, for the system has no "absolute knowledge" about itself that is localized somewhere within it. This collective knowledge resides in the social order of the system insofar as it is the "result of human action but not of human design" and cannot be appropriated by any individual consciousness. It is knowledge without a subject. It is embodied in norms, rules, conventions, institutions, which themselves are incorporated in individual minds in the form of abstract schemata: "The mind does not so much make rules as consist of rules of action. We can make use of so much experience, not because we possess such experience, but because, without our knowing it, it has become incorporated in the schemata of thought which guide us."

The prioritization of ethics and, as such, the possibility of creating positive social change, allowed subjects to feel as if they had some broader self-determination, even if only limitedly so, over their personal and collective destiny.

This creates, in turn, a new capitalist history for individual and communities to enter into and live within. Yet this represents a psychic history rather than a traditional temporal one. It affectively captures subjects within a repetitive cycle of struggling to be good and do good within a difficult neoliberal reality. It signifies, using Lacanian terminology, a perpetual history upon which ontological and psychic security is linked to the "love strife death" dynamic of fantasy. According to Lacan (2002: 262):

> This limit is present at every instant in what is finished in this history. It represents the past in its real form; it is not the physical past whose existence is abolished, nor the epic past as it has become perfected in the work of memory, nor the historical past in which man finds the guarantor of his future but rather the past which manifests itself in an inverted form in repetition.

The strive for and inevitable disappointment in achieving a moral and ethical free-market society in this respect anchors identity and forms a comfortable ongoing narrative for subjects to invest in and subjectively attach their sense of selfhood to. The realization of the fundamental contradictions and futility of such desires brings with it, however brief, an encounter with subjective death and, as such, a reignited longing to restart this safe present-day capitalist history.

Consequently, ethics stands as the actual end of neoliberal history. It is the discursive limit of concrete action, reasonable and realistic thinking, as well as more fundamentally contemporary subjectivity and social possibility. It critically grips people in a reoccurring and eternal history of trying to moralize the free market. It signifies an almost pathological obsession with balancing the economic needs of fiscal responsibility with the social longings for ethical responsibility. In a rather perverse manner, this echoes the vaunted virtuous cycle increasingly put forward by business experts and policymakers in which proper financial incentives and management can be used to encourage the creation of socially just and environmentally friendly markets. At a deeper level, neoliberalism represents the virtuous cycle of history—one whose good intentions lead societies even further down the path to ultimate social ruin.

The Ethics of Neoliberalism

This work has sought to critically reimagine the ethics of neoliberalism. It departs from established accounts that attribute the norms and ideals of neoliberalism to its underlying free-market ideology. Instead, it proposes that neoliberalism's social reproduction and survival is ironically dependent on the strategic promotion of nonmarket ethics. This durability is inexorably and even primarily associated with neoliberalism's ongoing moralization. Paradoxically, as actually existing neoliberalism reveals, the more capitalist an economy is, the more non-explicitly capitalist its ethics and populations often are and need to be.

This complex relationship between the free market and ethics allows for a potentially broader reconceptualization of modern capitalism's general historical development. Thus far, its history has been marked by the fluctuation between a free market and the progressive social response to regulate this capitalist economy for the sake of the public good. It reveals, in turn, the need for the broader moralization of capitalism. Indeed,

> [t]he maintenance of a market economy involves a basic paradox. For centuries writers such as Adam Smith have argued that the workings of the market should be based on the individual pursuit of self-interest. Yet, if the pursuit of self-interest goes too far in society, the very existence of the market may itself be endangered. If "opportunistic" behavior encompasses too many forms of social action, . . . a market economy may function very poorly. There is a wide range of behavior—including dishonest and "corrupt" transactions within the institutional framework of the market, "rent seeking" in government policy and administration, and actions that destroy trust in the legal system—that have the potential for undermining the efficient workings of the market.
>
> (Nelson, 2001: 1–2)

Polanyi (2001) refers, in this respect, to the "double movement" of capitalist history. More precisely,

> the extension of the market organization in respect to genuine commodities was accompanied by its restriction in respect to fictitious ones. While on the one hand markets spread all over the face of the globe and the amount of goods involved grew to unbelievable dimensions, on the other hand a network of measures and policies was integrated into powerful institutions designed to check the action of the market relative to labor, land, and money. While the organization of world commodity markets, world capital markets, and world currency markets under the aegis of the gold standard gave an unparalleled momentum to the mechanism of markets, a deep-seated movement sprang into being to resist the pernicious effects of a market-controlled economy Society protected itself against the perils inherent in a self-regulating market system—this was the one comprehensive feature in the history of the age.
>
> (80–81)

Expanding on this idea, this analysis introduces the third movement of capitalism associated with neoliberalism. Notably, the responsibility for ethically reigning in the excesses of the market is given not to the state but rather to individuals and communities. It, in essence, privatizes the regulation and moralization of capitalism. In doing so, neoliberalism has strategically combined an economic agenda of unrestrained capitalism with the cultural and political demand for social justice and collective welfare. The inevitable failures of such a project are attributed to the shortcomings of the subjects themselves, as opposed to a fundamental critique of the capitalist system itself.

The present era is threatening to replace social democracy with neoliberal ethics as a dominant progressive discourse. It is worth revisiting the beginnings of social democracy to understand better what is at risk of being lost. It is tempting to consider such politics as merely one variety of capitalism, a slightly less market-oriented alternative to the liberal models predominantly characterizing the Anglo world. This northern European model, with its greater regulation and public role of the state, is considered by the neoliberal paradigm and by committed neoliberals to be outdated and noncompetitive. Yet, at its root, it represents a potentially much more radical agenda (see Castles, 2009). Indeed, it is one in which economic equality will, over time, create the conditions necessary for the political and social equality necessary for evolving toward a more fully emancipated socialist society. It points to the potential for a more radical form of democracy (Mouffe, 1989), one in which individuals and groups can challenge social hegemonies and constantly discover "new revelations."

This may sound naive in the contemporary context; nonetheless, it highlights the danger of neoliberal ethics. One that goes beyond a lessening or full-scale rejection of the state's role in providing needed regulation and public good over the market. It is the inability to conceive and put into action a movement to evolve past capitalism that is in danger of fading away completely. More than simply the limiting of humanity's political imagination, it restricts their economic options to the continuation of marketization, privatization and financialization. The only available respite from this free-market condition is for subjects to find ways to better cope with this permanent capitalist reality.

This reading reframes the perceived purpose of neoliberalism popularly. It is no longer one of merely championing and spreading the free market to all spheres of human existence. It is now directly aimed at using hyper-capitalist techniques and methods to achieve a broader common good—one that is legitimate to non-capitalist values of cooperation, care for others, greater economic equality and even environmental sustainability. This is a subtle shift, perhaps, but it is not mere semantics either. It opens the way for capitalism itself to be legitimately challenged while also producing new means for disciplining subjects associated with these emergent ethical discourses.

Perhaps not surprisingly, the possibilities that have arisen from the subtle but potent discursive evolution of present-day neoliberalism have largely been disciplining rather than revolutionary. Far from its romanticized claims and predictions, neoliberalism has not eliminated bureaucracy and regulative collective norms—indeed, far from it. As Graeber (2015) notes, it can be more actually described as a "utopia of rules" whose market solutions produce "a nightmare fusion of the worst elements of bureaucracy and the worst elements of capitalism." A crucial component of this bureaucratization, this proliferation of ever more rules to direct and control human existence, is the increased moral and ethical responsibility placed on neoliberal subjects.

Critically, this emergent disciplining regime of ethical neoliberalism alters its operations and relations of power. Power is never simply repressive but instead primarily productive. "What makes power hold good, what makes power accepted," again quoting Foucault (1982: 225),

> is simply the fact that it doesn't only weigh on us as a force that says no, but that it traverses and produces things, it induces pleasures, forms of knowledge, produces discourse. It needs to be considered as a productive network that runs through the whole social body, much more than as a negative instance whose function is repression.

In its original conception and manifestation, neoliberal power took hold through producing market relations and largely marketized subjects. In the contemporary age these market pressures continue to exist but are augmented by moral and ethical discourses regulating subjects. Specifically, the present era is marked by a proliferation of moral and ethical forms of "empowerment," all done in the service of strengthening capitalism. It, moreover, nourishes the condition for disciplining and punishing individuals and groups who fail to live up to these daunting moral obligations.

The Business of Making Capitalism Moral

Capitalism is often portrayed primarily or even exclusively as an economic system. It is the buying and selling of goods and labor on the relatively free market. Its cultural components are usually considered, in turn, to be either a social reflection of this material foundation or a means of regulating and counteracting its most glaring injustices. Neoliberalism is perceived to be little different in this regard. Its ideology and ethics are ultimately a celebration and reification of the free market. In this view, capitalist exchange and the unrestrained striving for profit and economic growth reign supreme as the apex of human morality and progress. Yet a decidedly different reality has emerged in actually existing neoliberal systems, a reality marked by a resurgence of conventionally nonmarket values and social aspirations. The central aim of this book has, indeed, been to critically reveal the fundamental importance of such ideals for the contemporary survival and spread of global capitalism. This current hyper-capitalist order structurally and subjectively relies upon the social production of an ethical subject who can normatively fix the free market without threatening its underlying principles and hegemony. Individuals and communities are disciplined to invest their entrepreneurial spirit and creativity toward morally and ethically saving capitalism.

Significantly, this ethical paradox of neoliberalism retains the traditional language and ethos of the market, though in quite moralized ways. Individual accountability, initially a cultural mainstay of capitalism's original Protestant ethos, is currently translated into a profound political, institutional and personal ethical responsibility. This reading echoes and critically expands on the aforementioned "indebted man" proposed by Lazzarato (2012: 8–9):

> The series of financial crises has revealed a subjective figure that, while already present, now occupies the entirety of the public space: the "indebted man." The subjective achievements neoliberalism had promised ("everyone a shareholder, everyone an owner, everyone an entrepreneur") have plunged us into the existential condition of the indebted man, at once responsible and guilty for his particular fate.

Moreover, it speaks to Graeber's (2011) more encompassing linking of debt itself to a modern-day disciplining morality. At stake in this analysis is how subjects are regulated and made into duty-bound moral debtors to capitalism. It is an obligation to be a positive ethical force in order to counteract the system's immorality and injustice through one's own moral improvement and ethical commitment to improving society and the natural world.

Just as important, such ethics is a clear growth area for neoliberalism. At the purely economic level, ethical industries and renewable energy are undoubtedly big business. It is also vital to the reproduction and expansion of this hyper-capitalist order. Capitalism and capitalists have a seemingly unquenchable thirst for profit and the discovery of new markets to exploit. They are—to use Marx's famous description—vampiric in their desires for such gains (see Neocleous, 2003; Rasmussen, 2006). Likewise, in the present, the potential for neoliberal ethics to exploit subjects appears endless. The system and those who populate it, in this respect, can always be ethically renewed. Modern capitalism and capitalists are vampire-like in their hunger for strategically using the individual and collective desire to be good and do good—draining people of their highest ideals for their own fiscal and social profit.

Emerging, in turn, is a novel type of capitalist labor. Traditionally, of course, such labor was a rather straightforward economic proposition: one would work for a wage or entrepreneurially for one's own profit. As shown throughout this work and countless other historical analyses, this was far from obvious or natural evolution. Instead, to sell one's labor had to be a socially learned, promoted and culturally legitimized skill. Increasingly, scholars point to the appearance of what can be termed "moral labor" in the modern age, representing the intensification of work in the name of realizing a higher ideal (see Fechter, 2016; Johnson, 2015; Porter, 2009; Scheffler, 2005). It entails the accepting of longer hours and lower pay for achieving these laudatory aims. This study proposes a similar but expanded version of this ethical labor. Notably, it highlights the expectation that individuals work toward morally fixing themselves and ethically fixing society for the overall benefit of saving neoliberalism. Moreover, such labor mirrors the intensification associated with this contemporary hyper-capitalist order generally. Indeed, people are charged with constantly maximizing ethical labor for the ultimate advantage of the very capitalism and elites who have largely caused the problems they are responsible for solving.

At stake is the creation and evolution of an ownership ethos that remains so fundamental to the preservation of free-market societies (see Gates, 1999; Ong, 2006). A primary task of the capitalist state is to introduce and protect property rights. This notion extends socially, furthermore, to describe an ownership ethos fundamental to these capitalist cultures. To own something brings with it a sense of moral commitment, namely revolving around a sense of stewardship and more broadly caretaking. In a likewise fashion, neoliberal subjects are expected to take moral ownership for themselves and ethical ownership for their community, organization, and society at large. To this effect, there is an increased demand that they act as normative caretakers of modern capitalism so that it doesn't suffer a moral breakdown.

This follows in line with capitalism's general promotion of innovation rather than transformation. At the core of the free market lies a distinctive, though productive, contradiction. It is a system that demands economic improvement with little to no fundamental social political change. It desires, in this regard, perpetual dynamism in terms of product development and institutional processes within a stagnant and unbreakable commitment to maintaining an overarching market order. Schumpeter (1994) famously touches precisely on this tension in his theory of creative destruction. Here, the market constantly renews itself through a replacement of the old with the new, all within the relatively permanent confines of the capitalist system. Neoliberalism, despite its assumed naturalism and inherent "rationality," is furnished and made possible through such creative destruction. As Harvey presciently observes:

> The creation of this neoliberal system has entailed much destruction, not only of prior institutional frameworks and powers (such as the supposed prior state sovereignty over political-economic affairs) but also of divisions of labor, social relations, welfare provisions, technological mixes, ways of life, attachments to the land, habits of the heart, ways of thought, and the like.
>
> (Harvey, 2007)

Yet it also engages in a very specific and strategic form of continual ethical renewal. In the contemporary context of neoliberalism, there is an internal drive for ethical innovations—a constant cycle of creative destruction—that will not sacrifice in any substantial way the free market. Here, moreover, the constant destruction of social bonds and welfare creates the need for ever more creative ethical solutions—of which neoliberal subjects are almost solely responsible for providing.

This speaks to the deeper threat of neoliberal ethics. On the face of it, obviously, it appears quite strangely, perhaps, to pose otherwise welcome values such as cooperation and care for others as potentially dangerous. Indeed, as Foucault notes, everything is dangerous, as even the most high-minded ideals can in practice lead to oppressive outcomes. Accordingly, there is increased attention being rightly paid to the lack of moral hazards within present-day capitalist relations—specifically, hazards that would discourage those with the most economic and political power from moral impropriety and unethical behavior. The lack of punishment for such actions, it is said, creates no incentive for them or others with their influence or power to be just. This work points to a rather opposite problem—that of ethical hazard, in which the encouragement of people to be good and act good leads ultimately to adverse outcomes in the reinforcement of the very system responsible for these injustices in their own oppression.

There is, accordingly, a critical need to challenge and move past these hazardous desires to be ethical in the midst of these hyper-capitalist times. It is imperative to not be satisfied with the always disappointing desire to ethically fix the free market. Instead, there must be a renewed commitment to its replacement by a new order that takes human welfare and the potential for general transformation and liberation as its overriding priority. Ferguson (2009: 170) thus proposes the need to critically explore the "uses of neoliberalism":

> To understand the political possibilities and dangers that such emergent phenomena contain, we will surely need empirical research on the actual political processes at work. But beyond this, there is a palpable need for conceptual work, for new and better ways of thinking about practices of government and how they might be linked in new ways to the aspirations and demands of the economically and socially marginalized people who constitute the majority of the population in much of the world.

Politically and institutionally, this entails holding neoliberalism responsible for the social destruction it has wrought in a way that ideologically and concretely reveals the potential for the creation of a more ethical and emancipatory set of social and material relations. On a personal level, it involves going beyond (or traversing) the fantasy of a moral market for fresh visions of a different and better world, creating in its place a desire to supplant the drive for profit and the morality that sustains it with a more expansive drive for individual and collective individual liberty and collective possibility.

The focus of neoliberalism is on the lucrative business of making capitalism moral. It preaches a righteous agenda of holding subjects responsible for bringing justice to the free market. It holds up modern afflictions of political oligarchy, economic inequality and environmental devastation as an opportunity for moral improvement and ethical innovation—representing an eternal struggle to delicately balance the twin necessities of profit and welfare. It is increasingly absolutely clear that the survival and prosperity of capitalism and the survival and prosperity of humanity are mutually exclusive propositions. There is little time left or any real reason to continue ethically trying to fix neoliberalism. It is now well past the time to morally save the free market—it is instead now crucial to get down to the radically urgent business of actually solving the problem of capitalism for the sake of both the present and the future.

References

Aalbers, Manuel B., Ewald Engelen, and Anna Glasmacher. 2011. "'Cognitive closure' in the Netherlands: Mortgage securitization in a hybrid European political economy." *Environment and Planning A* 43 (8):1779–1795.
Abu-Lughod, Lila. 1990. "The romance of resistance: Tracing transformations of power through Bedouin women." *American Ethnologist* 17 (1):41–55.
Acharya, V. V., and M. P. Richardson (eds.). 2009. *Restoring financial stability: how to repair a failed system*. Hoboken, NJ: John Wiley & Sons.
Ackah, Carol, and Norma Heaton. 2004. "Human resource management careers: Different paths for men and women?" *Career Development International* 8 (3): 134–142.
Adler, Paul S. 2001. "Market, hierarchy, and trust: The knowledge economy and the future of capitalism." *Organization Science* 12 (2):215–234.
Adonis, Andrew. 1994. *A conservative revolution? The Thatcher-Reagan decade in perspective*. Manchester, UK: Manchester University Press.
Agger, B. 2007. "Time robbers, time rebels: Limits to fast capital." In R. Hassen (ed.) *24/7: Time and temporality in the network society*. Stanford, CA: Stanford University Press.
Aghion, Philippe, and Patrick Bolton. 1997. "A theory of trickle-down growth and development." *The Review of Economic Studies* 64 (2):151.
Ahmed, Sara. 2010. *The promise of happiness*. Durham, NC: Duke University Press.
Albrow, Martin. 1997. *Do organizations have feelings?* London, UK: Taylor & Francis.
Allen, D., and C. Berg. 2014. *The sharing economy*. The Institute for Public Affairs, December.
Allen, K. 2016. "Economic woes create anti-establishment movements around the world,"*The Guardian*, 5 November.
Allen, Marc. 1997. *A visionary life: Conversations on creating the life that you want*. Novato, CA: New World Library.
Allen, N., and S. Birch. 2010. *Ethics and integrity in British politics*. Novato, CA: New World Library.
Allier Montaño, Emilio, and Peter Bloom. 2014. "The closed promise: The authoritarian "grip" of democracy," *Theory & Event* 17 (3).
Alvesson, Mats. 2000. "Social identity and the problem of loyalty in knowledge-intensive companies." *Journal of Management Studies* 37 (8):1101–1124.
Alvesson, Mats, and A. Spicer. 2012. "Critical leadership studies: The case for critical performativity." *Human Relations* 65 (3):367–390.

Alvesson, Mats, and Hugh Willmott. 2002. "Identity regulation as organizational control: Producing the appropriate individual." *Journal of Management Studies* 39 (5):619–644.
Alvord, S. H., L. D. Brown, and C. W. Letts. 2004. Social entrepreneurship and societal transformation an exploratory study. *The Journal of Applied Behavioral Science*, 40 (3):260–282.
Anderson, A. R., and L. Warren. 2011. "The entrepreneur as hero and jester: Enacting the entrepreneurial discourse." *International Small Business Journal* 29 (6): 589–609.
Anderson, Benedict. 1991. *Imagined communities: Reflections on the origin and spread of nationalism*. 2nd ed. New York: Verso Books.
Anderson, P. 2012. "After the event," *New Left Review* 73:49–61.
Annan, K. 1998. "Secretary General of the United Nations: Annual report to the General Assembly on the work of the organisation." New York: United Nations.
Anon. 2014. "Study: US is an oligarchy, not a democracy." *BBC News*, 17 April.
Antonakakis, N., and Collins, A. 2014. The impact of fiscal austerity on suicide: on the empirics of a modern Greek tragedy. *Social Science & Medicine, 112*: 39–50.
Appleby, J. 2011. *The relentless revolution: A history of capitalism*. London, UK: W. W. Norton & Company.
Arndt, H. W. 1983. "The 'trickle-down' myth." *Economic Development and Cultural Change* 32 (1):1.
Arthington, P. 2016. "Mindfulness: A critical perspective." *Community Psychology in Global Perspective* 2 (1):87–104.
Arthur, Michael B., and Denise M. Rousseau. 1996. *The Boundaryless career: A new employment principle for a new organizational era*. New York: Oxford University Press.
Aune, James Arnt. 2002. *Selling the free market: The rhetoric of economic correctness*. New York: Guilford Press.
Austin, James, Howard Stevenson, and Jane Wei-Skillern. 2006. "Social and commercial entrepreneurship: Same, different, or both?" *Entrepreneurship Theory and Practice* 30 (1):1–22.
Babiak, Paul. 1995. "When psychopaths go to work: A case study of an industrial psychopath." *Applied Psychology* 44 (2):171–188.
Babiak, Paul, and Robert D. Hare. 2006. *Snakes in suits: When psychopaths go to work*. New York: ReganBooks.
Backhaus, K. B., B. A. Stone, and K. Heiner. 2002. "Exploring the relationship between corporate social performance and employer attractiveness." *Business & Society* 41 (3):292–318.
Bailyn, Lotte, Rhode Rapoport, and Joyce K. Fletcher. 1962. *Beyond work-family balance: Advancing gender equity and workplace performance*. San Francisco, CA: Jossey-Bass.
Bain, Peter, and Phil Taylor. 2000. "Entrapped by the 'electronic panopticon'? Worker resistance in the call centre." *New Technology, Work and Employment* 15 (1):2–18.
Baker, W. 1992. "The network organization in theory and practice." In N. Nohiria and R. Eccles (eds.). *Networks and organizations, structure, form and action* (pp. 57–91), Boston, MA: Harvard Business School Press.
Balogh, Brian. 2009. *A government out of sight: The mystery of national authority in nineteenth-century America*. Cambridge, UK: Cambridge University Press.

Banfe, Charles. 1991. *Entrepreneur: From zero to hero: How to be a blockbuster entrepreneur*. New York: Van Nostrand Reinhold.

Barber, Benjamin R. 1992. *Jihad vs. McWorld*. London, UK: Corgi Books.

Barber, B. B. 1997. Can Democracy Survive the War of Tribalism and Globalism. Rev. Const. Stud. 4 (1).

Barbera, R. 2009. *The cost of capitalism: Understanding market mayhem and stabilizing our economic future*. London, UK: McGraw-Hill Professional.

Barker, James R. 1993. "Tightening the iron cage: Concertive control in self-managing teams." *Administrative Science Quarterly* 38 (3):408.

Barley, Stephen R., and Gideon Kunda. 1992. "Design and devotion: Surges of rational and normative ideologies of control in managerial discourse." *Administrative Science Quarterly* 37 (3):363.

Barnes, R. D. 1992. "Standing guard for the PC Militia, or fighting hatred and indifference: Some thoughts on expressive hate-conduct and political correctness." *University of Illinois Law Review* 979.

Barnett, Clive. 2005. "The consolations of 'neoliberalism.'" *Geoforum* 36 (1):7–12.

Barnett, Clive. 2010. "Publics and markets: What's wrong with Neoliberalism?" In Susan Smith, Sallie Marston, Rachel Pain, and John Paul Jones III (eds.). *The SAGE handbook of social geographies*. London and New York: Sage, 269–296.

Barratt, Ed. 2002. "Foucault, Foucauldianism and human resource management." *Personnel Review* 31 (2):189–204.

Barry, A. 2004. "Ethical capitalism." In W. Larner and W. Walters (eds). *Global governmentality: Governing international*. London, UK: Routledge, 195.

Barry, A., T. Osbourne, and N. Rose. 1996. *Political reason: Liberalism, neoliberalism and rationalities of government*. Chicago, IL: University of Chicago Press.

Bartlett, Christopher A., and Sumantra Goshal. 1997. *The individualized corporation: A fundamentally new approach to management: Great companies are defined by purpose, process, and people*. New York: HarperBusiness.

Bass, Bernard M. 1985. *Leadership and performance beyond expectations*. New York: Collier Macmillan.

Bass, Jeff D., and Richard Cherwitz. 1978. "Imperial mission and manifest destiny: A case study of political myth in rhetorical discourse." *Southern Speech Communication Journal* 43 (3):213–232.

Bateman, T. S., and D. W. Organ. 1983. "Job satisfaction and the good soldier: The relationship between affect and employee 'citizenship.'" *Academy of Management Journal* 26 (4):587–595.

Bauman, Z. 2004. "To hope is human." *Tikkun, 19* (6):64–67.

Bayne, S., and J. Ross. 2007. "The 'digital native' and 'digital immigrant': A dangerous opposition." *Annual Conference of the Society for Research into Higher Education (SRHE)* 20.

Beckett, Francis. 2010. *What did the baby boomers ever do for us? Why the children of the sixties lived the dream and failed the future*. London, UK: Biteback Publishing.

Beeson, M., and A. Firth. 1998. "Neoliberalism as a political rationality: Australian public policy since the 1980s." *Journal of Sociology* 34 (3):215–231.

Bell, Daniel. 1976. *The cultural contradictions of capitalism*. 3rd ed. New York: Basic Books.

Bell, Duncan S. A. 2006. "Empire and international relations in Victorian political thought." *The Historical Journal* 49 (1):281.

Belleflamme, Paul, Thomas Lambert, and Armin Schwienbacher. 2014. "Crowdfunding: Tapping the right crowd." *Journal of Business Venturing* 29 (5):585–609.

Benaim, D., and P. Cammack. 2016. "How Trumpism went global." *New Republic*, 21 March.

Bennett, Jamie. 2008. "They hug Hoodies, don't they? Responsibility, irresponsibility and responsibilisation in conservative crime policy." *The Howard Journal of Criminal Justice* 47 (5):451–469.

Berend, Ivan T. 2006. *An economic history of twentieth-century Europe: Economic regimes from Laissez-Faire to globalization*. Cambridge, UK: Cambridge University Press (Virtual Publishing).

Bernstein, Henry. 1971. "Modernization theory and the sociological study of development." *The Journal of Development Studies* 7 (2):141–160.

Besmer, V. 2006. "The legal character of private codes of conduct: More than just a pseudo-formal gloss on corporate social responsibility." *Hastings Business Law Journal* 2 (1):279–306.

Best, Jacqueline. 2003. "Review article moralizing finance: The new financial architecture as ethical discourse." *Review of International Political Economy* 10 (3):579–603.

Beynon, Huw. 2014. "'Still too much socialism in Britain': The legacy of Margaret Thatcher." *Industrial Relations Journal* 45 (3):214–233.

Bhabha, Homi K. 2012. *The location of culture*. New York: Taylor & Francis.

Bhasin, Sanjay, and Peter Burcher. 2006. "Lean viewed as a philosophy." *Journal of Manufacturing Technology Management* 17 (1):56–72.

Biebricher, Thomas. 2015. "Neoliberalism and democracy." *Constellations* 22 (2):255–266.

Bieler, Andreas, and Adam David Morton. 2003. "Globalisation, the state and class struggle: A 'critical economy' engagement with open Marxism." *The British Journal of Politics and International Relations* 5 (4):467–499.

Binkley, Sam. 2011. "Happiness, positive psychology and the program of neoliberal governmentality." *Subjectivity* 4 (4):371–394.

Birch, S., and Allen, N. 2015. "Judging politicians: The role of political attentiveness in shaping how people evaluate the ethical behaviour of their leaders." *European Journal of Political Research* 54 (1):43–60.

Bishop, Matthew, John Kay, and C. Mayer, eds. 1994. *Privatization and economic performance*. Oxford, UK: Oxford University Press.

Blair, Tony. 2009. "Speech from Paris Summit 'New World, New Capitalism'." The Office of Tony Blair, 08 January.

Blanchard, O. J., and M. W. Watson. 1982. Bubbles, rational expectations and financial markets. In *Crisis in the economic and financial structure*, ed. P. Wachtel. Lexington, MA: Lexington Books, 295–316.

Block, F. 2010. "Crisis and renewal: The outlines of a twenty-first century new deal." *Socio-Economic Review* 9 (1):31–57.

Blomgren, Anna-Maria. 1997. Nyliberal politisk filosofi. En kritisk analys av Milton Friedman, Robert Nozick och F. A. Hayek. Nora: Bokförlaget Nya Doxa. Quoted in Thorsen, D. E. 2010. The neoliberal challenge. What is neoliberalism?. *Contemporary Readings in Law and Social Justice* 2 (2): 188–214.

Bloom, Peter. 2008. "Capitalism's cynical leviathan: Cynicism, totalitarianism, and Hobbes in modern capitalist regulation." *International Journal of Žižek Studies*, 2 (1).

Bloom, Peter. 2013. "The power of safe resistance." *Journal of Political Power* 6 (2): 219–239.

Bloom, Peter. 2014a. "Back to the capitalist future: Fantasy and the paradox of crisis." *Culture and Organization* 22 (2):158–177.

Bloom, Peter. 2014b. "We are all monsters now!" *Equality, Diversity and Inclusion: An International Journal* 33 (7):662–680.

Bloom, Peter. 2015. "Work as the contemporary limit of life: Capitalism, the death drive, and the lethal fantasy of 'work-life balance.'" *Organization*.

Bloom, Peter. 2016a. *Authoritarian capitalism in the age of globalization*. United Kingdom: Edward Elgar Publishing.

Bloom, Peter. 2016b. "Cutting off the king's head: The self disciplining fantasy of neoliberal sovereignty." *New Formations* 88 (88):8–29.

Bloom, Peter. 2016c. "Trump and the triumph of hopeful nihilism." *The Conversation*, 9 November.

Bloom, Peter. 2017. "How to resist the political rise of global nativism," *The Conversation*, 1 March.

Bloom, Peter, and Carl Cederstrom. 2009. "'The sky's the limit': Fantasy in the age of market rationality." *Journal of Organizational Change Management* 22 (2):159–180.

Bloom, Peter, and Sam Dallyn. 2011. "The paradox of order: Reimagining ideological domination." *Journal of Political Ideologies* 16 (1):53–78.

Bloom, P. and C. Rhodes. 2017. *The rise of CEO society: How the cult of corporate leadership is transforming the world*. Zed Books.

Bloom, Peter, and Paul J. White. 2016. "The moral work of subversion." *Human Relations* 69 (1):5–31.

Blouin, Michael J. 2016. *Magical thinking, fantastic film, and the illusions of neoliberalism*. New York: Palgrave Macmillan.

Blyth, M. 2013. *Austerity: The history of a dangerous idea*. Oxford University Press.

Boddy, Clive R. 2011. "Corporate Psychopaths, bullying and unfair supervision in the workplace." *Journal of Business Ethics* 100 (3):367–379.

Boggs, Carl. 2000. *The end of politics: Corporate power and the decline of the public sphere*. New York: Guilford Press.

Boje, David M. 1991. "The storytelling organization: A study of story performance in an office-supply firm." *Administrative Science Quarterly* 36 (1):106.

Boje, David M., and Robert D. Winsor. 1993. "The resurrection of Taylorism: Total quality management's hidden agenda." *Journal of Organizational Change Management* 6 (4):57–70.

Bolino, Mark C., and William H. Turnley. 2003. "Going the extra mile: Cultivating and managing employee citizenship behavior." *Academy of Management Executive* 17 (3):60–71.

Boltanski, Luc, and Eve Chiapello. 2005. "The new spirit of capitalism." *International Journal of Politics, Culture, and Society* 18 (3–4):161–188.

Bonefeld, W. 1992. "Social constitution and the form of the capitalist state." In W. Bonefeld, R. Gunn and K. Psychopedis (eds.), *Open Marxism, Vol. 1, Dialectics and History* (pp. 93–132), London, UK: Pluto Press.

Bonefeld, W., Gunn, R. and Psychopedis, K. 1992. Open Marxism: Dialectics and History, London, UK: Pluto Press.

Boreham, Paul. 1992. "The myth of post-Fordist management: Work organization and employee discretion in seven countries." *Employee Relations* 14 (2):13–24.

Boutillier, Sophie, and Dimitri Uzunidis. 2014. "The theory of the entrepreneur: From heroic to socialised entrepreneurship." *Journal of Innovation Economics* 14 (2):9.

Bowen, F., and J. A. Aragon-Correa. 2014. "Greenwashing in corporate environmentalism research and practice: The importance of what we say and do." *Organization & Environment* 27 (2):107–112.

Brammer, Stephen, Andrew Millington, and Bruce Rayton. 2007. "The contribution of corporate social responsibility to organizational commitment." *The International Journal of Human Resource Management* 18 (10):1701–1719.

Brantlinger, Patrick, and Donald Ulin. 1993. "Policing nomads: Discourse and social control in early Victorian England." *Cultural Critique* 25:33.

Brash, J. 2003. "Invoking fiscal crisis: Moral discourse and politics in New York City." *Social Text* 21 (3 76):59–83.

Brenner, Neil, Jamie Peck, and Nik Theodore. 2010. "Variegated neoliberalization: Geographies, modalities, pathways." *Global Networks* 10 (2):182–222.

Brenner, Neil, and Nik Theodore. 2002. "Cities and the geographies of 'actually existing neoliberalism.'" *Antipode* 34 (3):349–379.

Brocklehurst, M. 2001. "Power, identity and new technology homework: Implications for 'new forms' of organizing." *Organization Studies* 22 (3):445–466.

Broomhill, R. 2001. "Neoliberal globalism and the local state: a regulation approach." *The Journal of Australian Political Economy,* 48:115.

Brown, Wendy. 2005. "Neo-liberalism and the end of liberal democracy." *Theory & Event* 7 (1).

Brown, W. 2006. "American nightmare neoliberalism, neoconservatism, and de-democratization." *Political Theory 34* (6):690–714.

Brown, Wendy. 2015. *Undoing the Demos: Neoliberalism's stealth revolution.* Zone Books.

Brown, Wendy. 2016. "Sacrificial citizenship: Neoliberalism, human capital, and austerity politics." *Constellations* 23 (1):3–14.

Brückner, Hannah, and Karl Ulrich Mayer. 2005. "De-standardization of the life course: What it might mean? And if it means anything, whether it actually took place?" *Advances in Life Course Research* 9:27–53.

Budd, Leslie. 2007. "Post-bureaucracy and reanimating public governance." *International Journal of Public Sector Management* 20 (6):531–547.

Burack, Elmer H. 1999. "Spirituality in the workplace." *Journal of Organizational Change Management* 12 (4):280–292.

Burchell, Graham. 1993. "Liberal government and techniques of the self." *Economy and Society* 22 (3):267–282.

Burnham, P. 1994. "Open Marxism and vulgar international political economy." *Review of International Political Economy* 1 (2):221–231.

Bush, G. W. 2002. State of the Union Address. 30 January.

Butler, Judith P. 1997. *The psychic life of power: Theories in subjection.* Stanford, CA: Stanford University Press.

Byrne, U. 2005. "Work-life balance: Why are we talking about it at all?" *Business Information Review* 22 (1):53–59.

Cahill, Damien. 2010. "'Actually existing neoliberalism' and the global economic crisis." *Labour & Industry: A Journal of the Social and Economic Relations of Work* 20 (3):298–316.

Calavita, Kitty, Henry N. Pontell, and Robert Tillman. 1997. *Big money crime.* Berkeley, CA: University of California Press.

Caldwell, Cam. 2012. *Moral leadership: A transformative model for tomorrow's leaders.* New York: Business Expert Press.

Calhoun, Craig. 2013. "Occupy Wall Street in perspective." *The British Journal of Sociology* 64 (1):26–38.

Calomiris, C. W., and Mason, J. R. 2003. Fundamentals, panics, and bank distress during the depression. *The American Economic Review* 93 (5):1615–1647.

Campbell, Colin. 2005. *The romantic ethic and the spirit of modern consumerism.* 3rd ed. London, UK: Writersprintshop.

Campbell, John L., and Ove K. Pedersen, eds. 2001. *The rise of neoliberalism and institutional analysis.* Princeton, NJ: Princeton University Press.

Cannadine, David. 1984. "The present and the past in the English industrial revolution 1880–1980." *Past and Present* 103 (1):131–172.

Caproni, Paula J. 2004. "Work/life balance: You can't get there from here." *Journal of Applied Behavioral Science* 40 (2):208–218.

Carr, Adrian. 2000. "Critical theory and the management of change in organizations." *Journal of Organizational Change Management* 13 (3):208–220.

Carrette, J., and R. King. 2005. *Capitalist spirituality: The silent takeover of religion.* New York: Taylor & Francis.

Carroll, A. B. 1999. "Corporate social responsibility: Evolution of a definitional construct." *Business & Society* 38 (3):268–295.

Carroll, William K., and Colin Carson. 2010. *The making of a transnational capitalist class corporate power in the twenty-first century.* London, UK: Zed Books.

Casey, Catherine. 1995. *Work, self and society: After industrialism.* London, UK: Taylor & Francis.

Casey, Catherine. 1999. " 'Come, join our family': Discipline and integration in corporate organizational culture." *Human Relations* 52 (1):155–178.

Cashmore, Ellis. 2006. *Celebrity/culture.* New York: Taylor and Francis (Routledge).

Castles, F. 2009. *The social democratic image of society (Routledge Revivals): A Study of the achievements and origins of Scandinavian social democracy in comparative perspective.* Routledge.

Castro, Barry, ed. 1996. *Business and society: A reader in the history, sociology, and ethics of business.* New York: Oxford University Press.

Cederstrom, Carl. 2011. "Fit for everything: Health and the ideology of authenticity." *Ephemera: Theory and Politics in Organization* 11 (1):27–45.

Cederstrom, Carl, and R. Grassman. 2008. "The masochistic reflexive turn." *Ephemera: Theory and Politics in Organization* 8 (1):41–57.

Cederstrom, Carl, and Andrew Spicer. 2015. *The wellness syndrome.* Cambridge, UK: Polity Press.

Centeno, Miguel A., and Joseph N. Cohen. 2012. "The arc of neoliberalism." *Annual Review of Sociology* 38 (1):317–340.

Changfoot, Nadine. 2007. "Local activism and neoliberalism: Performing neoliberal citizenship as resistance." *Studies in Political Economy* 80 (1):129–149.

Chari, V. V., L. Christiano, and P. J. Kehoe. 2008. "Facts and myths about the financial crisis of 2008." *Federal Reserve Bank of Minneapolis Working Paper* 666.

Chaudhry, K. A. 1993. "The myths of the market and the common history of late developers." *Politics & Society* 21 (3):245–274.

Cherry, Miriam A., and Judd F. Sneirson. 2011. "Beyond profit: Rethinking corporate social responsibility and greenwashing after the BP oil disaster." *SSRN Electronic Journal.*

Chesnais, F. 1984. "Marx's crisis theory today." In *Design, innovation and long cycles in economic development*, ed. C. Freedman. London, UK: Frances Pinter, 186–193.

Chibba, Michael. 2009. "Financial inclusion, poverty reduction and the millennium development goals." *European Journal of Development Research* 21 (2): 213–230.

Cho, A. H. 2006. "Politics, values and social entrepreneurship: A critical appraisal." In *Social entrepreneurship*, eds. Johanna Mair, Jeffrey Robinson and Kai Hockerts. Basingstoke, UK: Palgrave Macmillan UK, 34–56.

Choat, Simon. 2010. *Marx through post-structuralism.* Lyotard, Derrida, Foucault, Deleuze, London: Continuum Books.

Chomsky, Noam. 1999. *Profit over people: Neoliberalism and global order.* New York: Seven Stories Press.

Choudhary, K. 2007. "Globalisation, governance reforms and development: An introduction." In *Globalisation, governance reforms and development in India*, ed. K. Choudhary. London, UK: Sage Publications.

Clarke, J. 2008. Living with/in and without neo-liberalism. *Focaal* 2008 (51): 135–147.

Clarke, John, and J. Newman. 2012. "The alchemy of austerity." *Critical Social Policy* 32 (3):299–319.

Clarke, Nick, Clive Barnett, Paul Cloke, and Alice Malpass. 2007. "Globalising the consumer: Doing politics in an ethical register." *Political Geography* 26 (3): 231–249.

Clarke, Simon. 1994. *Marx's theory of crisis.* New York: St. Martin's Press.

Clarke, Simon. 2005. "The neoliberal theory of society." In *Neoliberalism: A critical reader*, eds. A. Saad-Filho and D. Johnston. London, UK: Pluto Press, 50–59.

Cleaver, H. 1992. "The inversion of class perspective in Marxian theory: From valorisation to self-valorisation." Open Marxism, 2, 106–144.

Cohen, L. 2004. A consumers' republic: The politics of mass consumption in postwar America. *Journal of Consumer Research* 31 (1):236–239.

Cohen, Robin. 2006. *Migration and its enemies: Global capital, migrant labour and the nation-state.* Aldershot, UK: Ashgate Publishing.

Cohen, R. L. 2010. "Rethinking 'mobile work': Boundaries of space, time and social relation in the working lives of mobile hairstylists." *Work, Employment & Society* 24 (1):65–84.

Coleman, D. C. 1992. *Myth, history, and the Industrial Revolution.* Rio Grande, OH: Continuum International Publishing Group.

Collier, Jane, and Rafael Esteban. 2007. "Corporate social responsibility and employee commitment." *Business Ethics: A European Review* 16 (1):19–33.

Commons, John R. 1924. *Legal foundations of capitalism.* New Jersey: The Lawbook Exchange.

Congleton, Roger D. 2009. "On the political economy of the financial crisis and bailout of 2008–2009." *Public Choice* 140 (3–4):287–317.

Conley, John M., and Cynthia A. Williams. 2005. "Engage, embed, and embellish: Theory versus practice in the corporate social responsibility movement." *SSRN Electronic Journal* 31 (1):1–38.

Connolly, William E. 2008. *Capitalism and Christianity, American style*. Durham, NC: Duke University Press Books.

Connor, Stuart. 2010. "Promoting 'employ ability': The changing subject of welfare reform in the UK." *Critical Discourse Studies* 7 (1):41–54.

Contu, Alessia. 2008. "Decaf resistance: On misbehavior, cynicism, and desire in liberal workplaces." *Management Communication Quarterly* 21 (3):364–379.

Contu, Alessia, and Christopher Grey. 2003. "Against learning." *Human Relations* 56 (8):931–952.

Cook, Joanna. 2016. "Mindful in Westminster: The politics of meditation and the limits of neoliberal critique." *HAU: Journal of Ethnographic Theory* 6 (1): 141–161.

Cooper, Melinda. 2004. "Insecure times, tough decisions: The Nomos of neoliberalism." *Alternatives: Global, Local, Political* 29 (5):515–533.

Costas, Jana, and P. Fleming. 2009. "Beyond dis-identification: A discursive approach to self-alienation in contemporary organizations." *Human Relations* 62 (3):353–378.

Costas, Jana, and Christopher Grey. 2014. "The temporality of power and the power of temporality: Imaginary future selves in professional service firms." *Organization Studies* 35 (6):909–937.

Courpasson, D. and S. Clegg. 2006. "Dissolving the iron cages? Tocqueville, Michels, bureaucracy and the perpetuation of elite power." *Organization* 13 (3); 319–343.

Courpasson, D., and M. Reed. 2004. "Introduction: Bureaucracy in the age of enterprise." *Organization* 11 (1):5–12.

Covaleski, Mark A., Mark W. Dirsmith, James B. Heian, and Sajay Samuel. 1998. "The calculated and the avowed: Techniques of discipline and struggles over identity in Big Six public accounting firms." *Administrative Science Quarterly* 43 (2):293.

Crary, Jonathan. 2013. *24/7: Late capitalism and the ends of sleep*. New York: Verso Books.

Cremin, Colin. 2011. *Capitalism's new clothes: Enterprise, ethics and enjoyment in times of crisis*. London, UK: Palgrave Macmillan.

Critcher, C. 2003. *Moral panics and the media*. Philadelphia: Open University Press.

Cross, G. S. 2000. *An all-consuming century: Why commercialism won in modern America*. New York: Columbia University Press.

Crouch, Colin. 2011. *The strange non-death of neo-liberalism*. Cambridge, UK: John Wiley & Sons.

Cruikshank, Barbara. 1999. *The will to empower: Democratic citizens and other subjects*. Baltimore, MD: Cornell University Press.

Daft, Richard L. 2014. *The leadership experience*. 5th ed. Boston, MA: South-Western, Cengage Learning.

Daguerre, A. 2014. "New corporate elites and the erosion of the Keynesian social compact." *Work, Employment and Society*, 28 (2):323–334.

Davies, G. 2003. *Corporate reputation and competitiveness*. Abingdon, UK: Psychology Press.

Davies, William. 2014. *The limits of neoliberalism: Authority, sovereignty and the logic of competition*. London, UK: Sage Publications.

Davies, William. 2015. *The happiness industry: How the government and big business sold us well-being*. London, UK: Verso Books.

Davis, Allen F. 1984. *Spearheads for reform: The social settlements and the progressive movement, 1890–1914*. 3rd ed. New Brunswick, NJ: Rutgers University Press.

Dawson, G. 2013. *Soldier heroes: British adventure, empire and the imagining of masculinities*. London, UK: Routledge.
Day, K. 1993. *S & L hell: The people and the politics behind the $1 trillion savings and loan scandal*. London, UK: W. W. Norton & Company.
De Cock, C., Vachhani, S., and Murray, J. 2013. "Putting into question the imaginary of recovery: a dialectical reading of the global financial crisis and its aftermath." Culture and Organization, 19 (5):396–412.
Dean, Jodi. 2009. *Democracy and other neoliberal fantasies: Communicative capitalism and left politics*. Durham, NC: Duke University Press.
Dean, Mitchell. 1998. "Administrative ascetism." In *Governing Australia*, eds. M. Dean and B. Hindess. Cambridge, UK: Cambridge University Press.
Dean, Mitchell. 1994. " 'A social structure of many souls': Moral regulation, government, and self-formation." *Canadian Journal of Sociology/Cahiers canadiens de sociologie* 19 (2):145.
Dees, J. G. 2007. Taking social entrepreneurship seriously. *Society* 44 (3):24–31.
DeKeseredy, Walter S., Stephen L. Muzzatti, and Joseph F. Donnermeyer. 2013. "Mad men in bib overalls: Media's horrification and pornification of rural culture." *Critical Criminology* 22 (2):179–197.
Dekker, E. 2016. *The Viennese students of civilization*. Cambridge, UK: Cambridge University Press.
Delbridge, Rick. 1995. "Surviving JIT: Control and resistance in a Japanese transplant." *Journal of Management Studies* 32 (6):803–817.
Demmers, Joelle, and Jilberto, Alex Fernandez E. 2001. *Miraculous metamorphoses: The neoliberalization of Latin American populism*. London, UK: Zed Books.
Demmers, Jolle, Alex Fernandez E. Jilberto, and Barbara Hogenboom, eds. 2004. *Good governance in the era of global neoliberalism: Conflict and depolitisation in Latin America, Eastern Europe, Asia, and Africa*. London, UK: Taylor & Francis.
Derrida, Jacques. 1997. "Politics and friendship." An interview with Geoffrey Bennington, 1 December.
DeSante, Christopher D. 2013. "Working twice as hard to get half as far: Race, work ethic, and America's deserving poor." *American Journal of Political Science* 57 (2):342–356.
Deudney, D., and G. Ikenberry. 2009. "The myth of the autocratic revival." *Foreign Policy* 88 (1):77–93.
Diba, Behzad T., and Herschel I. Grossman. 1988. "The theory of rational bubbles in stock prices." *The Economic Journal* 98 (392):746.
Domagalski, T. A. 1999. "Emotion in organizations: Main currents." *Human Relations* 52 (6): 833–852.
Doornbos, M. 2003. "Good governance: The metamorphosis of a policy metaphor." *Journal of International Affairs* 57 (1):3.
Dore, R., W. Lazonick, and M. O'Sullivan. 1999. "Varieties of capitalism in the twentieth century." *Oxford Review of Economic Policy* 15 (4):102–120.
Dorrien, G. 2013. *Imperial designs: Neoconservatism and the New Pax Americana*. Routledge.
Drago, Robert, and Geoffrey K. Turnbull. 1988. "Individual versus group piece rates under team technologies." *Journal of the Japanese and International Economies* 2 (1):1–10.
Dupuy, Jean-Pierre. 2009. *On the origins of cognitive science: The mechanization of the mind*. Trans. M. B. DeBevoise. Cambridge, MA: MIT Press.

Dussel, E. 1997. *The invention of the Americas. Eclipse of 'the other' and the myth of modernity*. New York: Continuum Publishing.
Dyer, Richard. 1979. *Stars*. London, UK: British Film Inst.
Echtelt, P. E. van. 2006. "The new lumpiness of work: Explaining the mismatch between actual and preferred working hours." *Work, Employment & Society* 20 (3):493–512.
The Economist. 2009. "Virtual pleasures." *The Economist*, 5 February.
The Economist. 2016. "After Trump, Italy could be the next anti-establishment revolt," *The Economist*, 9 November.
Ehrenreich, Barbara. 2009. *Bright-sided [how the relentless promotion of positive thinking has undermined America]*. New York: Macmillan Audio.
Eikhof, D. R., C. Warhurst, and A. Haunschild. 2007. "Introduction: What work? What life? What balance?: Critical reflections on the work-life balance debate." *Employee Relations*, 29 (4): 325–333.
Elkington, John. 1997. *Cannibals with forks: The triple bottom line of 21st century business*. Oxford, UK: Capstone Publishing.
Elliott, J. 2013. "Suffering agency: Imagining neoliberal Personhood in North America and Britain." *Social Text* 31 (2 115):83–101.
Engelen, Ewald. 2015. "Don't mind the 'funding gap': What Dutch post crisis storytelling tells us about elite politics in financialized capitalism." *Environment and Planning A* 47 (8):1606–1623.
Engelen, Ewald, M. Konings, and R. Fernandez. 2008. "The rise of activist investors and patterns of political responses: Lessons on agency." *Socio-Economic Review* 6 (4):611–636.
Enriques, L. 2003. "Bad apples, bad oranges: A comment from old Europe on post-Enron corporate governance reforms." *Wake Forest Law Review* 38:911–934.
Esposito, L., and F. M. Perez. 2014. "Neoliberalism and the commodification of mental health." *Humanity & Society* 38 (4):414–442.
Eubanks, Philip. 2005. "Globalization, 'corporate rule,' and blended worlds: A conceptual-rhetorical analysis of metaphor, metonymy, and conceptual blending." *Metaphor and Symbol* 20 (3):173–197.
Falk, A., and Szech, N. 2013. Morals and markets. *Science*, *340* (6133):707–711.
Fechter, A.-M. 2016. "Aid work as moral labour." *Critique of Anthropology* 36 (3):228–243.
Feldman, Maryann P. 2014. "The character of innovative places: Entrepreneurial strategy, economic development, and prosperity." *Small Business Economics* 43 (1):9–20.
Felli, Romain. 2014. "The wrath of capital: Neoliberalism and climate change politics." *Social & Cultural Geography* 15 (5):588–590.
Ferguson, James. 2006. *Global shadows: Africa in the neoliberal world order*. 3rd ed. Durham, NC: Duke University Press.
Ferrell, O. C., and Larry G. Gresham. 1985. "A contingency framework for understanding ethical decision making in marketing." *Journal of Marketing* 49 (3):87.
Ferri-Reed, J. 2013. " 'Millennials-Generation' Screwed 'or Generation' Shrewd?" *The Journal for Quality and Participation* 36 (1):22.
Ferris, Kerry O. 2007. "The sociology of celebrity." *Sociology Compass* 1 (1): 371–384.
Field, F. 1997. "Re-inventing welfare: A response to Lawrence Mead." In *From welfare to work: Lessons from America* (pp. 59–64), ed. L. Mead. London, UK: Institute for Economic Affairs, Health and Welfare Unit.

Fineman, S. (ed.). 2000. *Emotion in organizations*. Thousand Oaks, CA: Sage.

Fisher, Mark. 2009. *Capitalist realism: Is there no alternative?* Hants, UK: John Hunt Publishing.

Fiss, Owen M. 1997. "Money and politics." *Columbia Law Review* 97 (8):2470.

Fleetwood, Steve. 2007. "Why work-life balance now?" *The International Journal of Human Resource Management* 18 (3):387–400.

Fleming, Peter, and Andre Spicer. 2003. "Working at a cynical distance: Implications for power, subjectivity and resistance." *Organization* 10 (1):157–179.

Fleming, Peter, and Andre Spicer. 2004. " 'You can checkout anytime, but you can never leave': Spatial boundaries in a high commitment organization." *Human Relations* 57 (1):75–94.

Fleming, Peter, and Andrew Sturdy. 2009. "Just be yourself!" *Employee Relations* 31 (6):569–583.

Flinders, Matthew V. 2012. "The demonisation of politicians: Moral panics, folk devils and MPs' expenses." *Contemporary Politics* 18 (1):1–17.

Folger, R., D. B. Ganegoda, D. B. Rice, R. Taylor, and D. X. Wo. 2013. "Bounded autonomy and behavioral ethics: Deonance and reactance as competing motives." *Human Relations* 66 (7):905–924.

Foner, Eric. 1999. *The story of American freedom*. New York: W. W. Norton & Company.

Ford, J., and D. Collinson. 2011. "In search of the perfect manager? Work-life balance and managerial work." *Work, Employment & Society* 25 (2):257–273.

Forkert, Kirsten. 2014. "The new moralism: Austerity, silencing and debt morality." *Soundings* 56:41–53.

Foster, John. 2003. *Class struggle and the industrial revolution: Early industrial capitalism in three English towns*. London, UK: Routledge.

Foucault, Michel. 1977. *Discipline and Punishment*. Penguin Books.

Foucault, Michel. 1982. "The subject and power." *Critical Inquiry* 8 (4):777.

Foucault, Michel. 1985. *Discourse and truth: The problematization of Parrhesia*. Evanston, IL: Northwestern University.

Foucault, Michel. 1988. "The ethic of care for the self as a practice of freedom: an interview with Michel Foucault on January 20, 1984 in the final Foucault: studies on Michel Foucault's last works." *Philosophy & Social Criticism*, 12 (2–3): 112–131.

Foucault, Michel. 1990. *Politics, Philosophy, Culture: Interviews and Other Writings, 1977–1984*. London, UK: Routledge.

Foucault, Michel. 1990 [1976]. *The will to knowledge: The history of sexuality, vol. 1* (trans. Hurley R). Penguin.

Fourcade, Marion, and Kieran Healy. 2007. "Moral views of market society." *Annual Review of Sociology* 33 (1):285–311.

Fraser, Benson P., and William J. Brown. 2002. "Media, celebrities, and social influence: Identification with Elvis Presley." *Mass Communication and Society* 5 (2): 183–206.

Fraser, Janet, and Michael Gold. 2001. " 'Portfolio workers': Autonomy and control amongst freelance translators." *Work, Employment & Society* 15 (4):679–697.

Freeden, Michael. 1996. *Ideologies and political theory: A conceptual approach*. Oxford, UK: Oxford University Press.

Freeden, Michael. 2009. "Editorial: what fails in ideologies?," *Journal of Political Ideologies* 14 (1):1–9.

Frenkel, S., M. Korczynski, L. Donoghue, and K. Shire. 1995. "Re-constituting work: Trends towards knowledge work and info-normative control." *Work, Employment & Society* 9 (4):773–796.

Fridman, Daniel. 2016. *Freedom from work: Embracing financial self-help in the United States and Argentina*. Stanford, CA: Stanford University Press.

Friedman, M. 1970. "The social responsibility of business is to increase its profits." *The New York Times*.

Friedman, A. 1977. "Responsible autonomy versus direct control over the labour process." *Capital & Class* 1 (1):43–57.

Friedman, M. 2009. *Capitalism and Freedom*. Chicago, IL: University of Chicago Press.

Froud, Julie, Adam Leaver, and Karel Williams. 2007. "New actors in a financialised economy and the remaking of capitalism." *New Political Economy* 12 (3): 339–347.

Fukuyama, Francis. 1989. *The end of history and the last man*. New York: Harper Perennial.

Fuller, S. 2006. "The market: Source or target of morality." Presented at International Conference on Moralization of the Markets. Essen, Germany May 6–9.

Furnham, Adrian, Michael Bond, Patrick Heaven, Denis Hilton, Thalma Lobel, John Masters, Monica Payne, R. Rajamanikam, Barrie Stacey, and H. Van Daalen. 1993. "A comparison of Protestant work ethic beliefs in thirteen nations." *The Journal of Social Psychology* 133 (2):185–197.

Galperin, Bella L., and Ronald J. Burke. 2006. "Uncovering the relationship between workaholism and workplace destructive and constructive deviance: An exploratory study." *The International Journal of Human Resource Management* 17 (2): 331–347.

Gamble, A. 1979. "The free economy and the strong state." *Socialist Register* 16:1–25.

Garland, D. 1996. "The limits of the sovereign state: Strategies of crime control in contemporary society." *British Journal of Criminology* 36 (4):445–471.

Garthwaite, Kayleigh. 2011. "'The language of shirkers and scroungers?' Talking about illness, disability and coalition welfare reform." *Disability & Society* 26 (3): 369–372.

Gates, Jeffrey R., and Stephan Schmidheiny. 1999. *The ownership solution: Toward a shared capitalism for the twenty-first century*. Reading, MA: Perseus Books.

Gerring, John, and Strom C. Thacker. 2005. "Do neoliberal policies deter political corruption?" *International Organization* 59 (1).

Giddens, Anthony. 1991. *Modernity and self-identity: Self and society in the late modern age*. Stanford, CA: Stanford University Press.

Giddens, Anthony. 1999. "Risk and responsibility." *Modern Law Review* 62 (1): 1–10.

Giddens, A. 2013. *The third way and its critics*. Hoboken, NJ: John Wiley & Sons.

Gideon, J. 1998. "The politics of social service provision through NGOs: A study of Latin America." *Bulletin of Latin American Research* 17 (3): 303–321.

Giger, Nathalie, and Moira Nelson. 2011. "The electoral consequences of welfare state retrenchment: Blame avoidance or credit claiming in the era of permanent austerity?" *European Journal of Political Research* 50 (1):1–23.

Gilbert, Jeremy. 2013. "What kind of thing is 'neoliberalism'?" *New Formations* 80 (1):7–22.

Gill, Stephen R., and David Law. 1989. "Global hegemony and the structural power of capital." *International Studies Quarterly* 33 (4):475.

Gilley, K. M., and A. Rasheed. 2000. "Making more by doing less: An analysis of outsourcing and its effects on firm performance." *Journal of Management* 26 (4): 763–790.

Gilson, M. 2016. *Sad guys at the Wise Owl Club: Hipsters, White nostalgia, & neoliberal optimism*. PhD thesis, George Washington University.

Gini, A. 1997. "Moral leadership and business ethics." *Journal of Leadership & Organizational Studies* 4 (4):64–81.

Giorgi, Liana, and Catherine Marsh. 1990. "The Protestant work ethic as a cultural phenomenon." *European Journal of Social Psychology* 20 (6):499–517.

Giroux, Henry. 2002. "Neoliberalism, corporate culture, and the promise of higher education: The university as a democratic public sphere." *Harvard Educational Review* 72 (4):425–464.

Giroux, Henry. 2003. "Utopian thinking under the sign of neoliberalism: Towards a critical pedagogy of educated hope." *Democracy & Nature* 9 (1):91–105.

Giroux, Henry. 2004. *The terror of neoliberalism: Authoritarianism and the eclipse of democracy*. Boulder, CO: Paradigm Publishers.

Giroux, Henry. 2015. *Against the terror of neoliberalism: Politics beyond the age of greed*. Boulder, CO: Paradigm Publishers.

Gitlin, Todd. 2012. *Occupy Nation: The roots, the spirit, and the promise of Occupy Wall Street*. New York: HarperCollins Publishers.

Glynos, Jason, and David Howarth. 2007. *Logics of critical explanation in social and political theory*. London, UK: Taylor & Francis.

Gofin, R., and Avitzour, M. 2012. "Traditional versus internet bullying in junior high school students." *Maternal and Child Health Journal* 16 (8):1625–1635.

GoGwilt, C. L. 1995. *The invention of the west: Joseph Conrad and the double-mapping of Europe and empire*. Stanford, CA: Stanford University Press.

Gond, J. P., Kang, N., and Moon, J. 2011. The government of self-regulation: On the comparative dynamics of corporate social responsibility. *Economy and Society*, *40*(4):640–671.

Goode, Erich, and Nachman Ben-Yehuda. 2009. *Moral panics: The social construction of deviance*. Cambridge, MA: Blackwell Publishers.

Gordon, Todd. 2005. "The political economy of law-and-order policies: Policing, class struggle, and neoliberal restructuring." *Studies in Political Economy* 75 (1): 53–77.

Gorman, Hugh S. 2004. "From apocalypse to way of life: Environmental crisis in the American century (review)." *Technology and Culture* 45 (3):638–639.

Gorton, G. 1988. "Banking panics and business cycles." *Oxford Economic Papers*, 40(4):751–781

Graeber, David. 2011. *Debt: The first 5,000 years*. New York: Melville House Publishing.

Graeber, David. 2015. *The utopia of rules: On technology, stupidity, and the secret joys of bureaucracy*. New York: Melville House Publishing.

Graham, David, and Ngaire Woods. 2006. "Making corporate self-regulation effective in developing countries." *World Development* 34 (5):868–883.

Gramsci, A. 1971. *Selections from the Prison Notebook*. London, UK: Smith, Lawrence and Wishart.

Gray, Andrew, Jane Broadbent, and Michaela Lavender. 2009. "Editorial: Crisis! What an opportunity!" *Public Money & Management* 29 (5):265–265.

Green, Stephen. 1999. "A plague on the panopticon: Surveillance and power in the global information economy." *Information, Communication & Society* 2 (1): 26–44.

Griffin, Emma. 2010. *A short history of the British Industrial Revolution*. Basingstoke: Palgrave Macmillan.

Griffin, Roger. 2007. *Modernism and fascism: The sense of a beginning under Mussolini and Hitler*. Basingstoke, Hampshire: Palgrave Macmillan.

Grindle, Merilee S. 2007. *Going local: Decentralization, democratization, and the promise of good governance*. 3rd ed. Princeton, NJ: Princeton University Press.

Groenhout, Ruth E. 2004. *Connected lives: Human nature and an ethics of care*. Maryland: Rowman & Littlefield Publishers.

Grugel, L. E. 1979. *Society and religion during the age of industrialization: Christianity in Victorian England*. Lanham, MD: The University Press of America.

Guerrera, F. 2010. "How wall street changed wall street." *Financial Times*.

Guldi, J. 2012. *Roads to power: Britain invents the infrastructure state*. Cambridge, MA: Harvard University Press.

Habermas, J. 1991. *The structural transformation of the public sphere: An inquiry into a category of bourgeois society*. Cambridge, MA: MIT Press.

Hager, Sandy Brian. 2015. "Corporate ownership of the public debt: Mapping the new aristocracy of finance." *Socio-Economic Review* 13 (3):505–523.

Haiven, Max. 2014. *Cultures of financialization: Fictitious capital in popular culture and everyday life*. Basingstoke, UK: Palgrave Macmillan.

Hales, C. 2000. "Management and empowerment programmes." *Work, Employment & Society* 14 (3):501–519.

Hall, Douglas T. 2001. *Careers in and out of organizations*. Thousand Oaks, CA: Sage Publications.

Hall, D. T. 2004. The protean career: A quarter-century journey. *Journal of Vocational Behavior* 65 (1):1–13.

Hall, Peter A., and David W. Soskice. 2001. *Varieties of capitalism: The institutional foundations of comparative advantage*. Oxford, UK: Oxford University Press.

Hall, T. J. 2011. "The triple bottom line: What is it and how does it work?" *Indian Business Review* 86 (1):4–8.

Halsall, R. 2013. "Signs of crisis and recovery: Geographical imaginaries in press coverage of the financial crisis in the UK and German press 2008–2009." *Culture and Organization* 19 (5):377–395.

Hamburg Coplan, Jill. 2009. "How positive psychology can boost your business." *Bloomberg Business*.

Hamer, John H. 1998. "Money and the moral order in late nineteenth and early twentieth-century American capitalism." *Anthropological Quarterly* 71 (3):138.

Hamann, T. H. 2009. Neoliberalism, governmentality, and ethics. *Foucault Studies*, 6:37–59.

Hampton, Keith N., Lauren F. Sessions, and Eun Ja Her. 2011. "Core networks, social isolation, and new media." *Information, Communication & Society* 14 (1):130–155.

Handler, Joel F., and Yeheskel Hasenfeld. 2006. *Blame welfare, ignore poverty and inequality*. Cambridge, UK: Cambridge University Press.

Handy, C. 1997. *The hungry spirit*. London, UK: Hutchinson.

Hanley, E. 1999. "Cadre capitalism in Hungary and Poland: Property accumulation among communist-era elites." *East European Politics & Societies* 14 (1): 143–178.

Hanlon, G. 2016a. "Total bureaucratisation, neo-liberalism, and Weberian oligarchy." *Ephemera: Theory and Politics in Organization* 16 (1):171–191.

Hanlon, G. 2016b. "The first neo-liberal science: Management and neo-liberalism." *Sociology*.

Hansen, Laura L., and Siamak Movahedi. 2010. "Wall Street scandals: The myth of individual greed." *Sociological Forum* 25 (2):367–374.

Hardt, M. 2010. "Two faces of the apocalypse: Letters from Copenhagen." *Polygraph* 22:265–274.

Harley, B. 1999. "The myth of empowerment: Work organisation, hierarchy and employee autonomy in contemporary Australian workplaces." *Work, Employment & Society* 13 (1):41–66.

Harris, A., Lin, M., and Selbin, J. 2007. "From 'The Art of War' to 'Being Peace': Mindfulness and community lawyering in a neoliberal age." *California Law Review* 95 (5):2073–2132.

Harris, B. 1996. "'American idols': Empire, war and the middling ranks in mid-eighteenth-century Britain." *Past & Present* 150 (1):111–141.

Harrison, Carol E. 1999. *The bourgeois citizen in nineteenth-century France: Gender, sociability, and the uses of emulation*. Oxford, UK: Oxford University Press.

Harvey, David. 2001. "Globalisation and the spatial fix." *Grographische tevue* 2 (3):23–31.

Harvey, David. 2005. *A brief history of neoliberalism*. Oxford: Oxford University Press.

Harvey, David. 2007. *A brief history of neoliberalism*. New York: Oxford University Press.

Hatier, Cécile. 2012. "'Them' and 'us': Demonising politicians by moral double standards." *Contemporary Politics* 18 (4):467–480.

Hawkins, David E. 2006. *Corporate social responsibility: Balancing tomorrow's sustainability and today's profitability*. Basingstoke: Palgrave Macmillan.

Hawkins, Mike, and Kingston. 1997. *Social Darwinism in European and American thought, 1860–1945: Nature as model and nature as threat*. Cambridge, UK: Cambridge University Press.

Hawkins, Peter. 1999. *The art of building windmills: Career tactics for the 21st century*. Liverpool: Graduate Into Employment Unit.

Hay, Colin. 2002. "Globalization as a problem of political analysis: Restoring agents to a process without a subject and politics to a logic of economic compulsion." *Cambridge Review of International Affairs* 15 (3):379–392.

Hayek, F. A. 2012. *Individualism and economic order*. Chicago, IL: University of Chicago Press.

Healy, Geraldine. 2004. "Work-life balance and family friendly policies—in whose interest?" *Work, Employment & Society* 18 (1):219–223.

Heckscher, C., C. Heckscher, and A. Donnellon. 1994. "Defining the post-bureaucratic type." In C. Heckscher and A. Donnellon (eds.) *The post-bureaucratic organization: New perspectives on organizational change*. Newbury Park, CA: Sage.

Heller, Kevin Jon. 1996. "Power, subjectification and resistance in Foucault." *Sub-Stance* 25 (1):78.

Henderson, H., and Sethi, S. 2006. *Ethical markets: Growing the green economy.* Chelsea Green Publishing.

Henig, Jeffrey R. 1989. "Privatization in the United States: Theory and practice." *Political Science Quarterly* 104 (4):649.

Henry, John F. 2008. "The ideology of the Laissez Faire program." *Journal of Economic Issues* 42 (1):209–224.

Herod, Andrew. 1997. "From a geography of labor to a labor geography: Labor's spatial fix and the geography of capitalism." *Antipode* 29 (1):1–31.

Hier, Sean P. 2011. "Tightening the focus: Moral panic, moral regulation and liberal government." *The British Journal of Sociology* 62 (3):523–541.

Hightower, J. 2002. "How Walmart is Remaking our World," Alternet 25 April

Hill, C. 1996. *Society and puritanism in pre-revolutionary England*, London: Secker & Warburg.

Hill, D., and Kumar, R. (eds.). 2009. *Global neoliberalism and education and its consequences.* Routledge.

Himmelstein, J. L., and J. A. McRae. 1984. "Social conservatism, New Republicans, and the 1980 election." *Public Opinion Quarterly* 48 (3):592–605.

Hirschman, A. O. 1982. "Rival interpretations of market society: Civilizing, destructive or feeble?" *Journal of Economic Literature* 20 (4):1463–1484.

Hitt, M. A., Ireland, R. D., Camp, S. M., and Sexton, D. L. 2001. Strategic entrepreneurship: Entrepreneurial strategies for wealth creation. *Strategic Management Journal*, 22 (6–7):479–491.

Hobsbawm, Eric J. 2010. *The age of empire, 1875–1914 (History of civilisation).* London, UK: Weidenfeld and Nicolson.

Hochschild, Arlie. 1997. "The time bind." *WorkingUSA* 1 (2):21–29.

Hodgson, G. M. 2004. *The evolution of institutional economics.* Routledge.

Hoedemaekers, Casper. 2016. "Work hard, play hard: Fantasies of nihilism and hedonism between work and consumption." *Ephemera: Theory and Politics in Organization* 16 (3):61–94.

Hofstadter, R. 1944. *Social Darwinism in American thought.* Beacon Press.

Holladay, Courtney L., and Miguel A. Quiñones. 2005. "Reactions to diversity training: An international comparison." *Human Resource Development Quarterly* 16 (4):529–545.

Hollander, Jocelyn A., and Rachel L. Einwohner. 2004. "Conceptualizing resistance." *Sociological Forum* 19 (4):533–554.

Holmstrom, B., and P. Milgrom. 1991. "Multitask principal-agent analyses: Incentive contracts, asset ownership, and job design." *Journal of Law, Economics, and Organization* 7 (special):24–52.

Holzmann, Robert, and Steen Jørgensen. 2001. "Social risk management: A new conceptual framework for social protection, and beyond'." *International Tax and Public Finance* 8 (4):529–556.

Honig, Benson, and Elizabeth Leslie Black. 2007. "The industrial revolution and beyond." *Journal of Management History* 13 (3):269–289.

Hoover, Kenneth R. 1987. "The rise of conservative capitalism: Ideological tensions within the Reagan and Thatcher governments." *Comparative Studies in Society and History* 29 (2):245.

Höpfl, H. M. 2006. "Post-bureaucracy and Weber's 'modern' bureaucrat." *Journal of Organizational Change Management* 19 (1):8–21.

Horowitz, Irving Louis. 2006. "The moral economy of meritocracy: Or, the unanticipated triumph of reform and the failure of revolution in the West." *The Political Quarterly* 77 (s1):127–133.

Houlihan, Maeve. 2002. "Tensions and variations in call centre management strategies." *Human Resource Management Journal* 12 (4):67–85.

Howarth, David J. 2000. *Discourse*. Buckingham: Open University Press.

Howe, N., and W. Strauss. 2009. *Millennials rising: The next great generation*. New York: Vintage.

Howitt, R. 2002. Preface. In R. Jenkins, R. Pearson, and G. Seyfang (eds), *Corporate responsibility and labour rights: Codes of conduct in the global economy*. London: Earthscan.

Hume, D. 1955. *Writings on Economics*. Madison, WI: University of Wisconsin Press.

Hunt, Lynn Avery. 2004. *Politics, culture, and class in the French Revolution*. Twentieth anniversary edition. Berkeley, CA: University of California Press.

Hutton, W. 2003. "I want to be my own time lord." *The Guardian*, 15 June.n.

Ilcan, Suzan. 2009. "Privatizing responsibility: Public sector reform under neoliberal government." *Canadian Review of Sociology/Revue canadienne de sociologie* 46 (3):207–234.

Illouz, Eva. 1997. *Consuming the romantic utopia: Love and the cultural contradictions of capitalism*. 2nd ed. Berkeley, CA: University of California Press.

Inglehart, Ronald F., and Pippa Norris. 2016. "Trump, Brexit and the rise of populism: Economic have-nots and cultural backlash." *HKS Working Paper Series*.

Isin, Engin F. 1997. "Who is the new citizen? Towards a genealogy." *Citizenship Studies* 1 (1):115–132.

Itoh, H. 1992. "Cooperation in hierarchical organizations: An incentive perspective." *The Journal of Law, Economics, and Organization* 8 (2):321–345.

Jackson, Paul, Hosein Gharavi, and Jane Klobas. 2006. "Technologies of the self: Virtual work and the inner panopticon." *Information Technology & People* 19 (3):219–243.

Jamali, Dima R., Ali M. El Dirani, and Ian A. Harwood. 2014. "Exploring human resource management roles in corporate social responsibility: The CSR-HRM co-creation model." *Business Ethics: A European Review* 24 (2):125–143.

Jamali, Dima R., Asem M. Safieddine, and Myriam Rabbath. 2008. "Corporate governance and corporate social responsibility synergies and interrelationships." *Corporate Governance: An International Review* 16 (5):443–459.

Jameson, F. 2003. "Future city." *New Left Review* 21:65–80.

Janssen, Marijn, and Elsa Estevez. 2013. "Lean government and platform-based governance—doing more with less." *Government Information Quarterly* 30:S1–S8.

Jefferess, D. 2012. "The 'Me to We' social enterprise: Global education as lifestyle brand." *Critical Literacy: Theories and Practices* 6 (1):18–30.

Jenkins, Rhys, Ruth Pearson, and Gill Seyfang, eds. 2002. *Corporate responsibility and labour rights: Codes of conduct in the global economy*. London, UK: Earthscan Publications.

Jensen, Tracey. 2014. "Welfare commonsense, poverty porn and doxosophy." *Sociological Research Online* 19 (3).

Jermier, J. M., Knights, D. E., and Nord, W. R. 1994. Resistance and power in organizations. Taylor & Francis/Routledge.

Jessop, B. 2002. "Liberalism, neoliberalism, and urban governance: A state–theoretical perspective." *Antipode* 34 (3):452–472.

Johnson, Craig E. 2013. *Meeting the ethical challenges of leadership: Casting light or shadow*. 5th ed. Thousand Oaks, CA: Sage Publications.
Johnson, Eleanor K. 2015. "The business of care: The moral labour of care workers." *Sociology of Health & Illness* 37 (1):112–126.
Jones, C., and A. Spicer. 2005. "The sublime object of entrepreneurship." *Organization* 12 (2):223–246.
Jones, Owen. 2012. *Chavs: The demonization of the working class*. New York: Verso Books.
Jones, Peter. 1991. *The 1848 revolutions*. 2nd ed. New York: Longman.
Kalinowski, T. 2013. "Regulating international finance and the diversity of capitalism." *Socio-Economic Review* 11 (3):471–496.
Kanter, R. M. 1977. *Work and family in the United States: A critical review and agenda for research and policy*. New York: Russell Sage.
Kanter, R. M. 1990. *When giants learn to dance*. New York: Simon & Schuster.
Kaplan, Nira. 2003. "Virtuous competition among citizens: Emulation in politics and pedagogy during the French Revolution." *Eighteenth-Century Studies* 36 (2):241–248.
Kapoor, Ilan. 2012. *Celebrity humanitarianism: The ideology of global charity*. New York: Taylor & Francis.
Karadag, R. 2010. *Neoliberal restructuring in Turkey: From state to oligarchic capitalism* (No. 10/7). MPIfG Discussion Paper.
Karisto, A. 2008. "Finnish baby boomers and the emergence of the third age." *International Journal of Ageing and Later Life* 2 (2):91–108.
Kasser, Tim. 2003. *The high price of materialism*. Cambridge, MA: Bradford Books.
Kaufman, D. 2015. "Corruption matters." *Finance & Development* 52 (3):20–23.
Kaul, Inge, P. Conceicao, K. Le Goulven, and R. Mendoza, eds. 2003. *Providing global public goods: Managing globalization*. New York: Published for the United Nations Development Programme [by] Oxford University Press.
Keller, D. 2001. "The sports spectacle, Michael Jordan and Nike: Unholy alliance?" In *Michael Jordan, Inc: Corporate sport, media culture, and late modern America*, ed. D. Andrews. Albany, NY: SUNY Press, 37–63.
Kelly, M., and C. Ó. Gráda. 2000. "Market contagion: Evidence from the Panics of 1854 and 1857." *American Economic Review* 90 (5):1110–1124.
Kenny, K., and E. Bell. 2014. "Irony as discipline: Self-help and gender in the knowledge economy." *Social Politics: International Studies in Gender, State & Society* 21 (4):562–584.
King, Barry. 1992. "Stardom and symbolic degeneracy: Television and the transformation of the stars as public symbols." *Semiotica* 92 (1–2):1–48.
Klerk, J. J. de. 2005. "Spirituality, meaning in life, and work wellness: A research agenda." *International Journal of Organizational Analysis* 13 (1):64–68.
Kline, N. "It was the Democrat's embrace of Neoliberalism that won it for Trump." *The Guardian*, 9 November.
Kloppenberg, James T. 1988. *Uncertain victory: Social democracy and progressivism in European and American thought, 1870–1920*. New York: Oxford University Press.
Koch, Anne. 2015. "Competitive charity: A neoliberal culture of 'giving back' in global yoga." *Journal of Contemporary Religion* 30 (1):73–88.
Kossek, E. E. and Friede, A. 2006. 'The business case: Managerial perspectives on work and the family," in M. Pitt-Catsouphes, E. E. Kossek and S. Sweet (eds) *The*

work-family handbook: Multidisciplinary perspectives methods and approaches, (pp. 611–628). Mahwah, NJ: Lawrence Erlbaum Associates Press.
Kossek, E., S. Lewis, S., and L. B. Hammer. 2010. "Work-life initiatives and organizational change: Overcoming mixed messages to move from the margin to the mainstream." *Human Relations* 63 (1):3–19.
Kotz, D. M. 2009. "The financial and economic crisis of 2008: A systemic crisis of neoliberal capitalism." *Review of Radical Political Economics* 41 (3):305–317.
Krahnke, Keiko, Robert A. Giacalone, and Carole L. Jurkiewicz. 2003. "Point-counterpoint: Measuring workplace spirituality." *Journal of Organizational Change Management* 16 (4):396–405.
Kriesi, Hanspeter. 2014. "The populist challenge." *West European Politics* 37 (2):361–378.
Krippner, Greta R. 2012. *Capitalizing on crisis: The political origins of the rise of finance*. Cambridge, MA: Harvard University Press.
Kuhner, Timothy K. 2014. *Capitalism v. Democracy: Money in politics and the free market constitution*. Stanford: Stanford Law Books.
Kunda, G. 1992. *Engineering culture: Control and commitment in a high-tech corporation*. Philadelphia: Temple University Press.
Kunda, Gideon, and G. Ailon-Souday. 2005. "Managers, markets, and ideologies." In *The Oxford handbook of work and organization*, eds. S. Ackroyd, R. Batt, P. Thompson, and P. Tolbert. Oxford, UK: Oxford University Press, 200–219.
Lacan, J. 200. *Ecrits* (trans B. Fink). New York: W.W. Norton and Company.
Laliberte Rudman, Debbie. 2015. "Embodying positive aging and neoliberal rationality: Talking about the aging body within narratives of retirement." *Journal of Aging Studies* 34:10–20.
Lanci, Y. 2014. "Zombie 2.0: Subjectivation in times of apocalypse." *Journal for Cultural and Religious Theory* 13 (2):25–37.
Land, C. and S. Taylor. 2010. "Surf's up: Work, life, balance and brand in a new age capitalist organization," *Sociology* 44 (3):395–413.
Larner, Wendy. 2000. "Neo-liberalismi policy, ideology, governmentality." *Studies in Political Economy* 63 (1):5–25.
Latham, Michael E. 2011. *The right kind of revolution: Modernization, development, and U.S. Foreign policy from the Cold War to the present*. Ithaca, NY: Cornell University Press.
Lau, Kimberly J. 2000. *New age capitalism: Making money east of Eden*. Philadelphia, PA: University of Pennsylvania Press.
Laufer, W. S. 2003. "Social accountability and corporate greenwashing." *Journal of Business Ethics* 43 (3):253–261.
Lavrence, Christine, and Kristin Lozanski. 2014. "' This is not your practice life': Lululemon and the neoliberal governance of self." *Canadian Review of Sociology/Revue canadienne de sociologie* 51 (1):76–94.
Law, D. M., J. D. Shapka, S. Hymel, B. F. Olson, and T. Waterhouse. 2012. "The changing face of bullying: An empirical comparison between traditional and internet bullying and victimization." *Computers in Human Behavior*, 28 (1):226–232.
Lawson, Victoria. 2007. "Geographies of care and responsibility." *Annals of the American Association of Geographers* 97 (1):1–11.
Lazear, Edward P. 1989. "Pay equality and industrial politics." *Journal of Political Economy* 97 (3):561–580.

Lazonick, William, and Mary O'Sullivan. 2000. "Maximizing shareholder value: A new ideology for corporate governance." *Economy and Society* 29 (1):13–35.

Lazzarato, Maurizio. 2009. "Neoliberalism in action: Inequality, insecurity and the reconstitution of the social." *Theory, Culture & Society* 26 (6):109–133.

Lazzarato, Maurizio. 2012. *The making of the indebted man: An essay on the neoliberal condition.* Cambridge, MA: Semiotext (E).

Leaviss, J. 2015. "How to end boom and bust: Make cash illegal." *The Telegraph*, 13 May.

Leclercq Vandelannoitte, Aurélie. 2011. "Organizations as discursive constructions: A Foucauldian approach." *Organization Studies* 32 (9):1247–1271.

Lee, C. W., K. McNulty, and S. Shaffer. 2012. " 'Hard times, hard choices': Marketing retrenchment as civic empowerment in an era of neoliberal crisis." *Socio-Economic Review* 11 (1):81–106.

Lee, Sang M., and Suzanne J. Peterson. 2001. "Culture, entrepreneurial orientation, and global competitiveness." *Journal of World Business* 35 (4):401–416.

Leidner, Robin. 1993. *Fast food, fast talk: Service work and the routinization of everyday life.* Berkeley, CA: University of California Press.

Lemke, Thomas. 2001. " 'The birth of bio-politics': Michel Foucault's lecture at the Collège de France on neo-liberal governmentality." *Economy and Society* 30 (2):190–207.

Leps, Marie-Christine. 1992. *Apprehending the criminal: The production of deviance in nineteenth-century discourse.* Durham, DC: Duke University Press.

Leslie, I. 2016. "In an age when even the powerful decry elites, we're all anti-establishment now." *The Guardian*, 29 May.

Levi-Faur, D., and Jordana, J. 2005. "The rise of regulatory capitalism: the global diffusion of a new order." *The annals of the American academy of political and social science*, 598 (1):200–217.

Levi-Faur, D., and Jordana, J. 2005. Preface: The making of a new regulatory order. *The Annals of the American Academy of Political and Social Science*, 598:6–9.

Levitas, Ruth. 2013. *Utopia as method: The imaginary reconstitution of society.* Basingstoke, UK: Palgrave Macmillan.

Lewis, Suzan, Richenda Gambles, and Rhona Rapoport. 2007. "The constraints of a 'work-life balance' approach: An international perspective." *The International Journal of Human Resource Management* 18 (3):360–373.

Lewis, Suzan, Rhona Rapoport, and Richenda Gambles. 2003. "Reflections on the integration of paid work and the rest of life." *Journal of Managerial Psychology* 18 (8):824–841.

Leys, Colin. 2001. *Market-driven politics: Neoliberal democracy and the public interest.* London, UK: Verso Books.

Lilla, M. 2016. "The end of identity liberalism." *New York Times*, 18 November.

Lindblom, Charles Edward E. 1982. *Politics and markets: The world's political-economic systems.* New York: Basic Books.

Lines, Gill. 2001. "Villains, fools or heroes? Sports stars as role models for young people." *Leisure Studies* 20 (4):285–303.

Lippe, T. van der. 2007. "Dutch workers and time pressure: Household and workplace characteristics." *Work, Employment & Society* 21 (4):693–711.

Lipset, Seymour Martin, and William Schneider. 1983. *The confidence gap: Business, labor, and government in the public mind.* New York: Free Press.

Littler, Craig R. 1978. "Understanding Taylorism." *The British Journal of Sociology* 29 (2):185.
Livne, Roi, and Yuval P. Yonay. 2015. "Performing neoliberal governmentality: An ethnography of financialized sovereign debt management practices." *Socio-Economic Review* 14 (2):339–362.
Logsdon, Jeanne M., and Donna J. Wood. 2002. "Business citizenship: From domestic to global level of analysis." *Business Ethics Quarterly* 12 (2):155.
Lowenthal, L. 1961. *Literature, popular culture and society.* California: Pacific Book Pub.
Lowi, Theodore J. 1972. "Four systems of policy, politics, and choice." *Public Administration Review* 32 (4):298.
Lugg, C. A. 1996. *For God and Country: Conservatism and American School Policy. Counterpoints.* New York: Peter Lang Publishing.
Lupton, Deborah. 2013. "The digitally engaged patient: Self-monitoring and self-care in the digital health era." *Social Theory & Health* 11 (3):256–270.
Luthans, Fred, and Tim R. V. Davis. 1979. "Behavioral self-management—the missing link in managerial effectiveness." *Organizational Dynamics* 8 (1):42–60.
Lyon, David. 1993. "An electronic panopticon? A sociological critique of surveillance theory." *The Sociological Review* 41 (4):653–678.
McCarty, Nolan M., Keith T. Poole, and Howard Rosenthal. 2013. *Political bubbles: Financial crises and the failure of American democracy.* Princeton, NJ: Princeton University Press.
McClintock, Anne. 2013. *Imperial leather: Race, gender, and sexuality in the colonial contest.* New York: Routledge.
McCloskey, Deirdre N. 2006. *The bourgeois virtues: Ethics for an age of commerce.* Chicago, IL: University of Chicago Press.
Mcdowell, Linda. 2004. "Work, workfare, work/life balance and an ethic of care." *Progress in Human Geography* 28 (2):145–163.
McGerr, Michael E. 2005. *A fierce discontent: The rise and fall of the Progressive movement in America, 1870–1920.* Oxford, UK: Oxford University Press.
McGregor J. B., Jr., 1993. "Toward a theory of public management success." In *Public management: The state of the art*, ed. B. Bozeman. New Jersey: Jossey-Bass Inc., 173–185.
McHoul, A. W., and Wendy Grace. 1997. *A Foucault primer: Discourse, power, and the subject.* New York: New York University Press.
Mackenzie, Craig, and Alan Lewis. 1999. "Morals and markets." *Business Ethics Quarterly* 9 (3):439–452.
McNamee, Stephen J., and Robert K. Miller. 2009. *The meritocracy myth.* Maryland: Rowman & Littlefield Publishers.
McNulty, Paul J. 1967. "A note on the history of perfect competition." *Journal of Political Economy* 75 (4, Part 1):395–399.
MacGregor Burns, J. 1978, *Leadership*, Harper & Row, New York, NY.
Maier, C. S. 1970. "Between Taylorism and technocracy: European ideologies and the vision of industrial productivity in the 1920s." *Journal of Contemporary History* 5 (2):27–61.
Mair, J., and Noboa, E. 2006. Social entrepreneurship: How intentions to create a social venture are formed. In Mair, J., Robinson, J., and Hockerts, K. (eds.) *Social Entrepreneurship* (pp. 121–135). London, UK: Palgrave Macmillan UK.

Maitland, Ian. 1997. "Virtuous markets: 'The market as school of the virtues.'" *Business Ethics Quarterly* 7 (1):17.

Malik, K. 2016. "Politics of disillusionment and the rise of Trump." *Al Jazeera*, 6 March.

Manz, Charles C., and Henry P. Sims. 1980. "Self-management as a substitute for leadership: A social learning theory perspective." *The Academy of Management Review* 5 (3):361.

Marable, Manning. 2015. *How capitalism underdeveloped Black America: Problems in race, political economy, and society*. New York: Haymarket Books.

Maravelias, Christian. 2003. "Post-bureaucracy—control through professional freedom." *Journal of Organizational Change Management* 16 (5):547–566.

Maravelias, Christian. 2007. "Freedom at work in the age of post-bureaucratic organization." *Ephemera: Theory and Politics in Organization* 7 (4):555–574.

Maravelias, Christian. 2009. "Health promotion and flexibility: Extending and obscuring power in organizations." *British Journal of Management* 20:S194–S203.

Marrewijk, Marcel van. 2003. "Concepts and definitions of CSR and corporate sustainability: Between agency and communion." *Journal of Business Ethics* 44 (2):95–105.

Marshall, David P. 1997. *Celebrity and power: Fame in contemporary culture*. 2nd ed. Minneapolis, MN: University of Minnesota Press.

Martin, Randy, Michael Rafferty, and Dick Bryan. 2008. "Financialization, risk and labour." *Competition & Change* 12 (2):120–132.

Marx, Karl. 1973. *Grundrisse*. New York: Penguin Books.

Mason, Paul. 2015. *Post-capitalism: A guide to our future*. New York: Farrar, Straus and Giroux.

Matthews, C. H., X. Qin, and G. M. Franklin. 1996. "Stepping toward prosperity: The development of entrepreneurial ventures in China and Russia." *Journal of Small Business Management* 34 (3):75–85.

Mattick, P. 1981. *Economic crisis and crisis theory*. London, UK: Merlin Press.

Mattson, M., Clair, R. P., Sanger, P. A., et al. 2000. "A feminist reframing of stress," in P. M. Buzzanell (ed.) Rethinking Organizational and Managerial Communication from Feminist Perspectives, pp. 157–76. Thousand Oaks, CA: Sage.

Maxwell, John W., Thomas P. Lyon, and Steven C. Hackett. 2000. "Self-regulation and social welfare: The political economy of corporate environmentalism." *SSRN Electronic Journal*.

Mearsheimer, J. J. 2005. "Hans Morgenthau and the Iraq War: Realism versus Neo-conservatism," *Open Democracy*, 18 May, retrieved 8 February 2009.

Mehta, L. 1999. "From darkness to light? Critical reflections on the world development report 1998/99." *The Journal of Development Studies*, 36 (1):151–161.

Merchant, B. 2015. "Fully automated luxury Communism," *The Guardian*, 18 March.

Meriläinen, Susan, Janne Tienari, Robyn Thomas, and Annette Davies. 2004. "Management consultant talk: A cross-cultural comparison of normalizing discourse and resistance." *Organization* 11 (4):539–564.

Meyer, D. S. 1995. "The challenge of cultural elites: Celebrities and social movements." *Sociological Inquiry*, 65 (2):181–206.

Micheletti, M., and D. Stoller, eds. 2004. *Politics, products, and markets: Exploring political consumerism past and present*. New Brunswick, NJ: Transaction Publishers.

Michelson, Grant, Nick Wailes, Sandra Van Der Laan, and Geoff Frost. 2004. "Ethical investment processes and outcomes." *Journal of Business Ethics* 52 (1):1–10.

Miller, J. 2014. " 'Millenials' cynical about politics: Harvard poll finds low interest in midterm elections." *Boston Globe*, 29 April.

Miller, Peter, and Nikolas Rose. 2008. *Governing the present: Administering economic, social and personal life*. Cambridge, UK: Polity Press.

Mills, C. W. 1959. *The sociological imagination*. Oxford, UK: Oxford University Press.

Mills, P. K. 1983. "Self-management: Its control and relationship to other organizational properties." *Academy of Management Review* 8 (3):445–453.

Mirowski, Philip, and Dieter Plehwe, eds. 2009. *The road from Mont Pèlerin: The making of the neoliberal thought collective*. Cambridge, MA: Harvard University Press.

Mirvis, P. H. 1994. "Human development or depersonalization? The company as total community." In *A fatal embrace? Assessing holistic trends in human resources programs*, eds. F. W. Heuberger and L. L. Nash. New Brunswick, NJ: Transaction Publishers.

Mitra, Sramana. 2015. "Is technology dehumanizing society?" *Huffington Post*, 3 July.

Mitrea, S. 2015. Evidence-Free Policy: Austerity, agnotology, and economies of knowledge in Canada. ECPR: 'Knowledge and Ignorance in Policy-Making' Panel Université de Montréal, Montréal, Quebec. 26–29 August.

Mitroff, I. I. and Denton, E. A. 1999. "A study of spirituality in the work place," *Sloan Management Review* 40(4):83–92.

Moffitt, Benjamin. 2016. *The global rise of populism: Performance, political style, and representation*. Stanford, CA: Stanford University Press.

Mollick, Ethan. 2014. "The dynamics of crowdfunding: An exploratory study." *Journal of Business Venturing* 29 (1):1–16.

Molloy, Andrea. 2005. *Stop living your job, start living your life: 85 simple strategies to achieve work/life balance*. Berkeley, CA: Ulysses Press.

Monaco, I. 1978. *Celebrity*. New York: Delta.

Monasterios, E. 2008. "Uncertain modernities: Amerindian epistemologies and the reorienting of culture." In S. Castro-Klaren (ed.), *Blackwell companions to literature and culture* (pp. 553–570). Malden, MA: Blackwell Publishing.

Monbiot, G. 2016. "Neoliberalism is creating loneliness: That's what's wrenching society apart." *The Guardian*, 12 October.

Moncrieff, J. 2006. "Psychiatric drug promotion and the politics of neoliberalism." *The British Journal of Psychiatry* 188 (4):301–302.

Morris, Richard B. 1962. "Class struggle and the American Revolution." *The William and Mary Quarterly* 19 (1):3.

Moscovici, Serge, Carol Sherrard, and Greta Heinz. 1976. *Social influence and social change [engl.]*. New York: Published in cooperation with European Association of Experimental Social Psychology by Academic Press.

Mosley, Paul, John Hudson, and Arjan Verschoor. 2004. "Aid, poverty reduction and the 'new conditionality'." *The Economic Journal* 114 (496):F217–F243.

Mosse, G. L. 1966. "Introduction: The genesis of fascism." *Journal of Contemporary History* 1 (1):14–26.

Mouffe, C., and P. Holdengräber. 1989. "Radical democracy: modern or postmodern?." *Social Text*, 21:31–45.

Mudde, C. 2016. "Europe's populist surge: A long time in the making," *Foreign Affairs*, November/December.

Mudge, S. L. 2008a. "What is neo-liberalism?" *Socio-Economic Review* 6 (4): 703–731.

Mudge, S. L. 2008b. "Neoliberalism's three faced." *EUI Work Papers* 34.

Muehlebach, A. 2012. *The moral neoliberal: welfare and citizenship in Italy*. Chicago, IL: University of Chicago Press.

Muller, J. 2015. "Rule-Breaking." *London Review of Books* 37 (16):3–7.

Mumby, D. K. 2005. "Theorizing resistance in organization studies: A dialectical approach." *Management Communication Quarterly* 19 (1):19–44.

Murray, R. 1971. "The internationalization of capital and the nation state." *New Left Review*, 67, 84–109.

Music, Graham. 2012. "The master and his emissary: The divided brain and the making of the Western world." *Journal of Child Psychotherapy* 38 (3):377–381.

Nadesan, Majia. 2010. *Governmentality, biopower, and everyday life*. New York: Taylor & Francis.

Nava, Mica. 1987. "Consumerism and its contradictions." *Cultural Studies* 1 (2): 204–210.

Navarro, V. 2007. "Neoliberalism as a class ideology; or, the political causes of the growth of inequalities." *International Journal of Health Services* 37 (1):47–62.

Neilson, Lisa A. 2010. "Boycott or buycott? Understanding political consumerism." *Journal of Consumer Behaviour* 9 (3):214–227.

Nelson, Robert H. 2001. *Economics as religion: From Samuelson to Chicago and beyond*. 3rd ed. Pennsylvania: Pennsylvania State University Press.

Neocleous, M. 2003. "The political economy of the dead Marx's Vampire." *History of Political Thought* 24 (4):668–684.

Nicholls, A. (Ed.). 2008. *Social entrepreneurship: New models of sustainable social change*. OUP Oxford.

Niebuhr, Reinhold, Cornel West, and Langdon B. Gilkey. 1963. *Moral man and immoral society: A study in ethics and politics*. 2nd ed. Louisville, KY: Westminster/John Knox Press.

Nisbet, Robert A. 1979. *History of the idea of progress*. New York: Basic Books.

Norman, Wayne, and Chris MacDonald. 2004. "Getting to the bottom of 'Triple bottom line.'" *Business Ethics Quarterly* 14 (2):243–262.

Obama, Barack. 2009. "New Conference for London Summit." Delivered at Excel Center in London, UK.

Office of the White House Press Secretary. 2008. "Declaration: Summit on financial markets and the world economy." *Financial Times*, 15 November.

O'Flynn, Gabrielle, and Eva Bendix Petersen. 2007. "The 'good life' and the 'rich portfolio': Young women, schooling and neoliberal subjectification." *British Journal of Sociology of Education* 28 (4):459–472.

Ogbor, John O. 2001. "Critical theory and the hegemony of corporate culture." *Journal of Organizational Change Management* 14 (6):590–608.

O'Malley, E., and J. Fitzgibbon. 2014. "Everywhere and nowhere: Populism and the puzzling non-reaction to Ireland's crises." In *Everywhere and nowhere: Populism and the puzzling non-reaction to Ireland's crises*, eds. H. Kriesi and T. S. Pappas. Colchester, UK: ECPR Press.

O'Malley, P. 1996. "Risk and responsibility." In *Foucault and political reason*, eds. A. Barry, T. Osbourne, and N. Rose. Chicago, IL: University of Chicago Press.

Ong, Aihwa. 2006. *Neoliberalism as exception: Mutations in citizenship and sovereignty*. Durham, NC: Duke University Press Books.

O'Riordan, Timothy. 2013. "Sustainability for wellbeing." *Environmental Innovation and Societal Transitions* 6:24–34.

O'Rourke, Anastasia. 2003. "The message and methods of ethical investment." *Journal of Cleaner Production* 11 (6):683–693.

Ortner, Sherry B. 1995. "Resistance and the problem of ethnographic refusal." *Comparative Studies in Society and History* 37 (01):173.

O'Toole, James. 1996. *Leading change: The argument for values-based leadership: Ballantine Books edition*. New York: Random House Publishing Group.

Ouellette, L. 2004. Take responsibility for yourself": Neoliberalism, citizenship, and reality television. In S. Murray and L. Oullette (eds.) *Reality TV: Remaking television culture*, New York: New York University Press, 231–250.

Overbeek, Henk, and Bastiaan van Apeldoorn, eds. 2012. *Neoliberalism in crisis*. Houndmills, Basingstoke, Hampshire [UK]: Palgrave Macmillan.

Palley, Thomas. 2007. "Macroeconomics and monetary policy: Competing theoretical frameworks." *Journal of Post Keynesian Economics* 30 (1):61–78.

Paluck, Elizabeth Levy. 2006. "Diversity training and intergroup contact: A call to action research." *Journal of Social Issues* 62 (3):577–595.

Pandey, S. K., and P. G. Scott. 2002. "Red tape: A review and assessment of concepts and measures." *Journal of Public Administration Research and Theory* 12 (4):553–580.

Papadopoulou, E. L., and G. Sakellaridis, eds. 2012. *The political economy of public debt and austerity in the EU*. Athens, Greece: Nissos Press.

Parker, Christine. 2002. *The open corporation: Effective self-regulation and democracy*. Cambridge, UK: Cambridge University Press.

Parker, Christopher, and Matt A. Barreto. 2014. *Change they can't believe in: The Tea Party and reactionary politics in America*. Princeton, NJ: Princeton University Press.

Partnoy, Frank. 2003. *Infectious greed: Enron and beyond—the story behind Enron and its wider implications*. London: Profile Books.

Passas, N. 2000. "Global anomie, dysnomie, and economic crime: Hidden consequences of globalization and neo-liberalism in Russia and around the world." *Social Justice* 27 (2):16–44.

Paul, Karen. 2004. "Business and society and business ethics journals: A citation and impact analysis." *Journal of Scholarly Publishing* 35 (2):103–117.

Peck, Jamie, and Adam Tickell. 1994. "Jungle law breaks out: Neoliberalism and global-local disorder." *Area* 26 (4):317–326.

Peck, Jamie, and Adam Tickell. 2007. "Conceptualizing neoliberalism, thinking Thatcherism." In *Contesting Neoliberalism: Urban Frontiers*, eds. H. Leitner, J. Peck and E.S. Sheppard. New York: Guilford Press, 26–50.

Peeples, J., Bsumek, P., Schwarze, S., and J. Schneider. 2014. "Industrial apocalyptic: Neoliberalism, coal, and the burlesque frame." *Rhetoric & Public Affairs* 17 (2):227–253.

Penman, D. 2012. "Meditate just like the U.S. Marines." *Psychology Today*, 3 July.

Peredo, Ana María, and Murdith McLean. 2006. "Social entrepreneurship: A critical review of the concept." *Journal of World Business* 41 (1):56–65.

Perez, C. 2009. "The double bubble at the turn of the century: Technological roots and structural implications." *Cambridge Journal of Economics* 33 (4):779–805.

Perlow, Leslie A. 1998. "Boundary control: The social ordering of work and family time in a high-tech corporation." *Administrative Science Quarterly* 43 (2):328.

Peters, Tom J. 1992. *Liberation management: Necessary disorganization for the nanosecond nineties*. 2nd ed. London, UK: Palgrave Macmillan.

Picchio, Antonella. 1992. *Social reproduction: The political economy of the labour market*. Cambridge, UK: Cambridge University Press.

Pierson, Paul. 1998. "Irresistible forces, immovable objects: Post-industrial welfare states confront permanent austerity." *Journal of European Public Policy* 5 (4):539–560.

Pilkington, M. 2016. "Well-being, happiness and the structural crisis of neoliberalism: an interdisciplinary analysis through the lenses of emotions." *Mind & Society, 15* (2):265–280.

Pitcher, Ben. 2012. "Race and capitalism redux." *Patterns of Prejudice* 46 (1):1–15.

Platman, Kerry. 2004. " 'Portfolio careers' and the search for flexibility in later life." *Work, Employment and Society* 18 (3):573–599.

Polanyi, Karl. 2001. *The great transformation: The political and economic origins of our time*. Boston, MA: Beacon Press.

Porter, Doug, and David Craig. 2004. "The third way and the third world: Poverty reduction and social inclusion in the rise of 'inclusive' liberalism." *Review of International Political Economy* 11 (2):387–423.

Porter, T. 2009. "The division of moral labour and the basic structure restriction." *Politics, Philosophy & Economics* 8 (2):173–199.

Portes, Alejandro. 1997. "Neoliberalism and the sociology of development: Emerging trends and unanticipated facts." *Population and Development Review* 23 (2):229.

Poulantzas, Nicos. 1978. *State, power, socialism*. London, UK: Verso Books.

Preuss, Lutz, Axel Haunschild, and Dirk Matten. 2009. "The rise of CSR: Implications for HRM and employee representation." *The International Journal of Human Resource Management* 20 (4):953–973.

Pritchard, K., and R. Whiting. 2014. "Baby boomers and the lost generation: On the discursive construction of generations at work." *Organization Studies* 35 (11): 1605–1626.

Putnam, Robert D. 2001. *Bowling alone: The collapse and revival of American community*. New York: Simon & Schuster.

Quinn, J. 1992. *Intelligent enterprise*. New York: Free Press.

Raddon, Mary-Beth. 2008. "Neoliberal legacies: Planned giving and the new philanthropy." *Studies in Political Economy* 81 (1):27–48.

Ragland, Ellie. 2013. *Essays on the pleasures of death: From Freud to Lacan*. New York: Taylor & Francis.

Ramakrishnan, R. T., and A. V. Thakor. 1991. "Cooperation versus competition in agency." *The Journal of Law, Economics, and Organization* 7 (2):248–283.

Ramsdell, L. 2004. "Dysfunctional family, dysfunctional nation: 'El cuarto mundo' by Diamela Eltit." In *The ties that bind: Questioning family dynamics and family discourse in Hispanic literature*, ed. Sara E. Cooper. Lanham, MD: University Press of America, 103–126.

Rapoport, R., Bailyn, L., Fletcher, J. K., et al. 2002. *Beyond work–family balance*. San Francisco, CA: Jossey-Bass.

Rasmussen, C. 2006. "Ugly and monstrous: Marxist aesthetics." Presented at Graduate Conference in the Humanities, University of Nebraska.

Redden, G. 2016. "Revisiting the spiritual supermarket: does the commodification of spirituality necessarily devalue it?" *Culture and Religion* 17 (2):231–249.

Rego, Arménio, and Miguel Pina e Cunha. 2008. "Workplace spirituality and organizational commitment: An empirical study." *Journal of Organizational Change Management* 21 (1):53–75.

Reich, R. B. 1987. "Reconsidered: The team as hero." *Harvard Business Review*, May–June:77–83.

Rhodes, C., and P. Bloom. 2013. "The cultural fantasy of hierarchy: Sovereignty and the desire for spiritual purity." *Research on the Sociology of Organisations* 35:141–169.

Roberts, John. 2003. "The manufacture of corporate social responsibility: Constructing corporate sensibility." *Organization* 10 (2):249–265.

Roberts, John. 2005. "The power of the 'imaginary' in disciplinary processes." *Organization* 12 (5):619–642.

Robinson, Brett T. 2013. *Appletopia: Media technology and the religious imagination of Steve Jobs*. Waco, TX: Baylor University Press.

Rodrik, D. 2011. *The globalization paradox: Democracy and the future of the world economy*. New York: W. W. Norton.

Rose, Nikolas. 1996. Inventing our selves: Psychology, power, and personhood. Cambridge: Cambridge University Press.

Rose, Nikolas. 1998. "Governing risky individuals: The role of psychiatry in new regimes of control." *Psychiatry, Psychology and Law* 5 (2):177–195.

Rose, Nikolas. 1999. *Powers of freedom, re-framing political thought*. Cambridge, UK: Cambridge University Press.

Rose, N., and P. Miller. 1992. "Political power beyond the state: Problematics of government." *British Journal of Sociology*, 173–205.

Rose, Nikolas, Pat O'Malley, and Mariana Valverde. 2006. "Governmentality." *Annual Review of Law and Social Science* 2 (1):83–104.

Rothbard, M. N. 2009. *Economic depressions: Their cause and cure*. Auburn, AL: Ludwig von Mises Institute.

Rothkopf, David J. 2008. *Superclass: The global power elite and the world they are making*. London, UK: Abacus.

Rottenberg, Catherine. 2013. "The rise of neoliberal feminism." *Cultural Studies* 28 (3):418–437.

Rotwein, E., and David Hume. 1955. *Writings on economics*. Freeport, NY: Books for Libraries Press.

Roubini, N., and S. Mihn. 2010. *Crisis economies*. New York: Penguin Books.

Rowe, J. K. 2005. Corporate social responsibility as business strategy. In R. D. Lipschutz, and J. K. Rowe, Globalization, governmentality and global politics: Regulation for the rest of us? (pp. 13070). Abingdon: Routledge.

Roy, A. 2001. Power politics. Cambridge, MA: South End Press.

Roy, R. K., A. T. Denzau, and T. D. Willett, eds. 2006. *Neo-liberalism: National and regional experiments with global ideas*. London, UK: Routledge.

Rueda, David, and Jonas Pontusson. 2000. "Wage inequality and varieties of capitalism." *World Politics* 52 (3):350–383.

Salsman, R. 2012. "Fiscal austerity and rational morality." *Forbes*, 17 May.

Samuel, R. 1992. "Mrs. Thatcher's return to Victorian values." In *Proceedings of the British Academy* 78:9–29.

Săndulescu, V. 2004. "Fascism and its quest for the 'New Man': the case of the Romanian legionary movement." *Studia Hebraica*, 4:349–361.
Sassatelli, R. 2009. "Promotional reflexivity, irony, de-fetishization and moralization in The Body Shop promotional rhetoric." *Prospettive degli studi culturali*:229–247.
Savage, Charles M. 1996. *Fifth generation management co-creating through virtual enterprising, dynamic teaming, and knowledge networking*. Boston, MA: Butterworth-Heinemann.
Savitz, Andrew W. 2012. *The triple bottom line: How today's best-run companies are achieving economic, social, and environmental success—and how you can too*. San Francisco, CA: John Wiley & Sons.
Scheffler, Samuel. 2005. "Egalitarian liberalism as moral pluralism." *Aristotelian Society* 79 (S1):229–253.
Shiller, R. 2015. "What good are economists," *Project Syndicalist*, 15 January.
Scholarios, Dora, and Abigail Marks. 2004. "Work-life balance and the software worker." *Human Resource Management Journal* 14 (2):54–74.
Schuh, Sebastian C., Xin-an Zhang, and Peng Tian. 2013. "For the good or the bad? Interactive effects of transformational leadership with moral and authoritarian leadership behaviors." *Journal of Business Ethics* 116 (3):629–640.
Schumpeter, Joseph A. 1994. *Capitalism, socialism and democracy*. Routledge.
Schumpeter, Joseph A. 2009. *Can capitalism survive? Creative destruction and the future of the global economy*. New York: HarperCollins Publishers.
Schuyler, R. L. 1921. "The climax of anti-imperialism in England." *Political Science Quarterly* 36 (4):537.
Scitovsky, Tibor. 1976. *The joyless economy: An inquiry into human satisfaction and consumer dissatisfaction*. New York: Oxford University Press.
Scott, A., A. Gilbert, and A. Gelan. 2007. *The urban-rural divide: Myth or reality?* Aberdeen, UK: Macaulay Institute.
Scott, James C. 1990. *Domination and the arts of resistance: Hidden transcripts*. New Haven, CT: Yale University Press.
Seabrooke, Leonard, and Rune Riisbjerg Thomsen. 2016. "Making sense of austerity: Everyday narratives in Denmark and the United Kingdom." *Politics* 36 (3):250–261.
Seaman, L. C. B. 2002. *Victorian England: Aspects of English and imperial history, 1837–1901*. London, UK: Routledge.
Seelos, Christian, and Johanna Mair. 2005. "Social entrepreneurship: Creating new business models to serve the poor." *Business Horizons* 48 (3):241–246.
Seib, G. 2016. "Behind the rise of populism, economic angst," *Wall Street Journal*, 20 January.
Semmel, B. 1970. *The rise of free trade imperialism, 1750–1850*. Cambridge, UK: Cambridge University Press.
Sen, S. 2004. "Liberal empire and illiberal trade: The political economy of responsible government in early British India." In *A new imperial history: Culture, identity and modernity in Britain and the Empire 1660–1840*, ed. K. Wilson. Cambridge, UK: Cambridge University Press, 136–154.
Sennett, Richard. 1998. *The corrosion of character: The personal consequences of work in the new capitalism*. New York: W. W. Norton & Company.
Shamir, Ronen. 2004. "Between self-regulation and the Alien Tort Claims Act: On the contested concept of corporate social responsibility." *Law and Society Review* 38 (4):635–664.

Shamir, Ronen. 2005a. "Corporate responsibility and the South African drug war." In A. Sarat and S. Scheingold (eds.) *The Worlds Cause Lawyers Make: Structure and Agency in Legal Practice*. Stanford University Press, p. 60.
Shamir, Ronen. 2005b. "Mind the gap: The commodification of corporate social responsibility." *Symbolic Interaction* 28 (2):229–253.
Shamir, Ronen. 2008. "The age of responsibilization: On market-embedded morality." *Economy and Society* 37 (1):1–19.
Sharma, Aradhana. 2008. *Logics of empowerment: Development, gender, and governance in neoliberal India*. Minneapolis, MN: University of Minnesota Press.
Sharma, Nandini, A. Venkat Raman, Sunita Dhaked, and Pawan Kumar. 2016. "Human resource challenges in Indian public health services." *South Asian Journal of Human Resources Management* 3 (2):173–190.
Sharma, Vineeta, Vic Naiker, and Barry Lee. 2009. "Determinants of audit committee meeting frequency: Evidence from a voluntary governance system." *Accounting Horizons* 23 (3):245–263.
Shelton, J. 2010. "Always on: Our 24/7 connectivity tips the work/life balance." *New Haven Register*.
Sidel, Ruth. 1996. "The enemy within: A commentary on the demonization of difference." *American Journal of Orthopsychiatry* 66 (4):490–495.
Sim, S., and Sumikawa, A. 2012. Challenges in Financial Services Transfer Pricing in 2012. *International Tax Review* 23:44.
Simola, Sheldene. 2010. "Anti-corporate anger as a form of care-based moral agency." *Journal of Business Ethics* 94 (S2):255–269.
Sklair, Leslie. 2002. *Globalization: Capitalism and its alternatives*. 3rd ed. New York: Oxford University Press.
Skocpol, Theda, and Kenneth Finegold. 1982. "State capacity and economic intervention in the early New Deal." *Political Science Quarterly* 97 (2):255.
Skocpol, Theda, and Vanessa Williamson. 2012. *The Tea Party and the remaking of Republican conservatism*. New York: Oxford University Press.
Slater, T. 2014. The myth of "Broken Britain": Welfare reform and the production of ignorance. *Antipode* 46 (4): 948–969.
Sloterdijk, Peter. 1984. "Cynicism: The twilight of false consciousness." *New German Critique* 33:190.
Smith, Iain Duncan. 2012. "2012 Conservative Party conference speech," 3 December.
Smith, R. 2011. "Green capitalism: The god that failed." *Real World Economics Review* 56:112–144.
Snyder, Richard. 1999. "After neoliberalism: The politics of reregulation in Mexico." *World Politics* 51 (2):173–204.
Soloway, R. A. 1990. *Demography and degeneration: Eugenics and the declining birthrate in twentieth-century Britain*. 2nd ed. North Carolina: The University of North Carolina Press.
Sparkes, Russell. 2001. "Ethical investment: Whose ethics, which investment?" *Business Ethics: A European Review* 10 (3):194–205.
Sparrow, Robert. 2002. "Talking sense about political correctness." *Journal of Australian Studies* 26 (73):119–131.
Spence, M. 2016. "The Inexorable Logic of the Sharing Economy" *Project Syndicalist*, 26 October.

Spicer, A, and Fleming, P. 2007. "Intervening in the inevitable: Contesting globalization in a public sector organization." *Organization* 14 (4):517–541.

Springer, Simon. 2010. *Cambodia's neoliberal order: Violence, authoritarianism, and the contestation of public space*. London, UK: Routledge.

Stanley, Liam. 2014. "'We're reaping what we sowed': Everyday crisis narratives and acquiescence to the age of austerity." *New Political Economy* 19 (6):895–917.

Stark, R. 2007. *The victory of reason: How Christianity led to freedom, capitalism, and Western success*. NY: Random House.

Stavrakakis, Yannis. 1999. *Lacan and the Political*. London: Routledge.

Stavrakakis, Yannis. 2007. *The Lacanian left: Psychoanalysis, theory, politics*. Edinburgh, UK: Edinburgh University Press.

Steger, Manfred B. 2005a. "From market globalism to imperial globalism: Ideology and American power after 9/11." *Globalizations* 2 (1):31–46.

Steger, Manfred B. 2005b. "Ideologies of globalization." *Journal of Political Ideologies* 10 (1):11–30.

Steger, Manfred B., and Ravi K. Roy. 2010. *Neoliberalism: A very short introduction (very short introductions)*. Oxford, UK: Oxford University Press.

Stehr, Nico. 2008. *Moral markets*. Boulder, CO: Paradigm Publishers.

Stehr, Nico, and M. Adolf. 2010. "Consumption between market and morals: A socio-cultural consideration of moralized markets." *European Journal of Social Theory* 13 (2):213–228.

Stein, Judith. 2011. *Pivotal decade: How the United States traded factories for finance in the seventies*. New Haven, CT: Yale University Press.

Stein, Judith. 2013. "Millennials: The me me me generation." *Time*, 9 May.

Stern, S. J. 2009. *Remembering Pinochet's Chile: On the eve of London 1998*. Durham, NC: Duke University Press.

Stigler, George J. 1957. "Perfect competition, historically contemplated." *Journal of Political Economy* 65 (1):1.

Stiglitz, J., and L. Bilmes. 2012. "The 1 percent's problem." *Vanity Fair*, 12 May.

Stoler, Ann Laura. 2002. *Carnal knowledge and imperial power: Race and the intimate in colonial rule*. Berkeley, CA: University of California Press.

Stone, D. 2008. "Biopower and modern genocide." In A. D. Moses (ed.) *Empire, colony, genocide: Conquest, occupation and subaltern resistance in world history* (pp. 162–182). New York: Berghahn Books.

Streeck, Wolfgang. 2014. *Buying time: The delayed crisis of democratic capitalism*. London, UK: Verso Books.

Strombeck, Andrew. 2006. "Invest in Jesus: Neoliberalism and the left behind novels." *Cultural Critique* 64 (1):161–195.

Sum, Shima, R. Mark Mathews, Ian Hughes, and Andrew Campbell. 2008. "Internet use and loneliness in older adults." *CyberPsychology & Behavior* 11 (2): 208–211.

Sundararajan, Arun. 2016. *The sharing economy: The end of employment and the rise of crowd-based capitalism*. Cambridge, MA: MIT Press.

Sunstein, Cass R. 1999. *Free markets and social justice*. New York: Oxford University Press.

Sutton, Robert. 2004. "Why these ideas work, but seem weird." *Design Management Review* 15 (1):43–49.

Szreter, Simon. 1988. "The importance of social intervention in Britain's mortality decline c.1850–1914: A re-interpretation of the role of public health." *Social History of Medicine* 1 (1):1–38.
Talbott, John R. 2011. *Obamanomics: How bottom-up economic prosperity will replace trickle-down economics*. New York: Seven Stories Press.
Tarnaff, B. 2016. "The future: Where borrowing is the norm and ownership is luxury." *The Guardian*, 17 October.
Tawney, R. H. 1926. *Religion and the rise of capitalism* (Vol. 23). Transaction Publishers.
Taylor, J.D. 2014. "Spent? Capitalism's growing problem with anxiety." *Roar Magazine*, 14 March.
Taylor, M. 2004. "Responding to neoliberalism in crisis: Discipline and empowerment in the World Bank's new development agenda." *Research in Political Economy* 21:3–30.
Teghtsoonian, Katherine. 2009. "Depression and mental health in neoliberal times: A critical analysis of policy and discourse." *Social Science & Medicine* 69 (1):28–35.
Teich, Mikulas, and Roy Porter, eds. 1996. *The industrial revolution in national context: Europe and the USA*. Cambridge, UK: Cambridge University Press.
Temperley, Howard. 1977. "Capitalism, slavery and ideology." *Past and Present* 75 (1):94–118.
Tett, Gillian. 2009. *Fool's gold: How the bold dream of a small tribe at J.P. Morgan was corrupted by Wall Street greed and unleashed a catastrophe*. New York: Simon & Schuster.
Thatcher, M. 1987. "Interview with Margaret Thatcher." *Women's Own*, 23 September.
Thompson, E. C. 2013. Urban Cosmopolitan Chauvinism and the Politics of Rural Identity. In Eric Thompson (ed.) *Cleavage, Connection and Conflict in Rural, Urban and Contemporary Asia* (pp. 161–179). Springer Netherlands.
Thompson, Helen. 2009. "The political origins of the financial crisis: The domestic and international politics of Fannie Mae and Freddie Mac." *The Political Quarterly* 80 (1):17–24.
Thompson, Kenneth. 2005. *Moral panics*. London, UK: Routledge.
Tie, W. 2004. "The psychic life of governmentality." *Culture, Theory and Critique*, 45 (2):161–176.
Tienhaara, Kyla. 2014. "Varieties of green capitalism: Economy and environment in the wake of the global financial crisis." *Environmental Politics* 23 (2):187–204.
Tingle, M. L. 1988. "Privatization and the Reagan administration." *Yale Law & Policy Review* 6 (1):229–257.
Tipps, Dean C. 1973. "Modernization theory and the comparative study of national societies: A critical perspective." *Comparative Studies in Society and History* 15 (2):199.
Topping A. 2014 Self-harm sites and cyberbullying: the threat to children from web's dark side. *The Guardian*, 10 March.
Trethewey, Angela. 1997. "Resistance, identity, and empowerment: A postmodern feminist analysis of clients in a human service organization." *Communication Monographs* 64 (4):281–301.
Trevino, Linda Klebe. 1986. "Ethical decision making in organizations: A person-situation interactionist model." *The Academy of Management Review* 11 (3):601.

Tudor, Deborah. 2012. "Selling nostalgia: Mad men, postmodernism and neoliberalism." *Society* 49 (4):333–338.

Turban, D. B., and D. W. Greening. 1997. "Corporate social performance and organizational attractiveness to prospective employees." *Academy of Management Journal* 40 (3):658–672.

Turken, S., E. Carlquist, and H. Allen. 2015. "Chasing happiness through personal debt: An example of neoliberal influence in Norwegian society." *Social and Psychological Dimensions of Personal Debt and the Debt Industry*:160–179.

Tyler, Imogen. 2013. *Revolting subjects: Social abjection and resistance in neoliberal Britain*. London, UK: Palgrave Macmillan.

Ubel, Peter A. 2013. *Free market madness: Why human nature is at odds with economics—and why it matters*. Boston, MA: Harvard Business Review Press.

United Nations. 2008. "Global financial crisis: Why corporate sustainability is more important than ever," United Nations Global Compact, 17 October.

Unger, Roberto Mangabeira. 2001. *False necessity: Anti-necessitarian social theory in the service of radical democracy: From* Politics, *a work in constructive social theory*. London, UK: Verso Books.

Van Echtelt, P. E., Glebbeek, A. C., & Lindenberg, S. M. 2006. "The new lumpiness of work explaining the mismatch between actual and preferred working hours." *Work, Employment & Society*, 20(3):493–512.

Velasquez, Manuel G. 2002. *Business ethics: Concepts and case*. Upper Saddle River, NJ: Prentice Hall.

Velasquez, Manuel G. 2006. *Business ethics: Concepts and cases*. 6th ed.

Velasquez, Manuel G. & Velazquez, M. 2006. *Business ethics: Concepts and cases* (Vol. 111). Upper Saddle River, NJ: Prentice Hall.

Venugopal, Rajesh. 2015. "Neoliberalism as concept." *Economy and Society* 44 (2):165–187.

Vertinsky, Ilan. 2009. "Debate: Public management in recessions and economic crises. Strategic options and political constraints." *Public Money & Management* 29 (5):265–267.

Vidaillet, B. 2007. "Lacanian theory's contribution to the study of workplace envy." *Human Relations* 60 (11):1669–1700.

Visser, W. 2011. *The age of responsibility: CSR 2.0 and the new DNA of business*. Sussex, UK: John Wiley & Sons.

Visser, Wayne. 2014. "The age of responsibility: CSR 2.0 and the new DNA of business." *Journal of Business Systems, Governance & Ethics* 5 (3).

Vogel, David. 2005. *The market for virtue: The potential and limits of corporate social responsibility*. Washington, DC: Hopkins Fulfillment Services.

Volberda, Henk Wijtze W. 1998. *Building the flexible firm: How to remain competitive*. New York: Oxford University Press.

Von Mises, L. 2012. *On the manipulation of money and credit*. Liberty Fund.

Vos, J. 2009. "Actions speak louder than words: Greenwashing in corporate America." *Notre Dame Journal of Law, Ethics, and Public Policy* 23:673.

Wacquant, Loïc. 2009. *Punishing the poor: The neoliberal government of social insecurity*. Durham, NC: Duke University Press.

Wacquant, Loïc. 2010. "Class, race & hyperincarceration in revanchist America." *Daedalus* 139 (3):74–90.

Wallerstein, Immanuel. 1995. *Historical capitalism; with capitalist civilization*. 10th ed. London, UK: Verso Books.

Wallis, Victor. 2009. "Beyond 'green capitalism.'" *Monthly Review* 61 (9):32.

Walsh, Z. 2016. "A meta-critique of mindfulness critiques: From mcMindfulness to critical mindfulness." In *Handbook of mindfulness*, eds. R. Purser, D. Forbes, and A. Burke. Switzerland: Springer International Publishing, 153–166.

Ward, L. 2015. "Caring for ourselves: Self-care and neoliberalism." In *Ethics of care: Critical advances in international perspective*, eds. M. Barnes, T. Brannelly, L. Ward, and N. Ward. Bristol, UK: Policy Press, 45–56.

Waring, Stephen P. 1994. *Taylorism transformed: Scientific management theory since 1945*. 2nd ed. New York: The University of North Carolina Press.

Warschauer, Mark. 2004. *Technology and social inclusion: Rethinking the digital divide*. Cambridge, MA: MIT Press.

Watts, M. 2002. "Preface: Green capitalism, green governmentality." *American Behavioral Scientist* 45 (9):1313–1317.

Weber, Eugen. 1976. *Peasants into Frenchmen: The modernization of rural France; 1870–1914*. Stanford, CA: Stanford University Press.

Weber, Max. 2002. *The Protestant ethic and the spirit of capitalism and other writings*. Penguin.

Weber, Max. 2007. *The Protestant ethic and the spirit of capitalism*. 3rd ed. New York: Oxford University Press.

Weeks, J. 2014. *Sex, politics and society: The regulations of sexuality since 1800*. Routledge.

Weick, Karl E. 1995. *Sensemaking in organizations*. Thousand Oaks, CA: Sage Publications.

Weiler, Michael, and Barnett W. Pearce. 1992. *Reagan and public discourse in America*. Tuscaloosa, AL: The University of Alabama Press.

Weiskopf, Richard, and Hugh Willmott. 2013. "Ethics as critical practice: The 'Pentagon Papers', deciding responsibly, truth-telling, and the unsettling of organizational morality." *Organization Studies* 34 (4):469–493.

Weiss, R. 2001. "Charitable choice as neoliberal social welfare strategy." *Social Justice* 28 (1):35–53.

Wernham, R. B. 1968. *The New Cambridge Modern History: Volume 3, Counter-Reformation and Price Revolution, 1559–1610*. Cambridge, UK: Cambridge University Press.

Whetstone, J. Thomas. 2002. "Personalism and moral leadership: The servant leader with a transforming vision." *Business Ethics: A European Review* 11 (4): 385–392.

Wiebe, Robert Huddleston. 1989. *Businessmen and reform*. Cambridge, MA: Harvard University Press.

Wiggan, J. 2012. "Telling stories of 21st century welfare: The UK coalition government and the neo-liberal discourse of worklessness and dependency." *Critical Social Policy* 32 (3):383–405.

Wilensky, Harold L. 1974. *The welfare state and equality: Structural and ideological roots of public expenditures*. Berkeley, CA: University of California Press.

Willard, Bob. 2012. *The new sustainability advantage: Seven business case benefits of a triple bottom line*. 10th ed. British Columbia, UK: New Society Publishers.

Williams, Colin C., and Sara J. Nadin. 2013. "Beyond the entrepreneur as a heroic figurehead of capitalism: Re-representing the lived practices of entrepreneurs." *Entrepreneurship & Regional Development* 25 (7–8):552–568.

Williams, Evan C. 2011. *Combined and uneven apocalypse: Luciferian Marxism*. Handt, UK: John Hunt Publishing.

Williams, Jeffrey, ed. 1999. *PC wars: Politics and theory in the academy*. New York: Routledge.

Williams, Leslie K., and Stephen J. McGuire. 2010. "Economic creativity and innovation implementation: The entrepreneurial drivers of growth? Evidence from 63 countries." *Small Business Economics* 34 (4):391–412.

Williamson, A. H. 1996. "Scots, Indians and empire: The Scottish politics of civilization 1519–1609." *Past & Present* (150):46–83.

Willmott, Hugh. 1993. "Strength is ignorance; slavery is freedom: Managing culture in modern organizations." *Journal of Management Studies* 30 (4):515–552.

Winslow, Luke. 2016. "Doing more with less: Modeling neoliberal labor relations in *Undercover Boss*." *The Journal of Popular Culture* 49 (6):1357–1374.

Winston, Morton. 2002. "NGO strategies for promoting corporate social responsibility." *Ethics & International Affairs* 16 (1):71–87.

Winter, J. 2007. "Greed is bad. Bad!" *Slate*, 25 September

Withorn, Ann. 1996. " 'Why do they hate me so much?' A history of welfare and its abandonment in the United States." *American Journal of Orthopsychiatry* 66 (4):496–509.

Wolfe, Alan. 1977. *The limits of legitimacy: Political contradictions of contemporary capitalism*. New York: Simon & Schuster.

Wolfensohn, J. D. 2001. *The challenges of globalization*. Address to Bundestag, Berlin Germany, 2 April.

Wood, Ellen Meiksins. 1991. *The pristine culture of capitalism: An historical essay on old regimes and modern states*. London, UK: Verso Books.

Wood, Sharon E. 2005. *The freedom of the streets: Work, citizenship, and sexuality in a gilded age city*. New York: The University of North Carolina Press.

Wren-Lewis, S. 2014. Ordoliberalism, neoliberalism and economics. *Online blog, January*, 22, 2014.

Wright, Christopher, and Alexis Rwabizambuga. 2006. "Institutional pressures, corporate reputation, and voluntary codes of conduct: An examination of the equator principles." *Business and Society Review* 111 (1):89–117.

Wyllie, I. 1954. *The self-made man in America: The myth of rags to riches*. New Jersey: Rutgers University Press.

Yaffe, David S. 1973. "The Marxian theory of crisis, capital and the state." *Economy and Society* 2 (2):186–232.

Yaffe, David S. 1975. "Inflation, the crisis and the post-war boom." *Revolutionary Communist* 3 (4):5–45.

Young, R. J. 2005. *Colonial desire: Hybridity in theory, culture and race*. Routledge.

Young, Stephen. 2003. *Moral capitalism: Reconciling private interest with the public good*. San Francisco, CA: Berrett-Koehler Publishers.

Zakaria, F. 2016. "Populism on the march." *Foreign Affairs*, November.

Žižek, A. 1997. "The seven veils of fantasy." In D. Bonus (ed.), *Key concepts of Lacanian psychoanalysis*. London, UK: Rebus Press.

Zižek, Slavoj. 1989. *The sublime object of ideology*. London, UK: Verso Books.
Zižek, Slavoj. 1993. *Tarrying with the negative: Kant, Hegel, and the critique of ideology*. 5th ed. Durham, NC: Duke University Press.
Žižek, Slavoj. 2001. *On belief*. London and New York: Verso Books.
Zižek, Slavoj. 2004. "What can psychoanalysis tell us about cyberspace?," *Psychoanalytic Review*, 91:801–830.
Zwan, N. van der. 2014. "Making sense of financialization." *Socio-Economic Review* 12 (1):99–129.

Index